More information about this series at http://www.springer.com/series/15875

# Moscow Lectures

## Volume 5

Valery A. Gritsenko • Vyacheslav P. Spiridonov
Editors

# Partition Functions and Automorphic Forms

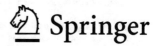

*Editors*

Valery A. Gritsenko
CNRS U.M.R. 8524
Université de Lille and IUF
Villeneuve d'Ascq Cedex, France

Vyacheslav P. Spiridonov
Laboratory of Theoretical Physics
Joint Institute for Nuclear Research
Dubna, c.Moscow, Russia

National Research University Higher
School of Economics
Laboratory of Mirror Symmetry
and Automorphic Forms
Moscow, Russia

National Research University Higher
School of Economics
Laboratory of Mirror Symmetry
and Automorphic Forms
Moscow, Russia

ISSN 2522-0314          ISSN 2522-0322   (electronic)
Moscow Lectures
ISBN 978-3-030-42402-2          ISBN 978-3-030-42400-8   (eBook)
https://doi.org/10.1007/978-3-030-42400-8

Mathematics Subject Classification (2020): 33-02, 81-02

Cover illustration: https://www.istockphoto.com/de/foto/panorama-der-stadt-moskau-gm490080014-75024685, with kind permission

This Springer imprint is published by the registered company Springer Nature Switzerland AG.
The registered company address is: Gewerbestrasse 11, 6330 Cham, Switzerland

# Preface to the Book Series *Moscow Lectures*

You hold a volume in a textbook series of Springer Nature dedicated to the Moscow mathematical tradition. Moscow mathematics has very strong and distinctive features. There are several reasons for this, all of which go back to good and bad aspects of Soviet organization of science. In the twentieth century, there was a veritable galaxy of great mathematicians in Russia, while it so happened that there were only few mathematical centers in which these experts clustered. A major one of these, and perhaps the most influential, was Moscow.

There are three major reasons for the spectacular success of Soviet mathematics:

1. Significant support from the government and the high prestige of science as a profession. Both factors were related to the process of rapid industrialization in the USSR.
2. Doing research in mathematics or physics was one of very few intellectual activities that had no mandatory ideological content. Many would-be computer scientists, historians, philosophers, or economists (and even artists or musicians) became mathematicians or physicists.
3. The Iron Curtain prevented international mobility.

These are specific factors that shaped the structure of Soviet science. Certainly, factors (2) and (3) are more on the negative side and cannot really be called favorable but they essentially came together in combination with the totalitarian system. Nowadays, it would be impossible to find a scientist who would want all of the three factors to be back in their totality. On the other hand, these factors left some positive and long lasting results.

An unprecedented concentration of many bright scientists in few places led eventually to the development of a unique "Soviet school". Of course, mathematical schools in a similar sense were formed in other countries too. An example is the French mathematical school, which has consistently produced first-rate results over a long period of time and where an extensive degree of collaboration takes place. On the other hand, the British mathematical community gave rise to many prominent successes but failed to form a "school" due to a lack of collaborations. Indeed, a school as such is not only a large group of closely collaborating individuals but also a group knit tightly together through student-advisor relationships. In the USA, which is currently the world leader in terms of the level and volume of mathematical research, the level of mobility is very high, and for this reason there are no US mathematical schools in the Soviet or French sense of the term. One can talk not only about the Soviet school of mathematics but also, more specifically, of the Moscow, Leningrad, Kiev, Novosibirsk, Kharkov, and other schools. In all these places, there were constellations of distinguished scientists with large numbers of students, conducting regular seminars. These distinguished scientists were often not merely advisors and leaders, but often they effectively became spiritual leaders in a very general sense.

A characteristic feature of the Moscow mathematical school is that it stresses the necessity for mathematicians to learn mathematics as broadly as they can, rather than focusing on a narrow field in order to get important results as soon as possible.

The Moscow mathematical school is particularly strong in the areas of algebra/algebraic geometry, analysis, geometry and topology, probability, mathematical physics and dynamical systems. The scenarios in which these areas were able to develop in Moscow have passed into history. However, it is possible to maintain and develop the Moscow mathematical tradition in new formats, taking into account modern realities such as globalization and mobility of science. There are three recently created centers—the Independent University of Moscow, the Faculty of Mathematics at the National Research University Higher School of Economics (HSE) and the Center for Advanced Studies at Skolkovo Institute of Science and Technology (SkolTech)—whose mission is to strengthen the Moscow mathematical tradition in new ways. HSE and SkolTech are universities offering officially licensed full-time educational programs. Mathematical curricula at these universities follow not only the Russian and Moscow tradition but also new global developments in mathematics. Mathematical programs at the HSE are influenced by those of the Independent University of Moscow (IUM). The IUM is not a formal university; it is rather a place where mathematics students of different universities can attend special topics courses as well as courses elaborating the core curriculum. The IUM was the main initiator of the HSE Faculty of Mathematics. Nowadays, there is a close collaboration between the two institutions.

While attempting to further elevate traditionally strong aspects of Moscow mathematics, we do not reproduce the former conditions. Instead of isolation and academic inbreeding, we foster global sharing of ideas and international cooperation. An important part of our mission is to make the Moscow tradition of mathematics at a university level a part of global culture and knowledge.

The "Moscow Lectures" series serves this goal. Our authors are mathematicians of different generations. All follow the Moscow mathematical tradition, and all teach or have taught university courses in Moscow. The authors may have taught mathematics at HSE, SkolTech, IUM, the Science and Education Center of the Steklov Institute, as well as traditional schools like MechMath in MGU or MIPT. Teaching and writing styles may be very different. However, all lecture notes are supposed to convey a live dialog between the instructor and the students. Not only personalities of the lecturers are imprinted in these notes, but also those of students.

We hope that expositions published within the "Moscow lectures" series will provide clear understanding of mathematical subjects, useful intuition, and a feeling of life in the Moscow mathematical school.

Moscow, Russia

Igor M. Krichever
Vladlen A. Timorin
Michael A. Tsfasman
Victor A. Vassiliev

# Preface

The international scientific school for mathematicians and physicists "**Partition Functions and Automorphic Forms**" took place in Dubna in 2018. It was organized by the International Laboratory of Mirror Symmetry and Automorphic Forms of the National Research University Higher School of Economics (NRU HSE) in Moscow, and the Bogoliubov Laboratory of Theoretical Physics of the Joint Institute for Nuclear Research (JINR) in Dubna, in the framework of the Dubna International Advanced School of Theoretical Physics (DIAS-TH). The program consisted of six minicourses and four survey lectures given by mathematicians and physicists and fourteen short scientific reports by the participants of the school. There were almost 80 participants from 15 countries, including postgraduate students, postdoctoral scholars, and young scientists. The school was advertised by the Russian central media (RIA News), the JINR, and by other local bulletins. All details about the school are available on the website https://indico.jinr.ru/event/304/. The scientific coordinators of the school were Valery A. Gritsenko and Vyacheslav P. Spiridonov.

This volume consists of eight sets of lecture notes representing all topics of the school. They are written by mathematicians and physicists and cover closely connected questions on automorphic forms, Feynman diagrams, algebraic geometry, infinite-dimensional Lie algebras, linear and nonlinear finite-difference equations, topological invariants, and supersymmetric quantum field theories. Written in a pedagogical style, they are targeted primarily at postdoctoral scholars, postgraduate students, and advanced undergraduate students, both in mathematics and in physics. We shall briefly outline their contents.

The introductory lecture notes by John F. R. Duncan, "*A Short Introduction to the Algebra, Geometry, Number Theory and Physics of Moonshine,*" are dedicated to the moonshine phenomenon. This emerged in the 1970s as a collection of coincidences connecting modular functions to the monster simple group. Attempts to explain these relations resulted in new algebraic structures, including vertex algebras and generalized Kac–Moody algebras, and fruitful interactions between mathematics and modern theoretical physics. In the twenty-first century, moonshine has continued to develop, particularly with the discovery of deeper connections to geometry and number theory. Professor J. Duncan has made significant contributions to the subject. The section subtitles of his beautiful notes precisely describe their content, covering various types of complex manifolds, modular forms, infinite-dimensional Lie algebras, and exceptional finite groups.

The elliptic genus of Calabi–Yau manifolds, on the one hand, is an extension of the Witten genus. On the other hand, it is an elliptic generalization of the Hirzebruch $\chi_y$-genus. In string theory, the elliptic genus emerges in the context of two-dimensional $\mathcal{N} = 2$ superconformal field theories. In the lecture notes by Valery Gritsenko, "*Modified Elliptic Genus*," a construction is described which matches an automorphic invariant—more precisely, a (possibly meromorphic) Jacobi form—to each Hermitian bundle on a complex manifold. This construction generalizes the Witten genus of Spin$^c$-manifolds, the elliptic genus of Calabi–Yau manifolds, and the partition functions of (0, 2)-theories. It is based on the idea of automorphic corrections of Jacobi forms which has been successfully applied by Gritsenko in the theory of Borcherds products. In particular, the second quantization of the elliptic genus due to Dijkgraaf–Moore–Verlinde–Verlinde leads to the construction of Lorentzian Kac–Moody algebras, as suggested by Gritsenko–Nikulin. The notes of V. Gritsenko serve as a geometric complement to his popular internet course "Jacobi modular forms: 30 ans après," which has been available on Coursera for the last three years.

In 2002, Nekrasov computed the instanton partition function in four-dimensional $\mathcal{N} = 2$ supersymmetric gauge theory on the omega-deformed space-time $\mathbb{R}^4$. In 2005, Römelsberger and Kinney–Maldacena–Minwalla–Raju have defined super-conformal indices of the field theories in space-time of dimensions 3, 4, 5, and 6. These indices represent a generalization of the Witten index counting BPS states in supersymmetric theories on compact curved spaces. The Nekrasov instanton sums can be represented in the form of an infinite series, the $k$-th term of which is described by the generalized Witten index of $k$-instanton sector. In some special limit, it defines the Seiberg–Witten prepotential. This formula has been investigated in detail both from the physical and mathematical points of view, as well as generalized to field theories in space-times of higher dimensions. Lectures by Seok Kim, "*Superconformal Indices and Instanton Partition Functions*," are dedicated to a survey of results dealing with instanton sums in five- and six-dimensional theories and their connection with superconformal indices of these theories on curved spaces. All these functions obey certain automorphic properties which were not investigated sufficiently deeply yet. Professor S. Kim is a leading expert in this field. In particular, he has pioneering results on superconformal indices of three-dimensional field theories.

As to the superconformal indices of four-dimensional field theories, they are partially considered in the lecture notes by Vyacheslav Spiridonov, "*Introduction to the Theory of Elliptic Hypergeometric Integrals*." Elliptic hypergeometric integrals were discovered by Spiridonov in 2000, and in 2008, Dolan–Osborn showed that these integrals coincide with superconformal indices of four-dimensional supersymmetric field theories. It appeared that the elliptic beta integrals admitting exact evaluation describe the confinement phenomenon, and the symmetries of more general integrals describe the electric–magnetic dualities of Seiberg, which may be regarded as analogues of mirror symmetry in four-dimensional theories. From the automorphic point of view, elliptic hypergeometric integrals are related to automorphic forms for the $SL(3, \mathbb{Z})$ modular group. An elementary introduction to the theory of these integrals is given in these notes, and a number of their applications are outlined. In particular, the properties of the elliptic gamma function and an elliptic analogue of the Euler–Gauss hypergeometric function are described, including the elliptic hypergeometric equation that is satisfied by the latter. An important conjecture of Gaiotto–Kim states that eigenfunctions of the elliptic Fourier transformation, which is partially described in these notes, are given by the Nekrasov partition functions of five-dimensional theories. In addition to this, the univariate elliptic beta integral describes the star-triangle relation and thus leads to the most complicated known solution of the Yang–Baxter equation. This means that there exists a direct relation between four-dimensional superconformal indices and partition functions of lattice spin models of the Ising type.

The contribution by Axel Kleinschmidt, Hermann Nicolai, and Adriano Vigano *"On Spinorial Representations of Involutory Subalgebras of Kac–Moody Algebras,"* describes some parts of the lectures read by H. Nicolai. Namely, it contains basics of the theory of representations of involutive (or "maximally compact") subalgebras of the infinite-dimensional Kac–Moody algebras. This theory by large remains a terra incognita, especially in respect to the fermionic representations. In the lectures, the recent efforts on the systematic investigation of spinorial representations are summarized, especially for one of the most important hyperbolic Kac–Moody algebra $E_{10}$ for which it is expected that the spinors of the involutive subalgebra $K(E_{10})$ will play an important role in the algebraic description of the fermionic sector of $D = 11$ supergravity and $M$-theory. Although these results remain incomplete, they indicate a beginning of possible explanation of the structure of fermions observed in the standard model of particle physics.

In the lectures of Du Pei, *"BPS Spectra and Invariants for Three- and Four-Manifolds,"* it is described how partition functions and superconformal indices of the field theories in two, three, and four dimensions describe topological invariants of the corresponding spaces, such as the Witten–Reshetikhin–Turaev invariants. The six-dimensional space-time equipped by different ways of the compactification serves as a key basis of considerations. During these compactifications, there emerge partition functions related to each other due to the duality of corresponding theories. Among other things, lectures contain discussions of the elliptic genus, topological modular forms, and the lens space. Dr. D. Pei is a young highly qualified expert in the intersections of field theories and topology having interesting results in this area.

The series of lectures by Pierre Vanhove, *"Feynman Integrals and Mirror Symmetry,"* contains a description of general Feynman integrals in spaces of arbitrary dimension from the point of view of algebraic geometry. The sum of all Feynman integrals describes path integrals in the perturbative regime or partition functions in the Euclidean space-time. In the suggested approach to the investigation of these integrals, the toric geometry is used for derivation of differential equations satisfied by the imaginary parts of the Feynman integrals. The differential equations for the whole integrals are discussed as well. Additionally, a survey is given of an interesting connection of one concrete class of Feynman integrals with the mirror symmetry discovered by Bloch and Vanhove in 2013. This class includes into itself the two-dimensional "sunset" diagram which is expressed in terms of the Bloch's elliptic dilogarithm function obeying certain automorphic properties. Professor P. Vanhove is a world leader in investigations of Feynman integrals who has discovered important applications of automorphic forms in the theory of strings and field theory.

The lectures of Yasuhiko Yamada, *"Theory and Applications of the Elliptic Painlevé Equation,"* are devoted to the most general known nonlinear finite-difference equation describing discrete analogue of the Painlevé equations. This equation was constructed by Sakai in 2001 and it is associated with the affine root system $E_8$. Professor Y. Yamada is a well-known specialist both in quantum field theory and integrable systems. He has found the Lax pair for the elliptic Painlevé equation and investigated different properties of the corresponding tau-function. In particular, in collaboration with Kajiwara, Masuda, Noumi, and Ohta, he has shown that special tau-function of this equation satisfies the elliptic hypergeometric equation, which emerges after some reduction of the elliptic Painlevé equation. This result is analogous to the statement that the most general explicit solution of the ordinary Painlevé-VI equation is given by the Euler–Gauss hypergeometric function. Also, a beautiful interpretation of the elliptic Painlevé equation is suggested as a nonautonomous dynamical system based on the group law of addition of points on the elliptic curve. In this way, one can trace a connection with the general theory of elliptic hypergeometric functions, superconformal indices, and gauge theories in space-times of different dimensions. General automorphic properties of the tau-function of the elliptic Painlevé equation are not investigated yet.

In conclusion, we would like to acknowledge the financial support for the school from the Laboratory of Mirror Symmetry and Automorphic Forms NRU HSE (RF government grant, ag. no. 14.641.31.0001), the Bogoliubov Laboratory of Theoretical Physics JINR, the Heisenberg–Landau program, and the Russian Foundation for Basic Research. Their support was crucial for the success of the school.

Moscow, Russia                                                                               Valery A. Gritsenko
Dubna, Russia                                                                        Vyacheslav P. Spiridonov

# Contents

**A Short Introduction to the Algebra, Geometry, Number Theory and Physics of Moonshine** .................................................... 1
John F. R. Duncan

**Modified Elliptic Genus** ........................................................ 87
Valery Gritsenko

**Superconformal Indices and Instanton Partition Functions** ................. 121
Seok Kim

**On Spinorial Representations of Involutory Subalgebras of Kac–Moody Algebras** .......................................................... 179
Axel Kleinschmidt, Hermann Nicolai, and Adriano Viganò

**BPS Spectra and Invariants for Three- and Four-Manifolds** .............. 217
Du Pei

**Introduction to the Theory of Elliptic Hypergeometric Integrals** .......... 271
Vyacheslav P. Spiridonov

**Feynman Integrals and Mirror Symmetry** .................................... 319
Pierre Vanhove

**Theory and Applications of the Elliptic Painlevé Equation** ................. 369
Yasuhiko Yamada

# A Short Introduction to the Algebra, Geometry, Number Theory and Physics of Moonshine

John F. R. Duncan

## Contents

1   Elliptic Curves . . . . . . . . . . . . . . . . . . . . . . . . . . . . . . . . . . . . . . . . . . . . . . . . . . . . . . . . . . . . 2
2   Supersingular Elliptic Curves . . . . . . . . . . . . . . . . . . . . . . . . . . . . . . . . . . . . . . . . . . . . . 4
3   Ogg's Observation . . . . . . . . . . . . . . . . . . . . . . . . . . . . . . . . . . . . . . . . . . . . . . . . . . . . . . . 5
4   Complex Elliptic Curves . . . . . . . . . . . . . . . . . . . . . . . . . . . . . . . . . . . . . . . . . . . . . . . . . . 7
5   Monstrous Moonshine . . . . . . . . . . . . . . . . . . . . . . . . . . . . . . . . . . . . . . . . . . . . . . . . . . . 9
6   Elliptic Curves with Level . . . . . . . . . . . . . . . . . . . . . . . . . . . . . . . . . . . . . . . . . . . . . . . 11
7   Affine Lie Algebras . . . . . . . . . . . . . . . . . . . . . . . . . . . . . . . . . . . . . . . . . . . . . . . . . . . . . 14
8   Modular Forms . . . . . . . . . . . . . . . . . . . . . . . . . . . . . . . . . . . . . . . . . . . . . . . . . . . . . . . . . 17
9   Vertex Operators . . . . . . . . . . . . . . . . . . . . . . . . . . . . . . . . . . . . . . . . . . . . . . . . . . . . . . . 20
10 Conformal Field Theory . . . . . . . . . . . . . . . . . . . . . . . . . . . . . . . . . . . . . . . . . . . . . . . . . 24
11 The Monster Module . . . . . . . . . . . . . . . . . . . . . . . . . . . . . . . . . . . . . . . . . . . . . . . . . . . . 26
12 Affinization . . . . . . . . . . . . . . . . . . . . . . . . . . . . . . . . . . . . . . . . . . . . . . . . . . . . . . . . . . . . 30
13 Vertex Algebras . . . . . . . . . . . . . . . . . . . . . . . . . . . . . . . . . . . . . . . . . . . . . . . . . . . . . . . . 31
14 Vertex Operator Algebras . . . . . . . . . . . . . . . . . . . . . . . . . . . . . . . . . . . . . . . . . . . . . . . 34
15 Orbifolds . . . . . . . . . . . . . . . . . . . . . . . . . . . . . . . . . . . . . . . . . . . . . . . . . . . . . . . . . . . . . . 36
16 The Genus Zero Problem . . . . . . . . . . . . . . . . . . . . . . . . . . . . . . . . . . . . . . . . . . . . . . . . 38
17 Super Structure . . . . . . . . . . . . . . . . . . . . . . . . . . . . . . . . . . . . . . . . . . . . . . . . . . . . . . . . 42
18 Super Moonshine . . . . . . . . . . . . . . . . . . . . . . . . . . . . . . . . . . . . . . . . . . . . . . . . . . . . . . 45
19 Superconformal Structure . . . . . . . . . . . . . . . . . . . . . . . . . . . . . . . . . . . . . . . . . . . . . . . 48
20 Conway Moonshine . . . . . . . . . . . . . . . . . . . . . . . . . . . . . . . . . . . . . . . . . . . . . . . . . . . . 51
21 The Mathieu Groups . . . . . . . . . . . . . . . . . . . . . . . . . . . . . . . . . . . . . . . . . . . . . . . . . . . . 53
22 K3 Surfaces . . . . . . . . . . . . . . . . . . . . . . . . . . . . . . . . . . . . . . . . . . . . . . . . . . . . . . . . . . . 56
23 Complex Elliptic Genera . . . . . . . . . . . . . . . . . . . . . . . . . . . . . . . . . . . . . . . . . . . . . . . . 58
24 Mathieu Moonshine . . . . . . . . . . . . . . . . . . . . . . . . . . . . . . . . . . . . . . . . . . . . . . . . . . . . 61
25 Conway Returns . . . . . . . . . . . . . . . . . . . . . . . . . . . . . . . . . . . . . . . . . . . . . . . . . . . . . . . . 66
26 Forward from Here . . . . . . . . . . . . . . . . . . . . . . . . . . . . . . . . . . . . . . . . . . . . . . . . . . . . . 70
References . . . . . . . . . . . . . . . . . . . . . . . . . . . . . . . . . . . . . . . . . . . . . . . . . . . . . . . . . . . . . . . . . 75

J. F. R. Duncan (✉)
Department of Mathematics, Emory University, Atlanta, GA, USA
e-mail: john.duncan@emory.edu

© Springer Nature Switzerland AG 2020
V. A. Gritsenko, V. P. Spiridonov (eds.), *Partition Functions and Automorphic Forms*, Moscow Lectures 5,
https://doi.org/10.1007/978-3-030-42400-8_1

**Abstract** Moonshine arose in the 1970s as a collection of coincidences connecting modular functions to the monster simple group, which was newly discovered at that time. The effort to elucidate these connections led to new algebraic structures, and fruitful cross-fertilization between mathematics and physics. In this century the field has been further enriched, with the discovery of new roles for finite groups in geometry, and new relations to number theory. We offer an introduction and invitation to this theory in these notes.

# 1 Elliptic Curves

Since they are almost always present in our story we begin with elliptic curves. An *elliptic curve* over a field $K$ is a pair $(E, O)$ where $E$ is a non-singular projective algebraic curve over $K$ of genus one, and $O$ is a point of $E$ that is defined over $K$. In elementary terms (see e.g. Chapter III of [1]) this means that $E$ is given by the solutions in $\mathbb{P}^2(K)$ of a *Weierstrass equation*

$$E \ : \ y^2 + a_1 xy + a_3 y = x^3 + a_2 x^2 + a_4 x + a_6, \tag{1}$$

for some $a_1, a_2, a_3, a_4, a_6 \in K$ such that the *discriminant*

$$\Delta = \Delta(E) := -b_2^2 b_8 - 8b_4^3 - 27b_6^2 + 9b_2 b_4 b_6 \tag{2}$$

is non-zero, where $b_2 := a_1^2 + 4a_2$ and $b_4 := 2a_4 + a_1 a_3$ and $b_6 := a_3^2 + 4a_6$ and $b_8 := a_1^2 a_6 + 4a_2 a_6 - a_1 a_3 a_4 + a_2 a_3^2 - a_4^2$. Strictly, we should use the homogeneous cubic $Y^2 Z + a_1 XYZ + \ldots$ corresponding to (1) in order to specify points in $\mathbb{P}^2(K)$. In practice the *affine form* (1) is usually given. The distinguished point is $O = [0, 1, 0]$, and is often suppressed from notation.

It is natural to ask for a description of the set of elliptic curves over $K$. We can address this quite concretely when $K$ is algebraically closed by considering the *j-invariant*. This is the $K$-valued function on elliptic curves over $K$ which is defined on $E$ as in (1) by setting

$$j(E) := \frac{c_4^3}{\Delta} \tag{3}$$

where $c_4 := b_2^2 - 24b_4$. The next result shows that if $K$ is algebraically closed then elliptic curves over $K$ are parameterized, via the $j$-invariant, by $K$.

**Proposition 1** *Let $\overline{K}$ be an algebraic closure of $K$.*

1. *If $E$ and $E'$ are elliptic curves defined over $K$ then $E$ is isomorphic to $E'$ over $\overline{K}$ if and only if $j(E) = j(E')$.*
2. *For every $j_0 \in \overline{K}$ there exists an elliptic curve $E_0$ defined over $K(j_0)$ such that $j(E_0) = j_0$.*

See e.g. §1 in Chapter III of [1] for a proof of Proposition 1.

Another useful fact is that if the characteristic of $K$ is not 2 or 3 then for $c_6 := -b_2^3 + 36b_2b_4 - 216b_6$ we have $\Delta = \frac{1}{1728}(c_4^3 - c_6^2)$, and (1) can be brought into the simpler form

$$E \; : \; y^2 = x^3 + Ax + B. \tag{4}$$

Tracing through the required change of variables (cf. e.g. §2 in Chapter 3 of [2]) one sees that $c_4 = -48A$ and $c_6 = -864B$ so that $\Delta = -16(4A^3 + 27B^2)$ and

$$j(E) = 1728 \frac{4A^3}{4A^3 + 27B^2} \tag{5}$$

for $E$ as in (4).

A fundamental problem in the theory of elliptic curves is to determine the set $E(K)$ of *K-rational points* of $E$, which are the $[x, y, 1]$ where $(x, y) \in K^2$ solves (1) together with $O = [0, 1, 0]$. There is a million dollar bounty on a solution to this (cf. [3]) already in the case that $K = \mathbb{Q}$.

The significance of the requirement that an elliptic curve have genus one is that it leads naturally to an abelian group structure on $E(K)$ (see e.g. §2 in Chapter III of [1]). It turns out that this group is finitely generated when $K$ is a number field (i.e. a finite-dimensional field extension of $\mathbb{Q}$).

**Theorem 1 (Mordell [4], Weil [5])** *If $K$ is a number field then $E(K)$ is finitely generated.*

The $K = \mathbb{Q}$ case of Theorem 1 was proved by Mordell in 1922. Weil extended it to number fields (and higher dimensional abelian varieties) in his doctoral thesis of 1928. A proof of Theorem 1 is explained in Chapter VIII of [1].

Of course maps between elliptic curves play an essential role in their theory. An *isogeny* of elliptic curves $(E, O) \to (E', O')$ is a morphism $\phi : E \to E'$ of algebraic varieties such that $\phi(O) = O'$. It turns out that isogenies of elliptic curves automatically preserve their group structure (cf. Theorem 4.8 in Chapter III of [1]) so we may write $\mathrm{Hom}(E, E')$ for the group of isogenies from $E$ and $E'$, and $\mathrm{End}(E)$ for the ring of isogenies on $E$.

An isogeny $\phi : E \to E'$ of elliptic curves over $K$ naturally defines an injection $\phi^* : \overline{K}(E') \to \overline{K}(E)$ of function fields for $\overline{K}$ an algebraic closure of $K$. The *degree* of $\phi$ is the degree $[\overline{K}(E) : \overline{K}(E')]$ of the corresponding field extension. One obtains the following result by using the fact that the degree, so defined, is multiplicative on $\mathrm{End}(E)$ (cf. Proposition 4.2 in Chapter III of [1]).

**Proposition 2** *For $E$ an elliptic curve $\mathrm{End}(E)$ is an integral domain of characteristic zero.*

## 2  Supersingular Elliptic Curves

The ring $\text{End}(E)$ of isogenies on $E$ may be regarded as a measure of the symmetry of $E$. In order to describe this symmetry more precisely we now review some facts from algebra.

Recall that a number field is called *quadratic imaginary* if it takes the form $K = \mathbb{Q}(\sqrt{D})$ for $D$ a negative integer. Say that $D \in \mathbb{Z}$ is a *fundamental discriminant* if $D \equiv 1 \bmod 4$ and $D$ is squarefree, or if $D = 4d$ where $d \equiv 2, 3 \bmod 4$ and $d$ is squarefree. Then every quadratic imaginary field arises as $K = \mathbb{Q}(\sqrt{D})$ for a unique fundamental discriminant $D < 0$, and

$$\mathcal{O} := \mathbb{Z}\left[\frac{D + \sqrt{D}}{2}\right] \tag{6}$$

is the ring of integers in $K$ (i.e. the integral closure of $\mathbb{Z}$ in $K$).

The ring of integers $\mathcal{O}$ in a number field $K$ is an example of an order in $K$. More generally, for $\mathcal{K}$ an algebra over $\mathbb{Q}$ an *order* in $\mathcal{K}$ is a subring $\mathcal{R} < \mathcal{K}$ that is finitely generated as a $\mathbb{Z}$-module and satisfies $\mathcal{R} \otimes_{\mathbb{Z}} \mathbb{Q} = \mathcal{K}$. For $K$ quadratic imaginary the orders in $K$ are exactly the subrings of the form $\mathcal{R} = \mathbb{Z} + f\mathcal{O}$ for $f \in \mathbb{Z}^{+}$.

Number fields are perhaps the most prominent examples of $\mathbb{Q}$-algebras in number theory. Another important class of examples is the *quaternion algebras* over $\mathbb{Q}$, which are the $\mathbb{Q}$-algebras of the form

$$\mathcal{K} = \mathbb{Q} + \mathbb{Q}i + \mathbb{Q}j + \mathbb{Q}k \tag{7}$$

where $i^2$ and $j^2$ are non-zero elements of $\mathbb{Q}$ and $ij = k = -ji$. Probably the most familiar of these is *Hamilton's quaternions*, which we obtain by taking $i^2 = j^2 = -1$. We obtain the *split* quaternion algebra by taking $i^2 = j^2 = 1$. For a comprehensive introduction to quaternion algebras we refer to [6].

Observe that $\mathcal{K} \otimes_{\mathbb{Q}} \overline{\mathbb{Q}} \simeq M_2(\overline{\mathbb{Q}})$ for $\mathcal{K}$ a quaternion algebra and $\overline{\mathbb{Q}}$ an algebraic closure of $\mathbb{Q}$ where $M_2(K)$ is the ring of $2 \times 2$ matrices with coefficients in $K$. Indeed if $i^2 = a$ and $j^2 = b$ then an explicit isomorphism is given by

$$i \mapsto \frac{1}{\sqrt{a}}\begin{pmatrix} 1 & 0 \\ 0 & -1 \end{pmatrix}, \quad j \mapsto \begin{pmatrix} 0 & b \\ 1 & 0 \end{pmatrix}. \tag{8}$$

So in particular the split quaternion algebra is a copy of $M_2(\mathbb{Q})$. With this as inspiration say that $\mathcal{K}$ is *split* at $p$, for $p$ a prime or $p = \infty$, if

$$\mathcal{K} \otimes_{\mathbb{Q}} \mathbb{Q}_p \simeq M_2(\mathbb{Q}_p) \tag{9}$$

where $\mathbb{Q}_p$ is the completion of $\mathbb{Q}$ with respect to the $p$-adic metric for $p > 0$, and $\mathbb{Q}_\infty = \mathbb{R}$. Say that $\mathcal{K}$ is *ramified* at $p$ otherwise. A quaternion algebra that is ramified at $\infty$ is called *definite*.

For $N$ prime there is a unique definite quaternion algebra over $\mathbb{Q}$ that is ramified only at $\infty$ and $N$. For example Hamilton's quaternions are ramified at $\infty$ and $N = 2$. For $N \equiv 3 \bmod 4$ we obtain an example ramified only at $\infty$ and $N$ by taking $i^2 = -1$ and $j^2 = -N$. For $N \equiv 5 \bmod 8$ take $i^2 = -2$ and $j^2 = -N$, and for $N \equiv 1 \bmod 8$ choose a prime $\ell \equiv 3 \bmod 4$ which is a quadratic residue mod $N$ and take $i^2 = -\ell$ and $j^2 = -N$.

The endomorphism ring of an elliptic curve $E$ is a free $\mathbb{Z}$-module of positive rank according to Proposition 2. The next result shows that the possibilities are very restricted. See Theorem 9.3 in Chapter III of [1] for a proof.

**Theorem 2** *For $E$ an elliptic curve either* $\mathrm{End}(E) \simeq \mathbb{Z}$ *or* $\mathrm{End}(E)$ *is an order in a quadratic imaginary field or* $\mathrm{End}(E)$ *is an order in a definite quaternion algebra over* $\mathbb{Q}$.

So in particular $\mathrm{End}(E)$ has rank 1, 2 or 4 as a $\mathbb{Z}$-module. In the extreme case, that $\mathrm{End}(E)$ has rank 4 and $\mathrm{End}(E) \otimes_{\mathbb{Z}} \mathbb{Q}$ is a quaternion algebra, the elliptic curve $E$ is called *supersingular*. Deuring made a comprehensive study [7] of $\mathrm{End}(E)$ for supersingular $E$ and obtained the following result (amongst others). To state it we let $\mathbb{F}_q$ denote the unique field with $q$ elements for $q$ a power of a prime.

**Theorem 3 (Deuring [7])** *If $E$ is a supersingular elliptic curve defined over $K$ then* $\mathrm{char}(K) > 0$, *the quaternion algebra* $\mathcal{K} = \mathrm{End}(E) \otimes_{\mathbb{Z}} \mathbb{Q}$ *is ramified only at $\infty$ and $p = \mathrm{char}(K)$, and* $\mathrm{End}(E)$ *is a maximal order in $\mathcal{K}$. Also,* $j(E) \in \mathbb{F}_{p^2}$.

It might be regarded as curious that the $j$-invariants of supersingular elliptic curves are so restricted as to lie in $\mathbb{F}_{p^2}$. Are there primes $p$ for which the $j$-invariants of supersingular curves in characteristic $p$ always lie in the prime subfield $\mathbb{F}_p$? Ogg answered this question, for $K$ an algebraic closure of $\mathbb{F}_p$, in 1975.

**Theorem 4 (Ogg [8])** *Let $p$ be a prime. Then every supersingular elliptic curve $E$ over $\overline{\mathbb{F}}_p$ has $j(E) \in \mathbb{F}_p$ if and only if $p$ belongs to the following list.*

$$2, 3, 5, 7, 11, 13, 17, 19, 23, 29, 31, 41, 47, 59, 71. \tag{10}$$

## 3 Ogg's Observation

We now turn to a topic which at first glance seems unrelated to elliptic curves, viz. the classification of finite simple groups. This 100 year program (see [9, 10] for detailed accounts) entered a state of high activity in the 1960s. This was motivated not just by the development of new methods but also by the unearthing of new examples. Particularly, Janko's unexpected discovery [11] in 1965 of a finite simple group $J_1$ with an element $z$ of order 2 whose centralizer $C_{J_1}(z) := \{g \in J_1 \mid zg = gz\}$ takes the form

$$C_{J_1}(z) \simeq \mathbb{Z}/2\mathbb{Z} \times PSL_2(\mathbb{F}_5) \simeq \mathbb{Z}/2\mathbb{Z} \times A_5 \tag{11}$$

initiated a decade of exploration during which 20 further examples of *sporadic simple groups*—i.e. non-abelian finite simple groups that are not alternating groups or simple groups of Lie type—would be revealed. (See e.g. [12–14] for background on groups of Lie type.)

The largest of these sporadic simple groups came to light in 1973, contemporaneously with Ogg's Theorem 4, when Fischer and Griess independently produced evidence (see [15] and §15 of [16]) for a finite simple group $\mathbb{M}$ with an involution centralizer $C_{\mathbb{M}}(z)$ that participates in a (non-split) short exact sequence

$$1 \to 2^{1+24} \to C_{\mathbb{M}}(z) \to Co_1 \to 1. \tag{12}$$

Here $2^{1+24}$ denotes a kind of finite Heisenberg group, itself satisfying a (non-split) short exact sequence

$$1 \to \mathbb{Z}/2\mathbb{Z} \to 2^{1+24} \to (\mathbb{Z}/2\mathbb{Z})^{24} \to 1, \tag{13}$$

and $Co_1$ is the largest sporadic simple group of Conway (which we will meet again later on, see (98)).

A proof of the existence of $\mathbb{M}$, now known as the *Fischer–Griess monster*, was announced by Griess in 1980 (see §1 of [16]) but many properties of $\mathbb{M}$ were deduced in advance of this. In particular it was determined that the minimal $d$ for which $\mathbb{M}$ maps non-trivially to $GL_d(\mathbb{C})$ should be $d = 196883$,

$$\mathbb{M} \to GL_{196833}(\mathbb{C}), \tag{14}$$

and the order of $\mathbb{M}$ was predicted to be

$$\#\mathbb{M} = 2^{46}.3^{20}.5^9.7^6.11^2.13^3.17.19.23.29.31.41.47.59.71$$
$$= 808017424794512875886459904961710757005754368000000000 \tag{15}$$

(hence the name).

Jacques Tits reported on conjectural properties of the monster group, including its order, in his inaugural lecture at the Collège de France on 14 January 1975. Andrew Ogg was present, and observed that the primes (10) arising in his theorem on $j$-invariants of supersingular elliptic curves are exactly the primes that appear in (15), as divisors of the order of $\mathbb{M}$! Ogg offered a bottle of Jack Daniels[1] for an explanation of this coincidence. (See Remarque 1 in [8]. Also see [17] for recent work that puts Ogg's observation in a more general context.)

---

[1] Jack Daniels is the brand name of a certain Tennessee whiskey.

We may now recognize Ogg's observation as the initiating step in the development of moonshine. For the next step we return to elliptic curves but specialize to $K = \mathbb{C}$.

## 4 Complex Elliptic Curves

We now consider elliptic curves over the complex number field $\mathbb{C}$. For $E$ such a curve the set $E(\mathbb{C})$ of $\mathbb{C}$-rational points inherits a complex manifold structure from its embedding in $\mathbb{P}^2(\mathbb{C})$, and the group law on $E$ equips $E(\mathbb{C})$ with product and inverse maps that are given locally by complex analytic functions. That is, $E(\mathbb{C})$ is naturally a complex Lie group of (complex) dimension 1. On the other hand if $\Lambda$ is a discrete subgroup of $\mathbb{C}$ with the property that $\Lambda \otimes_{\mathbb{Z}} \mathbb{R} = \mathbb{C}$ then the quotient $\mathbb{C}/\Lambda$ is also a one-dimensional complex Lie group. One obtains an elliptic curve $E$ and a complex analytic isomorphism $\mathbb{C}/\Lambda \to E(\mathbb{C})$ by defining the *Weierstrass* $\wp$-*function*

$$\wp(\Lambda, z) := \frac{1}{z^2} + \sum_{\substack{\omega \in \Lambda \\ \omega \neq 0}} \frac{1}{(z - \omega)^2} - \frac{1}{\omega^2}, \tag{16}$$

and checking that

$$\wp'(\Lambda, z)^2 = 4\wp(\Lambda, z)^3 - g_2(\Lambda)\wp(\Lambda, z) - g_3(\Lambda) \tag{17}$$

for $z \in \mathbb{C} \setminus \Lambda$ where $\wp'(\Lambda, z) := \frac{d}{dz}\wp(\Lambda, z)$ and

$$g_2(\Lambda) := 60 \sum_{\substack{\omega \in \Lambda \\ \omega \neq 0}} \frac{1}{\omega^4}, \quad g_3(\Lambda) := 140 \sum_{\substack{\omega \in \Lambda \\ \omega \neq 0}} \frac{1}{\omega^6}. \tag{18}$$

Equation (17) shows that if $E$ is given by $y^2 = x^3 + Ax + B$ (cf. (4)) where $A = -\frac{1}{4}g_2(\Lambda)$ and $B = -\frac{1}{4}g_3(\Lambda)$ then the map $\mathbb{C}/\Lambda \to \mathbb{P}^2(\mathbb{C})$ which sends $z + \Lambda$ to $[\wp(\Lambda, z), 2\wp'(\Lambda, z), 1]$ for $z \notin \Lambda$, and sends the trivial coset to $O = [0, 1, 0]$, defines the desired isomorphism $\mathbb{C}/\Lambda \to E(\mathbb{C})$. The *uniformization theorem* for elliptic curves (see Proposition 5 in Chapter VII of [18], or §4.2 of [19]) states that every elliptic curve over $\mathbb{C}$ arises in this way. (For a verification of this fact that does not use the uniformization theorem see Chapter VI of [2].)

So to understand complex elliptic curves we may focus our attention on the discrete subgroups $\Lambda \subset \mathbb{C}$ such that $\Lambda \otimes_{\mathbb{Z}} \mathbb{R} = \mathbb{C}$. When do two such subgroups define the same elliptic curve? Certainly $\mathbb{C}/\Lambda \simeq \mathbb{C}/\Lambda'$ if $\Lambda = \alpha\Lambda'$ for some nonzero $\alpha \in \mathbb{C}$. For this reason every complex elliptic curve takes the form

$$E_\tau := \mathbb{C}/\Lambda_\tau \tag{19}$$

for some $\tau$ in the *upper half-plane* $\mathbb{H} := \{\tau \in \mathbb{C} \mid \Im(\tau) > 0\}$ where $\Lambda_\tau := \mathbb{Z}\tau + \mathbb{Z}$. Indeed if $\Lambda = \mathbb{Z}\omega_1 + \mathbb{Z}\omega_2$ for some $\omega_1, \omega_2 \in \mathbb{C}$ then the quotient $\frac{\omega_1}{\omega_2}$ has non-zero imaginary part (otherwise $\Lambda \otimes_{\mathbb{Z}} \mathbb{R} = \mathbb{R}\omega_1 \neq \mathbb{C}$) so that $\Lambda = \omega_2 \Lambda_\tau$ for $\tau = \pm\frac{\omega_1}{\omega_2}$ (for one or other choice of sign).

Now we may ask when $E_\tau$ and $E_{\tau'}$ are isomorphic for given $\tau, \tau' \in \mathbb{H}$. To answer this question observe that $PSL_2(\mathbb{C})$ acts naturally on $\mathbb{P}^1(\mathbb{C})$ according to

$$
\begin{bmatrix} a & b \\ c & d \end{bmatrix} \begin{bmatrix} \omega_1 \\ \omega_2 \end{bmatrix} = \begin{bmatrix} a\omega_1 + b\omega_2 \\ c\omega_1 + d\omega_2 \end{bmatrix}
\tag{20}
$$

so $SL_2(\mathbb{R})$ acts naturally on $\mathbb{P}^1(\mathbb{C})$ via the composition of natural maps $SL_2(\mathbb{R}) \to PSL_2(\mathbb{R}) \to PSL_2(\mathbb{C})$. Now identify $\mathbb{H}$ as a subset of $\mathbb{P}^1(\mathbb{C})$ via

$$
\tau \mapsto \begin{bmatrix} \tau \\ 1 \end{bmatrix}.
\tag{21}
$$

Then this copy of $\mathbb{H}$ forms a single orbit for the action of $SL_2(\mathbb{R})$ on $\mathbb{P}^1(\mathbb{C})$ with point stabilizer isomorphic to $SO_2(\mathbb{R})$. (In particular the homogeneous space $SL_2(\mathbb{R})/SO_2(\mathbb{R})$ has the atypical property that it admits a complex analytic structure.)

So for given $\tau, \tau' \in \mathbb{H}$ there exist $a, b, c, d \in \mathbb{R}$ such that $ad - bc = 1$ and

$$
\tau' = \begin{pmatrix} a & b \\ c & d \end{pmatrix} \tau := \frac{a\tau + b}{c\tau + d}.
\tag{22}
$$

Set $\alpha = c\tau + d$. Then $\alpha\Lambda_{\tau'} = \mathbb{Z}(a\tau + b) + \mathbb{Z}(c\tau + d)$, which is contained in $\Lambda_\tau$ if $a, b, c, d \in \mathbb{Z}$, and then actually equal to $\Lambda_\tau$ since $ad - bc = 1$. So $E_{\tau'} \simeq E_\tau$ if there exists $\gamma \in SL_2(\mathbb{Z})$ such that $\tau' = \gamma\tau$, and in this case multiplication by $\alpha = c\tau + d$ gives an explicit isomorphism $E_{\tau'} \to E_\tau$ when $\gamma = \begin{pmatrix} * & * \\ c & d \end{pmatrix}$. This establishes the "if" part of the following result.

**Proposition 3** *We have $E_{\tau'} \simeq E_\tau$ for $\tau, \tau' \in \mathbb{H}$ if and only if there exists $\gamma \in SL_2(\mathbb{Z})$ such that $\tau' = \gamma\tau$.*

The "only if" part of Proposition 3 may be proved by noting that any isomorphism $\mathbb{C}/\Lambda_{\tau'} \to \mathbb{C}/\Lambda_\tau$ lifts to an isomorphism $\mathbb{C} \to \mathbb{C}$ of covering spaces, which is necessarily multiplication by $\alpha$ for some $\alpha \in \mathbb{C}^*$. (Cf. e.g. Theorem 5.2 in Chapter V of [20].)

Proposition 3 tells us that isomorphism classes of complex elliptic curves are in correspondence with the orbits of the *modular group* $SL_2(\mathbb{Z})$ on $\mathbb{H}$ where the action is as in (22). Invoking Proposition 1 we conclude that the $j$-invariant (3) defines a bijective map

$$
SL_2(\mathbb{Z}) \backslash \mathbb{H} \xrightarrow{\sim} \mathbb{C}.
\tag{23}
$$

Alternatively we may regard the $j$-invariant as defining an $SL_2(\mathbb{Z})$-invariant function $j(\tau) := j(E_\tau)$ on $\mathbb{H}$.

We obtain that $j(\tau)$, the *elliptic modular invariant*, is complex analytic by observing that

$$j(\tau) = 1728 \frac{g_2(\tau)^3}{g_2(\tau)^3 - 27g_3(\tau)^2} \tag{24}$$

where $g_2(\tau) := g_2(\Lambda_\tau)$ and $g_3(\tau) := g_3(\Lambda_\tau)$ are complex analytic by construction (cf. (18)). Because $\mathbb{Z}(\tau+1)+\mathbb{Z} = \mathbb{Z}\tau+\mathbb{Z}$ we have $j(\tau+1) = j(\tau)$ for all $\tau$ so $j(\tau)$ admits a Fourier expansion of the form $j(\tau) = \sum_{n \in \mathbb{Z}} c(n)e^{2\pi i n \tau}$. It is traditional to set $q = e^{2\pi i \tau}$ and write

$$j(\tau) = \sum_{n \in \mathbb{Z}} c(n)q^n. \tag{25}$$

The fact that $j(\tau)$ is injective on $SL_2(\mathbb{Z})\backslash\mathbb{H}$ then forces $c(n) = 0$ for $n < -1$. It turns out that with the given choice of scaling (24) all the remaining coefficients are positive integers,

$$j(\tau) = q^{-1} + 744 + 196884q + 21493760q^2 + 864299970q^3 + \ldots \tag{26}$$

## 5 Monstrous Moonshine

We have noted (14) that 196883 has a special significance for the monster. John McKay observed, sometime in the 1970s (cf. [21, 22]), that

$$196884 = 1 + 196883 \tag{27}$$

is the coefficient of $q$ in the Fourier expansion (26) of the elliptic modular invariant (24). Could this be merely a coincidence?

By 1979 the character table of the monster had been worked out, assuming the existence of the embedding (14), by Fischer, Livingstone and Thorne (cf. [21, 23]). With this at hand Thompson observed analogues of (27) for higher coefficients of $j(\tau)$ such as

$$21493760 = 1 + 196883 + 21296876, \tag{28}$$

$$864299970 = 2 \times 1 + 2 \times 196883 + 21296876 + 842609326, \tag{29}$$

and suggested [22] a possible explanation. Namely, Thompson suggested that there is an infinite-dimensional graded module

$$V = \bigoplus_{n \geq -1} V_n \tag{30}$$

for $\mathbb{M}$ such that dim $V_n$ is the coefficient of $q^n$ in the Fourier expansion (25–26) of $j(\tau)$ (at least for $n \neq 0$).

The notion that the $j$-invariant for complex elliptic curves could have any meaningful relationship to the monster group could have been passed off as lunacy but Thompson formulated a substantially stronger relationship between complex elliptic curves and the monster [24], and significant evidence in support of this was presented by Conway–Norton [21]. Taken together their work led to the following amazing *monstrous moonshine* conjecture.

*Conjecture 1 (Thompson–Conway–Norton [21, 22, 24])* There exists an $\mathbb{M}$-module $V = \bigoplus_{n \geq -1} V_n$ such that

$$\sum_{n \geq -1} \dim V_n q^n = j(\tau) - 744. \tag{31}$$

More generally, for each $g \in \mathbb{M}$ the *McKay–Thompson series*

$$T_g(\tau) := \sum_{n \geq -1} \mathrm{tr}(g|V_n)q^n \tag{32}$$

is the normalized principal modulus for a genus zero group $\Gamma_g < SL_2(\mathbb{R})$.

To appreciate Conjecture 1 we should explain the terms "normalized principal modulus" and "genus zero group". For this note first that $\mathbb{P}^1(\mathbb{R}) \subset \mathbb{P}^1(\mathbb{C})$ is another orbit for the natural action of $SL_2(\mathbb{R})$ on $\mathbb{P}^1(\mathbb{C})$, and is naturally identified with the boundary of $\mathbb{H} \subset \mathbb{P}^1(\mathbb{C})$ (cf. (21)). Also $\mathbb{P}^1(\mathbb{Q}) \subset \mathbb{P}^1(\mathbb{R})$ is an orbit for the subgroup $SL_2(\mathbb{Z})$. We have $j(\tau) = q^{-1} + O(1)$ as $\Im(\tau) \to \infty$ (i.e. $|j(\tau) - e^{-2\pi i \tau}|$ remains bounded as $\Im(\tau) \to \infty$). So by modular invariance (i.e. $SL_2(\mathbb{Z})$-invariance) $j(\tau) \to \infty$ as $\tau$ tends to any point of $\mathbb{P}^1(\mathbb{Q})$ (regarded as a subset of $\partial\mathbb{H}$). For this reason $j(\tau)$ induces an isomorphism

$$SL_2(\mathbb{Z}) \backslash \widehat{\mathbb{H}} \xrightarrow{\sim} \widehat{\mathbb{C}} \tag{33}$$

of compact complex curves where $\widehat{\mathbb{H}} := \mathbb{H} \cup \mathbb{P}^1(\mathbb{Q})$ and $\widehat{\mathbb{C}} := \mathbb{C} \cup \{\infty\}$ is the one point compactification of $\mathbb{C}$. In particular there is a natural complex manifold structure on the orbit space $SL_2(\mathbb{Z}) \backslash \widehat{\mathbb{H}}$. We say that the modular group $SL_2(\mathbb{Z})$ is *genus zero* because the (real) orientable surface underlying $SL_2(\mathbb{Z}) \backslash \widehat{\mathbb{H}} \simeq \widehat{\mathbb{C}}$ is a sphere. We say that $j(\tau)$ is a *principal modulus* for $SL_2(\mathbb{Z})$ because it induces an isomorphism (33) that witnesses this fact. We say that $J(\tau) := j(\tau) - 744$ is the *normalized* principal modulus for $SL_2(\mathbb{Z})$ because $J(\tau) = q^{-1} + O(q)$ as $\Im(\tau) \to$ $\infty$ (i.e. $|J(\tau) - e^{-2\pi i \tau}| < Ce^{-2\pi \Im(\tau)}$ for some $C > 0$, for $\Im(\tau)$ sufficiently large).

For a more general statement suppose that $\Gamma < SL_2(\mathbb{R})$ is *commensurable* with $SL_2(\mathbb{Z})$ in the sense that $\Gamma \cap SL_2(\mathbb{Z})$ has finite index in both $\Gamma$ and $SL_2(\mathbb{Z})$. Then the action of $\Gamma$ on $\mathbb{P}^1(\mathbb{R})$ preserves the subset $\mathbb{P}^1(\mathbb{Q})$ (cf. Proposition 2.13 in [25]) so it makes sense to consider $\Gamma \backslash \widehat{\mathbb{H}}$. This space admits a compact complex manifold structure in a similar fashion to the case that $\Gamma = SL_2(\mathbb{Z})$ (cf. §1.5 of [19]), so the underlying real surface has a well-defined genus. We say that $\Gamma$ has *genus zero* if $\Gamma \backslash \widehat{\mathbb{H}}$ has genus zero, i.e. if there exists an isomorphism

$$\Gamma \backslash \widehat{\mathbb{H}} \xrightarrow{\sim} \widehat{\mathbb{C}} \tag{34}$$

of complex curves. A $\Gamma$-invariant function on $\mathbb{H}$ that induces such an isomorphism (34) is called a *principal modulus* (or *Hauptmodul*) for $\Gamma$. A principal modulus $T$ for $\Gamma$ is called *normalized* if it satisfies $T(\tau) = q^{-1} + O(q)$ as $\Im(\tau) \to \infty$. There are no non-constant bounded analytic functions on a complex curve so each genus zero group $\Gamma < SL_2(\mathbb{R})$ has a uniquely determined normalized principal modulus $T_\Gamma$.

Since principal moduli only exist for genus zero groups the condition that $T_g$ be a normalized principal modulus is often referred to as the *genus zero property* of monstrous moonshine. Because normalized principal moduli are unique the genus zero property imbues Conjecture 1 with impressive predictive power. For example, assuming the conjecture we can compute the trace of a given $g \in \mathbb{M}$ on any of the putative $\mathbb{M}$-modules $V_n$ of (30) as soon as we know the associated group $\Gamma_g$. Since there are 194 conjugacy classes in the monster this reduces the entire $\mathbb{M}$-module structure (30) to the determination of 194 groups $\Gamma_g < PSL_2(\mathbb{R})$. We may say that the explicit specification of these groups (it turns out that 171 distinct groups arise) is one of the main features of [21]. On the flipside it is hard to imagine how such a specification might have been possible without the genus zero hypothesis.

As we will review in Sect. 16 the monstrous moonshine conjecture was verified by Borcherds [26] in 1992 (see Theorem 9). He developed new algebraic structures and techniques for this purpose which have since found applications in many areas of mathematics and physics.

## 6  Elliptic Curves with Level

Conjecture 1 says that there is a non-trivial graded infinite-dimensional $\mathbb{M}$-module for which the McKay–Thompson series $T_e(\tau) = j(\tau) - 744$ (cf. (31 and 32)) attached to the identity element parametrizes isomorphism classes of complex elliptic curves. It is natural to ask if there are geometric interpretations for the other monstrous trace functions $T_g$ (32).

We can answer this question by shifting our focus from elliptic curves to isogenies of elliptic curves (cf. Proposition 2). Then a natural counterpart to the setup of Sect. 4 is to consider equivalence classes of isogenies $\phi : E_1 \to E_2$ with the property that $\ker \phi$ has a certain fixed structure, where $\phi$ and $\phi' : E_1' \to E_2'$ are regarded as equivalent if there exist isomorphisms $\psi_i : E_i \to E_i'$ such that $\psi_2 \circ \phi = \phi' \circ \psi_1$. This gives an example of what is known as *level structure* for elliptic curves. (See the introduction of [27] for a general discussion.)

Considering isogenies $\phi$ for which $\ker \phi = \{O\}$ is trivial we recover the problem of describing elliptic curves up to isomorphism, which is solved for $K = \mathbb{C}$ by $j(\tau)$ (24) or $T_e(\tau) = j(\tau) - 744$. To interpret more general trace functions $T_g$ (32) we consider the case that $\ker \phi$ is cyclic with some fixed order.

So choose a positive integer $N$ and consider isogenies of complex elliptic curves with kernel isomorphic to $\mathbb{Z}/N\mathbb{Z}$. Working up to equivalence, any such isogeny must take the form

$$\phi : E_\tau \to E_\tau / C \tag{35}$$

for some $\tau \in \mathbb{H}$ and some subgroup $C \simeq \mathbb{Z}/N\mathbb{Z}$ of $E_\tau$, according to the discussion in Sect. 4. Under what conditions is this isogeny equivalent to another one of the same form?

Given $\gamma \in SL_2(\mathbb{Z})$, a point $\tau \in \mathbb{H}$ and a subgroup $C < E_\tau$ define $\gamma^* C$ to be the kernel of the composition

$$E_{\gamma\tau} \to E_\tau \to E_\tau / C \tag{36}$$

where the first map is multiplication by $c\tau + d$ when $\gamma = \left(\begin{smallmatrix} * & * \\ c & d \end{smallmatrix}\right)$ (cf. Proposition 3), and the second map is the natural projection. Then $\gamma^* C$ is also cyclic of order $N$ since the first map in (36) is an isomorphism so we obtain an action of $SL_2(\mathbb{Z})$ on isogenies as in (35) by defining $\gamma\phi$ to be the natural projection

$$\gamma\phi : E_{\gamma\tau} \to E_{\gamma\tau} / \gamma^* C. \tag{37}$$

Any $\mathbb{Z}/N\mathbb{Z}$ subgroup of $E_\tau$ takes the form $C = \langle \frac{c}{N}\tau + \frac{d}{N} \rangle$ for some $c, d \in \mathbb{Z}$ with $(c, d) = 1$. Given such $c, d \in \mathbb{Z}$ choose $a, b \in \mathbb{Z}$ such that $ad - bc = 1$. Then $\gamma^* C = \langle \frac{1}{N} \rangle$ when $\gamma = \left(\begin{smallmatrix} a & b \\ c & d \end{smallmatrix}\right)$. Thus the analogue of the modular group for the isogenies under consideration is the subgroup of $SL_2(\mathbb{Z})$ composed of the $\gamma$ such that $\gamma^* \langle \frac{1}{N} \rangle = \langle \frac{1}{N} \rangle$. In this way we arrive at the *Hecke congruence subgroup*

$$\Gamma_0(N) := \left\{ \begin{pmatrix} a & b \\ c & d \end{pmatrix} \in SL_2(\mathbb{Z}) \mid c \equiv 0 \bmod N \right\} \tag{38}$$

and the following generalization of Proposition 3.

**Proposition 4** *Equivalence classes of isogenies* $\phi : E_1 \to E_2$ *with* $\ker \phi \simeq \mathbb{Z}/N\mathbb{Z}$ *are in natural correspondence with the orbits of* $\Gamma_0(N)$ *on* $\mathbb{H}$.

We may now ask for an analogue of the $j$-function—satisfying the conditions of Proposition 1—for isogenies of complex elliptic curves with kernel isomorphic to $\mathbb{Z}/N\mathbb{Z}$. In fact the second condition of Proposition 1 can be achieved only for $N = 1$. The first condition amounts to the existence of a holomorphic function which embeds the orbit space $\Gamma_0(N)\backslash\mathbb{H}$ into the complex plane (cf. (23)) and for most $N$ this condition cannot be satisfied either. The $N$ for which it can be satisfied are exactly those for which the orbit space $\Gamma_0(N)\backslash\widehat{\mathbb{H}}$ has genus zero (34), and for such $N$ the normalized principal modulus of $\Gamma_0(N)$ is the natural analogue of $j - 744$. For example $\Gamma_0(2)\backslash\widehat{\mathbb{H}}$ has genus zero and the corresponding normalized principal modulus is given explicitly by

$$T_{2B}(\tau) = \frac{\eta(\tau)^{24}}{\eta(2\tau)^{24}} + 24 = q^{-1} + 276q - 2048q^2 + 11202q^3 + \dots. \qquad (39)$$

(Cf. (58).) Sure enough this function occurs in monstrous moonshine as the McKay–Thompson series $T_g$ (32) for any $g \in \mathbb{M}$ that is conjugate to the involution $z$ of (12). (This conjugacy class is labelled $2B$ in [21, 23].)

More generally, Conway–Norton hypothesized that $\Gamma_g$ always lies between $\Gamma_0(N_g)$ and the normalizer of this group in $SL_2(\mathbb{R})$ for certain integers $N_g = o(g)h_g$ where $o(g)$ is the order of $g$ and $h_g$ divides $\gcd(o(g), 24)$. This greatly reduced the problem of determining the $T_g$ (32), especially in the case that $o(g)$ is divisible by large primes. For example the group $\Gamma_0(p)$ has index two in its normalizer $N(\Gamma_0(p))$ if $p$ is prime. So for $g \in \mathbb{M}$ such that $o(g) = p$ is prime with $p > 3$ we have $h_g = 1$, so $\Gamma_g = \Gamma_0(p)$ or $\Gamma_g = N(\Gamma_0(p))$. For $p$ prime the genus of $\Gamma_0(p)$ is zero only for $p \in \{2, 3, 5, 7, 13\}$, so if $p$ is a prime dividing $\#\mathbb{M}$, and $p = 11$ or $p > 13$, then $T_g$ must be the normalized principal modulus for $N(\Gamma_0(p))$.

We can now explicate a sense in which monstrous moonshine explains "one half" of Ogg's observation that a prime $p$ divides the order of the monster if and only if the $j$-invariant of every supersingular elliptic curve over $\overline{\mathbb{F}}_p$ lies in $\mathbb{F}_p$ (cf. Sect. 3). For this note that Ogg obtained Theorem 4 by using the fact that the genus of $N(\Gamma_0(p))$ is $\frac{1}{2}$ times the number of supersingular $j$-invariants that lie outside of $\mathbb{F}_p$ (cf. (15) in [8]). So Ogg's list (10) may also be described as the list of primes $p$ for which $N(\Gamma_0(p))$ has genus zero. Now if $p$ is a prime dividing $\#\mathbb{M}$ then $\mathbb{M}$ necessarily has an element $g$ of order $p$, and Conjecture 1 attaches a genus zero group $\Gamma_g$ to this element. According to the discussion above $\Gamma_g$ is either $\Gamma_0(p)$ or $N(\Gamma_0(p))$ (at least if $p > 3$). So according to monstrous moonshine, if $p$ divides the order of the monster then $p$ must appear in Ogg's list (10).

What remains mysterious is why a prime having the distinguished connection to supersingular elliptic curves exhibited in Theorem 4 should necessarily arise as a divisor of the order of the monster. Is there some construction, using supersingular elliptic curves or the normalizer of $\Gamma_0(p)$, that exhibits order $p$ automorphisms of an object whose automorphism group is provably the monster, for each prime $p$ in Ogg's list?

Although this question is currently beyond our reach an explicit construction of the monster module of Conjecture 1, together with an algebraic structure for which the monster is the automorphism group, is known. It is this topic that we turn to next.

## 7 Affine Lie Algebras

The period of rapid development in group theory that was initiated by Janko's discovery of $J_1$ (cf. (11)), and which led to the discovery of the monster group $\mathbb{M}$ (cf. (12)), was contemporaneous with the emergence of a transformative new idea in theoretical physics, namely that fundamental particles may be modeled by strings rather than points.

The earliest manifestations of this idea appeared in the dual resonance models which developed from Veneziano's work [28] in 1968. The significance of this for us is that the problem of modeling interactions of particles in these early models of string theory was solved, in part, by certain generating functions of operators on polynomial algebras in infinitely many variables (introduced by various authors, see e.g. Fubini–Veneziano [29]) called vertex operators. As we will see in Sect. 12 these vertex operators played an essential role in the Frenkel–Lepowsky–Meurman construction [30–32] of the monster module that would ultimately solve (cf. Theorems 8 and 9) the monstrous moonshine conjecture, Conjecture 1.

The methods of Frenkel–Lepowsky–Meurman are rooted in the theory of affine Lie algebras. These are special cases of Kac–Moody algebras, which were introduced (independently) by Kac [33], Kantor [34] and Moody [35] around the same time the dual resonance model was discovered by Veneziano. Indeed generating functions of operators on polynomial algebras in infinitely many variables were first utilized in the construction of affine Lie algebras by Lepowsky–Wilson in [36], and the similarity of these operators to the vertex operators of dual resonance theory was noticed by Howard Garland. Vertex operators became manifest in mathematics when the so-called homogeneous construction of basic representations of simple affine Lie algebras were obtained (independently) by Frenkel–Kac [37] and Segal [38]. (We will see a special case of this construction in Sect. 9.)

Affine Lie algebras also have a relationship to the basic idea of string theory. To explain this suppose that $\mathfrak{g}$ is a *complex Lie algebra*. This means that $\mathfrak{g}$ is a complex vector space equipped with an antisymmetric bilinear *Lie bracket* $[\cdot, \cdot] : \mathfrak{g} \times \mathfrak{g} \to \mathfrak{g}$ satisfying

$$[x, [y, z]] = [[x, y], z] + [y, [x, z]] \tag{40}$$

for all $x, y, z \in \mathfrak{g}$. At least if $\mathfrak{g}$ is finite-dimensional there is a complex Lie group $G$ whose tangent space at the identity is naturally identified with $\mathfrak{g}$, and the Lie bracket on $\mathfrak{g}$ derives in a natural way from the group structure on $G$. (See e.g. [39, 40] for background on Lie theory.)

The *loop algebra* of $\mathfrak{g}$ is the space $\bar{\mathfrak{g}} := \mathfrak{g}[t, t^{-1}]$ of Laurent polynomials with coefficients in $\mathfrak{g}$, equipped with the *pointwise* Lie bracket $[xt^m, yt^n] := [x, y]t^{m+n}$.

Regarding $t = e^{i\theta}$ as a coordinate on the circle $S^1 = \{e^{i\theta} \mid \theta \in \mathbb{R}\}$ we see that the loop algebra $\bar{\mathfrak{g}}$ serves a model for the Lie algebra of the *loop group* $\bar{G}^\infty$ of smooth maps $S^1 \to G$. So we may view $\bar{\mathfrak{g}}$ as a Lie algebra of "(closed) strings on $G$".

The symmetry of the circle $S^1$ opens a door to richer structure. To make this manifest we choose a non-degenerate symmetric bilinear form $(\cdot, \cdot)$ on $\mathfrak{g}$ that is *invariant* in the sense that $([x, y], z) = (x, [y, z])$ for $x, y, z \in \mathfrak{g}$. Then the associated *affine Lie algebra* is

$$\tilde{\mathfrak{g}} := \mathfrak{g}[t, t^{-1}] \oplus \mathbb{C}\mathbf{c} \oplus \mathbb{C}\mathbf{d} \tag{41}$$

where $\mathbf{c}$ is *central* (i.e. all brackets $[\mathbf{c}, *]$ vanish), and now

$$[xt^m, yt^n] := [x, y]t^{m+n} + (x, y)m\delta_{m+n,0}\mathbf{c},$$
$$[\mathbf{d}, xt^m] := mxt^m, \tag{42}$$

for $x, y \in \mathfrak{g}$ and $m, n \in \mathbb{Z}$. Somewhat confusingly the subalgebra

$$\hat{\mathfrak{g}} := \mathfrak{g}[t, t^{-1}] \oplus \mathbb{C}\mathbf{c} \tag{43}$$

is also referred to as the affine Lie algebra associated to $\mathfrak{g}$.

Say that a complex Lie algebra $\mathfrak{g}$ is *simple* if $\dim \mathfrak{g} > 1$ and $[\mathfrak{g}, \mathfrak{k}] \subset \mathfrak{k}$ for $\mathfrak{k}$ a subspace of $\mathfrak{g}$ implies $\mathfrak{k} = \{0\}$ or $\mathfrak{k} = \mathfrak{g}$. In the case that $\mathfrak{g}$ is simple the choice of invariant bilinear form on $\mathfrak{g}$ is unique up to scale. For $\mathfrak{g}$ simple $\tilde{\mathfrak{g}}$ (41) and $\hat{\mathfrak{g}}$ (43) are called *simple affine* Lie algebras.

The simplest example of a simple complex Lie algebra is obtained by equipping the three-dimensional space

$$\mathfrak{sl}_2 := \left\{ \begin{pmatrix} a & b \\ c & d \end{pmatrix} \in M_2(\mathbb{C}) \,\middle|\, a + d = 0 \right\} \tag{44}$$

with the *commutator* $[A, B] := AB - BA$. For an invariant bilinear form on $\mathfrak{sl}_2$ we may take $(A, B) := \operatorname{tr} AB$. In case $\mathfrak{g} = \mathfrak{sl}_2$ the affine Lie algebras $\tilde{\mathfrak{g}}$ and $\hat{\mathfrak{g}}$ are also called the *affine $A_1$ Lie algebras*.

To motivate the adjective "affine" let $\bar{\mathfrak{g}}^\infty$ be the Lie algebra of smooth maps $S^1 \to \mathfrak{g}$, equipped with the pointwise bracket, and observe that for fixed $a \in \mathbb{C}$ we obtain a twist of the conjugation action of $\bar{G}^\infty$ on the Lie algebra $\bar{\mathfrak{g}}^\infty$ by setting $g \cdot x := gxg^{-1} - ag'g^{-1}$ for $g \in \bar{G}^\infty$ and $x \in \bar{\mathfrak{g}}^\infty$ where $g' := \frac{dg}{d\theta}$. This action, called a *gauge transformation* in physics, is reminiscent of an *affine automorphism* of a vector space, i.e. the composition of a linear automorphism and a translation. (See [41] for a fuller discussion of this.)

For a richer example of a simple affine Lie algebra we consider the case that $\mathfrak{g} = \mathfrak{e}_8$ is the $E_8$ Lie algebra. For this we first consider the $E_8$ *lattice*, which is the unique (up to isomorphism) unimodular even positive-definite lattice of rank 8. A *lattice* is a free $\mathbb{Z}$-module $L$ of finite rank together with a symmetric bilinear form

$$(\cdot,\cdot) : L \times L \to \mathbb{Z}. \tag{45}$$

A lattice $L$ is called *positive-definite* if the natural extension of the bilinear form makes $L \otimes_{\mathbb{Z}} \mathbb{R}$ a Euclidean space, and *even* if $(\lambda, \lambda) \in 2\mathbb{Z}$ for all $\lambda \in L$. The *dual* of $L$ is defined by

$$L^* := \{\mu \in L \otimes_{\mathbb{Z}} \mathbb{R} \mid (\lambda, \mu) \in \mathbb{Z} \text{ for all } \lambda \in L\}, \tag{46}$$

and $L$ is called *unimodular* if $L^* = L$. See e.g. [42] for background on lattices.

To realize the $E_8$ lattice explicitly we may take $L$ to be the free $\mathbb{Z}$-module, in an eight-dimensional Euclidean space with orthonormal basis $\{e_1, \ldots, e_8\}$, generated by $\pm e_i \pm e_j$ for $1 \le i, j \le 8$ and $\frac{1}{2}(e_1 + \cdots + e_8)$. Then the $E_8$ *Lie algebra* is the 248-dimensional simple Lie algebra that may be described as

$$\mathfrak{g} = \mathfrak{h} \oplus \bigoplus_{\alpha \in \Delta} \mathbb{C}e_\alpha \tag{47}$$

for $\mathfrak{h} \simeq \mathbb{C}^8$ the dual space of $L \otimes_{\mathbb{Z}} \mathbb{C}$, and $\Delta$ the set of vectors $\alpha \in L$ satisfying $(\alpha, \alpha) = 2$. Write $\langle \cdot, \cdot \rangle$ for the natural pairing between $\mathfrak{h}$ and $L \otimes_{\mathbb{Z}} \mathbb{C}$. For $\alpha \in L$ let $h_\alpha$ be the unique element of $\mathfrak{h}$ such that $\langle h_\alpha, \lambda \rangle = (\alpha, \lambda)$ for all $\lambda \in L$. Also choose a $\mathbb{Z}$-basis $\{\lambda_i\}$ for $L$, and let $\varepsilon$ be the function $L \times L \to \{\pm 1\}$ determined by bimultiplicativity (i.e. $\varepsilon(\lambda + \lambda', \mu) = \varepsilon(\lambda, \mu)\varepsilon(\lambda', \mu)$ and $\varepsilon(\lambda, \mu + \mu') = \varepsilon(\lambda, \mu)\varepsilon(\lambda, \mu')$) and

$$\varepsilon(\lambda_i, \lambda_j) = \begin{cases} 1 & \text{if } i < j, \\ (-1)^{\frac{1}{2}(\lambda_i, \lambda_i)} & \text{if } i = j, \\ (-1)^{(\lambda_i, \lambda_j)} & \text{if } i > j. \end{cases} \tag{48}$$

Then a suitable Lie bracket on $\mathfrak{g}$ may be defined by requiring that for any $\alpha, \beta \in \Delta$ we have

$$[h_\alpha, h_\beta] = 0,$$

$$[h_\alpha, e_\beta] = (\alpha, \beta)e_\beta,$$

$$[e_\alpha, e_\beta] = \begin{cases} \varepsilon(\alpha, \beta)h_\alpha & \text{if } \alpha + \beta = 0, \\ \varepsilon(\alpha, \beta)e_{\alpha+\beta} & \text{if } \alpha + \beta \in \Delta, \\ 0 & \text{if } \alpha + \beta \notin \Delta \cup \{0\}, \end{cases} \tag{49}$$

and an invariant symmetric bilinear form on $\mathfrak{g}$ is obtained by requiring, for $\alpha, \beta \in \Delta$, that

$$(h_\alpha, h_\beta) = (\alpha, \beta),$$

$$(h_\alpha, e_\beta) = 0,$$

$$(e_\alpha, e_\beta) = \begin{cases} \varepsilon(\alpha, \beta) & \text{if } \alpha + \beta = 0, \\ 0 & \text{if } \alpha + \beta \neq 0. \end{cases} \tag{50}$$

For $\mathfrak{g}$ simple the *basic representation* of $\tilde{\mathfrak{g}}$ (cf. (41)) is characterized by the existence of a generating vector $\mathbf{v}$, called the *vacuum vector*, such that $\mathbf{c} \cdot \mathbf{v} = \mathbf{v}$ and $xt^m \cdot \mathbf{v} = \mathbf{d} \cdot \mathbf{v} = 0$ for $x \in \mathfrak{g}$ and $m \geq 0$. For each negative integer $n$ let $\mathfrak{h}(n)$ be a vector space isomorphic to $\mathfrak{h}$, choose an isomorphism $\mathfrak{h} \to \mathfrak{h}(n)$ and denote it $h \mapsto h(n)$. Set

$$\check{\mathfrak{h}} := \bigoplus_{n<0} \mathfrak{h}(n). \tag{51}$$

The basic representation of the simple affine $E_8$ Lie algebra (i.e. of $\tilde{\mathfrak{g}}$ for $\mathfrak{g} = \mathfrak{e}_8$ as in (47–50)) was realized in the *homogeneous* form

$$V_L := S(\check{\mathfrak{h}}) \otimes \mathbb{C}L \tag{52}$$

in [37, 38]. Here $S(\check{\mathfrak{h}})$ denotes the symmetric algebra of $\check{\mathfrak{h}}$ and $\mathbb{C}L = \bigoplus_{\lambda \in L} \mathbb{C}e_\lambda$ is the group algebra of $L$. The vacuum vector of $V_L$ is $\mathbf{v} := 1 \otimes e_0$.

Note that the construction (47–50) associates a complex Lie algebra to any even lattice. For example we obtain $\mathfrak{g} = \mathfrak{sl}_2$ (44) by replacing the $E_8$ lattice with the $A_1$ *lattice* $\sqrt{2}\mathbb{Z}$. In the case that (47–50) yields a simple complex Lie algebra the construction (52) also furnishes the basic representation of the corresponding simple affine Lie algebra. We will explain the action of $\tilde{\mathfrak{g}}$ on $V_L$ in Sect. 9. In Sect. 8 we will exhibit a connection between $V_L$ (for $L$ the $E_8$ lattice) and the elliptic modular invariant (24).

## 8 Modular Forms

In Sect. 9 we will explain how vertex operators equip $V_L$ (52) with a $\tilde{\mathfrak{g}}$-module structure (cf. (41–42)). For now we content ourselves to define the operator that will ultimately represent $\mathbf{d}$. For this define $L(0) \in \text{End } V_L$ by setting

$$L(0)(p \otimes e_\lambda) := \tfrac{1}{2}(\lambda, \lambda)L(0)p \otimes e_\lambda \tag{53}$$

for $p \in S(\check{\mathfrak{h}})$ and $\lambda \in L$ where $p \mapsto L(0)p$ is determined by requiring that in End $S(\check{\mathfrak{h}})$ we have

$$[L(0), h(-n)] = nh(-n) \tag{54}$$

for $h \in \mathfrak{h}$ and $n \in \mathbb{Z}^+$ (cf. (51)). Here $[A, B]$ is the usual commutator in the endomorphism ring of a vector space (cf. (44)) and $h(-n)$ acts on $S(\check{\mathfrak{h}})$ by left-multiplication.

Observe that $L(0)$ acts semisimply on $V_L$ with non-negative integer eigenvalues (since $L$ is positive-definite and even). Let $(V_L)_n$ be the $L(0)$-eigenspace with eigenvalue $n$. Then $(V_L)_n$ is finite-dimensional for all $n$ so it is natural to consider the *graded dimension* of $V_L$, which is the generating function

$$\operatorname{gdim} V_L := \sum_{n \in \mathbb{Z}} \dim (V_L)_n \, q^n. \tag{55}$$

Using the fact that $\operatorname{gdim}(U \otimes V) = (\operatorname{gdim} U)(\operatorname{gdim} V)$ we compute

$$\operatorname{gdim} V_L = \frac{\sum_{\lambda \in L} q^{\frac{1}{2}(\lambda,\lambda)}}{\prod_{n>0}(1 - q^n)^8} = 1 + 248q + 4124q^2 + 34752q^3 + \dots. \tag{56}$$

(As we will see in Sect. 9 it is no accident that $\dim(V_L)_1 = \dim \mathfrak{g}$.)

A priori $\operatorname{gdim} V_L$ is a formal series but in fact it may be interpreted as a holomorphic function on the complex upper half-plane $\mathbb{H}$ (cf. (19)). Indeed if we take $q = e^{2\pi i \tau}$ for $\tau \in \mathbb{H}$ as in (25) then the numerator in (56) becomes the *theta series*

$$\theta_L(\tau) := \sum_{\lambda \in L} q^{\frac{1}{2}(\lambda,\lambda)} \tag{57}$$

of $L$, and the denominator recovers $q^{-\frac{1}{3}} = e^{-\frac{2}{3}\pi i \tau}$ times the eighth power of the *Dedekind eta function*, which is

$$\eta(\tau) := q^{\frac{1}{24}} \prod_{n>0} (1 - q^n). \tag{58}$$

These functions (57 and 58) are examples of modular forms. Specifically, $\theta_L$ is a *modular form* of *weight* 4 for $SL_2(\mathbb{Z})$, which means that $\theta_L(\tau) = O(1)$ as $\Im(\tau) \to \infty$ and

$$\theta_L \left( \frac{a\tau + b}{c\tau + d} \right) \frac{1}{(c\tau + d)^4} = \theta_L(\tau) \tag{59}$$

for $\left(\begin{smallmatrix} a & b \\ c & d \end{smallmatrix}\right) \in SL_2(\mathbb{Z})$ and $\tau \in \mathbb{H}$. The Dedekind eta function (58) is a *cuspidal modular form* of weight $\frac{1}{2}$ with *multiplier* for $SL_2(\mathbb{Z})$, which means that $\eta(\tau) \to 0$ as $\Im(\tau) \to \infty$ and

$$\epsilon \begin{pmatrix} a & b \\ c & d \end{pmatrix} \eta \left( \frac{a\tau + b}{c\tau + d} \right) \frac{1}{\sqrt{c\tau + d}} = \eta(\tau) \tag{60}$$

for $\left(\begin{smallmatrix} a & b \\ c & d \end{smallmatrix}\right) \in SL_2(\mathbb{Z})$ and $\tau \in \mathbb{H}$ for a certain function $\epsilon : SL_2(\mathbb{Z}) \to \mathbb{C}$ called the *multiplier system* of $\eta$.

The proof of Theorem 6 in Chapter VII of [18] verifies (60) and shows that the multiplier system of $\eta$ takes values in the 24-th roots of unity. This statement is true for $\sqrt{\cdot}$ any branch of the square root. For the sake of concreteness we take the principal branch,

$$\sqrt{re^{i\theta}} := \sqrt{r} e^{\frac{1}{2}i\theta} \tag{61}$$

for $r > 0$ and $-\pi < \theta \leq \pi$. Then the multiplier system of $\eta$ is determined by the particular values

$$\epsilon \begin{pmatrix} 1 & 1 \\ 0 & 1 \end{pmatrix} = e^{-\frac{1}{12}\pi i}, \quad \epsilon \begin{pmatrix} 0 & -1 \\ 1 & 0 \end{pmatrix} = e^{\frac{1}{4}\pi i}, \tag{62}$$

and the fact that $SL_2(\mathbb{Z})$ is generated by the matrices appearing in (62).

The modularity (59) of $\theta_L$ is part of a more general picture. For $k \in \frac{1}{2}\mathbb{Z}$ define the *weight $k$ action* of $SL_2(\mathbb{R})$ on functions $f : \mathbb{H} \to \mathbb{C}$ by setting

$$\left( f \Big|_k \begin{pmatrix} a & b \\ c & d \end{pmatrix} \right)(\tau) := f \left( \frac{a\tau + b}{c\tau + d} \right) \frac{1}{(c\tau + d)^k}. \tag{63}$$

Next, given $\Gamma < SL_2(\mathbb{R})$ that is commensurable with $SL_2(\mathbb{Z})$ (cf. Sect. 5) say that a holomorphic function $f : \mathbb{H} \to \mathbb{C}$ is a *modular form of weight $k$ for $\Gamma$* if $f|_k\gamma = f$ for all $\gamma \in \Gamma$, and if the growth condition $(f|_k\gamma)(\tau) = O(1)$ as $\Im(\tau) \to \infty$ is satisfied for all $\gamma \in SL_2(\mathbb{Z})$. (Say that $f$ is *cuspidal* if the stronger condition $(f|_k\gamma)(\tau) \to 0$ as $\Im(\tau) \to \infty$ is satisfied for all $\gamma \in SL_2(\mathbb{Z})$, and say that $f$ is *weakly holomorphic* in case the weaker growth condition $(f|_k\gamma)(\tau) = O(e^{C\Im(\tau)})$ is satisfied for all $\gamma \in SL_2(\mathbb{Z})$, for some $C > 0$.) Then we have the following result.

**Theorem 5** *For $L$ a positive-definite lattice the expression (57) defines a modular form $\theta_L$ of weight $k = \frac{1}{2} \operatorname{rank} L$ for some finite index subgroup $\Gamma$ of $SL_2(\mathbb{Z})$. If $L$ is even and unimodular then $\operatorname{rank} L \equiv 0 \bmod 8$ and $\Gamma = SL_2(\mathbb{Z})$.*

See e.g. §6 in Chapter VII of [18] for a proof of Theorem 5.

Taking $L$ to be the $E_8$ lattice again, as in Sect. 7, we deduce from (59 and 60) and the fact that $3 \times 248 = 744$ (cf. (56)) that

$$\frac{\theta_L(\tau)^3}{\eta(\tau)^{24}} = q^{-1}(\text{gdim } V_L)^3 \tag{64}$$

is an $SL_2(\mathbb{Z})$-invariant holomorphic function on $\mathbb{H}$ satisfying $q^{-1} + 744 + O(q)$ as $\Im(\tau) \to \infty$. Comparing with (26) we conclude that

$$q^{-1}(\text{gdim } V_L)^3 = j(\tau). \tag{65}$$

That is, the graded dimension of the basic representation of the simple affine Lie algebra of type $E_8$ defines a cube root of the elliptic modular invariant (24). In fact it may be checked that $\theta_L = \frac{3}{4\pi^4} g_2$ and $\eta^{24} = \frac{1}{(2\pi)^{12}}(g_2^3 - 27g_3^2)$ where $g_2$ and $g_3$ are as in (24).

## 9  Vertex Operators

Our next objective is to define vertex operators and explain how they furnish a $\tilde{\mathfrak{g}}$-module structure on the space $V_L$ as realized in (52).

The structure of $V_L$ as a module for the subalgebra $\hat{\mathfrak{h}} < \tilde{\mathfrak{g}}$ (cf. (41), (47)) is relatively easy to describe. (This is the reason that the construction we about to present was dubbed homogeneous in [37].) For this we define derivations $\partial_{h(-n)}$ of the associative algebra $S(\check{\mathfrak{h}})$ for $h \in \mathfrak{h}$ and $n \in \mathbb{Z}^+$ by requiring $\partial_{h(-n)} h'(-n') = n(h, h')\delta_{n,n'}$ for $h' \in \mathfrak{h}$ and $n' \in \mathbb{Z}^+$. Next define $h(n) \in \text{End } V_L$ for $h \in \mathfrak{h}$ and $n \in \mathbb{Z}$ by setting

$$h(n)(p \otimes e_\lambda) := \begin{cases} h(n)p \otimes e_\lambda & \text{if } n < 0, \\ \langle h, \lambda \rangle p \otimes e_\lambda & \text{if } n = 0, \\ \partial_{h(-n)} p \otimes e_\lambda & \text{if } n > 0, \end{cases} \tag{66}$$

for $p \in S(\check{\mathfrak{h}})$ and $\lambda \in L$. Then (66) agrees with (54), and the linear embedding $\hat{\mathfrak{h}} \to \text{End } V_L$ determined by $ht^n \mapsto h(n)$, $\mathbf{c} \mapsto I$ and $\mathbf{d} \mapsto -L(0)$ (cf. (41)) is an embedding of Lie algebras, and in particular realizes an $\hat{\mathfrak{h}}$-module structure on $V_L$. That is, we have

$$[h(n), h'(n')] = (h, h')n\delta_{n+n',0},$$
$$[-L(0), h(n)] = nh(n), \tag{67}$$

in $\text{End } V_L$ for $h, h' \in \mathfrak{h}$ and $n, n' \in \mathbb{Z}$ (cf. (42), (54)).

One of the fundamental observations of [36] is that it is profitable to consider generating functions of operators in the context of affine Lie algebras. For example given $h \in \mathfrak{h}$ we may consider the generating function $h(z) \in (\mathrm{End}\, V_L)[[z, z^{-1}]]$, for $z$ a formal variable, defined by setting

$$h(z) := \sum_{n \in \mathbb{Z}} h(n) z^{-n-1}. \tag{68}$$

Then the commutation relations (67) satisfied by the $h(n) \in \mathrm{End}\, V_L$ translate into the identity

$$[h(z), h'(w)] = (h, h') \delta^{(2)}(z - w) \tag{69}$$

in $(\mathrm{End}\, V_L)[[z, w, z^{-1}, w^{-1}]]$ where $\delta^{(n+1)}(z - w) := \frac{1}{n!} \frac{\partial^n}{\partial w^n} \delta(z - w)$ for $n \geq 0$ and $\delta(z - w) := \sum_{n \in \mathbb{Z}} z^{-n-1} w^n$. To get a feeling for (69) observe that

$$(z - w)^n \delta^{(n)}(z - w) = 0 \tag{70}$$

in $\mathbb{C}[[z, w, z^{-1}, w^{-1}]]$ for $n \geq 1$, and in particular $z\delta(z - w) = w\delta(z - w)$. So the $\delta^{(n)}(z - w)$ are *algebraic delta functions*, which we can think of as "supported on the *diagonal* $z = w$". We may regard (69) as saying that $h(z)$ and $h'(w)$ "commute" away from the diagonal. Indeed from (69) and (70) we have that

$$(z - w)^2 [h(z), h'(w)] = 0 \tag{71}$$

in $(\mathrm{End}\, V_L)[[z, w, z^{-1}, w^{-1}]]$. This analogue (71) of commutativity is called *locality* for $h(z)$ and $h'(z)$.

Given formal series $A(z), B(z) \in (\mathrm{End}\, V)[[z, z^{-1}]]$ write $A(z) = \sum_n A_n z^{-n-1}$ and define the *normally ordered product* of $A(z)$ and $B(w)$ by setting

$$: A(z)B(w) :\, := A(z)_{\mathrm{rgl}} B(w) + B(w)A(z)_{\mathrm{sgl}} \tag{72}$$

where $A(z)_{\mathrm{rgl}} := \sum_{n<0} A_n z^{-n-1}$ and $A(z)_{\mathrm{sgl}} := \sum_{n \geq 0} A_n z^{-n-1}$. Then it turns out that (69) is equivalent to the identity

$$h(z)h'(w) = \frac{(h, h')}{(z - w)^2} + :h(z)h'(w): \tag{73}$$

in $(\mathrm{End}\, V_L)[[z, w, z^{-1}, w^{-1}]]$ so long as we interpret $(z - w)^{-2}$ as the formal series $\sum_{n \geq 0}(n + 1) z^{-n-2} w^n$, i.e. the formal expansion of $(z - w)^{-2}$ in the domain $|z| < |w|$. (Note that $\delta^{(n)}(z - w)$ is the difference between the expansions of $(z - w)^{-n}$ in $|z| < |w|$ and $|w| < |z|$.) For this reason the formal expression

$$h(z)h'(w) \sim \frac{(h, h')}{(z - w)^2}, \tag{74}$$

called the *operator product expansion (OPE)* of $h(z)$ with $h'(w)$, is often used. Via the equivalence of (73) with (69) the OPE (74) encodes the commutators of all the component operators of $h(z)$ and $h'(z)$.

The space $(\text{End } V_L)[[z, z^{-1}]] = (\text{End } V_L) \otimes \mathbb{C}[[z, z^{-1}]]$ is awkward to work with because $\mathbb{C}[[z, z^{-1}]] := \{\sum_{n \in \mathbb{Z}} c_n z^n \mid c_n \in \mathbb{C}\}$ is not naturally a ring. But $\mathbb{C}((z)) := \mathbb{C}[[z]][z^{-1}]$ is naturally a ring, being the field of fractions of $\mathbb{C}[[z]] := \{\sum_{n \geq 0} c_n z^n \mid c_n \in \mathbb{C}\}$. We have $h(n)v = 0$ for large enough $n$, for any fixed $v \in V_L$, so the generating function $h(z)$ (cf. (68)) defines a *field* on $V_L$. That is, the assignment $v \mapsto h(z)v$ defines a linear map

$$V_L \to V_L((z)). \tag{75}$$

Note that the naive product $h(z)h'(z)$ is not a well-defined series (let alone a field on $V_L$) even though $h(z)$ and $h'(z)$ are fields and $\mathbb{C}((z))$ is a ring. The normally ordered product ameliorates this. Indeed if $A(z)$ and $B(z)$ are fields on a vector space $V$ then the specialization $: A(z)B(z):$ of (72) at $z = w$ is a field on $V$ too. Note however that the normally ordered product of fields is in general neither commutative nor associative.

Given $\lambda \in L$ we now consider the corresponding *vertex operator* on $V_L$, which may be defined by setting

$$Y(e_\lambda, z)(p \otimes e_\mu) :=$$

$$\varepsilon(\lambda, \mu) \left( \exp\left( \sum_{n>0} \frac{1}{n} h_\lambda(-n) z^n \right) \exp\left( \sum_{n>0} -\frac{1}{n} h_\lambda(n) z^{-n} \right) p \right) \otimes e_{\lambda+\mu} z^{(\lambda, \mu)} \tag{76}$$

for $p \in S(\check{\mathfrak{h}})$ and $\mu \in L$. Define $e_\lambda(n) \in \text{End } V_L$ for $n \in \mathbb{Z}$ by requiring that

$$Y(e_\lambda, z) = \sum_{n \in \mathbb{Z}} e_\lambda(n) z^{-n-1}. \tag{77}$$

Then again we have $e_\lambda(n)v = 0$ for sufficiently large $n$, for any fixed $v \in V_L$, so $Y(e_\lambda, z)$ also defines a field on $V_L$ (cf. (75)).

Let $\pi$ be the linear map $\tilde{\mathfrak{g}} \to \text{End } V_L$ determined by

$$ht^n \mapsto h(n),$$

$$e_\alpha t^n \mapsto e_\alpha(n),$$

$$\mathbf{c} \mapsto I,$$

$$\mathbf{d} \mapsto -L(0), \tag{78}$$

for $h \in \mathfrak{h}$, $\alpha \in \Delta$ and $n \in \mathbb{Z}$ (cf. (41), (47), (53), (66), (76)). The $E_8$ case of the main result of [37] may now be expressed as follows.

**Theorem 6** *The map* $\pi : \tilde{\mathfrak{g}} \to \operatorname{End} V_L$ *is an embedding of Lie algebras which realizes $V_L$ as a copy of the basic representation of the simple affine Lie algebra of type $E_8$.*

The operator (76) is a modification of the original vertex operator of the dual resonance theory. The main idea is that $Y(e_\lambda, z)$ should serve as the "exponential of an antiderivative of $h_\lambda(z)$" so that $\frac{\partial}{\partial z} Y(e_\lambda, z) = :h_\lambda(z)Y(e_\lambda, z):$ (cf. (72)). Indeed $\exp\left(\int h_\lambda(z)\right)$, or better $:\exp\left(\int h_\lambda(z)\right):$, can serve as a mnemonic for (76) where the colons remind us to let the $h_\lambda(n)$ with $n \geq 0$ act before those with $n < 0$.

We can now explain the significance of the function $\varepsilon$ of (48). The reader may check that

$$Y(e_\lambda, z)Y(e_\mu, w) = (z - w)^{(\lambda,\mu)} :Y(e_\lambda, z)Y(e_\mu, w): \tag{79}$$

for $\lambda, \mu \in L$ (cf. (72)). From our prescription (48) for $\varepsilon$ it follows that

$$e_\lambda e_\mu = (-1)^{(\lambda,\mu)} e_\mu e_\lambda \tag{80}$$

for $\lambda, \mu \in L$. This in turn leads us to

$$:Y(e_\lambda, z)Y(e_\mu, w): = (-1)^{(\lambda,\mu)} :Y(e_\mu, z)Y(e_\lambda, w): \tag{81}$$

for $\lambda, \mu \in L$ so we have $(z - w)^N[Y(e_\lambda, z), Y(e_\mu, w)] = 0$ for $N := -(\lambda, \mu)$. That is, the choice (48) of $\varepsilon$ ensures that the vertex operators of (76) define mutually local fields on $V_L$ (cf. (71)).

The fields $h(z)$ (68) and $Y(e_\lambda, z)$ (76) are also mutually local. We have the OPE

$$h(z)Y(e_\lambda, w) \sim \frac{h(\lambda)Y(e_\lambda, w)}{z - w} \tag{82}$$

(cf. (74)) so $(z - w)[h(z), Y(e_\lambda, w)] = 0$ (cf. (71)).

We refer to [43–45] for reviews of the dual resonance models and their vertex operators from the physical point of view. Historical mathematical perspectives on vertex operators may be found in the introductions to [37] and [32]. See [41] or the introduction to [32] for some historical perspectives on affine Lie algebras. A standard reference for affine Lie algebras, and Kac–Moody algebras more generally, is [46].

## 10  Conformal Field Theory

Given our notational choices it is natural to write $x(n)$ for the image of $xt^n$ under $\pi$ (cf. (78)) for $x \in \mathfrak{g}$ and $n \in \mathbb{Z}$. Then from (42) and Theorem 6 we see that the map

$$\mathfrak{g} \to \text{End } V_L$$
$$x \mapsto x(0)$$
(83)

is an embedding of Lie algebras so $V_L$ is naturally a $\mathfrak{g}$-module. By considering the subgroup of $GL(V_L)$ generated by the exponentials $e^{tx(0)}$ for $t \in \mathbb{C}$ and $x \in \mathfrak{g}$ we obtain an action of the complex Lie group

$$G := \text{Aut}(\mathfrak{g}) \simeq E_8(\mathbb{C})$$
(84)

on $V_L$. This action commutes with $L(0)$ according to (53) so actually each homogeneous subspace $(V_L)_n < V_L$ (cf. (55)) is a $G$-module.

The first five numbers that arise as dimensions of irreducible $E_8(\mathbb{C})$-modules are 1, 248, 3875, 27000 and 30380 (cf. e.g. [47]). Comparing with (56) we note the identities

$$\dim(V_L)_1 = 248 = 248,$$
(85)

$$\dim(V_L)_2 = 4124 = 1 + 248 + 3875,$$
(86)

$$\dim(V_L)_3 = 34752 = 1 + 2 \times 248 + 3875 + 30380,$$
(87)

which are reminiscent of (27–29). (These coincidences (85–87) were noticed by McKay [48] but Queen confirmed [49] that there is no analogue of the genus zero property of Sect. 5 for $E_8(\mathbb{C})$.)

To explain why $248 = \dim \mathfrak{g}$ appears in each of (85–87) observe that the map $x \mapsto x(-n)\mathbf{v}$ (cf. (52)) defines an embedding $\mathfrak{g} \to (V_L)_n$ (cf. (55)) of $\mathfrak{g}$-modules for each $n > 0$. This map is a linear isomorphism precisely for $n = 1$ so it is natural that $\dim(V_L)_1 = \dim \mathfrak{g}$.

Actually more is true. Let $\mathfrak{F}(V_L)$ denote the space of fields on $V_L$ (cf. (75)) and define a linear map

$$Y : (V_L)_1 \to \mathfrak{F}(V_L)$$
(88)

by setting $Y(h(-1)\mathbf{v}, z) := h(z)$ for $h \in \mathfrak{h}$ (cf. (68)) and letting $Y(e_\lambda, z)$ be as in (76) for $\lambda \in \Delta$. Then $Y(v, z)$ is a field for every $v \in (V_L)_1$ and the component operators of these $Y(v, z)$ generate the action of $\hat{\mathfrak{g}}$ (43) on $V_L$. Also we have

$$[x(0), Y(v, z)] = Y(x(0)v, z)$$
(89)

for $x \in \mathfrak{g}$ and $v \in (V_L)_1$ so $\mathfrak{g}$ acts naturally on $V_L$, preserving the association (88), and a directly similar statement holds for $G$ (84).

What about the 1s in (86 and 87)? They represent one-dimensional subspaces that are fixed by the action of $G$. For $1 \leq i \leq 8$ let $h_i$ be the element of $\mathfrak{h}$ that is dual to $e_i$ (cf. (47)) so that $\langle h_i, e_j \rangle = \delta_{i,j}$. For $m, n \in \mathbb{Z}$ define $: h(m)h(n):$ to be $h(m)h(n)$ in case $n \geq m$, and $h(n)h(m)$ otherwise. Then one may check that the one-dimensional $G$-fixed subspaces of (86 and 87) are spanned by

$$L(-2)\mathbf{v} = \frac{1}{2} \sum_i h_i(-1)^2 \mathbf{v} \tag{90}$$

and $L(-3)\mathbf{v} = \sum_i h_i(-2)h_i(-1)\mathbf{v}$, respectively, where $L(n) \in \text{End } V_L$ is defined for $n \in \mathbb{Z}$ by

$$L(n) := \frac{1}{2} \sum_i \sum_{m \in \mathbb{Z}} :h_i(m)h_i(n-m): . \tag{91}$$

The operators (91) were introduced originally in the dual resonance theory (cf. [43, 45]). They are called *Virasoro operators* because they satisfy the commutation relations of the *Virasoro algebra* [50]. That is, we have

$$[L(m), L(n)] = (m - n)L(m + n) + \frac{m^3 - m}{12}\delta_{m+n,0}\mathbf{c} \tag{92}$$

where $\mathbf{c}$ is central (cf. (41)). For us $\mathbf{c} = 8I$ and we may recognize $L(n)$ as the coefficient of $z^{-n-2}$ in the field

$$L(z) := \frac{1}{2} \sum_i :h_i(z)^2: \tag{93}$$

(cf. (68), (72)).

The importance of the Virasoro algebra in physics was developed significantly by Belavin–Polyakov–Zamolodchikov [51]. To catch a glimpse of why consider the case that $\mathbf{c}$ acts as 0 in (92). Then (92) reduces to $[L(m), L(n)] = (m-n)L(m+n)$, and these relations are realized in a concrete way by letting $L(n)$ act on $\mathbb{C}[t, t^{-1}]$ via

$$L(n)p := -t^{n+1}\frac{dp}{dt} \tag{94}$$

for $n \in \mathbb{Z}$ and $p \in \mathbb{C}[t, t^{-1}]$. That is, the quotient of the Virasoro algebra by its center $\mathbb{C}\mathbf{c}$ is the Lie algebra of (polynomial) vector fields on the circle $S^1$. The Virasoro algebra (92) is a central extension of this Lie algebra and its representations are projective representations of the Lie algebra of vector fields on $S^1$. In [51] Belavin–Polyakov–Zamolodchikov introduced conformal (quantum) field theory, and in so doing abstracted many of the features of string theory (cf. the introduction to [32]). They explained that the Virasoro algebra serves as an algebra of symmetries for conformal field theories in two-dimensions.

To appreciate the relationship between the Virasoro algebra and the affine Lie algebra $\tilde{\mathfrak{g}}$ (41) observe that

$$[L(m), h(n)] = -nh(m+n),$$

$$[L(m), e_\lambda(n)] = \left(\frac{m+1}{2}\langle\lambda,\lambda\rangle - m - 1\right)e_\lambda(m+n) \tag{95}$$

(cf. (66), (77)). Taking $m = 0$ in (95) and noting that $L(0)\mathbf{v} = 0$ and $e_\lambda = e_\lambda(-1)\mathbf{v}$ we see that the operator $L(0)$ defined by (91) agrees with the $L(0)$ we defined earlier in (53–54). Taking $n = 0$ and $\lambda \in \Delta$ (cf. (47)) in (95) we see that the Virasoro action on $V_L$ commutes with that of $\mathfrak{g}$ (cf. (83), (89)) so it also commutes with the action of the complex Lie group $G$ (84). In addition to $L(-2)\mathbf{v}$ and $L(-3)\mathbf{v}$ (cf. (90)) the space of $G$-fixed points in $(V_L)_n$ includes $L(-n)\mathbf{v}$ for every $n \geq 2$ (cf. (91)).

## 11    The Monster Module

We now describe the construction by Frenkel–Lepowsky–Meurman [30–32] of an infinite-dimensional graded monster module that solves Conjecture 1.

The reader may have noticed that the construction (52) of $V_L$ as an $\tilde{\mathfrak{h}}$-module (cf. (41), (47), (52), (66)), and in particular as a graded vector space, makes sense for any lattice $L$ if we take $\mathfrak{h}$ to be the dual of $L \otimes_{\mathbb{Z}} \mathbb{C}$ and equip it with the *trivial Lie bracket*, $[h, h'] = 0$ for $h, h' \in \mathfrak{h}$. Then (56) gets replaced by

$$\operatorname{gdim} V_L = q^{\frac{c}{24}}\frac{\theta_L(\tau)}{\eta(\tau)^c} \tag{96}$$

where $c = \operatorname{rank} L$ and $\theta_L$ and $\eta$ are as in (57–58). Now observe that $\theta_{L\oplus L'} = \theta_L\theta_{L'}$ for lattices $L$ and $L'$. This means that the identity (65) may be rewritten as

$$q^{-1}\operatorname{gdim} V_N = j(\tau) \tag{97}$$

where $N = L^{\oplus 3}$ is the unimodular even positive-definite lattice of rank 24 (cf. (45 and 46)) obtained by taking the direct sum of 3 copies of the $E_8$ lattice $L$ constructed in Sect. 7. So $N$ could be a starting point for the construction of a graded vector space whose graded dimension is $j(\tau) - 744$.

There is however another unimodular even positive-definite lattice $\Lambda$ of rank 24 called the *Leech lattice* (cf. [52]) which is more promising. To explain this we first note the following characterization of $\Lambda$ due to Conway.

**Theorem 7 (Conway [53])** *The Leech lattice* $\Lambda$ *is the unique unimodular even positive-definite lattice with rank* 24 *and no vectors* $\lambda \in \Lambda$ *such that* $(\lambda, \lambda) = 2$.

Next we note that the automorphism group $Co_0 := \mathrm{Aut}(\Lambda)$ of the Leech lattice—henceforth referred to as *the Conway group* (cf. [54, 55])—satisfies a (non-split) short exact sequence

$$1 \to \mathbb{Z}/2\mathbb{Z} \to Co_0 \to Co_1 \to 1 \tag{98}$$

where $Co_1 := \mathrm{Aut}(\Lambda)/\{\pm I\}$ is the *largest sporadic simple Conway group*, which is involved in an involution centralizer for the monster according to (12). (There are two further sporadic simple Conway groups $Co_2$ and $Co_3$. See [23, 55, 56] for more on these.)

Moreover if we take a $\mathbb{Z}$-basis $\{\lambda_i\}$ for $\Lambda$ and define a bimultiplicative function $\varepsilon : \Lambda \times \Lambda \to \{\pm 1\}$ as in (48) then the set $\hat{\Lambda} := \{\pm e_\lambda \mid \lambda \in \Lambda\}$ becomes a group—a *double cover* of $\Lambda$—once we define $e_\lambda e_\mu = \varepsilon(\lambda, \mu) e_{\lambda + \mu}$. As we see from (76–81) it is really $\hat{\Lambda}$, rather than $\Lambda$, that we use to construct the vertex operators on $V_\Lambda$.

Observe that any automorphism of the group structure on $\hat{\Lambda}$ naturally defines an automorphism of the group structure on $\Lambda$. Since $\mathrm{Aut}(\Lambda)$ denotes the automorphisms of $\Lambda$ that preserve the bilinear form $(\cdot, \cdot)$ let us write $\mathrm{Aut}(\hat{\Lambda})$ for the group of automorphisms of $\hat{\Lambda}$ that project to $\mathrm{Aut}(\Lambda)$. Then from the discussion above it is $\mathrm{Aut}(\hat{\Lambda})$, rather than $\mathrm{Aut}(\Lambda)$, that we should expect to act naturally on $V_\Lambda$. Encouragingly, for $\mathrm{Aut}(\hat{\Lambda})$ we have

$$1 \to (\mathbb{Z}/2\mathbb{Z})^{24} \to \mathrm{Aut}(\hat{\Lambda}) \to Co_0 \to 1 \tag{99}$$

which is tantalizingly close to (12).

However (99) is not the same as (12), and the challenge (cf. (31)) of removing the constant term 744 from $j(\tau)$ (cf. (26)) is not solved by $V_\Lambda$ either. Indeed by the hypothesis on vectors of square-length 2 in Theorem 7 we must have $\theta_\Lambda(\tau) = 1 + O(q^2)$ (cf. (57)) so

$$q^{-1} \operatorname{gdim} V_\Lambda = \frac{\theta_\Lambda(\tau)}{\eta(\tau)^{24}} = q^{-1} + 24 + 196884q + \dots . \tag{100}$$

Applying (60) and Theorem 5 to (100) we may conclude that

$$q^{-1} \operatorname{gdim} V_\Lambda = j(\tau) - 720. \tag{101}$$

Frenkel–Lepowsky–Meurman solved both the symmetry problem (99) and the constant term problem (100) simultaneously [30, 31] by introducing a companion space $V_{\Lambda,\mathrm{tw}}$ to $V_\Lambda$ and defining

$$V^{\natural} := V_\Lambda^0 \oplus V_{\Lambda,\mathrm{tw}}^1 \tag{102}$$

where $V_\Lambda = V_\Lambda^0 \oplus V_\Lambda^1$ is the eigenspace decomposition for the action of a certain involution $z_\Lambda$ on $V_\Lambda$, and similarly for $V_{\Lambda,\text{tw}}^1$. Specifically, the action of $z_\Lambda$ on $V_\Lambda$ is given by

$$z_\Lambda(p \otimes e_\lambda) := z_\Lambda(p) \otimes z_\Lambda(e_\lambda) \tag{103}$$

for $p \in S(\check{\mathfrak{h}})$ (cf. (51 and 52)) and $\lambda \in \Lambda$ where $p \mapsto z_\Lambda(p)$ is the unique algebra automorphism such that $z_\Lambda(v) = -v$ for $v \in \check{\mathfrak{h}}$, and the $z_\Lambda$ in $z_\Lambda(e_\lambda)$ denotes the element of $\text{Aut}(\hat{\Lambda})$ defined by $z_\Lambda(e_\lambda) := e_{-\lambda}$. In particular this $z_\Lambda \in \text{Aut}(\hat{\Lambda})$ maps to the central involution of $Co_0$ under the projection of (99). So $\text{Aut}(\hat{\Lambda})/\langle z_\Lambda \rangle$ acts naturally on $V_\Lambda^0$. Upon comparing (12) with the short exact sequence

$$1 \to (\mathbb{Z}/2\mathbb{Z})^{24} \to \text{Aut}(\hat{\Lambda})/\langle z_\Lambda \rangle \to Co_1 \to 1 \tag{104}$$

we might hope that $\text{Aut}(\hat{\Lambda})/\langle z_\Lambda \rangle$ is isomorphic to the quotient of $C_{\mathbb{M}}(z)$ by the central involution in its normal subgroup $2^{1+24}$ (cf. (12 and 13)). This turns out to be true (cf. [16, 57] or §10.4 of [32]).

So $C_{\mathbb{M}}(z)$ acts naturally on $V_\Lambda^0$ but not faithfully. So in order to recover an action of the monster on $V^\natural$ (102) we require $C_{\mathbb{M}}(z)$ to act faithfully on $V_{\Lambda,\text{tw}}^1$. In particular we require a faithful action of the normal subgroup $2^{1+24}$. Let $K$ be the subgroup of $\hat{\Lambda}$ composed of the elements $e_{2\lambda}$ for $\lambda \in \Lambda$. The quotient $\hat{\Lambda}/K$ has order $2^{25}$ and turns out to be isomorphic to the subgroup of $C_{\mathbb{M}}(z)$ denoted $2^{1+24}$ in (12).

The group $\hat{\Lambda}/K$ has a unique faithful irreducible module up to equivalence. Let $T_\Lambda$ be such a module. Then $\dim T_\Lambda = 2^{12} = 4096$ and the space $V_{\Lambda,\text{tw}}$ may now be described as

$$V_{\Lambda,\text{tw}} = S(\check{\mathfrak{h}}_{\text{tw}}) \otimes T_\Lambda \tag{105}$$

where $\check{\mathfrak{h}}_{\text{tw}} := \bigoplus_{n<0} \mathfrak{h}(n + \frac{1}{2})$ (cf. (51)).

How should $z_\Lambda$ act on $V_{\Lambda,\text{tw}}$? Observe that our prescription (48) ensures that

$$\varepsilon(\lambda, \lambda) = (-1)^{\frac{1}{2}(\lambda,\lambda)} \tag{106}$$

for $\lambda \in \Lambda$. So $e_\lambda^{-1} = (-1)^{\frac{1}{2}(\lambda,\lambda)} e_{-\lambda}$ and $K$ coincides with the set of elements of $\hat{\Lambda}$ of the form $a^{-1}z_\Lambda(a)$ for $a \in \hat{\Lambda}$. In particular $K < \hat{\Lambda}$ is invariant under $z_\Lambda$, and the induced action of $z_\Lambda$ on $\hat{\Lambda}/K$ is trivial. So it is natural to define

$$z_\Lambda(p \otimes t) := z_\Lambda(p) \otimes t \tag{107}$$

for $p \in S(\check{\mathfrak{h}}_{\text{tw}})$ and $t \in T_\Lambda$ where $p \mapsto z_\Lambda(p)$ is, as before (cf. (103)), the unique algebra automorphism such that $z_\Lambda(v) = -v$ for $v \in \check{\mathfrak{h}}_{\text{tw}}$.

If $V^\natural$ is to solve the monstrous moonshine conjecture, Conjecture 1, it should admit a $\mathbb{Z}$-grading $V^\natural = \bigoplus_n V_n^\natural$ for which the corresponding *graded dimension* function

$$\text{gdim } V^\natural := \sum_n \dim V_n^\natural q^n \tag{108}$$

recovers $j(\tau) - 744$. The space $V_\Lambda$ is naturally graded by the action of $L(0)$ (cf. (53)) so it is natural to ask for an extension of this action to $V_{\Lambda,\text{tw}}$. We will see momentarily (cf. (112)) that a good choice is made by setting

$$L(0)(p \otimes t) := \tfrac{3}{2}L(0)p \otimes t \tag{109}$$

(cf. (53)) for $p \in S(\check{\mathfrak{h}}_{\text{tw}})$ and $t \in T_\Lambda$ where the endomorphism $p \mapsto L(0)p$ is determined in direct analogy with (54) by requiring that

$$[L(0), h(-n + \tfrac{1}{2})] = (n - \tfrac{1}{2})h(-n + \tfrac{1}{2}). \tag{110}$$

As in (55) let us define $(V_{\Lambda,\text{tw}})_n$ to be the $L(0)$-eigenspace of $V_{\Lambda,\text{tw}}$ with eigenvalue $n$, now for $n \in \tfrac{1}{2}\mathbb{Z}$. Since the actions of $z_\Lambda$ on $V_\Lambda$ and $V_{\Lambda,\text{tw}}$ commute with those of $L(0)$ we may make similar definitions for $V_\Lambda^j$ and $V_{\Lambda,\text{tw}}^j$, and thus obtain an action of $L(0)$ on $V^\natural$ (102), and a corresponding decomposition $V^\natural = \bigoplus_n V_n^\natural$. Then from the explicit description (103), (107) of $z_\Lambda$ we obtain

$$q^{-1}\text{gdim } V_\Lambda^0 = \frac{1}{2}\left(\frac{\theta_\Lambda(\tau)}{\eta(\tau)^{24}} + \frac{\eta(\tau)^{24}}{\eta(2\tau)^{24}}\right),$$

$$q^{-1}\text{gdim } V_\Lambda^1 = \frac{1}{2}\left(\frac{\theta_\Lambda(\tau)}{\eta(\tau)^{24}} - \frac{\eta(\tau)^{24}}{\eta(2\tau)^{24}}\right),$$

$$q^{-1}\text{gdim } V_{\Lambda,\text{tw}}^0 = 2^{11}\left(\frac{\eta(\tau)^{24}}{\eta(\tfrac{1}{2}\tau)^{24}} + \frac{\eta(\tfrac{1}{2}\tau)^{24}\eta(2\tau)^{24}}{\eta(\tau)^{48}}\right), \tag{111}$$

$$q^{-1}\text{gdim } V_{\Lambda,\text{tw}}^1 = 2^{11}\left(\frac{\eta(\tau)^{24}}{\eta(\tfrac{1}{2}\tau)^{24}} - \frac{\eta(\tfrac{1}{2}\tau)^{24}\eta(2\tau)^{24}}{\eta(\tau)^{48}}\right).$$

At this point an amazing identity intervenes. Namely $\text{gdim } V_\Lambda^1 = \text{gdim } V_{\Lambda,\text{tw}}^1 + 24$. Thus for the graded dimension (108) of $V^\natural$ we have

$$q^{-1}\text{gdim } V^\natural = q^{-1}\text{gdim } V_\Lambda^0 + q^{-1}\text{gdim } V_{\Lambda,\text{tw}}^1$$

$$= q^{-1}\text{gdim } V_\Lambda^0 + q^{-1}\text{gdim } V_\Lambda^1 - 24 \tag{112}$$

$$= q^{-1}\text{gdim } V_\Lambda - 24$$

which is $j(\tau) - 744$ according to (101) in perfect agreement with (31).

## 12 Affinization

We have not yet described the full monster module structure on $V^\natural$ but in a sense we do not need to because much more than this is obtained in [30, 31]. Indeed vertex operators are constructed for $V^\natural$ which furnish a parallel treatment for the monster group to that which we described for the complex $E_8$ Lie group (84) in Sect. 10. The monster turns out to be the automorphism group of the resulting algebraic structure on $V^\natural$ (cf. Theorem 8).

An important antecedent to this is the proof by Griess [16] that the monster group $\mathbb{M}$ exists. Armed with the character table of $\mathbb{M}$ one can compute that there is a unique up to scale $\mathbb{M}$-invariant symmetric bilinear form $(\cdot, \cdot)$ and a unique up to scale $\mathbb{M}$-invariant totally symmetric trilinear form $(\cdot, \cdot, \cdot)$ on any non-trivial 196883-dimensional $\mathbb{M}$-module (cf. (14)). Given this, one can—in principle—define an algebra structure $*$ on such an $\mathbb{M}$-module $B'$ by letting $a * b$ for $a, b \in B'$ be the unique element of $B'$ such that

$$(a, b, c) = (a * b, c) \tag{113}$$

for all $c \in B'$. Because the forms are $\mathbb{M}$-invariant we have $ga * gb = g(a * b)$ for $a, b \in B'$ and $g \in \mathbb{M}$. Because they are (totally) symmetric we have commutativity $a * b = b * a$.

In practice this is difficult because the monster is large (15) and complicated. It is also difficult because the algebra structure on $B'$ is not associative, and satisfies no obvious analogue of associativity, so the number of structure constants that have to be specified by hand in a ground-up construction is greater than we might otherwise hope. For this reason there is no obvious module theory for non-associative algebras such as $B'$. Nonetheless Griess managed to give a concrete construction of a certain one-dimensional extension $B$ of the algebra $B'$ in [16] (see also [58] for a reinterpretation by Conway) and moreover managed to verify that the subgroup of $GL(B)$ such that $g(a * b) = ga * gb$ for $a, b \in B$ has all the properties expected of $\mathbb{M}$.

In the approach of Frenkel–Lepowsky–Meurman [30, 31] the challenges of working directly with $B'$ are overcome by affinization. That is, by proceeding in line with the discussion in Sect. 9 an association

$$Y : V_2^\natural \to \mathfrak{F}(V^\natural) \tag{114}$$

of fields on $V^\natural$ (102) to elements of the 196,884-dimensional subspace $V_2^\natural$ (cf. (102), (112)) is obtained which naturally generalizes the association (88) that realized the affine $E_8$ Lie algebra in Sect. 10. A modification $\mathcal{B}$ of the algebra $B$ of Griess is then recovered by defining $\mathcal{B} := V_2^\natural$ and

$$a * b := a_1 b \tag{115}$$

for $a, b \in \mathcal{B}$ where $v \mapsto a_n v$ for $a \in V_2^\natural$ and $n \in \mathbb{Z}$ is the endomorphism of $V^\natural$ determined by requiring that

$$Y(a, z)v = \sum_n a_n v z^{-n-1}. \tag{116}$$

These endomorphisms $a_n$ realize the action of an *affinization* $\hat{\mathcal{B}}$ of $\mathcal{B}$ on $V^\natural$, and the monster may be recovered as the set of $g \in GL(V^\natural)$ such that $gY(a, z)v = Y(ga, z)gv$ for $a \in \mathcal{B}$ and $v \in V^\natural$. Moreover, as we will see in Sect. 15, there is a rich module theory for objects such as $V_L$ (52) and $V^\natural$ (102), and the subspaces of these that arise as fixed points with respect to automorphisms.

Just as in Sect. 10 we obtain a representation of the Virasoro algebra (92) on $V^\natural$ by letting $L(n)$ be defined as in (91), (93) but taking $\{h_i\}$ to be an orthonormal basis for the dual, $\mathfrak{h}$, of $\Lambda \otimes_{\mathbb{Z}} \mathbb{C}$. As in Sect. 10 this action commutes with that of $\mathbb{M}$. So in particular the vectors $L(-n)\mathbf{v} \in V_n^\natural$ are $\mathbb{M}$-invariant (and non-zero for $n \geq 2$). In this context the vector

$$\omega := L(-2)\mathbf{v} = \frac{1}{2} \sum_i h_i(-1)^2 \mathbf{v} \tag{117}$$

(cf. (90)) takes on a special significance. Namely we have $\omega_1 b = L(0)b = 2b$ for every $b \in \mathcal{B} = V_2^\natural$ so $\frac{1}{2}\omega$ is a unit for the algebra structure (115) on $\mathcal{B}$. (This is one difference between $\mathcal{B}$ and the algebra defined by Griess in [16]. The latter has no unit.)

## 13 Vertex Algebras

The construction of [30, 31] associates a field on $V_L$ to any element in $(V_L)_n$ for $n = 1$ and $n = 2$ for any even positive-definite lattice $L$ (45). Given this it is natural to ask for an association

$$Y : V_L \to \mathfrak{F}(V_L) \tag{118}$$

of fields on $V_L$ to arbitrary elements of $V_L$. Even better, one would like to axiomatize the resulting algebraic structure on $V_L$. What identities should the fields $Y(v, z)$ for $v \in V_L$ satisfy? This question was answered by Borcherds with his introduction of the notion of vertex algebra in [59].

In op. cit. Borcherds prescribed an extension (118) of (88), for an arbitrary even lattice $L$ (45), and formulated identities for the resulting fields $Y(v, z)$ that are independent of the choice of $L$. He then defined a vertex algebra to be a triple $(V, Y, \mathbf{v})$, where $Y : V \to \mathfrak{F}(V)$ and $\mathbf{v} \in V$ is a distinguished element, for which these *Borcherds identities* are satisfied. It develops (cf. [32]) that the association (114) also extends so as to furnish a vertex algebra structure on $V^\natural$.

The Borcherds identities were originally formulated in terms of the component endomorphisms of the fields $Y(v, z)$. It turns out that we may reformulate these identities as the requirement that the *vertex product* $V \otimes V \to V((z))$ defined by $u \otimes v \mapsto Y(u, z)v$ is commutative and associative in the following sense. Observe that there are naturally defined linear maps

$$
V[[z, w]][z^{-1}, w^{-1}, (z - w)^{-1}] \to
\begin{cases}
\nearrow V((z))((w)) \\
V((w))((z)) \\
\searrow V((w))((z - w))
\end{cases}
\tag{119}
$$

given by formal power series expansions in the domains $|z| > |w|$, $|w| > |z|$ and $|w| > |z - w|$, respectively (cf. (73)). Specifically, the first map in (119) is determined by the requirement that

$$
z^m w^n (z - w)^k \mapsto \sum_{\ell \geq 0} \binom{k}{\ell} z^{m+k-\ell} w^{n+\ell}
\tag{120}
$$

for $m, n, k \in \mathbb{Z}$ where $\binom{k}{\ell} := \frac{k(k-1)(k-2)\cdots(k-\ell+1)}{\ell!}$ for $\ell$ non-negative. The second map in (119) is similar but with the roles of $z$ and $w$ reversed so that

$$
z^m w^n (z - w)^k \mapsto (-1)^k \sum_{\ell \geq 0} \binom{k}{\ell} w^{n+k-\ell} z^{m+\ell}.
\tag{121}
$$

For the third map in (119) we write $z = w + (z - w)$ and regard $w$ and $z - w$ as independent variables, thus obtaining

$$
z^m w^n (z - w)^k \mapsto \sum_{\ell \geq 0} \binom{m}{\ell} (-1)^\ell w^{m+n-\ell} (z - w)^{k+\ell}.
\tag{122}
$$

With these conventions in place *commutativity* for a linear map $Y : V \to \mathfrak{F}(V)$ is the requirement that $Y(a, z)Y(b, w)c$ and $Y(b, w)Y(a, z)c$ are the formal series expansions in $|z| > |w|$ and $|w| > |z|$, respectively, of a common element $f_{a,b,c}$ of $V[[z, w]][z^{-1}, w^{-1}, (z - w)^{-1}]$ for all $a, b, c \in V$. By a similar token *associativity* for $Y$ is the requirement that $Y(a, z)Y(b, w)$ and $Y(Y(a, z-w)b, w)c$ are the formal series expansions in $|z| > |w|$ and $|w| > |z-w|$, respectively, of a common element $f_{a,b,c}$ of $V[[z, w]][z^{-1}, w^{-1}, (z - w)^{-1}]$ for $a, b, c \in V$.

We may now define a *vertex algebra* to be a triple $(V, Y, \mathbf{v})$ where the *vertex operator map* $Y : V \to \mathfrak{F}(V)$ is commutative and associative in the sense that the compositions

$$Y(a, z)Y(b, w)c \in V((z))((w))$$

$$Y(b, w)Y(a, z)c \in V((w))((z)) \qquad (123)$$

$$Y(Y(a, z - w)b, w)c \in V((w))((z - w))$$

are the formal series expansions (120–122) of some

$$f_{a,b,c} \in V[[z, w]][z^{-1}, w^{-1}, (z - w)^{-1}] \qquad (124)$$

for $a, b, c \in V$, and where the *vacuum element* $\mathbf{v} \in V$ satisfies the *identity axioms*

$$Y(\mathbf{v}, z)a = a,$$
$$Y(a, z)\mathbf{v} = a + O(z). \qquad (125)$$

By $O(z)$ here we mean some unspecified element of $zV[[z]]$.

It develops (cf. Proposition 3.1 of [60] or Theorem 4.4.1 of [61]) that the aforementioned associations $V_L \to \mathfrak{F}(V_L)$ and $V^\natural \to \mathfrak{F}(V^\natural)$ are the unique extensions of (88) and (114), respectively, which satisfy the vertex algebra axioms (123–125).

It is instructive to consider the special case that the vertex operator map satisfies $Y(a, z) \in V[[z]]$ for all $a \in V$. Then the compositions of (123) all lie in $V[[z, w]]$, which maps to itself under each of the formal series expansions (120–122) of (119). So commutativity and associativity in this case is the requirement that the compositions of (123) coincide as elements of $V[[z, w]]$. In particular then the product $(a, b) \mapsto ab$ on $V$ that we obtain by requiring that

$$Y(a, z)b = ab + O(z) \qquad (126)$$

equips $V$ with a commutative associative algebra structure in the classical sense, and the identity axioms (125) imply that $\mathbf{v}$ is a unit element for this algebra structure.

To recover the higher order terms in the series $Y(a, z)b$ we consider the identity

$$Y(a, z)Y(b, w)c = Y(Y(a, z - w)b, w)c \qquad (127)$$

in $V[[z, w]]$. Taking $b = \mathbf{v}$ and comparing the coefficients of $z^n w^0$ on each side of (127) we obtain that $a_{-n-1}c = (a_{-n-1}\mathbf{v})c$ for $n \geq 0$ where $a_{-n-1}$ is the endomorphism of $V$ defined by requiring that $Y(a, z)c = \sum_{n \geq 0} a_{-n-1}cz^n$ (cf. (116)). That is, the coefficient of $z^n$ in $Y(a, z) \in \text{End}(V)[[z]]$ acts as left-multiplication by the element $a_{-n-1}\mathbf{v}$ of $V$. To ease notation set $a^{(n)} := a_{-n-1}\mathbf{v} \in V$. Then the $a^{(n)}$ are related to each other in a nice way. To see this define a linear map $T : V \to V$ by setting $Ta := a^{(1)}$ so that

$$Y(a, z)b = ab + (Ta)bz + O(z^2) \qquad (128)$$

for $a, b \in V$. Then by taking $c = \mathbf{v}$ and comparing the coefficients of $z^n w^l$ on each side of (127) we obtain the identity

$$[T, Y(a, z)] = \frac{\mathrm{d}}{\mathrm{d}z} Y(a, z) \tag{129}$$

in $\mathrm{End}(V)[[z]]$. Finally, applying (129) to $\mathbf{v}$ yields $Ta^{(n)} = (n + 1)a^{(n+1)}$, or equivalently $Y(Ta, z) = \frac{\mathrm{d}}{\mathrm{d}z} Y(a, z)$, which in turn implies that $a^{(n)} = \frac{1}{n!} T^n a$ for $a \in V$.

So we may write $Y(a, z)b = (e^{zT} a)b$ and with this notation the identity (127) becomes

$$e^{wT} ((e^{(z-w)T} a)b)c = (e^{zT} a)(e^{wT} b)c \tag{130}$$

for $a, b, c \in V$. If $T$ acts *locally nilpotently* on $V$ in the sense that $e^{zT} a \in V[z]$ (i.e. $T^N a = 0$ for sufficiently large $N = N_a$) for every $a \in V$ then we may take $z$ to be a complex variable, and $z \cdot a := e^{zT} a$ defines an action of the additive group of $\mathbb{C}$ on $V$. So we may think of the notion of vertex algebra as a concrete axiomatization of the idea of a "commutative associative algebra with an action by $\mathbb{C}$ that is allowed to be singular at the origin".

Just as commutative associative algebras have modules, so too do vertex algebras. Such an object is a pair $(M, Y)$ where $Y : V \to \mathfrak{F}(M)$ satisfies commutativity and associativity in the sense of (123) and (124) but with $M$ in place of $V$ on the right-hand side. We also require the identity axiom $Y(\mathbf{v}, z)v = v$ for $v \in M$ (cf. (125)).

We refer to [61, 62] for fuller accounts of vertex algebra theory. The idea of regarding vertex algebras as algebras with "singular" actions by $\mathbb{C}$ was developed in greater generality by Borcherds in [63]. See also [64] for related work.

## 14  Vertex Operator Algebras

Taking motivation from representation theory, number theory and mathematical physics, Frenkel–Lepowsky–Meurman introduced the notion of vertex operator algebra in [32]; a specialization of the notion of vertex algebra just discussed.

Perhaps the most important distinction is that a vertex operator algebra comes equipped with an action of the Virasoro algebra (92) by definition. Specifically, a *vertex operator algebra* is a quadruple $(V, Y, \mathbf{v}, \omega)$ where $(V, Y, \mathbf{v})$ is a vertex algebra in the sense of Sect. 13, and the *Virasoro element* $\omega \in V$ has the property that if $L(n)$ is the endomorphism of $V$ determined by requiring that

$$Y(\omega, z) = \sum_{n \in \mathbb{Z}} L(n) z^{-n-2} \tag{131}$$

then the Virasoro algebra relations (92) are satisfied by the $L(n)$, with $\mathbf{c}$ acting as $cI$ for some $c \in \mathbb{C}$. The particular value of $c$ arising is called the *central charge* or *rank* of $V$. It is also required that

$$[L(-1), Y(a, z)] = Y(L(-1)a, z) = \frac{d}{dz} Y(a, z) \tag{132}$$

for $a \in V$ (cf. (129)), and $L(0)$ should act semisimply on $V$ with integer eigenvalues bounded from below, and with finite-dimensional eigenspaces. So in particular if $V$ is a vertex operator algebra and $V = \bigoplus_n V_n$ is its decomposition into eigenspaces for $L(0)$ then the *graded dimension*

$$\mathrm{gdim}\, V := \sum_n \dim V_n q^n \tag{133}$$

of $V$ (cf. (55), (108)) is a well-defined $q$-series with at most finitely many terms that are polar in $q$.

Note that some vertex algebras admit many different vertex operator algebra structures and some admit none. In certain circumstances vertex operator algebras define two-dimensional conformal field theories (cf. Sect. 10 and e.g. [65, 66]).

For $L$ an even positive-definite lattice (cf. (45)) the vertex algebra structure on $V_L$ (cf. (52)) described by Borcherds in [59] becomes a vertex operator algebra structure with central charge $c = \mathrm{rank}\, L$ once we set

$$\omega := \frac{1}{2} \sum_i h_i(-1)^2 \mathbf{v} \tag{134}$$

(cf. (90), (117)) for $\{h_i\}$ an orthonormal basis of the dual of $L \otimes_{\mathbb{Z}} \mathbb{C}$. Indeed the prescription of [59] may be recovered by defining $Y(1 \otimes e_\lambda, z) := Y(e_\lambda, z)$ where $Y(e_\lambda, z)$ is as in (76), and requiring that

$$Y(h(-n-1)p \otimes e_\lambda, z) = :h^{(n)}(z)Y(p \otimes e_\lambda, z): \tag{135}$$

(cf. (72)) for $p \in S(\check{\mathfrak{h}})$ (cf. (52)) where we set $h^{(n)}(z) := \frac{1}{n!} \left(\frac{d}{dz}\right)^n h(z)$ and $h(z)$ is as in (68). Thus for $\omega$ as in (134) we have

$$Y(\omega, z) = \frac{1}{2} \sum_i :h_i(z)^2: \tag{136}$$

and the coefficients of $Y(\omega, z)$ generate a Virasoro action on $V_L$ just as in Sect. 10 (cf. (93)).

In physical terms the vertex operator algebra structure on $V_L$ manifests the *lattice* conformal field theory associated to $L$. The particular example $V_\Lambda$ for $\Lambda$

the Leech lattice (cf. Theorem 7) may be regarded as a realization of the two-dimensional conformal field theory that results when the 26-dimensional bosonic string is compactified on the 24-dimensional torus

$$\mathbb{R}^{24}/\Lambda. \tag{137}$$

Given the above description (134 and 135) of the vertex operator algebra structure on $V_\Lambda$ we may check that action (103) of the involution $z_\Lambda$ on $V_\Lambda$ defines an *automorphism* of this structure in the sense that $z_\Lambda$ preseves the Virasoro element $z_\Lambda \omega = \omega$, preserves the vacuum $z_\Lambda \mathbf{v} = \mathbf{v}$, and satisfies

$$z_\Lambda Y(a, z)b = Y(z_\Lambda a, z)z_\Lambda b, \tag{138}$$

or equivalently $z_\Lambda(a_n b) = (z_\Lambda a)_n(z_\Lambda b)$ for $n \in \mathbb{Z}$ (cf. (116)), for $a, b \in V_L$. From this it follows that the vertex operator algebra structure on $V_\Lambda$ restricts to a vertex operator algebra structure with the same central charge on the subspace $V_\Lambda^0$ of $z_\Lambda$-fixed points in $V_\Lambda$ (cf. (102)). In [32] Frenkel–Lepowsky–Meurman solved the problem of extending this structure to $V^\natural$ (102) in such a way that the resulting vertex operator map

$$Y : V^\natural \to \mathfrak{F}(V^\natural) \tag{139}$$

extends the association (114) they arrived at earlier for $V_2^\natural$. Thus the following result was obtained.

**Theorem 8 (Frenkel–Lepowsky–Meurman [30–32], Borcherds [59])** *There exists a vertex operator algebra $V^\natural$ such that* $\mathrm{Aut}(V^\natural) \simeq \mathbb{M}$ *and* $q^{-1} \mathrm{gdim}\, V^\natural = j(\tau) - 744$.

## 15 Orbifolds

We mentioned in Sect. 12 that, in contrast to the situation for the Griess algebra (113), there is a theory of modules for vertex operator algebras. Indeed a module for a vertex operator algebra is a module $M$ for the underlying vertex algebra (cf. Sect. 13) such that the action of $L(0)$ on $M$ is semisimple with finite-dimensional eigenspaces and eigenvalues bounded from below (cf. (133)).

With this definition it develops (cf. [67]) that if $L$ is an even positive-definite lattice then the isomorphism classes of irreducible modules for $V_L$ are in correspondence with the cosets $L^*/L$ of $L$ in its dual $L^*$ (46). So in particular $V_\Lambda$ has a single irreducible module—namely itself—since $\Lambda$ is unimodular (cf. (46)), and the same is true for the vertex operator algebra $V_L$ in (52) that realizes (78) the basic representation of the affine $E_8$ Lie algebra. Frenkel–Lepowsky–Meurman showed in [32] that this property extends to $V^\natural$. That is, $V^\natural$ is *holomorphic* in the sense that any irreducible module for $V^\natural$ is isomorphic to $V^\natural$. (This terminology is motivated by the conventions of conformal field theory.)

Recall Conway's result, Theorem 7, that the Leech lattice $\Lambda$ is the unique unimodular even positive-definite lattice with rank 24 and no vectors $\lambda \in \Lambda$ such that $(\lambda, \lambda) = 2$. Frenkel–Lepowsky–Meurman proposed the following conjecture in [32] which, if proven, would furnish a directly analogous characterization of $V^\natural$.

*Conjecture 2 (Frenkel–Lepowsky–Meurman [32])* The vertex operator algebra $V^\natural$ is the unique holomorphic vertex operator algebra with central charge 24 and no non-zero vectors $v \in V^\natural$ such that $L(0)v = v$.

Perhaps the principal appeal of Conjecture 2 is that it promises a beautiful conceptual definition of the monster group. To this day the monster is most commonly defined as the largest of the sporadic simple groups (cf. Sect. 3), but this takes the classification of finite simple groups—the so-called enormous theorem (cf. [10])—as a prerequisite. If Conjecture 2 were confirmed we would be able to define the monster as the automorphism group of the unique holomorphic vertex operator algebra that has central charge 24 and no non-zero vectors $v$ with $L(0)v = v$. This in particular motivates the notation $V^\natural$. The superscript $\natural$ is the *natural sign* from musical notation. One of the key ideas of Frenkel–Lepowsky–Meurman is that the monster group, despite being finite, is most naturally represented on the infinite-dimensional space $V^\natural$. We may also note the similarity of Conjecture 2 to Theorem 7.

Recently there has been much progress on the classification of holomorphic vertex operator algebras (see in particular [68–75] and references therein) but Conjecture 2 remains open.

Part of what makes $V^\natural$ natural is that it admits a string theoretic interpretation. In this vein we note that the extension of the vertex operator algebra structure from $V_\Lambda^0 < V^\natural$ to $V^\natural = V_\Lambda^0 \oplus V_{\Lambda,\mathrm{tw}}^1$ (102) involves *twisted* vertex operators, which in the case at hand are fields of the form

$$Y_{\mathrm{tw}}(v, z^{\frac{1}{2}}) : V_{\Lambda,\mathrm{tw}} \to V_{\Lambda,\mathrm{tw}}((z^{\frac{1}{2}})) \tag{140}$$

(cf. (75), (88)) for $v \in V_\Lambda$. The main difference between these twisted vertex operators and the fields (75) introduced in Sect. 9 is the relationship

$$Y_{\mathrm{tw}}(z_\Lambda v, z^{\frac{1}{2}}) = Y_{\mathrm{tw}}(v, -z^{\frac{1}{2}}) \tag{141}$$

with the automorphism $z_\Lambda$ of $V_\Lambda$. In physical terms such twisted vertex operators are used to describe conformal field theory on *orbifolds* (cf. [76–78]), which are quotients of manifolds by automorphisms that act with fixed points. The construction of $V^\natural$ by Frenkel–Lepowsky–Meurman was later recognized as the first example of a conformal field theory on an orbifold which is not a torus. In the context of string theory $V^\natural$ manifests the compactification of the 26-dimensional bosonic string on the canonical 24-dimensional orbifold

$$\left(\mathbb{R}^{24}/\Lambda\right)/\{\pm I\} \tag{142}$$

of the torus (137) defined by the Leech lattice.

A more abstract motivation for the twisted fields of (140) is the fact that while each eigenspace for the action of $z_\Lambda$ on $V_{\Lambda,tw}$ is a module for the vertex operator algebra $V_\Lambda^0$ of $z_\Lambda$-fixed points in $V_\Lambda$, the full space $V_{\Lambda,tw}$ is naturally a *twisted module* for $V_\Lambda$ where the notion of twisted module for a vertex operator algebra is obtained by formulating a suitable analogue of the main vertex algebra axiom (123 and 124) for twisted fields as in (140), and putting suitable conditions on the action of $L(0)$ (cf. (133)). The hope then is that, given an automorphism $g$ of a vertex operator algebra $V$, ordinary modules for the $g$-fixed subspace of $V$ may be understood in terms of $g$-twisted modules for the original space $V$. This idea stems from the foundational works [79, 80] on orbifold conformal field theory. The full realization of this idea is an ongoing goal of the *orbifold theory* of vertex operator algebras.

We refer to §IV of the introduction to [32] for a fuller discussion of the relationship between $V^\natural$ and string theory. In addition to the vertex algebra references mentioned in Sect. 13 we refer to [32, 81] for introductory accounts that focus on vertex operator algebra theory. For a recent review of the vertex algebraic orbifold theory see §2 of [82].

## 16   The Genus Zero Problem

Armed with the $\mathbb{M}$-module structure on $V^\natural$ (cf. Theorem 8) we may consider the *graded trace functions*

$$T_g^\natural(\tau) := \sum_n \operatorname{tr}(g|V_n^\natural)q^{n-1} \tag{143}$$

for $g \in \mathbb{M}$. With monstrous moonshine Sect. 5 in mind one now hopes to find

$$T_g^\natural = T_g \tag{144}$$

(cf. (32)) for all $g \in \mathbb{M}$ for this would verify the monstrous moonshine conjecture, Conjecture 1. Somewhat frustratingly the description of $V^\natural$ given in Sect. 11 does not allow to verify (144) directly because it is difficult to compute $T_g^\natural$ (143) for elements $g \in \mathbb{M}$ that fall outside of the involution centralizer $C_\mathbb{M}(z)$ (cf. (12), (104)). For this reason the monstrous moonshine conjecture was only partially verified by the work [30–32] of Frenkel–Lepowsky–Meurman.

The full verification of Conjecture 1 was achieved by Borcherds in [26] following his introduction [83] of another new algebraic structure. Namely, Borcherds introduced *generalized Kac–Moody (Lie) algebras*, also known as *Borcherds–Kac–Moody (Lie) algebras*, in op. cit., and used $V^\natural$ to construct a particularly distinguished example called the *monster Lie algebra* in [26]. In analogy with simple finite-dimensional Lie algebras, and simple affine Lie algebras as described in Sect. 7, a generalized Kac–Moody algebra admits a *Weyl–Kac denominator formula*, which is a product-sum identity of the form

$$e^\rho \prod_{\alpha>0}(1 - e^\alpha)^{\text{mult}(\alpha)} = \sum_{w\in W} \det(\alpha)w\left(e^\rho \sum_\alpha \varepsilon(\alpha)e^\alpha\right). \tag{145}$$

We refer to §4 of [26] for the precise definition of the symbols appearing in (145). For the affine $A_1$ Lie algebra (cf. (44)) the Weyl–Kac denominator formula is

$$\prod_{n>0}(1 - u^{n-1}v^n)(1 - u^n v^{n-1})(1 - u^n v^n) = \sum_{n\in\mathbb{Z}}(-1)^n u^{\frac{1}{2}n(n+1)}v^{\frac{1}{2}n(n-1)} \tag{146}$$

(see e.g. (12.1.5) in [46]), which is a form of the famous *Jacobi triple product identity*. (Substitute $x$ for $uv$ and $-\zeta^{-1}$ for $v$ in (146) to recover the version of (146) that appears on p.377 of [84], for example.)

In the case of the monster Lie algebra (145) specializes to

$$p^{-1}\prod_{m>0}(1 - p^m q^n)^{c(mn)} = j(\sigma) - j(\tau) \tag{147}$$

where $p = e^{2\pi i\sigma}$ for $\sigma \in \mathbb{H}$ and $c(n)$ is as in (25). The amazing fact that the right-hand side of (147) involves no terms $p^m q^n$ with $m$ and $n$ both non-zero implies a system of quadratic identities for the $c(n)$. For example it follows from (147) that

$$c(4k + 2) = c(2k + 2) + \sum_{1\le j\le k} c(j)c(2k + 1 - j) \tag{148}$$

for $k \geq 1$. By using the natural action of the monster on the monster Lie algebra that it inherits from $V^\natural$ Borcherds showed [26] that (147) generalizes to

$$p^{-1}\exp\left(-\sum_{\substack{k,m,n\in\mathbb{Z}\\k,m>0}} \frac{1}{k}c^\natural_{g^k}(mn)p^{mk}q^{nk}\right) = T^\natural_g(\sigma) - T^\natural_g(\tau) \tag{149}$$

for $g \in \mathbb{M}$ where $c^\natural_g(n) := \text{tr}(g|V^\natural_{n+1})$ is the coefficient of $q^n$ in $T^\natural_g$ (143). It develops (see §9 of [26]) that the consequent quadratic identities for the $c^\natural_g(n)$ are strong enough to show that the functions $T_g^\natural$ are all principal moduli (cf. (34)) for genus zero groups. This fact reduces the verification of (144) to a short computation.

**Theorem 9 (Borcherds [26])** *We have $T_g^\natural = T_g$ for all $g \in \mathbb{M}$.*

As we explained in Sect. 5 the main content of the monstrous moonshine conjecture is the genus zero property, i.e. the statement that the McKay–Thompson series $T_g$ are all principal moduli for genus zero groups (cf. (34)). The proof of Theorem 9 given by Borcherds confirms the genus zero property of monstrous moonshine by reducing the verification to a surprisingly succinct calculation, but it does not yield a conceptual explanation of the phenomenon. The *genus zero problem* of monstrous moonshine is to provide such an explanation.

To give this problem some context let us consider the weaker *modularity property* of monstrous moonshine, being the statement that the McKay–Thompson series are invariant—but not necessarily principal moduli—for some subgroups of $SL_2(\mathbb{R})$. This problem was solved conceptually by work of Zhu [85] and Dong–Li–Mason [86], which demonstrates that the trace functions associated to the action of a finite group on a vertex operator algebra define modular invariant functions under general conditions. From the physical perspective this is expected given the conformal invariance property of conformal field theory (cf. Sect. 10), which manifests algebraically via the action of the Virasoro algebra (cf. (131)). Motivated by this Zhu developed a rigorous method [85] for showing that the graded dimension functions of irreducible modules for a vertex operator algebra define modular functions. This *modularity theorem* was extended and refined by Dong–Li–Mason [86] so as to establish similar results for the trace functions arising from the action of a finite group. In both cases certain technical conditions on the vertex operator algebra in question are required. (Some of these conditions are not expected to be necessary. More recent works on the modularity of vertex operator algebras include [87–90].)

So it is a general phenomenon that the trace functions arising from a vertex operator algebra are modular invariant. In fact just by considering lattice vertex operator algebras for suitable lattices we can realize any finite group as a group of automorphisms of a holomorphic vertex operator algebra satisfying the hypotheses of Zhu and Dong–Li–Mason, so vertex operator algebra theory attaches modular invariant functions to every finite group. This reinforces the claim that it is the genus zero property of monstrous moonshine that makes it special. What we are missing, so far, is a class of algebraic objects for which the associated trace functions are principal moduli (cf. (34)) under general conditions.

We do however have some results that hint at where such a class might be found. One such is an argument of Tuite, given in physical terms [91, 92] (see also [93, 94]), that the genus zero property of monstrous moonshine follows from the uniqueness conjecture, Conjecture 2, of Frenkel–Lepowsky–Meurman. This has been revisited and refined recently [95, 96] by Paquette–Persson–Volpato in the context of string theory.

Another such result is a connection between the genus zero property of monstrous moonshine and certain speculative properties of three-dimensional quantum gravity. This connection is developed in [97], wherein it is proven that the McKay–Thompson series $T_g$ (32) of monstrous moonshine may be expressed as regularized averages over their invariance groups. Specifically, for each $g \in \mathbb{M}$ we have

$$T_g(\tau) = R_{\Gamma_g}(\tau) - \frac{1}{2}C_{\Gamma_g} \tag{150}$$

for a certain naturally defined complex number $C_\Gamma$ called the Rademacher constant of $\Gamma$ (cf. §6 of [98]) where $R_\Gamma$ denotes the *Rademacher sum*

$$R_\Gamma(\tau) := \exp(-2\pi i \tau) + \lim_{K \to \infty} \sum_{\gamma \in (\Gamma_\infty \backslash \Gamma)_K^\times} \exp(-2\pi i \gamma \tau) - \exp(-2\pi i \gamma \infty). \tag{151}$$

In (151) we may take $\Gamma$ to be any subgroup of $SL_2(\mathbb{R})$ that is commensurable with the modular group (in the sense of Sect. 5). We write $\Gamma_\infty$ for the subgroup of $\Gamma$ composed of the matrices $\pm\left(\begin{smallmatrix} 1 & n \\ 0 & 1 \end{smallmatrix}\right)$ for $n \in \mathbb{Z}$, we write $(\Gamma_\infty \backslash \Gamma)_K^\times$ for the set of cosets of $\Gamma_\infty$ in $\Gamma$ with representatives $\left(\begin{smallmatrix} a & b \\ c & d \end{smallmatrix}\right)$ such that $0 < c < K$ and $|d| < K^2$, and $\gamma\infty$ means $\lim_{\tau \to \infty} \gamma\tau = \frac{a}{c}$ in case $\gamma = \left(\begin{smallmatrix} a & b \\ c & d \end{smallmatrix}\right)$. For simplicity we are assuming that $\Gamma_\infty$ is precisely the set of upper triangular matrices in $\Gamma$ but the definition can easily be modified (see [97]) if this is not the case. We call $R_\Gamma$ a Rademacher sum in honor of Rademacher's introduction [99] of the $\Gamma = SL_2(\mathbb{Z})$ case of (151) in 1939, and his proof (op. cit.) that

$$j(\tau) - 744 = R_{SL_2(\mathbb{Z})}(\tau) - 12. \tag{152}$$

Rademacher probably did not anticipate connections to quantum gravity or the monster, but we may recognize (152) as the $g = e$ case of (150) (cf. (26), (31), (32)).

The right-hand side of (151) should be regarded as a regularization of the naive average

$$\sum_{\gamma \in \Gamma_\infty \backslash \Gamma} q^{-1}\Big|_0 \gamma \tag{153}$$

(cf. (63)) which is manifestly $\Gamma$-invariant but does not converge. By contrast the right-hand side of (151) converges locally uniformly, and thus defines a holomorphic function $R_\Gamma$ on $\mathbb{H}$, but it is not at all clear that this function $R_\Gamma$ is $\Gamma$-invariant. Indeed it develops that the $\Gamma$-invariance of $R_\Gamma$ depends crucially upon the geometry of $\Gamma$.

**Theorem 10 ([97])** *For* $\Gamma < SL_2(\mathbb{R})$ *commensurable with* $SL_2(\mathbb{Z})$ *we have* $R_\Gamma(\gamma\tau) = R_\Gamma(\tau)$ *for all* $\gamma \in \Gamma$ *if and only if* $\Gamma$ *has genus zero. Moreover if* $\Gamma$ *has genus zero then* $R_\Gamma$ *is a principal modulus for* $\Gamma$.

Observe that Theorem 10 essentially proves (150). The pertinence to gravity is that there are physical arguments (see [97]) that suggest that the partition functions of an appropriately defined three-dimensional theory of quantum gravity should be Rademacher sums by construction.

So it is possible that either a rigorous reworking of the string theoretic arguments of Paquette–Persson–Volpato, or the construction of a suitably defined three-dimensional theory of quantum gravity, would establish the genus zero property in a conceptual way. So far there is no clear relationship between these two approaches. Uncovering such a relationship may be the key to progress on this problem.

## 17  Super Structure

As mentioned in Sect. 15 the vertex operator algebra $V^\natural$ (102) that realizes monstrous moonshine (cf. Conjecture 1, (143 and 144), Theorem 9) is a manifestation of bosonic string theory in 26 dimensions (cf. (142)). In principle string theory can be formulated in any dimension but for a generic choice there is a *conformal anomaly*, meaning a failure of the theory to be invariant under rescalings of metrics on the *worldsheets*, being the surfaces which model the propagation of strings through spacetime.

The fact that bosonic string theory is optimally formulated in dimension 26—the *critical dimension* of the bosonic string—was observed by Lovelace [100] in 1971. Soon after this a precursor to supersymmetric string theory was introduced by Gervais–Sakita [101] following Ramond [102] and Neveu–Schwarz [103, 104]. Schwarz observed [105] that the critical dimension reduces to 10 in this setting.

So it was natural for Frenkel–Lepowsky–Meurman to consider a super analogue $V^{f\natural}$ of the moonshine module in [31] in which bosonic and fermionic fields are present, and an unimodular even lattice of rank $8 = 10 - 2$ takes on the role played by the Leech lattice (cf. Theorem 7, (142)) in the construction (102) of $V^\natural$. In fact the $E_8$ lattice $L$ we introduced in Sect. 7 (cf. (47)) is the unique unimodular even positive-definite lattice of rank 8 so $V^{f\natural}$ should be related to the basic representation (52) of the affine $E_8$ Lie algebra. The main difference is the inclusion of fermionic fields, which is achieved by considering

$$V_L^f := A(\mathfrak{e}) \otimes V_L \tag{154}$$

where $\mathfrak{e}$ is a copy of $L \otimes \mathbb{C}$, also equipped with the bilinear form $(\,\cdot\,,\,\cdot\,)$ (cf. (45)), we set

$$\check{\mathfrak{e}} := \bigoplus_{n<0} \mathfrak{e}(n + \tfrac{1}{2}) \tag{155}$$

(cf. (51)), and $A(\mathfrak{e}) := \bigwedge(\check{\mathfrak{e}})$ is the exterior algebra of $\check{\mathfrak{e}}$.

An exterior algebra $\bigwedge(V)$ is an example of a *superalgebra* because we have $ab \in \bigwedge(V)_{\bar{j}+\bar{k}}$ for $a \in \bigwedge(V)_{\bar{j}}$ and $b \in \bigwedge(V)_{\bar{k}}$, where

$$\bigwedge(V)_{\bar{j}} := \bigoplus_n \bigwedge^{2n+j}(V) \tag{156}$$

and $\bar{j}$ denotes the residue class of $j \in \mathbb{Z}$ modulo 2. More generally a *superspace* is a vector space $A$ with a $\mathbb{Z}/2\mathbb{Z}$-grading

$$A = A_{\bar{0}} \oplus A_{\bar{1}}. \tag{157}$$

Say that $a \in A_{\bar{j}}$ is *even* or *odd* according as $\bar{j}$ is $\bar{0}$ or $\bar{1}$, and call $\bar{j}$ the *parity* of $a$. Define the *canonical involution* of a superalgebra or superspace $A$ to be the linear map $z_A : A \to A$ that satisfies

$$z_A(a) = (-1)^j a \tag{158}$$

for $a \in A_{\bar{j}}$ homogeneous, and in this case define $|a| := \bar{j}$.

A superalgebra $A = A_{\bar{0}} \oplus A_{\bar{1}}$ becomes a *Lie superalgebra* when equipped with the *supercommutator*

$$[a, b] := ab - (-1)^{|a||b|}ba, \tag{159}$$

but this Lie superalgebra structure is trivial for $A = \bigwedge(V)$. So $\bigwedge(V)$ is called *supercommutative*. The endomorphism ring of a superalgebra $A$ is also naturally a superalgebra,

$$\text{End } A = (\text{End } A)_{\bar{0}} \oplus (\text{End } A)_{\bar{1}}. \tag{160}$$

A homogeneous endomorphism $x \in (\text{End } A)_{|x|}$ satisfies $xb \in A_{|x|+|b|}$ for $b \in A_{|b|}$.

With the bilinear form $(\cdot, \cdot)$ in hand we may define odd endomorphisms $\partial_{e(-r)}$ of $A(\mathfrak{e}) = \bigwedge(\check{\mathfrak{e}})$ for $e \in \mathfrak{e}$ and $r \in \mathbb{Z} + \frac{1}{2}$ with $r > 0$ by requiring that

$$\partial_{e(-r)}(e'(-r')b) = -(e, e')\delta_{r,r'}b - e'(-r')(\partial_{e(-r)}b) \tag{161}$$

for $e' \in \mathfrak{e}$ and $r' \in \mathbb{Z} + \frac{1}{2}$ (with $r' > 0$) and $b \in A(\mathfrak{e})$. Then $\partial_{e(-r)}$ is a *superderivation* of $A(\mathfrak{e})$ because we have $\partial_{e(-r)}(ab) = (\partial_{e(-r)}a)b + (-1)^{|a|}a(\partial_{e(-r)}b)$ for $a \in A(\mathfrak{e})^{|a|}$.

Now define $e(r) \in \text{End } A(\mathfrak{e})$ for $e \in \mathfrak{e}$ and $r \in \mathbb{Z} + \frac{1}{2}$ by setting

$$e(r)b := \begin{cases} e(r)b & \text{if } r < 0, \\ \partial_{e(-r)}b & \text{if } r > 0, \end{cases} \tag{162}$$

for $b \in A(\mathfrak{e})$ (cf. (66)). Then we obtain a *fermionic field* $e(z)$ on $A(\mathfrak{e})$ for $e \in \mathfrak{e}$ by defining

$$e(z) := \sum_{n \in \mathbb{Z}} e(n + \tfrac{1}{2})z^{-n-1} \tag{163}$$

(cf. (68)). The action (162) implies the identity

$$[e(z), e'(w)] = -(e, e')\delta(z - w) \tag{164}$$

in $\text{End } A(\mathfrak{e})[[z, z^{-1}, w, w^{-1}]]$ for $e, e' \in \mathfrak{e}$ (cf. (69)) where $[\cdot, \cdot]$ is the supercommutator in $\text{End } A(\mathfrak{e})$. So the fields (163) are *superlocal* in the sense that we have

$$(z - w)[e(z), e'(w)] = 0 \tag{165}$$

in $\text{End } A(\mathfrak{e})[[z, z^{-1}, w, w^{-1}]]$ (cf. (71)).

We may equip $A(\mathfrak{e})$ with a $\tfrac{1}{2}\mathbb{Z}$-grading by defining $L(0) \in \text{End } A(\mathfrak{e})$ so that $L(0)\mathbf{v} = 0$ for $\mathbf{v}$ the unit element of $A(\mathfrak{e}) = \bigwedge(\check{\mathfrak{e}})$ and

$$[L(0), e(-n + \tfrac{1}{2})] = (n - \tfrac{1}{2})e(-n + \tfrac{1}{2}) \tag{166}$$

for $e \in \mathfrak{e}$ and $n > 0$ (cf. (54), (110)). Then $L(0)$ acts semisimply on $A(\mathfrak{e})$ and the decomposition into eigenspaces furnishes a non-negative grading $A(\mathfrak{e}) = \bigoplus_{n \geq 0} A(\mathfrak{e})_{\frac{n}{2}}$. The resulting graded dimension function (cf. (55)) satisfies

$$\begin{aligned} q^{-\frac{1}{6}} \text{gdim } A(\mathfrak{e}) &= q^{-\frac{1}{6}} \prod_{n > 0}(1 + q^{n-\frac{1}{2}})^8 \\ &= \frac{\eta(\tau)^{16}}{\eta(\tfrac{1}{2}\tau)^8 \eta(2\tau)^8} = q^{-\frac{1}{6}} + 8q^{\frac{1}{3}} + 28q^{\frac{5}{6}} + 64q^{\frac{4}{3}} + \dots \end{aligned} \tag{167}$$

where $\eta$ is as in (58).

Note that $A(\mathfrak{e})_0$ is spanned by $\mathbf{v}$ and $A(\mathfrak{e})_{\frac{1}{2}}$ is composed of the $e(-\tfrac{1}{2})\mathbf{v}$ for $e \in \mathfrak{e}$. So we may define a linear map

$$Y : A(\mathfrak{e})_{\frac{1}{2}} \to \mathfrak{F}(A(\mathfrak{e})) \tag{168}$$

(cf. (88), (114)) by setting $Y(e(-\tfrac{1}{2})\mathbf{v}, z) := e(z)$ for $e \in \mathfrak{e}$ (cf. (163)). Then, similar to the discussion of fields on $V_L$ and $V^\natural$ in Sect. 13, it develops that there is a unique extension $Y : A(\mathfrak{e}) \to \mathfrak{F}(A(\mathfrak{e}))$ of (168) such that $(A(\mathfrak{e}), Y, \mathbf{v})$ satisfies the axioms of a *vertex superalgebra*. That is, the identity axioms (125) hold, and there is a $\mathbb{Z}/2\mathbb{Z}$-grading $A(\mathfrak{e}) = A(\mathfrak{e})_{\bar{0}} \oplus A(\mathfrak{e})_{\bar{1}}$ for which the vertex operator map $Y$ is *supercommutative* and associative in the sense that

$$Y(a, z)Y(b, w)c \in V((z))((w))$$

$$(-1)^{|a||b|}Y(b, w)Y(a, z)c \in V((w))((z)) \tag{169}$$

$$Y(Y(a, z - w)b, w)c \in V((w))((z - w))$$

are the formal expansions (120–122) of some

$$f_{a,b,c} \in A(\mathfrak{e})[[z, w]][z^{-1}, w^{-1}, (z - w)^{-1}] \tag{170}$$

(cf. (123)) for each $a, b, c \in A(\mathfrak{e})$, for $a \in A(\mathfrak{e})_{|a|}$ and $b \in A(\mathfrak{e})_{|b|}$ homogeneous. In our case $A(\mathfrak{e})_{\bar{j}} = \sum_n A(\mathfrak{e})_{n+\frac{j}{2}}$.

We obtain a *vertex operator superalgebra* structure (cf. Sect. 14) of central charge $4 = \frac{1}{2} \dim \mathfrak{e}$ on $A(\mathfrak{e})$ by taking

$$\omega := -\frac{1}{2} \sum_i e_i(-\tfrac{3}{2})e_i(\tfrac{1}{2})\mathbf{v} \tag{171}$$

to be the Virasoro element (cf. (131), (132), (134)) where $\{e_i\}$ is an orthonormal basis for $\mathfrak{e}$. Then $V_L^f = A(\mathfrak{e}) \otimes V_L$ (154) becomes a vertex operator superalgebra of central charge 12 when we take $\mathbf{v} := \mathbf{v} \otimes \mathbf{v}$ to be the vacuum, let

$$\omega := \omega \otimes \mathbf{v} + \mathbf{v} \otimes \omega \tag{172}$$

be the Virasoro element, and define the vertex operator map $Y : V_L^f \to \mathfrak{F}(V_L^f)$ by setting

$$Y(a \otimes v, z) := Y(a, z) \otimes Y(v, z) \tag{173}$$

for $a \in A(\mathfrak{e})$ and $v \in V_L$. The $\mathbb{Z}/2\mathbb{Z}$-grading $V_L^f = (V_L^f)_{\bar{0}} \oplus (V_L^f)_{\bar{1}}$ (cf. (157)) is given by $(V_L^f)_{\bar{j}} = A(\mathfrak{e})_{\bar{j}} \otimes V_L$ (cf. (155)).

## 18 Super Moonshine

From (172 and 173) it follows that the graded dimension (cf. (55)) of $V_L^f$ is the product of gdim $A(\mathfrak{e})$ (167) and gdim $V_L$ (56). So the $\frac{1}{2}\mathbb{Z}$-grading on $V_L^f$ determined by the vertex operator superalgebra structure specified in Sect. 17 satisfies

$$q^{-\frac{1}{2}} \text{ gdim } V_L^f = \frac{\eta(\tau)^{16}}{\eta(\frac{1}{2}\tau)^8 \eta(2\tau)^8} \frac{\theta_L(\tau)}{\eta(\tau)^8} = q^{-\frac{1}{2}} + 8 + 276q^{\frac{1}{2}} + 2048q + \dots . \tag{174}$$

This is striking because 276 is the minimal dimension of a non-trivial irreducible module for the sporadic simple Conway group $Co_1$ (cf. (12), (98)), and 2048 may be interpreted in terms of dimensions of irreducible $Co_1$-modules as

$$2048 = 1 + 276 + 1771. \tag{175}$$

(cf. (27–29), (85–87)). Moreover the function $q^{-\frac{1}{2}} \operatorname{gdim} V_L^f$ is analogous to the elliptic modular invariant (24) in that it is a principal modulus (cf. (34)) for the index 3 subgroup of $SL_2(\mathbb{Z})$ generated by $\left(\begin{smallmatrix} 0 & -1 \\ 1 & 0 \end{smallmatrix}\right)$ and $\left(\begin{smallmatrix} 1 & 2 \\ 0 & 1 \end{smallmatrix}\right)$.

In fact we can relate $V_L^f$ quite directly to the discussion of Sect. 6, for the super structure $V_L^f = (V_L^f)_{\bar{0}} \oplus (V_L^f)_{\bar{1}}$ invites us to consider the *graded superdimension*

$$\operatorname{gsdim} V_L^f := \operatorname{gdim}(V_L^f)_{\bar{0}} - \operatorname{gdim}(V_L^f)_{\bar{1}}. \tag{176}$$

Let $z_A$ denote the canonical involution of $A(\mathfrak{e})$ (cf. (158)), satisfying $z_A(a) = (-1)^j a$ for $a \in A(\mathfrak{e})_{\bar{j}}$. Then we have

$$
\begin{aligned}
q^{-\frac{1}{6}} \operatorname{gsdim} A(\mathfrak{e}) &= \sum_{n \geq 0} \operatorname{tr}(z_A | A(\mathfrak{e})_{\frac{n}{2}}) q^{\frac{n}{2} - \frac{1}{6}} \\
&= q^{-\frac{1}{6}} \prod_{n > 0} (1 - q^{n - \frac{1}{2}})^8 \\
&= \frac{\eta(\frac{1}{2}\tau)^8}{\eta(\tau)^8} = q^{-\frac{1}{6}} - 8q^{\frac{1}{3}} + 28q^{\frac{5}{6}} - 64q^{\frac{4}{3}} + \dots
\end{aligned}
\tag{177}
$$

(cf. (167)). Putting together (177) and gsdim $V_L = \operatorname{gdim} V_L$ (56) we obtain that the graded superdimension of $V_L^f$ satisfies

$$q^{-\frac{1}{2}} \operatorname{gsdim} V_L^f = \frac{\eta(\frac{1}{2}\tau)^8}{\eta(\tau)^8} \frac{\theta_L(\tau)}{\eta(\tau)^8} = q^{-\frac{1}{2}} - 8 + 276q^{\frac{1}{2}} - 2048q + \dots. \tag{178}$$

It turns out that this function (178) coincides with $T_{2B}(\frac{1}{2}\tau) - 8$ where $T_{2B}$ (cf. (39)) is the normalized principal modulus for $\Gamma_0(2)$ (and the McKay–Thompson series of a $2B$ element of the monster, such as the $z$ in (12)).

However, similar to the situation in (100), there is no natural $Co_1$-module interpretation for the constant term 8 in (174). So Frenkel–Lepowsky–Meurman suggested [31] to consider

$$V^{f\natural} := (V_L^f)^0 \oplus (V_{L,\text{tw}}^f)^1 \tag{179}$$

(cf. (102)) where $V_L^f = (V_L^f)^0 \oplus (V_L^f)^1$ and $V_{L,\mathrm{tw}}^f = (V_{L,\mathrm{tw}}^f)^0 \oplus (V_{L,\mathrm{tw}}^f)^1$ are the eigenspace decompositions for the action of an involution $z_L^f := z_A \otimes z_L$, and

$$V_{L,\mathrm{tw}}^f := A(\mathfrak{e})_{\mathrm{tw}} \otimes V_{L,\mathrm{tw}}. \tag{180}$$

Here $z_A$ is as in (177) and $z_L$ is as in (103), we define $V_{L,\mathrm{tw}} := S(\check{\mathfrak{h}}_{\mathrm{tw}}) \otimes T_L$ in direct analogy with $V_{\Lambda,\mathrm{tw}}$ (105), and describe the action of $z_L^f$ on $V_{L,\mathrm{tw}}$ momentarily. Note that $\dim T_L = 2^4 = 16$.

To describe the space $A(\mathfrak{e})_{\mathrm{tw}}$ we require a *polarization* of $\mathfrak{e}$, which is a decomposition

$$\mathfrak{e} = \mathfrak{e}^- \oplus \mathfrak{e}^+ \tag{181}$$

such that the symmetric bilinear form $(\cdot, \cdot)$ vanishes identically on each of $\mathfrak{e}^-$ and $\mathfrak{e}^+$. Given this datum we may write

$$A(\mathfrak{e})_{\mathrm{tw}} := \bigwedge(\mathfrak{e}^-(0) \oplus \check{\mathfrak{e}}_{\mathrm{tw}}) \tag{182}$$

where $\check{\mathfrak{e}}_{\mathrm{tw}} := \bigoplus_{n<0} \mathfrak{e}(n)$ (cf. (105)). As the reader may anticipate by now $A(\mathfrak{e})_{\mathrm{tw}}$ is naturally a $z_A$-twisted module for $A(\mathfrak{e})$ (cf. Sect. 15) and $V_{L,\mathrm{tw}}$ is naturally a $z_L$-twisted module for $V_L$, and it follows that $V_{L,\mathrm{tw}}^f$ is naturally a $z_L^f$-twisted module for $V^{f\natural}$. We obtain an action of $z_L^f$ on $V_{L,\mathrm{tw}}^f$ by letting $z_L$ act on $V_{L,\mathrm{tw}}$ in direct analogy with (107), and letting $z_A$ act as the canonical involution (158) associated to the super structure (156) on the exterior algebra $A(\mathfrak{e})_{\mathrm{tw}}$ (182).

To compute the graded dimension of $V^{f\natural}$ we require an action of $L(0)$ on $V_{L,\mathrm{tw}}^f$. It develops that the natural analogue of (109) is

$$L(0)(a \otimes p \otimes t) = (L(0)a) \otimes p \otimes t + a \otimes (L(0)p) \otimes t + \tfrac{1}{2} a \otimes p \otimes t \tag{183}$$

where $p \mapsto L(0)p$ is defined as in (110) for $p \in S(\check{\mathfrak{h}}_{\mathrm{tw}})$, and $a \mapsto L(0)a$ for $a \in A(\mathfrak{e})_{\mathrm{tw}} = \bigwedge(\check{\mathfrak{e}}_{\mathrm{tw}})$ is determined by requiring that $L(0)\mathbf{v}_{\mathrm{tw}} = \tfrac{1}{2}\mathbf{v}_{\mathrm{tw}}$ for $\mathbf{v}_{\mathrm{tw}}$ the unit element of the exterior algebra $\bigwedge(\check{\mathfrak{e}}_{\mathrm{tw}})$, and

$$[L(0), e(-n)] = ne(-n) \tag{184}$$

for $n \in \mathbb{Z}$ with $n \geq 0$ (cf. (54), (110), (166)). We may now compute

$$q^{-\frac{1}{2}} \operatorname{gdim}(V_L^f)^0 = \frac{1}{2} \left( \frac{\eta(\tau)^{16}}{\eta(\frac{1}{2}\tau)^8 \eta(2\tau)^8} \frac{\theta_L(\tau)}{\eta(\tau)^8} + \frac{\eta(\frac{1}{2}\tau)^8}{\eta(\tau)^8} \frac{\eta(\tau)^8}{\eta(2\tau)^8} \right),$$

$$q^{-\frac{1}{2}} \operatorname{gdim}(V_L^f)^1 = \frac{1}{2} \left( \frac{\eta(\tau)^{16}}{\eta(\frac{1}{2}\tau)^8 \eta(2\tau)^8} \frac{\theta_L(\tau)}{\eta(\tau)^8} - \frac{\eta(\frac{1}{2}\tau)^8}{\eta(\tau)^8} \frac{\eta(\tau)^8}{\eta(2\tau)^8} \right), \qquad (185)$$

$$q^{-\frac{1}{2}} \operatorname{gdim}(V_{L,\mathrm{tw}}^f)^0 = q^{-\frac{1}{2}} \operatorname{gdim}(V_{L,\mathrm{tw}}^f)^1 = 2^7 \frac{\eta(2\tau)^8}{\eta(\tau)^8} \frac{\eta(\tau)^8}{\eta(\frac{1}{2}\tau)^8},$$

(cf. (111), (174), (177)). The last identity in (185) holds because the graded trace of $z_A$ on $A(\mathfrak{e})_{\mathrm{tw}}$ vanishes identically.

For the construction of $V^\natural$ the identity $\operatorname{gdim} V_\Lambda^1 = \operatorname{gdim} V_{\Lambda,\mathrm{tw}}^1 + 24$ played a crucial role (cf. (112)). The analogue of this in the present situation is perhaps more remarkable. We have

$$\operatorname{gdim}(V_L^f)^1 = \operatorname{gdim}(V_{L,\mathrm{tw}}^f)^1 + 8 = \operatorname{gdim}(V_{L,\mathrm{tw}}^f)^0 + 8 \qquad (186)$$

so there are actually two natural counterparts to $V^\natural$ that we may consider. Indeed from (178 and 179) and (186) we see that $V^{f\natural}$ has the graded superdimension we seek,

$$q^{-\frac{1}{2}} \operatorname{gsdim} V^{f\natural} = T_{2B}(\tfrac{1}{2}\tau) = q^{-\frac{1}{2}} + 276q^{\frac{1}{2}} - 2048q + \dots, \qquad (187)$$

and if we replace $(V_{L,\mathrm{tw}}^f)^1$ with $(V_{L,\mathrm{tw}}^f)^0$ in (179) then we obtain (187) again.

## 19  Superconformal Structure

The coincidence (186) may be derived from a phenomenon known as triality for the $D_4$ root system, which in turn may be explained in terms of the vertex operator algebra $A(\mathfrak{e})^0$ (cf. (155), (158)) and the $A(\mathfrak{e})^0$-module $A(\mathfrak{e}) \oplus A(\mathfrak{e})_{\mathrm{tw}}$ (cf. (182)). To prepare for such an explanation let us first ease notation by setting

$$U_0 := A(\mathfrak{e})^0, \quad U_1 := A(\mathfrak{e})^1, \quad U_\sigma := A(\mathfrak{e})_{\mathrm{tw}}^0, \quad U_{\bar{\sigma}} := A(\mathfrak{e})_{\mathrm{tw}}^1. \qquad (188)$$

Also set $U_\star := U_0 \oplus U_1 \oplus U_\sigma \oplus U_{\bar{\sigma}}$. Then *triality* for $D_4$ is the statement that there is an order 3 automorphism $\sigma$ of $U_0$ that intertwines the $U_0$-module structures on $U_x$ for $x \in \{1, \sigma, \bar{\sigma}\}$. That is, the action of $\sigma$ on $U_0$ extends to $U_\star$ in such a way that

$$\sigma(U_1) \subset U_\sigma, \quad \sigma(U_\sigma) \subset U_{\bar{\sigma}}, \quad \sigma(U_{\bar{\sigma}}) \subset U_1, \qquad (189)$$

and $\sigma Y(a, z)b = Y(\sigma a, z)\sigma b$ (cf. (138)) for $a \in U_0$ and $b \in U_\star$. In particular the $U_0$-modules in (189) all have the same graded dimension. This entails the eta-product identity

$$\frac{1}{2}\left(\frac{\eta(\tau)^{16}}{\eta(\frac{1}{2}\tau)^8\eta(2\tau)^8} - \frac{\eta(\frac{1}{2}\tau)^8}{\eta(\tau)^8}\right) = 2^3\frac{\eta(2\tau)^8}{\eta(\tau)^8} \tag{190}$$

(cf. (58)) because the left-hand side of (190) is $q^{-\frac{1}{6}}$ times the expression we get from (167) and (177) for the graded dimension of $U_1 = A(\mathfrak{e})^1$, and the right-hand side of (190) is $q^{-\frac{1}{6}}$ times the expression we get via the same calculation (cf. (185)) for the graded dimension of both $U_\sigma = A(\mathfrak{e})^0_{\mathrm{tw}}$ and $U_{\bar\sigma} = A(\mathfrak{e})^1_{\mathrm{tw}}$ (since $\mathrm{tr}(z_A|(A(\mathfrak{e})_{\mathrm{tw}})_n) = 0$ for all $n$).

From the classical relationship between Clifford algebras and orthogonal groups (cf. e.g. [39]) it develops that $U_0 = A(\mathfrak{e})^0$ serves as a natural analogue of (52) for the $D_4$ Lie algebra. That is to say $A(\mathfrak{e})^0$ realizes the basic representation of the simple affine Lie algebra (cf. Sect. 7) associated to $\mathfrak{so}_8(\mathbb{C})$. This hints at a closer connection between $A(\mathfrak{e})$ and $V_L$. Indeed from the explicit description (47) we may check that the simple complex $E_8$ Lie algebra contains a commuting pair of sub Lie algebras isomorphic to $\mathfrak{so}_8(\mathbb{C})$, and from this it follows that we may express $V_L$ in terms of modules for $U_0$. It develops that we have

$$V_L \simeq U_0 \otimes U_0 \oplus U_1 \otimes U_1 \oplus U_\sigma \otimes U_\sigma \oplus U_{\bar\sigma} \otimes U_{\bar\sigma} \tag{191}$$

as modules for $U_0 \otimes U_0$, and in fact there is a similar expression for $V_{L,\mathrm{tw}}$. Namely in §6.4 of [106] it is shown that

$$V_{L,\mathrm{tw}} \simeq U_0 \otimes U_\sigma \oplus U_1 \otimes U_{\bar\sigma} \oplus U_\sigma \otimes U_0 \oplus U_{\bar\sigma} \otimes U_1 \tag{192}$$

as modules for $U_0 \otimes U_0$.

A practical consequence of (191 and 192) is that we may re-express $V^{f\natural}$ purely in terms of the $U_0$-modules $U_x$. In [106] this point of view, together with the triality isomorphism (189), is used to show that $V^{f\natural}$ admits an alternative description as

$$V^{f\natural} = A(\mathfrak{l})^0 \oplus A(\mathfrak{l})^0_{\mathrm{tw}} \tag{193}$$

(cf. (179)) where $A(\mathfrak{l}) = A(\mathfrak{l})^0 \oplus A(\mathfrak{l})^1$ and $A(\mathfrak{l})_{\mathrm{tw}} = A(\mathfrak{l})^0_{\mathrm{tw}} \oplus A(\mathfrak{l})^1_{\mathrm{tw}}$ are defined just as in Sects. 17 and 18, respectively, but with the 24-dimensional space $\mathfrak{l} := \Lambda \otimes_{\mathbb{Z}} \mathbb{C}$ (cf. (98)) in place of the eight-dimensional $\mathfrak{e} = L \otimes_{\mathbb{Z}} \mathbb{C}$ (cf. (47), (154)).

Armed with (193) it is relatively easy to show that the vertex operator algebra structure on $A(\mathfrak{l})^0$ naturally extends to a vertex operator superalgebra structure on $V^{f\natural}$ (179), and the fact that the Conway group $Co_0$ (cf. (98)) acts naturally on $\mathfrak{l} = \Lambda \otimes_{\mathbb{Z}} \mathbb{C}$ makes it relatively easy to identify an action of the sporadic Conway group $Co_1$ on $V^{f\natural}$ by vertex operator superalgebra automorphisms. But $Co_1$ is not the full automorphism group of $V^{f\natural}$. Indeed any orthogonal transformation of $\mathfrak{l} \simeq \mathbb{C}^{24}$ with determinant 1 admits a lift to an automorphism of $V^{f\natural}$.

However it is not true that every lift to $\mathrm{Aut}(V^{f\natural})$ of every $g \in SO(\mathfrak{l})$ fixes the unique up to scale $Co_1$-invariant vector in the degree $\frac{3}{2}$ subspace of $V^{f\natural}$ that is represented by the 1 in (175). So it natural to ask if this invariant vector has some abstract significance.

In the case of the monster we found (cf. (117), (131)) that the 1 in McKay's original observation (27) equips $V^{\natural}$ with a representation of the Virasoro algebra (92), which is a crucial ingredient for the conformal field theory interpretation of $V^{\natural}$ (cf. Sects. 10 and 14). In [106] it is shown that the 1 for $Co_1$ in (175) plays a directly analogous role in connecting $V^{f\natural}$ to superconformal field theory (cf. Sect. 17). Moreover this superconformal structure solves the problem of abstractly identifying $Co_1$ within the full group of vertex operator superalgebra automorphisms of $V^{f\natural}$.

Indeed it is proven in [106] that $Co_1$ is the automorphism group of an $N = 1$ *structure* on $V^{f\natural}$, which means a choice of odd vector $\tau \in V^{f\natural}$ with the property that if operators $G(n + \frac{1}{2}) \in \mathrm{End}(V^{f\natural})$ are defined for $n \in \mathbb{Z}$ by

$$Y(\tau, z)v = \sum_{n \in \mathbb{Z}} G(n + \tfrac{1}{2})v z^{-n-2} \tag{194}$$

for $v \in V^{f\natural}$ then we have

$$\left[ G(m + \tfrac{1}{2}), G(n - \tfrac{1}{2}) \right] = 2L(m + n) + \frac{m^2 + m}{3} \delta_{m+n,0} cI,$$
$$\left[ G(m + \tfrac{1}{2}), L(n) \right] = \left( m + \tfrac{1}{2} - \tfrac{n}{2} \right) G(m + n + \tfrac{1}{2}), \tag{195}$$

in $\mathrm{End}(V^{f\natural})$ for all $m, n \in \mathbb{Z}$. Here $L(n) \in \mathrm{End}(V^{f\natural})$ is determined in the usual way (131) by the Virasoro element (cf. (171), (193)) that defines the vertex operator superalgebra structure on $V^{f\natural}$, and $c$ is the central charge of $V^{f\natural}$ (i.e. $c = 12$, cf. (172)).

Note that the $G(n + \frac{1}{2})$ of (194) are odd endomorphisms (cf. (160)) since $\tau$ belongs to the odd part of $V^{f\natural}$ while the $L(n)$ are even. So according to (159) we have

$$[G(m + \tfrac{1}{2}), G(n - \tfrac{1}{2})] = G(m + \tfrac{1}{2})G(n - \tfrac{1}{2}) + G(n - \tfrac{1}{2})G(m + \tfrac{1}{2}),$$
$$[G(m + \tfrac{1}{2}), L(n)] = G(m + \tfrac{1}{2})L(n) - L(n)G(m + \tfrac{1}{2}). \tag{196}$$

If we replace $cI$ in (195) with an abstract central element $\mathbf{c}$ as in (92) then the resulting super commutation relations, taken together with (92), define the *Neveu–Schwarz algebra*, which is sometimes also referred to as the $N = 1$ *superconformal algebra*. As we have hinted already this a Lie superalgebra that serves as a natural replacement for the Virasoro algebra (92) in superconformal field theory (cf. Sects. 10 and 17). A representation with $\mathbf{c}$ acting trivially may be realized explicitly in terms of (polynomial) vector fields on a super analogue $S^{1|1}$ of the circle $S^1$ (cf. (94)).

Define an $N = 1$ *vertex operator superalgebra* to be a pair $(V, \tau)$ where $V$ is a vertex operator superalgebra and $\tau$ is an $N = 1$ structure on $V$, and define Aut$(V, \tau)$ to be the subgroup of Aut$(V)$ composed of automorphisms that fix $\tau$. Then we have the following counterpart to Theorem 8.

**Theorem 11 ([106])** *There exists an $N = 1$ vertex operator superalgebra $(V^{f\natural}, \tau)$ such that* Aut$(V^{f\natural}, \tau) \simeq Co_1$ *and* $q^{-\frac{1}{2}}$ gdim $V^{f\natural} = T_{2B}(\frac{1}{2}\tau)$.

Given the similarity of Theorems 11 to 8 it is natural to ask for a counterpart to Conjecture 2. Such a result is actually proven, modulo some technical conditions, in op. cit.

**Theorem 12 ([106])** *The $N = 1$ vertex operator superalgebra $(V^{f\natural}, \tau)$ is the unique holomorphic $C_2$-cofinite $N = 1$ vertex operator superalgebra of CFT type with central charge 12 and no non-zero vectors $v \in V^{f\natural}$ such that $L(0)v = \frac{1}{2}v$.*

Here we say that a vertex operator superalgebra $V = \bigoplus_n V_{\frac{n}{2}}$ is of *CFT type* if $V_0$ is spanned by the vacuum and $V_{\frac{n}{2}} = 0$ for $n < 0$. More generally a vertex superalgebra $V$ is called *$C_2$-cofinite* if the *$C_2$ space*

$$C_2(V) := \{a_{-2}b \mid a, b \in V\} \tag{197}$$

has finite codimension in $V$ where $a_n \in \text{End}(V)$ is defined for $a \in V$ by requiring that $Y(a, z)b = \sum_n a_n b z^{-n-1}$ for $b \in V$ (cf. (116)). The notion of $C_2$-cofiniteness was introduced by Zhu in the course of proving his modularity theorem [85] (cf. Sect. 16).

We refer to [107–109] for general results on $N = 1$ vertex operator superalgebras, and to [110–112] for more on the supergeometry of these objects. For an $N = 1$ counterpart to the discussion in Sect. 7 see [109]. A detailed analysis of the relationship between $A(\mathfrak{e})$ and $V_L$ is given in [113]. Notions of triality also play an important role in the original analyses [30–32] of $V^{\natural}$.

## 20 Conway Moonshine

The reader may have anticipated already that the graded dimensions of $A(\mathfrak{l})^0_{\text{tw}}$ and $A(\mathfrak{l})^1_{\text{tw}}$ coincide, for the same reason that we have the last identity in (185). Namely, the trace of the canonical involution on $A(\mathfrak{l})_{\text{tw}}$ vanishes identically. In light of (193) this means that we obtain another space with the same graded dimension as $V^{f\natural}$, and the same suggestive relationship to the Conway group (cf. (175)), by setting

$$V^{s\natural} := A(\mathfrak{l})^0 \oplus A(\mathfrak{l})^1_{\text{tw}}. \tag{198}$$

Indeed we saw already in Sect. 18 (cf. (186)) that by replacing $(V_{L,\text{tw}}^f)^1$ with $(V_{L,\text{tw}}^f)^0$ in the original prescription (179) for $V^{f\natural}$ we obtain a graded space with the same graded dimension. The description (198) is precisely what results when we adopt the approach of Sect. 19, applying triality (189) and the isomorphisms (191) and (192) to $(V_L^f)^0 \oplus (V_{L,\text{tw}}^f)^0$ (cf. (179), (193)).

So what is the significance of $V^{s\natural}$? Just as for $V^{f\natural} = A(\mathfrak{l})^0 \oplus A(\mathfrak{l})_{\text{tw}}^0$ it is relatively easy to show that the vertex operator algebra structure on $A(\mathfrak{l})^0$ extends naturally to a vertex operator superalgebra structure on $V^{s\natural}$, and the fact that $Co_0$ acts naturally on $\mathfrak{l} = \Lambda \otimes_{\mathbb{Z}} \mathbb{C}$ (cf. (98)) again leads naturally to an action of $Co_0$ on $V^{s\natural}$. However, in contrast to $V^{f\natural}$ the action of $Co_0$ on $V^{s\natural}$ is faithful because $Co_0$ acts faithfully on the odd component $V_{\bar{1}}^{s\natural} = A(\mathfrak{l})_{\text{tw}}^1$. Indeed the action of the non-trivial central involution of $Co_0$ on $V^{s\natural}$ recovers the canonical involution (cf. (158)) of $V^{s\natural}$ so none of the homogeneous subspaces of $V_{\bar{1}}^{s\natural} = A(\mathfrak{l})_{\text{tw}}^1$ can involve irreducible representations of $Co_0$ that are not faithful. It develops that the $V^{s\natural}$ counterpart to (175) is

$$2048 = 24 + 2024 \tag{199}$$

where the numbers on the right are dimensions of faithful irreducible representations of $Co_0$. In particular the degree $\frac{3}{2}$ subspace of $V^{s\natural}$ has no non-trivial $Co_0$-invariant vectors, and $V^{s\natural}$ admits no $Co_0$-invariant $N = 1$ structure.

Nonetheless the $Co_0$-module $V^{s\natural}$ is of considerable interest in moonshine, for it is shown in [114] that it shares with $V^\natural$ (102) the genus zero property we explained in Sect. 5. That is, if we define McKay–Thompson series (cf. (32))

$$T_g^s(\tau) := \sum_{n \geq -1} \text{str}(g|V_{\frac{n}{2}}^{s\natural})q^{\frac{n}{2}} \tag{200}$$

for each $g \in Co_0$, where the *supertrace* of an even endomorphism $g \in \text{End}(V)_{\bar{0}}$ (cf. (160)) on a superspace $V = V_{\bar{0}} \oplus V_{\bar{1}}$ is defined by setting

$$\text{str}(g|V) := \text{tr}(g|V_{\bar{0}}) - \text{tr}(g|V_{\bar{1}}) \tag{201}$$

(cf. (176)), then we have the following result.

**Theorem 13 ([114])** *For every $g \in Co_0$ the function $T_g^s$ is the normalized principal modulus for a genus zero group $\Gamma_g^s < SL_2(\mathbb{R})$.*

Note that this genus zero property is not held, in general, by the graded trace functions that derive from the $Co_1$-module structure on $V^{f\natural}$.

Recall that $V^\natural$ is conjectured to be the unique holomorphic vertex operator algebra with central charge 24 and no non-zero vectors satisfying $L(0)v = v$ (see Conjecture 2), and analogous statements have been proven for the Leech lattice (Theorem 7) and $V^{f\natural}$ (Theorem 12). The latter result is strengthened to the following in op. cit.

**Theorem 14 ([114])** *There is a unique holomorphic $C_2$-cofinite vertex operator superalgebra of CFT type with central charge $12$ and no non-zero vectors $v$ such that $L(0)v = \frac{1}{2}v$.*

It follows from Theorem 14 that $V^{f\natural}$ and $V^{s\natural}$ are actually isomorphic as vertex operator superalgebras. The key difference then—crucial for moonshine in light of Theorem 13—is the module structure. Indeed, whereas $V^{f\natural}$ admits an action by the simple group $Co_1$, and this action is precisely determined by a choice of $N = 1$ structure on $V^{f\natural}$, the double cover $Co_0$ acts naturally and faithfully on $V^{s\natural}$. Whereas the McKay–Thompson series (200) associated to the action of $Co_0$ on $V^{s\natural}$ are normalized principal moduli for genus zero groups, this is generally not true for the graded trace functions arising from the action of $Co_1$ on $V^{f\natural}$.

This raises the question of how to identify $Co_0$ as the relevant subgroup of automorphisms for $V^{s\natural}$. Any vertex operator superalgebra admits a canonically defined involution (cf. (158)) so it is natural to consider its *canonically twisted modules*, being those twisted modules (cf. Sect. 15) for the which the twisted fields are as in (140–141) but with the canonical involution taking the place of $z_\Lambda$ in (141). It develops that $V^{s\natural}$ has a unique irreducible canonically twisted module $V^{s\natural}_{\mathrm{tw}}$, and it may be realized explicitly as

$$V^{s\natural}_{\mathrm{tw}} := A(\mathfrak{l})^1 \oplus A(\mathfrak{l})^0_{\mathrm{tw}}. \tag{202}$$

In particular $V^{s\natural}_{\mathrm{tw}}$ includes the vector $\tau$ that defines the $Co_0$-invariant $N = 1$ structure (cf. (175), (194)) on $V^{f\natural}$. So we may identify $Co_0$ as the subgroup of $\mathrm{Aut}(V^{s\natural})$ that fixes an $N = 1$ element in the canonically twisted module $V^{s\natural}_{\mathrm{tw}}$.

## 21 The Mathieu Groups

A notable role in the analysis of [106] is played by the *largest sporadic Mathieu group* $M_{24}$. This is a sporadic simple group (cf. Sect. 3) that was discovered by Émile Mathieu more than 150 years ago [115, 116] as a permutation group on 24 points. In Sect. 3 we mentioned that 21 of the sporadic simple groups were discovered within the period 1965 to 1975. There are 26 sporadic simple groups[2] in total (cf. [14, 23]), and the remaining five are all visible inside $M_{24}$. We obtain the sporadic simple *Mathieu groups* $M_{23}$ and $M_{22}$ by fixing any one or two points, respectively, in the degree 24 permutation representation of $M_{24}$. The stabilizer of a certain configuration of 12 points is a sporadic simple group denoted $M_{12}$, and the stabilizer of any point in the resulting degree 12 permutation representation of $M_{12}$ is the smallest sporadic simple group, $M_{11}$.

---

[2] Although the coincidence is striking it is not expected that there is any direct connection to the critical dimension of string theory (cf. Sect. 17).

We may identify $M_{24}$ simultaneously as a subgroup of $Co_0 = \mathrm{Aut}(\Lambda)$ (cf. (98)) and as a subgroup of the symmetric group of degree 24 by choosing 24 vectors $\lambda_i \in \Lambda$ such that

$$(\lambda_i, \lambda_j) = 8\delta_{ij}. \tag{203}$$

(For a beautiful proof of the existence of such vectors see [53].) Then the subgroup of $Co_0$ that preserves the set $\{\lambda_i\}$ is a copy of $M_{24}$. If we let $N$ denote the subgroup of $Co_0$ that preserves the larger set $\{\pm\lambda_i\}$—this is a *coordinate-frame* for $\Lambda$ in the language of [56]—then $N$ turns out to be a splitting extension

$$1 \to E \to N \to M_{24} \to 1 \tag{204}$$

where $E \simeq (\mathbb{Z}/2\mathbb{Z})^{12}$ is elementary abelian of order $2^{12}$ and acts by sign changes on the $\lambda_i$.

The significance of this for $V^{f\natural}$ is that the action of $E$ on $A(\mathfrak{l})_{\mathrm{tw}}$ allows us to give an alternative construction of this space, not involving a polarization $\mathfrak{l} = \mathfrak{l}^- \oplus \mathfrak{l}^+$ (cf. (181–182)), which makes the 1 in (175) manifest. To explain this we first note that for each character $\chi \in E^* := \hom(E, \mathbb{C}^*)$ of $E$ there is a unique up to scale vector $e_\chi \in A(\mathfrak{l})_{\mathrm{tw}}$ with $L(0)e_\chi = \frac{3}{2}e_\chi$ such that $ge_\chi = \chi(g)e_\chi$ for all $g \in E$. Next, noting the coincidence $\dim(A(\mathfrak{l})_{\mathrm{tw}})_{\frac{3}{2}} = 4096 = \#E^*$ we conclude that the action of $E$ defines an identification of the group algebra

$$\mathbb{C}E^* = \bigoplus_{\chi \in E^*} \mathbb{C}e_\chi \tag{205}$$

with the degree $\frac{3}{2}$ subspace of $A(\mathfrak{l})_{\mathrm{tw}}$. In [106] this fact is used to give a realization of $A(\mathfrak{l})_{\mathrm{tw}}$ as a canonically twisted module for $A(\mathfrak{l})$ (cf. (193)) in the form

$$A(\mathfrak{l})_{\mathrm{tw}} = \mathbb{C}E^* \otimes \bigwedge(\check{\mathfrak{l}}_{\mathrm{tw}}) \tag{206}$$

where $\check{\mathfrak{l}}_{\mathrm{tw}} := \bigoplus_{n<0} \mathfrak{l}(n)$ (cf. (182)).

To see why this manifests the superconformal structure on $V^{f\natural}$ observe that the trivial character of $E$ defines a distinguished vector in $\mathbb{C}E^* \simeq (A(\mathfrak{l})_{\mathrm{tw}})_{\frac{3}{2}}$. Namely, it defines the unique up to scale vector that is fixed by $E$. We know from (175) and (199) that the $N = 1$ element $\tau \in A(\mathfrak{l})^0_{\mathrm{tw}} < V^{f\natural}$ (cf. (194)) is the unique up to scale $Co_0$-invariant vector in $(A(\mathfrak{l})_{\mathrm{tw}})_{\frac{3}{2}}$, so—up to a matter of scale—these two distinguished vectors must coincide. That is, given the realization (206) the trivial character of $E$ explicitly realizes the $N = 1$ structure on $V^{f\natural}$.

There is an interesting relationship between the subgroup $E < Co_0$ (204) and subsets of the coordinate-frame $\{\pm\lambda_i\}$ (cf. (203)) that we used to identity it. To explain this let us take $\Omega := \{1, \ldots, 24\}$ to be the index set of the $\lambda_i$, and given a subset $C \subset \Omega$ define a function $\chi_C : \Omega \to \mathbb{C}^*$ by requiring that $\chi_C(i)$ is 1 or $-1$ according as $i$ belongs to $C$ or not. Then we may associate a subset $C = C(g)$ of $\Omega$

to each $g \in E$ by requiring that

$$g(\lambda_i) = \chi_C(i)\lambda_i \qquad (207)$$

for each $i \in \Omega$.

Now let $C$ be the set of subsets of $\Omega$ so obtained, $C := \{C(g) \mid g \in E\}$, and define another set $C^*$ of subsets of $\Omega$ by requiring that

$$C^* := \{D \subset \Omega \mid \#C \cap D \equiv 0 \bmod 2 \text{ for all } C \in C\}. \qquad (208)$$

Then it develops that $C$ is closed under *symmetric difference*, $C + D := C \cup D - C \cap D$, we have $\#C = 0 \bmod 4$ for every $C \in C$, the set $C^*$ coincides with $C$, and there are no $C \in C$ with $\#C = 4$.

A subset of a power set that is closed under symmetric difference is commonly referred to as a *linear binary code*. Its elements are called *codewords*, and the cardinality of a codeword is called its *weight*. A binary linear code is called *even* if the weights of all of its codewords are even, and *doubly even* if these weights are all 0 mod 4. The prescription (208) defines the *dual* of the binary linear code $C$, and $C$ is called *self-dual* if $C^* = C$. So we may reformulate the conclusions above as follows.

**Proposition 5** *The subgroup $E < Co_0$ defined by the coordinate-frame $\{\pm\lambda_i\}$ defines a self-dual doubly even binary linear code $C$ of length 24 that has no codewords of weight* 4.

Notice the similarity of this description to the hypotheses in Theorem 7, Conjecture 2, Theorems 12 and 14. As the reader may guess by now, there is exactly one binary linear code that satisfies these properties. It is called the *(extended binary) Golay code*, and the full group of coordinate permutations that preserves such a code is a copy of the largest sporadic Mathieu group, $M_{24}$.

Most analyses of the Mathieu groups $M_{24}$, $M_{23}$ and $M_{22}$ take the Golay code as a starting point, and there is a ternary Golay code, naturally regarded as a subspace of $\mathbb{F}_3^{12}$, that plays a similar role for $M_{12}$ and $M_{11}$ (cf. e.g. [56] and Chapter 11 of [42]). By a similar token the Leech lattice $\Lambda$ is the most common starting point for analyses of $Co_0$. Indeed it served such a role for us in Sect. 11. As one may guess from Proposition 5 the Golay code may be used, and often is used, as a tool for constructing the Leech lattice (see e.g. [56]). In [106] the Golay code serves as the main ingredient for the realization (206) of $A(\mathfrak{l})_{tw}$. From this perspective the fact that $Co_0$ arises naturally as the stabilizer of a distinguished vector in the space $\mathbb{C}E^* < A(\mathfrak{l})_{tw}$ may be regarded as furnishing a code-based lattice-free construction of this group. (Another lattice-free construction of $Co_0$ has appeared recently in [117]. It would be interesting to know how it is related to the construction just described.)

## 22   K3 Surfaces

We may summarize the content of Sects. 5 and 6 very briefly by saying that
monstrous moonshine witnesses a curious connection between the monster simple
group and complex elliptic curves. In 1988 Mukai noticed a curious connection
between the Mathieu group $M_{23}$ (cf. Sect. 21) and certain higher dimensional
analogues of elliptic curves called K3 surfaces.

To explain what a K3 surface is, and to see an analogy with elliptic curves, we
note first that the *Fermat cubic*

$$X^3 + Y^3 + Z^3 = 0 \tag{209}$$

defines a complex elliptic curve $C \subset \mathbb{P}^2(\mathbb{C})$. Indeed, after making the substitutions
$\frac{X}{Z} = \frac{36+y}{-6x}$ and $\frac{Y}{Z} = \frac{36-y}{-6x}$ we obtain a Weierstrass equation $y^2 = x^3 - 432$ as in (4)
with $\Delta \neq 0$. (This transformation may be derived from an application of Nagell's
algorithm [118]. Cf. also [119], for example.)

Every complex elliptic curve is isomorphic to one of the form $E_\tau$ (cf. (19))
according to Sect. 4. (The evident three-fold symmetry of $C$ forces $C \simeq E_\omega$ for
$\omega$ a cube root of unity.) So in particular every complex elliptic curve is the same
topologically. The flip side of this is that we may regard the general complex elliptic
curve as a choice of complex structure on the two-dimensional (real) manifold
$S^1 \times S^1$ that underlies $C$.

Next let us consider the complex surface in $\mathbb{P}^3(\mathbb{C})$ that is defined by the *Fermat
quartic*

$$X^4 + Y^4 + Z^4 + W^4 = 0. \tag{210}$$

This is an example of a complex K3 surface. All complex K3 surfaces are the same
topologically so for a general definition we may say that a *complex K3 surface* is
a choice of complex structure on the four-dimensional manifold that underlies this
particular example.

To formulate Mukai's result on $M_{23}$ note that, just as a complex elliptic curve
admits a unique up to scale non-vanishing holomorphic 1-form, a complex K3
surface has a unique up to scale non-vanishing holomorphic 2-form. In other
words, both complex elliptic curves and complex K3 surfaces have *trivial canonical
bundles*. (Indeed, it is more standard to define a complex K3 surface to be a simply
connected two-dimensional complex manifold with trivial canonical bundle.)

Say that an automorphism $g$ of a complex K3 surface $S$ is *symplectic* if the
induced action of $g$ on the holomorphic 2-forms on $S$ is trivial. Mukai's 1988 result
on $M_{23}$ and K3 surfaces may now be stated as follows.

**Theorem 15 (Mukai [120])** *A finite group occurs as a group of symplectic automorphisms of a complex K3 surface if and only if it is isomorphic to a subgroup of
$M_{23}$ that has at least 5 orbits in its action on 24 points.*

So the sporadic simple group $M_{23}$ plays a governing role for symmetries of complex K3 surfaces. Why should this be so? Although Mukai's proof of Theorem 15 is constructive, and a later proof [121] by Kondo is more uniform, neither method seems to provide a conceptual explanation of the coincidence.

Every complex K3 surface has trivial canonical bundle as we have mentioned. Siu proved [122] that every complex K3 surface is *Kähler*. This means that any such surface $S$ admits a Riemannian metric $g$ that is compatible with the complex manifold structure in the sense that $g(iu, iv) = g(u, v)$ in the tangent space $T_x S$ at each point $x \in S$, and for any $y \in S$ the complex vector space structure on $T_y S$ coincides with that obtained by parallel transport with respect to $g$ from $x$. (See e.g. [123] for more on Kähler manifolds.) So complex K3 surfaces are *Calabi–Yau*. They are in some sense the simplest "non-trivial" examples of Calabi–Yau manifolds because any other example in (complex) dimension 2 or less is a torus. Indeed, complex elliptic curves are the only Calabi–Yau manifolds of dimension 1, and any connected but not simply connected compact complex manifold of dimension 2 with trivial canonical bundle is a quotient $\mathbb{C}^2/L$ for some subgroup $L \simeq \mathbb{Z}^4$ of $\mathbb{C}^2$.

Calabi–Yau manifolds in general, and complex K3 surfaces in particular, have come to play an important role in string theory because they facilitate compact-ifications of the many dimensions of the superstring (cf. Sect. 17) that preserve desirable supersymmetry. So it is natural to look to physics for an explanation of the connection between $M_{23}$ and complex K3 surfaces. As we will see presently the physical perspective has shed light on Mukai's result by furnishing a richer context for it (cf. Sect. 25). However it has also revealed a new aspect to the relationship between Mathieu groups and K3 surfaces which is, for now at least, even more mysterious (cf. Sects. 23 and 24).

Just as elliptic curves may be defined over arbitrary fields, there is a notion of algebraic K3 surface which makes sense for any field. It develops that the connection between symplectic symmetries of complex K3 surfaces and the sporadic simple Mathieu groups persists in positive characteristic. Indeed larger subgroups of $M_{23}$ can appear, having as few as 3 orbits on 24 points. A positive-characteristic counterpart to the classification of Theorem 15 has not yet been obtained but it is known that there are examples of automorphism groups that cannot be subgroups of $M_{23}$ (or $M_{24}$). See [124–126] for some precise results along these lines.

It also develops that there is a counterpart to Theorem 15 for the smaller Mathieu group $M_{12}$. In this setting the role of complex K3 surfaces is taken up by *Enriques surfaces*, which may be obtained as the quotients of complex K3 surfaces by fixed-point-free automorphisms of order 2. See [127] for details.

The classical introductory references for complex K3 surfaces are [123, 128]. A more recent reference that also covers algebraic K3 surfaces is [129]. For a nice survey on symplectic automorphisms of K3 surfaces see [130]. For a more detailed introduction to the role that K3 surfaces play in string theory see [131].

## 23   Complex Elliptic Genera

In 2010 a new aspect of the enigmatic connection between Mathieu groups and
K3 surfaces was exposed by Eguchi–Ooguri–Tachikawa with the announcement
[132] of their *Mathieu moonshine* observation. To describe this we introduce a
holomorphic function $Z : \mathbb{H} \times \mathbb{C} \to \mathbb{C}$ by setting

$$Z(\tau, z) := \frac{24}{4\pi^2} \wp(\tau, z) \frac{\theta_1(\tau, z)^2}{\eta(\tau)^6} \tag{211}$$

where $\wp(\tau, z) := \wp(\Lambda_\tau, z)$ for $\wp$ as in (16) and $\Lambda_\tau$ as in (19) (cf. (24)), the
Dedekind eta function $\eta$ is as in (58), and $\theta_1$ denotes the *Jacobi theta function*

$$\theta_1(\tau, z) := -iq^{\frac{1}{8}} y^{\frac{1}{2}} \prod_{n>0} (1 - y^{-1}q^{n-1})(1 - yq^n)(1 - q^n). \tag{212}$$

In (212) we take $q = e^{2\pi i \tau}$ for $\tau \in \mathbb{H}$ as usual (cf. (25), (57), (58)) and also use
$y = e^{2\pi i z}$ for $z \in \mathbb{C}$. Notice that for any fixed $\tau \in \mathbb{H}$ the function $z \mapsto \wp(\tau, z)$
is meromorphic with a pole of order two at every lattice point $z \in \Lambda_\tau = \mathbb{Z}\tau + \mathbb{Z}$.
However $z \mapsto \theta_1(\tau, z)^2$ vanishes to order two at these points so $Z$ is holomorphic
as claimed.

The geometric significance of $Z$ (211) is that it recovers the complex elliptic
genus (see below) of any complex K3 surface. The significance of $Z$ for $M_{24}$, and
the observation [132] of Eguchi–Ooguri–Tachikawa, is that if $H : \mathbb{H} \to \mathbb{C}$ is defined
by requiring that

$$Z(\tau, z) = 24\mu(\tau, z) \frac{\theta_1(\tau, z)^2}{\eta(\tau)^3} + H(\tau) \frac{\theta_1(\tau, z)^2}{\eta(\tau)^3}, \tag{213}$$

where $\mu : \mathbb{H} \times \mathbb{C} \to \mathbb{C}$ is the *Appell–Lerch sum*

$$\mu(\tau, z) := \frac{-iy^{\frac{1}{2}}}{\theta_1(\tau, z)} \sum_{n \in \mathbb{Z}} (-1)^n \frac{y^n q^{\frac{1}{2}n(n+1)}}{(1 - yq^n)}, \tag{214}$$

then we have

$$H(\tau) = -2q^{-\frac{1}{8}} + 90q^{\frac{7}{8}} + 462q^{\frac{15}{8}} + 1540q^{\frac{23}{8}} + 4554q^{\frac{31}{8}} + 11592q^{\frac{39}{8}} + \dots \tag{215}$$

What is striking about this is that each of the non-polar coefficients in (215) is 2
times the dimension of a non-trivial irreducible representation of $M_{24}$ (cf. [23]). (We
may regard the polar coefficient as $-2$ times the dimension of the trivial irreducible
representation.)

For $X$ a compact complex manifold the *complex elliptic genus* of $X$ is a formal series in $q$ and $y$ whose coefficients are Euler characteristics of certain vector bundles on $X$ (cf. e.g. [133, 134]). This series defines a holomorphic function $EG_X(\tau, z)$ on $\mathbb{H} \times \mathbb{C}$ once we specialize to $q = e^{2\pi i \tau}$ and $y = e^{2\pi i z}$ as in (212). If $X$ has trivial canonical bundle (or more generally if the first Chern class of $X$ is trivial or torsion) then it develops that $EG_X$ has good modular properties. Specifically $EG_X$ is a *weak Jacobi form* of *weight* 0 and *index* $m$ for $SL_2(\mathbb{Z})$ where

$$m = \frac{1}{2} \dim_\mathbb{C} X. \qquad (216)$$

To explain what this means define the weight $k$ and index $m$ *modular action* of $SL_2(\mathbb{Z})$ on functions $\phi : \mathbb{H} \times \mathbb{C} \to \mathbb{C}$ by setting

$$\left( \phi \Big|_{k,m} \begin{pmatrix} a & b \\ c & d \end{pmatrix} \right)(\tau, z) := \phi\left( \frac{a\tau + b}{c\tau + d}, \frac{z}{c\tau + d} \right) \frac{1}{(c\tau + d)^k} e^{-2\pi i m \frac{cz^2}{c\tau + d}} \qquad (217)$$

for $k, m \in \mathbb{Z}$ and $\begin{pmatrix} a & b \\ c & d \end{pmatrix} \in SL_2(\mathbb{Z})$ (cf. (63)), and define the index $m$ *elliptic action* of $\mathbb{Z}^2$ on functions $\phi : \mathbb{H} \times \mathbb{C} \to \mathbb{C}$ by setting

$$(\phi|_m (a, b))(\tau, z) := e^{2\pi i m(a^2\tau + 2az)} \phi(\tau, z + a\tau + b) \qquad (218)$$

for $(a, b) \in \mathbb{Z}^2$. Then for $\gamma \in SL_2(\mathbb{Z})$ and $\lambda \in \mathbb{Z}^2$ we have

$$EG_X|_{0,m} \gamma = EG_X|_m \lambda = EG_X \qquad (219)$$

and the adjective "weak" is applied because $\tau \mapsto EG_X(\tau, z)$ is bounded as $\Im(\tau) \to \infty$ for any fixed $z \in \mathbb{C}$.

The invariance (219) of $EG_X$ under the modular action (217) of $SL_2(\mathbb{Z})$ implies that $\tau \mapsto EG_X(\tau, 0)$ is a modular form of weight 0 (cf. (63)), but there are no non-constant modular forms of weight 0 that are bounded as $\Im(\tau) \to \infty$ because there are no non-constant bounded holomorphic functions on the orbit space $SL_2(\mathbb{Z})\backslash\mathbb{H}$ (cf. (23)). So $EG_X(\tau, 0)$ must be constant. The complex elliptic genus is defined in such a way that

$$EG_X(\tau, 0) = \chi(X) \qquad (220)$$

is none other than the usual Euler characteristic of $X$.

The invariance (219) of $EG_X$ under the elliptic action (218) of $\mathbb{Z}^2$ implies that $EG_X$ admits a *theta decomposition*

$$EG_X(\tau, z) = \sum_{r \bmod 2m} h_r(\tau) \theta_{m,r}(\tau, z) \qquad (221)$$

for some holomorphic functions $h_r : \mathbb{H} \to \mathbb{C}$, called the *theta coefficients* of $EG_X$, where

$$\theta_{m,r}(\tau, z) := \sum_{\ell \equiv r \bmod 2m} y^\ell q^{\frac{\ell^2}{4m}} \tag{222}$$

for $m, r \in \mathbb{Z}$.

Observe that $\theta_{m,0}(\tau, 0) = \theta_L(\tau)$ in the notation of (57) for $L = \sqrt{2m}\mathbb{Z}$, and the *Thetanullwerte* $\theta_{m,r}(\tau, 0)$ for general $r \in \mathbb{Z}$ are obtained in a similar way from the cosets of $L$ in its dual (cf. (46)). So Theorem 5 suggests that the $\theta_{m,r}$ of (222) should satisfy a weight $\frac{1}{2}$ analogue of (219). Indeed this is true (see e.g. [135, 136]), and thus another consequence of the modular invariance of $EG_X$ is that the theta coefficients $h_r$ in (221) constitute the components of a weakly holomorphic vector-valued modular form of weight $-\frac{1}{2}$ for $SL_2(\mathbb{Z})$ (cf. (63)). More generally, the theta coefficients of a weak Jacobi form of weight $k$ transform with weight $k - \frac{1}{2}$.

It develops (see e.g. [137]) that the space of weak Jacobi forms of weight 0 and index 1 is one dimensional and $Z$ (211) is an example. From this we may conclude that

$$EG_S = Z \tag{223}$$

for any complex K3 surface $S$ because the Euler characteristic of any such surface is $24 = Z(\tau, 0)$ (cf. (220)).

It follows from the defining expression (16) that the Weierstrass $\wp$-function $\wp(\tau, z) := \wp(\Lambda_\tau, z)$ satisfies

$$\wp|_{2,0} \gamma = \wp|_0 \lambda = \wp \tag{224}$$

for $\gamma \in SL_2(\mathbb{Z})$ and $\lambda \in \mathbb{Z}^2$. Thus $\wp$ would furnish an example of a Jacobi form of weight 2 and index 0 (cf. (217 and 218)) except that $z \mapsto \wp(\tau, z)$ has poles at lattice points $z \in \Lambda_\tau$. So we call $\wp$ a *meromorphic Jacobi form* of weight 2 and index 0. Note that a theta decomposition (221) does not generally exist for meromorphic Jacobi forms (but see [138] for an interesting analogue).

From (211), (219), (223 and 224) we may now deduce that $\theta_1^2$ is a *Jacobi form* of weight 1 and index 1 with multiplier. More precisely, we have

$$\epsilon(\gamma)^6 \theta_1^2\big|_{1,1} \gamma = \theta_1^2\big|_1 \lambda = \theta_1^2 \tag{225}$$

for $\gamma \in SL_2(\mathbb{Z})$ and $\lambda \in \mathbb{Z}^2$ where $\epsilon$ denotes the multiplier system of the Dedekind eta function (60). The theta decomposition (221) for $\theta_1^2$ takes the form

$$\theta_1(\tau, z)^2 = \theta_{1,1}(\tau, 0)\theta_{1,0}(\tau, z) - \theta_{1,0}(\tau, 0)\theta_{1,1}(\tau, z). \tag{226}$$

We call $\theta_1^2$ a Jacobi form, rather than a weak Jacobi form, because the theta coefficients in (226) remain bounded as $\Im(\tau) \to \infty$ (cf. (63)).

The invariance (225) suggests of course that $\theta_1$ (212) is a Jacobi form of weight $\frac{1}{2}$ and index $\frac{1}{2}$ with multiplier. This statement can be confirmed once we have a suitable definition of $\theta_{m,r}$ for $m \in \frac{1}{2}\mathbb{Z}$. (See e.g. (2.4) in [139]. The multiplier of $\theta_1$ works out to be the same as that of $\eta^3$.)

If $\phi(\tau, z)$ is a Jacobi form of index $m$ and some weight then $\phi(\tau, hz)$ is a Jacobi form of index $h^2 m$. So $\theta_1(\tau, 2z)$ is a Jacobi form of index 2. This fact manifests in the identity

$$i\theta_1(\tau, z) = \theta_{2,1}(\tau, \tfrac{1}{2}z) - \theta_{2,-1}(\tau, \tfrac{1}{2}z) \tag{227}$$

(cf. (222)) which is another form of the Jacobi triple product identity (146), and thus a consequence of the representation theory of the affine $A_1$ Lie algebra (cf. (44)).

The standard reference for Jacobi forms is [137]. See [138] for a more concise introduction that also discusses meromorphic Jacobi forms. Certain foundational results on Jacobi forms were obtained in [140], in a setting not far removed from the topic of Sect. 7.

The complex elliptic genera of complex manifolds were first studied by Höhn [141] and Krichever [142]. See [133] for some history of the notion, and a proof of the Jacobi form property (219–221) (amongst other results). See [143] for a generalization to manifolds with non-vanishing first Chern class. Note that there are other related but different notions of elliptic genus in the literature, cf. e.g. [144–146]. As we will see in Sect. 24, elliptic genera play an important role in physics.

Counterparts to the Mathieu moonshine observation (215) that relate $M_{12}$ (cf. Sect. 21) to Enriques surfaces (cf. Sect. 22) have been investigated by Eguchi–Hikami [147] and Govindarajan (cf. [148]).

## 24 Mathieu Moonshine

Regarding (215) in the light of monstrous moonshine (cf. Sect. 5) we are inexorably led to the hypothesis that there exists a graded infinite-dimensional $M_{24}$-module

$$K = \bigoplus_{n>0} K_{n-\frac{1}{8}} \tag{228}$$

with $-2q^{-\frac{1}{8}} + \operatorname{gdim} K = H(\tau)$ (cf. (213), (215)). We also expect the corresponding graded trace functions

$$H_g(\tau) := -2q^{-\frac{1}{8}} + \sum_{n>0} \operatorname{tr}\left(g \,\middle|\, K_{n-\frac{1}{8}}\right) q^{n-\frac{1}{8}} \tag{229}$$

for $g \in M_{24}$ to serve as distinguished analogues of $H$.

How should we identify these $H_g$? In contrast to the situation in monstrous moonshine the function $H$ of (213) and (215) is not a principal modulus for a genus zero group (cf. (34)). In fact we may deduce from (213) that $H$ is not even a modular form because $\theta_1^2 \eta^{-3}$ satisfies a modular transformation rule according to (60) and (225) but the Appell–Lerch sum (214) is not modular according to the original (independent) investigations of Appell [149] and Lerch [150, 151] (see also [152–154]). So at first glance at least it is not clear that there is any analogue of the powerful genus zero property of monstrous moonshine (cf. Sects. 5 and 16) that we can expect the $H_g$ in (229) to satisfy.

However there is another contrast to the situation with the monster in that some hints come directly to us from geometry by virtue of the work of Mukai (see Sect. 22). Indeed if $g \in M_{24}$ has a fixed point in the permutation representation on 24 points (i.e. belongs to a subgroup $M_{23}$) and has order not exceeding 8 then $g$ acts faithfully and symplectically on a complex K3 surface $S$ according to Theorem 15. For such an automorphism there is a naturally defined $g$-*equivariant* variant $EG_{S,g}$ (cf. e.g. [155]) of the complex elliptic genus $EG_S$ of (219) and (221), and it is natural to guess that the corresponding $H_g$ are determined via a corresponding analogue of (213).

This idea was implemented and verified by Cheng in [156]. Indeed the $EG_{S,g}$ for $g \in M_{23}$ with $o(g) \in \{2, 3, 5, 7\}$ had been computed earlier by Sen [157]. Cheng defined $H_g : \mathbb{H} \to \mathbb{C}$ for such $g \in M_{24}$ by requiring that

$$EG_{S,g}(\tau, z) = \frac{24}{p+1}\mu(\tau, z)\frac{\theta_1(\tau, z)^2}{\eta(\tau)^3} + H_g(\tau)\frac{\theta_1(\tau, z)^2}{\eta(\tau)^3} \tag{230}$$

where $p = o(g)$ is the order of $g$. She then observed that the resulting $H_g$ exhibit compatibility with the hypothesis that (229) is satisfied for a graded $M_{24}$-module $K$ as in (228). For example, for $n \in \{1, 2, 3, 4, 5\}$ there is a unique (up to isomorphism) $M_{24}$-module $K_d$ for $d = n - \frac{1}{8}$ that

1. has dimension equal to the coefficient of $q^d$ in $H(\tau)$ (cf. (215)),
2. is the direct sum of two irreducible modules for $M_{24}$, and
3. is isomorphic as an $M_{24}$-module to its dual.

When defined—purely geometrically—by (230) the coefficient of $q^d$ in $H_g(\tau)$, for $0 < d < 5$, turns out to be precisely the trace of $g \in M_{23}$ on the $M_{24}$-module $K_d$ characterized above. Moreover it follows from (230) that $E_{S,g}(\tau, 0) = \frac{24}{p+1}$. This is both the $g$-equivariant Euler characteristic of $S$ (cf. (220)), for $S$ a complex K3 surface and $g$ an order $p$ symplectic automorphism of $S$, and also the trace of $g \in M_{23}$ on the 24-dimensional permutation module for $M_{24}$ (cf. Sect. 21).

It also follows from (230) that the $H_g$ so defined fail to be modular for the same reason that $H$ does, but the situation is better for $E_{S,g}$. Indeed $E_{S,g}$ is a weak Jacobi form of weight 0 and index 1 for $\Gamma_0(o(g))$, meaning that (219) holds with $m = 1$ and $EG_{S,g}$ in place of $EG_X$ but only for $\gamma \in \Gamma_0(o(g))$ (cf. (38)).

We cannot expect to identify $H_g$ (229) via an equivariant elliptic genus $EG_{S,g}$ for every $g \in M_{24}$ because Theorem 15 excludes many such $g$ from acting symplectically on a complex K3 surface. So Cheng proposed to identify these more general $H_g$ by finding suitable weak Jacobi forms $Z_g$, for subgroups $\Gamma_0(o(g)) < SL_2(\mathbb{Z})$, and requiring that

$$Z_g(\tau, z) = \chi_g \mu(\tau, z) \frac{\theta_1(\tau, z)^2}{\eta(\tau)^3} + H_g(\tau) \frac{\theta_1(\tau, z)^2}{\eta(\tau)^3} \tag{231}$$

for $\tau \in \mathbb{H}$ and $z \in \mathbb{C}$ where $g \mapsto \chi_g$ denotes the character of the 24-dimensional permutation representation of $M_{24}$ (i.e. $\chi_g$ is the number of fixed points of $g$ as a permutation on 24 points, cf. Sect. 21). Here "suitable" includes the hypothesis that $Z_g = E_{S,g}$ for $g \in M_{23}$ with $o(g) \in \{2, 3, 5, 7\}$, and compatibility with the proposal sketched following (230) for the $M_{24}$-module structure on the $K_d$ for $0 < d < 5$.

This Jacobi form approach to $H_g$ was independently formulated by Gaberdiel–Hohenegger–Volpato in [158]. In both works [156, 158] a number of the $Z_g$ were determined beyond those given by the equivariant elliptic genera $E_{S,g}$ of Sen. With subsequent work of Gaberdiel–Hohenegger–Volpato [159] and Eguchi–Hikami [160] a full list of $Z_g$ was obtained, and in due course Gannon established [82] that these $Z_g$ are indeed compatible with (229). That is, there exists a graded $M_{24}$-module $K$ such that if $H_g$ is defined by (229) and $Z_g$ is as determined by Cheng, Eguchi–Hikami, and Gaberdiel–Hohenneger–Volpato in [156, 158–160] then (231) is satisfied for all $g \in M_{24}$. Note however that this does not really qualify as a counterpart to Theorems 8 or 11 as it does not include any $M_{24}$-invariant algebraic structure, such as vertex operator (super)algebra structure or similar (cf. Sects. 9, 12–14, 17, and 19). Indeed there has as yet been only limited success in defining such a structure on $K$.

Why was it natural to expect $Z_g$ as in (231) to be modular? We will take this on trust for now and consider it more carefully in Sect. 25. A related natural question is whether this really deserves to be called moonshine. As we saw in Sect. 16 it is not difficult to attach modular forms to a finite group in such a way that the Fourier coefficients are recovered by taking traces on some graded infinite-dimensional module. By the results of [161], for example, the same is true if we replace modular forms with Jacobi forms. It is the genus zero property (cf. Sects. 5 and 16) that distinguishes monstrous moonshine from the generic examples, and the genus zero property extends to Conway moonshine according to Theorem 13.

So is there a counterpart to the genus zero property for Mathieu moonshine? It turns out that the answer is yes. That is, it is possible to arrive at the $H_g$ (229) in a natural way, for each $g \in M_{24}$, with very little a priori knowledge of $g$. To explain this we recall from Sect. 16 that the McKay–Thompson series $T_g$ of monstrous moonshine (32) may be realized (150) as Rademacher sums. More concretely, we have

$$T_g(\tau) + \frac{1}{2}C_{\Gamma_g} = \text{Reg}\left(\sum_{\gamma \in \Gamma_\infty \backslash \Gamma_g} q^{-1}\Big|_0 \gamma\right) \tag{232}$$

for $g \in \mathbb{M}$ where the right-hand side of (232) is a shorthand for (151) with $\Gamma = \Gamma_g$. Moreover the analogous formula

$$T_g^s(\tau) + \frac{1}{2}C_{\Gamma_g^s} = \text{Reg}\left(\sum_{\gamma \in \Gamma_\infty \backslash \Gamma_g^s} q^{-\frac{1}{2}}\Big|_0 \gamma\right) \tag{233}$$

holds for $g \in Co_0$ (cf. (200)) by essentially the same proof on the strength of Theorem 13.

The formulas (232 and 233) constitute a concrete and computable incarnation of the statement that the normalized principal modulus for a genus zero group is uniquely determined by that group (cf. Sect. 5). We call it computable because the expression (232), for example, leads naturally to the formula

$$\text{tr}\left(g\Big|V_{n+1}^\natural\right) = \lim_{K \to \infty} \sum_{\left(\begin{smallmatrix} a & b \\ c & d \end{smallmatrix}\right) \in (\Gamma_\infty \backslash \Gamma / \Gamma_\infty)_K^\times} \exp\left(2\pi i\frac{nd-a}{c}\right)\frac{2\pi}{c\sqrt{n}}I_1\left(\frac{4\pi\sqrt{n}}{c}\right) \tag{234}$$

for the coefficient of $q^n$ in $T_g$ for $g \in \mathbb{M}$ (cf. (32), (143 and 144)). Here $(\Gamma_\infty \backslash \Gamma / \Gamma_\infty)_K^\times$ denotes the set of double coset representatives $\left(\begin{smallmatrix} a & b \\ c & d \end{smallmatrix}\right)$ such that $0 < c < K$, and $I_1$ denotes the *modified Bessel function of the first kind*,

$$I_1(x) := \sum_{n \geq 0} \frac{1}{n!(n+1)!}\left(\frac{x}{2}\right)^{2n+1}. \tag{235}$$

(See e.g. [162–164] for background on Bessel functions.)

As is demonstrated in [165] Mathieu moonshine admits direct counterparts to (232–233). Namely we have

$$H_g(\tau) = -2\,\text{Reg}\left(\sum_{\gamma \in \Gamma_\infty \backslash \Gamma_g} q^{-\frac{1}{8}}\Big|_{\psi_g,\frac{1}{2}} \gamma\right) \tag{236}$$

for each $g \in M_{24}$ where $\Gamma_g := \Gamma_0(o(g))$ and $\psi_g : \Gamma_g \to \mathbb{C}^*$ is a certain weight $\frac{1}{2}$ multiplier system (cf. (60–62)), and $f|_{\psi,k}\gamma := \psi(\gamma)f|_k\gamma$ for $\gamma \in \Gamma$ in the case that $\psi$ is a $\mathbb{C}$-valued function on $\Gamma$ (cf. (63)). This leads to an analogue of (234) for the coefficients of $H_g$ by essentially the same method (see [165, 166]). Now the multiplier system $\psi_g$ is given explicitly by

$$\psi_g \begin{pmatrix} a & b \\ c & d \end{pmatrix} := \epsilon \begin{pmatrix} a & b \\ c & d \end{pmatrix}^{-3} \exp\left(-2\pi i \frac{cd}{nh}\right) \tag{237}$$

where $\epsilon$ is the multiplier system (62) of the Dedekind eta function, we let $n := o(g)$ be the order of $g$, and $h := h(g)$ denotes the minimal cycle length in a cycle decomposition of $g$ (when regarded as a permutation on 24 points, cf. Sect. 21). So (236–237) say that $H_g$ is determined in a concrete and directly computable way as soon as we know the order and minimal cycle length of $g$. In particular (236) imbues Mathieu moonshine with a predictive power that is directly parallel to that enjoyed by monstrous moonshine and Conway moonshine by virtue of the genus zero property (cf. Sects. 5 and 16).

One point of contrast between (232 and 233) and (236) is that the former regularized averages are invariant for the groups over which they average whereas the latter is not. However modularity fails in a controlled way in the latter case. Namely, it follows (see [165, 166]) from (236) that we have

$$\left(H_g\big|_{\psi_g, \frac{1}{2}} \gamma\right)(\tau) = H_g(\tau) + \chi_g C \int_{-\gamma^{-1}\infty}^{\infty} (\tau + \tau')^{-\frac{1}{2}} \eta(\tau')^3 d\tau' \tag{238}$$

for $\gamma \in \Gamma_0(o(g))$ for each $g \in M_{24}$, where $\chi_g$ is as in (231), for a certain constant $C$ that is independent of $g$. (In particular $H_g$ is modular in case $\chi_g = 0$, or equivalently if $g$ does not belong to a subgroup $M_{23}$, as we may expect given (231)).

Functions satisfying expressions of the form (238) have been the focus of much activity in recent years both in mathematics and in physics. They are called *mock modular forms* following the 2002 doctoral thesis [153] of Zwegers, which introduced the general definition. A main significance of op. cit. is that it furnished a theoretical framework for the so-called mock theta functions of Ramanujan—first presented by him in his "last letter to Hardy" in 1920—by demonstrating that they are all examples.

The literature on mock modular forms is already extensive. For mathematical introductions see [152, 167–169]. A physically motivated development is given in [138]. For the text of Ramanujan's last letter to Hardy see pp. 220–224 of [170]. Part of what is surprising about $H_g$, and the examples produced a century ago by Ramanujan, is that they have rational integer Fourier coefficients. As explained in [171] most mock modular forms have transcendental coefficients. A connection between rational integer mock modular forms and genus zero groups is established in [135]. (We will say more about this in Sect. 26.)

## 25 Conway Returns

We return now to the modularity of the $Z_g$ (231) for $g \in M_{24}$. Why was this natural to expect? A key principle underlying Sen's determination [157] of $E_{S,g}$ for $g \in M_{23}$ with $o(g) \in \{2, 3, 5, 7\}$ is Witten's interpretation [172] of the complex elliptic genus in terms of superconformal field theory (cf. Sects. 10, 17, and 19). Witten uses the theory of non-linear sigma models to attach a graded Hilbert space Aut($X$)-module $\mathcal{H} = \bigoplus_{\ell,n,\bar{n}} \mathcal{H}_{\ell,n,\bar{n}}$ to $X$ and argues that

$$EG_{X,g}(\tau, z) = \sum_{\ell,n,\bar{n} \in \mathbb{Z}} \mathrm{tr}\left(g(-1)^F \middle| \mathcal{H}_{\ell,n,\bar{n}}\right) y^\ell q^n \bar{q}^{\bar{n}} \tag{239}$$

for $g \in \mathrm{Aut}(X)$ where $(-1)^F$ is a semisimple grading preserving operator on $\mathcal{H}$ that has $\pm 1$ for eigenvalues. Significantly for modularity, $\mathcal{H}$ comes equipped with two commuting actions—called *left-moving* and *right-moving*—of the $N = 2$ superconformal algebra in the case that $X$ is Calabi–Yau (cf. Sect. 22). These actions allow for a physical argument (see [173], and e.g. [156, 158]) for the modularity of the physical realization (239) of the equivariant elliptic genus $EG_{X,g}$ so long as $g$ is symplectic (i.e. acts trivially on the canonical bundle of $X$, cf. Sect. 22).

The $N = 2$ superconformal algebra, first introduced in [174], provides a further enhancement of the Virasoro algebra (92) beyond the $N = 1$ superconformal structure (195) we encountered in Sect. 19. (In particular, it is actually a Lie superalgebra.) The 2 in $N = 2$ reflects the fact that it is typically presented in terms of 2 odd generating fields,

$$Y(\tau^\pm, z) = \sum_n G^\pm(n + \tfrac{1}{2}) z^{-n-2} \tag{240}$$

(cf. (194)). As the reader may guess from the notation we recover a generating field $Y(\tau, z)$ for the $N = 1$ superconformal algebra (cf. (194 and 195)) by taking $\tau = \tau^- + \tau^+$.

Note that the right-hand side of (239) is often written in the form

$$\mathrm{tr}\left(g(-1)^F y^{J(0)} q^{L(0)-\frac{c}{24}} \bar{q}^{\bar{L}(0)-\frac{\bar{c}}{24}} \middle| \mathcal{H}\right). \tag{241}$$

Here $J(0)$ and $L(0)$ represent commuting semisimple operators on $\mathcal{H}$ belonging to the left-moving action of the $N = 2$ algebra, and $c$ is the *left-moving central charge*. These are determined by the operators $G^\pm(n + \tfrac{1}{2})$ (cf. (240)) according to

$$[G^+(n + \tfrac{1}{2}), G^-(-n - \tfrac{1}{2})] = L(0) + (n + \tfrac{1}{2})J(0) + \frac{n^2 + n}{6} cI \tag{242}$$

(cf. (195 and 196)), and $\bar{L}(0)$ and $\bar{c}$ are the counterparts to $L(0)$ and $c$ for the right-moving action. We may translate between (239) and (241) by interpreting $\mathcal{H}_{\ell,n,\bar{n}}$ as the $(\ell, n + \frac{c}{24}, \bar{n} + \frac{\bar{c}}{24})$-eigenspace for the triple $(J(0), L(0), \bar{L}(0))$.

The motivation for the decomposition (213) of $Z = EG_S$ (cf. (223)) may now be explained. It develops that any Calabi–Yau manifold of complex dimension 2 is automatically *hyper-Kähler*, meaning that it admits a 2-sphere of complex structures that all admit a common Kähler metric (cf. Sect. 22). So in particular complex K3 surfaces are hyper-Kähler.

In the case that $X$ is hyper-Kähler the two commuting $N = 2$ actions on the Hilbert space $\mathcal{H}$ in (239) are promoted to unitary actions of the (small) $N = 4$ superconformal algebra. (As the reader may guess this is an infinite-dimensional Lie superalgebra, also introduced in [174], that further extends the Virasoro algebra beyond the $N = 2$ structure mentioned above.) Given this structure it is natural to ask for a decomposition of the Hilbert space $\mathcal{H}$ of (239) into irreducible representations of the $N = 4$ superconformal algebra. At present this question is too hard to answer because there is as yet no general method for constructing the Hilbert space attached to a hyper-Kähler manifold $X$ by a non-linear sigma model on $X$, even in the K3 case, except for a few special examples.

However the K3 elliptic genus $EG_S$ is independent of the choice of complex K3 surface as we have seen in Sect. 23. Moreover $EG_X$ is independent of $\bar{q}$ (cf. (239), (241)) so we should be able to write $EG_S$ as a linear combination of the $N = 4$ *characters*

$$\mathrm{ch}_{m,h,\ell}^{N=4}(\tau, z) := \mathrm{tr}\left((-1)^F y^{J(0)} q^{L(0) - \frac{c}{24}} \,\middle|\, \mathcal{V}_{m,h,\ell}\right) \tag{243}$$

where $(m, h, \ell)$ indexes irreducible unitary modules $\mathcal{V}_{m,h,\ell}$ for the $N = 4$ superconformal algebra.

The irreducible unitary $N = 4$ characters (243) were determined by Eguchi–Taormina in [175, 176]. The module $\mathcal{V}_{m,h,\ell}$ is uniquely determined amongst those with central charge $c = 6m$ (cf. (242)) by the property that it admits a generating vector $\mathbf{v}_{m,h,\ell}$ such that $L(0)\mathbf{v}_{m,h,\ell} = h\mathbf{v}_{m,h,\ell}$ and $J(0)\mathbf{v}_{m,h,\ell} = 2\ell\mathbf{v}_{m,h,\ell}$ where $h$ is the minimal eigenvalue for $L(0)$ on $\mathcal{V}_{m,h,\ell}$. For the K3 elliptic genus we have $m = 1$ (cf. (216)) and the possible values of $\ell$ are 0 and $\frac{1}{2}$. If $\ell = 0$ then $h = \frac{1}{4}$ and we have

$$\mathrm{ch}_{1,\frac{1}{4},0}^{N=4}(\tau, z) = \mu(\tau, z)\frac{\theta_1(\tau, z)^2}{\eta(\tau)^3}, \tag{244}$$

whereas if $\ell = \frac{1}{2}$ then $h$ can be any real number greater than or equal to $\frac{1}{4}$. In case $\ell = \frac{1}{2}$ and $h > \frac{1}{4}$ we have

$$\mathrm{ch}_{1,\frac{1}{4},\frac{1}{2}}^{N=4}(\tau, z) = q^{h - \frac{3}{8}}\frac{\theta_1(\tau, z)^2}{\eta(\tau)^3}, \tag{245}$$

and the remaining character $\mathrm{ch}^{N=4}_{1,\frac{1}{4},\frac{1}{2}}$ is determined by the identity

$$2\,\mathrm{ch}^{N=4}_{1,\frac{1}{4},0}(\tau,z) + \mathrm{ch}^{N=4}_{1,\frac{1}{4},\frac{1}{2}}(\tau,z) = \lim_{h\to\frac{1}{4}} \mathrm{ch}^{N=4}_{1,h,\frac{1}{2}}(\tau,z) = q^{-\frac{1}{8}} \frac{\theta_1(\tau,z)^2}{\eta(\tau)^3}. \tag{246}$$

The characters with $h = \frac{1}{4}$ are called *BPS* or *short* or *massless*, and those with $h > \frac{1}{4}$ are called *non-BPS* or *long* or *massive*.

Comparing (211) and (223) with (244–246) one may compute (cf. [177–179]) that

$$EG_S = 20\,\mathrm{ch}^{N=4}_{1,\frac{1}{4},0} - 2\,\mathrm{ch}^{N=4}_{1,\frac{1}{4},\frac{1}{2}} + \sum_{n>0} c(n - \tfrac{1}{8})\,\mathrm{ch}^{N=4}_{1,n+\frac{1}{4},\frac{1}{2}} \tag{247}$$

for certain integers $c(n - \frac{1}{8})$ for $S$ a complex K3 surface. The decomposition (213) now follows from (244–247), for $c(d)$ the coefficient of $q^d$ in the Fourier expansion (215) of $H$. So (213) is, essentially, the decomposition of the K3 elliptic genus into irreducible unitary characters of the $N = 4$ superconformal algebra, and the coefficients (215) of $H$ may be interpreted as multiplicities of these irreducible characters.

One feature of the non-linear sigma model approach (239) to complex elliptic genera is that it suggests a new route to explaining the Mathieu moonshine observation (cf. (215), (228–229)) of Eguchi–Ooguri–Tachikawa. Namely we may consider the supersymmetry preserving symmetry groups of the non-linear sigma models with complex K3 surface targets. The Hilbert space $\mathcal{H}$ associated to a non-linear sigma model with target $X$ naturally admits a supersymmetry preserving action by the group of symplectic automorphisms of $X$, but generally admits other automorphisms in addition, and these more general automorphisms also define equivariant elliptic genera $EG_{X,g}$ as in (239), (241). The symplectic automorphisms of a complex K3 surface are restricted by $M_{23}$ according to Mukai's Theorem 15, but perhaps there is a non-linear sigma model with K3 target that admits $M_{24}$ as automorphisms? If so then we may, after all, be able to identify $Z_g$ (231) as $EG_{S,g}$ for every $g \in M_{24}$, for a suitable choice of $S$.

This question was taken up by Gaberdiel–Hohenegger–Volpato in [180]. (See also [181].) Despite the absence of a general method for constructing non-linear sigma models they were able to adapt Kondo's proof [121] of Theorem 15 so as to arrive—modulo certain natural conjectures about K3 sigma models—at the following "stringy" generalization of that result.

**Theorem 16 (Gaberdiel–Hohenegger–Volpato [180])** *A finite group occurs as a group of supersymmetry preserving automorphisms of a non-linear sigma model with complex K3 surface target if and only if it is isomorphic to a subgroup of $Co_0$ that fixes a space of dimension at least 4 in its action on 24-dimensional space.*

Note that the only non-trivial 24-dimensional action of $Co_0$ is that defined (cf. (98)) by the Leech lattice.

Theorem 16 is disappointing from the point of view of Mathieu moonshine because although $M_{24}$ embeds in $Co_0$ (cf. Sect. 21) there is no such embedding which fixes a four-dimensional space in the representation $\Lambda \otimes_{\mathbb{Z}} \mathbb{C}$ defined by the Leech lattice. So there is no K3 sigma model that admits $M_{24}$ as automorphisms, and we cannot hope to recover every $Z_g$ (231) as an equivariant K3 elliptic genus $EG_{S,g}$. On the other hand Theorem 16 demonstrates a strong connection between K3 surfaces and the Conway group. We might be forgiven for wondering if this connection has any kind of manifestation in the context of super moonshine (cf. Sects. 18–20). Perhaps surprisingly, it turns out that it does, according to [182].

To explain this we note that, as detailed in [180], the data of a K3 sigma model defines a four-dimensional subspace $\Pi < \Lambda \otimes_{\mathbb{Z}} \mathbb{C}$, and the subgroup of $Co_0$ that acts on the sigma model in question according to Theorem 16 is precisely the stabilizer

$$\operatorname{Aut}(\Lambda)_\Pi := \{g \in Co_0 \mid g|_\Pi = I_\Pi\} \tag{248}$$

of $\Pi$ in $Co_0 = \operatorname{Aut}(\Lambda)$. It is shown in [182] that such a 4-space $\Pi < \Lambda \otimes_{\mathbb{Z}} \mathbb{C}$ naturally defines a semisimple grading preserving operator $J_\Pi(0)$ on $V_{\mathrm{tw}}^{s\natural}$ (cf. (202)) such that

$$-\operatorname{tr}\left((-1)^F y^{J_\Pi(0)} q^{L(0)-\frac{c}{24}} \,\Big|\, V_{\mathrm{tw}}^{s\natural}\right) = EG_S(\tau, z) \tag{249}$$

is the K3 elliptic genus (cf. (239), (241)) so long as we interpret $(-1)^F$ as the non-trivial central element of $Co_0$ (cf. (98)) and take $c = 12$ to be the central charge of $V^{s\natural}$ (cf. Theorem 14). Moreover the bigrading on $V_{\mathrm{tw}}^{s\natural}$ defined by $J_\Pi(0)$ and $L(0) - \frac{c}{24}$ is preserved by any $g \in Co_0$ that fixes $\Pi$. So for $g \in \operatorname{Aut}(\Lambda)_\Pi$ (248) we may consider the graded trace function

$$\phi_g(\tau, z) := -\operatorname{tr}\left(g(-1)^F y^{J_\Pi(0)} q^{L(0)-\frac{c}{24}} \,\Big|\, V_{\mathrm{tw}}^{s\natural}\right). \tag{250}$$

If $\phi_e$ is the K3 elliptic genus according to (249 and 250) what is the meaning of $\phi_g$ for more general $g \in Co_0$? At the time of publication of [182] equivariant K3 elliptic genera $EG_{S,g}$ (239) had been computed explicitly for K3 sigma model symmetries $g$ corresponding, via Theorem 16, to about half the relevant conjugacy classes of $Co_0$ (i.e. about half of the classes represented by elements fixing a space of dimension at least 4 in $\Lambda \otimes_{\mathbb{Z}} \mathbb{C}$). In all of these cases it was found that

$$\phi_g = EG_{S,g}. \tag{251}$$

In other words the single vertex operator superalgebra module $V_{\mathrm{tw}}^{s\natural}$ allows to compute equivariant K3 elliptic genera arising from many different K3 sigma models. Perhaps all of the $\phi_g$ are equivariant K3 elliptic genera for suitable K3 sigma model symmetries?

A systematic analysis of equivariant K3 elliptic genera was subsequently accomplished in [183, 184]. These works confirmed that, indeed, (251) holds for suitable $S$ and $g$, for every $g$ in $Co_0$ that belongs to $\mathrm{Aut}(\Lambda)_\Pi$ for some 4-space $\Pi < \Lambda \otimes_{\mathbb{Z}} \mathbb{C}$. Curiously the converse is not quite true. More specifically there are four equivariant K3 elliptic genera $EG_{S,g}$ that are not realized as $\phi_g$ (250) for the corresponding $g \in Co_0$. So $V_{\mathrm{tw}}^{s\natural}$ serves—to a certain extent—as a universal object for K3 sigma models, but this "service" is not as complete as we might naively have hoped.

Linear and non-linear sigma models were introduced by Gell-Mann–Lévy in [185]. For a detailed reference on non-linear sigma models see [186]. For a survey of non-linear sigma models with complex K3 surface target we refer to [187]. (See also [188, 189].)

The elliptic genus $EG_X$ fails to be modular in a controlled way in the case that $X$ is Kähler but not Calabi–Yau. See [143] or [173] for precise descriptions.

As the notation of (240) suggests it is natural to consider vertex operator superalgebras $V = V_{\bar{0}} \oplus V_{\bar{1}}$ (cf. (157)) equipped with odd elements $\tau^\pm \in V_{\bar{1}}$ such that the $N = 2$ superconformal algebra relations (242) are satisfied for $G^\pm(n + \frac{1}{2})$ as in (240). We call such data an $N = 2$ *structure* on $V$ (cf. Sect. 19). Structural results on vertex operator superalgebras with $N = 2$ structure can be found in [108, 190], and the supergeometry of these objects is developed in [112, 191] (cf. also [192, 193]). Given the discussion above we may expect to recover Jacobi forms from vertex operator superalgebras with $N = 2$ structure via analogues of (239), (241). This expectation is confirmed in [194]. (See also [195].)

We refer to [196] for background on hyper-Kähler manifolds. The decomposition (247) was given first by Ooguri in [179].

The validity of Theorem 16 is subject to certain conjectures on non-linear sigma models with K3 target. It is expected that an equivalent theory may be obtained by considering stability conditions on suitable derived categories of coherent sheaves (cf. e.g. [197]). A mathematically rigorous derived category counterpart to Theorem 16 is established in [198]. The main result of [182] is formulated in a similar language. Theorem 16 is refined to some extent in [181].

# 26  Forward from Here

We hope these notes have served as both an introduction and an invitation to the reader, to involve themselves deeper in the plexus of ideas that moonshine represents. In this final section we summarize some more recent results in the field, and thereby indicate some further directions the reader might next pursue.

### Generalized Monstrous Moonshine

Monstrous moonshine, as described in Sect. 5, associates principal moduli (cf. (34)) to elements of the monster group $\mathbb{M}$ in a systematic way. Analogues of this for centralizer subgroups $C_{\mathbb{M}}(g)$ (cf. (11 and 12)) of the monster appeared in [21, 49, 199], and Norton was prompted by this to propose *generalized monstrous moonshine*, wherein principal moduli are attached to commuting pairs of elements of $\mathbb{M}$. The corresponding generalization of Conjecture 1 was first formulated in Norton's appendix to [200], and later revised in [201]. In this setting the original monstrous moonshine is recovered by considering pairs $(e, g)$ where $g \in \mathbb{M}$ and $e$ is the identity.

It turns out that there are just two conjugacy classes of involutions in $\mathbb{M}$, with that containing the $z$ of (12) being the larger of these. If $t \in \mathbb{M}$ belongs to the smaller class of involutions then the centralizer $C_{\mathbb{M}}(t)$ participates in a (non-split) short exact sequence

$$1 \to \mathbb{Z}/2\mathbb{Z} \to C_{\mathbb{M}}(t) \to \mathbb{B} \to 1 \tag{252}$$

where $\mathbb{B}$ is a sporadic simple group (cf. Sect. 3) called the *baby monster*. Having

$$
\begin{aligned}
\#\mathbb{B} &= 2^{41}.3^{13}.5^6.7^2.11.13.17.19.23.31.47 \\
&= 4154781481226426191177580544000000
\end{aligned}
\tag{253}
$$

for its order (cf. (15)) it qualifies as the second largest of the sporadic simple groups.

Generalized monstrous moonshine for pairs $(t, g)$ where $g \in C_{\mathbb{M}}(t)$ and $t$ is as in (253) was first established by Höhn [202], following his earlier work [203] (see also [204]) on a vertex operator superalgebra $VB^{\natural}$ that admits an action by the baby monster $\mathbb{B}$ as automorphisms. The full generalized monstrous moonshine conjecture was recently resolved in a tour de force [205] by Carnahan. (Certain preliminaries to op. cit. appear in [206, 207]. See [208, 209] for some refinements.) In Carnahan's approach the irreducible twisted modules (cf. Sect. 15) for $V^{\natural}$ take on the role played by $V^{\natural}$ (cf. Sect. 16) in monstrous moonshine.

## Umbral Moonshine

Mathieu moonshine, as described in Sect. 24, turns out to be a special case of a more general phenomenon called *umbral moonshine* that attaches mock modular forms (cf. (238)) to automorphisms of unimodular even positive-definite lattices (cf. (45 and 46)) of rank 24. This is explained in [136, 210, 211]. Gannon's result [82] that the mock modular forms of Mathieu moonshine may be recovered from a graded module for $M_{24}$ was extended to umbral moonshine in [212]. The fact (236) that the mock modular forms of Mathieu moonshine may be realized as Rademacher sums also extends to umbral moonshine according to the main result of [136].

As mentioned in Sect. 24 a connection between genus zero groups and mock modular forms with rational integer Fourier coefficients is established in [135]. This connection actually provides a second kind of genus zero property for Mathieu moonshine, and umbral moonshine more generally. Namely we may recover the mock modular forms of umbral moonshine in a concrete way from principal moduli for certain genus zero groups. This is explained in §4.5 of op. cit. Partial results in this direction were obtained earlier in [213].

So far there is no known analogue of $V^\natural$ for Mathieu moonshine but counterparts for other cases of umbral moonshine have been constructed in [139, 214–216]. Given the relationship between the orbifold theory of $V^\natural$ and generalized monstrous moonshine mentioned above it counts as evidence in support of the existence of a vertex algebraic realization of the Mathieu moonshine module (228) that generalized monstrous moonshine admits a natural analogue for Mathieu moonshine. This *generalized Mathieu moonshine* is studied in [217] (see also [218]). An extension of generalized Mathieu moonshine to umbral moonshine has also been formulated. This is the main focus of [219].

### Supersingular Elliptic Curves with Level

Generalized monstrous moonshine for commuting pairs $(g, h)$ with fixed $g$ furnishes an analogue of monstrous moonshine for the subgroup $C_M(g)$. In fact many sporadic groups arise as quotients of such subgroups (see e.g. [14, 23]) so it is natural to ask if Ogg's observation (cf. Sect. 3) generalizes to monstrous sporadic groups too. This was confirmed recently by Aricheta [17] via an analysis of supersingular elliptic curves with level structure (cf. Sects. 2 and 6). In particular a higher-level generalization of Theorem 4 is obtained in op. cit., and coincidences with the prime spectra of sporadic simple groups beyond the monster are observed. Interestingly the results of op. cit. also demonstrate a connection between supersingular elliptic curves and umbral moonshine.

### Vertex Operator Superalgebras and Sigma Models

As we alluded to in Sect. 25 the relationship between $V_{tw}^{s\natural}$ and K3 sigma models incapsulated in (251) suggests a role for vertex operator superalgebra theory as a source of universal objects for families of non-linear sigma models. This theme is developed systematically in [65] and [220].

See also [221] for some related discussion and a conjectural interpretation of the Jacobi forms $\phi_g$ (250) in the context of enumerative geometry for complex K3 surfaces. Some concrete results on the conjectures of op. cit. are obtained in [222].

### Symmetry Surfing

Taormina–Wendland have developed an approach to Mathieu moonshine Sect. 24 called *symmetry surfing*, wherein one seeks to construct a suitable action of $M_{24}$ by "collecting" symmetries from different K3 sigma models in a systematic way. See [223–227] and [228] for details on this.

So far there is no extension of the symmetry surfing program to umbral moonshine but see [229] (and also [183, 184]) for the development of a relationship between umbral moonshine and K3 surfaces.

### Thompson Moonshine

Mathieu moonshine attaches mock modular forms of weight $\frac{1}{2}$ to elements of the Mathieu group $M_{24}$ according to Sect. 24. An interesting analogue of this that associates (weakly holomorphic non-mock) modular forms of weight $\frac{1}{2}$ (cf. (63)) to the Thompson group was introduced by Harvey–Rayhaun in [230].

The *Thompson group*, *Th*, is a sporadic simple group that was discovered by Thompson as a result of investigations [231] into the $E_8$ Lie algebra (47). It has order

$$\# Th = 2^{15}.3^{10}.5^3.7^2.13.19.31$$
$$= 90745943887872000 \tag{254}$$

and was first constructed by Smith and Thompson (see [232]) as the automorphism group of an unimodular even positive-definite lattice (cf. (45–46)) of rank 248.

Similar to monstrous, Conway and umbral moonshine the modular forms identified by Harvey–Rayhaun in [230] may be recovered as graded traces from a graded module for *Th* (cf. (143), (200), (229)). This was verified by Griffin–Mertens in [233]. The analysis of op. cit. confirms that the weight $\frac{1}{2}$ modular forms arising are Rademacher sums, in direct parallel with monstrous, Conway and umbral moonshine (cf. (232), (233), (236)).

It is no accident that 248 is the dimension of the $E_8$ Lie algebra (47). Although there is no direct relationship between *Th* and the Lie group $E_8(\mathbb{C})$ of (84), there is a finite simple group $E_8(\mathbb{F}_3)$—a version of the $E_8$ Lie group defined over the field with 3 elements—into which *Th* embeds. (For a vertex algebraic development of this fact see [234, 235].) The Thompson group also embeds in the monster. More specifically it may be recovered from the monster via a split exact sequence

$$1 \to \mathbb{Z}/3\mathbb{Z} \to C_{\mathbb{M}}(w) \to Th \to 1 \tag{255}$$

(cf. (12), (252)) for $w \in \mathbb{M}$ a suitably chosen element of order 3.

From (255) we may conclude that generalized monstrous moonshine for the pairs $(w, g)$ with $w$ as in (255) and $g \in C_{\mathbb{M}}(w)$ defines an association of principal moduli to elements of the Thompson group. (See [199] for some details on the principal moduli arising.) So the Thompson simple group is in the privileged situation of enjoying two distinct forms of moonshine. One can begin to compare the two directly using product formulas that Borcherds introduced in [236]. The results of such a comparison are the focus of the forthcoming work [237].

The product formulas of [236] are now known as *Borcherds products*. Their theory was greatly extended by Borcherds in [238], and has since been contextualized and extended further by many authors. See [239] for an introduction to this theory and see [240] for a thorough treatment.

**Moonshine and Arithmetic**

A basic motivating problem in arithmetic geometry is to understand the rational, or rational integer solutions to a polynomial equation. The determination of $E(K)$, for example, for $E$ an elliptic curve and $K$ a number field is an open problem of this type (cf. Theorem 1). Recently some results in this direction have emerged from moonshine.

The pioneering example of this is *O'Nan moonshine*, which is detailed in [241, 242]. In these works (weakly holomorphic) modular forms of weight $\frac{3}{2}$

(cf. (63)) are assigned to elements of the sporadic simple group *ON* of O'Nan, and they are shown to arise as the graded traces defined by a graded (virtual) *ON*-module. Also, these forms are shown to have a close relationship to Rademacher sums, similar to that (cf. (232), (233), (236)) which we have discussed for the other manifestations of moonshine that have appeared in this work. But more than this, the graded *ON*-module is used to formulate obstructions to the vanishing of arithmetic invariants of some elliptic curves defined over $\mathbb{Q}$ (cf. Sect. 1). As explained in some detail in [243], part of the appeal of this is that it reduces statements about bivariate polynomials of degree 3—viz. those (1), (4) that define elliptic curves—to statements about the univariate polynomials of degree 2 that define quadratic imaginary number fields (cf. Sect. 2). These latter statements are in principle—cf. e.g. [244–246]—easier to solve.

Up to this point all of the sporadic simple groups (cf. Sect. 3) we have encountered are *monstrous* in the sense that they may realized as quotients $H/K$ for suitable chains $K \triangleleft H < \mathbb{M}$ (cf (12), (252), (255)). It is notable that the O'Nan group is one of the six sporadic simple groups that does not arise in this way (cf. §14 of [16]). It was discovered by O'Nan [247] around 1974 and constructed on a computer shortly afterwards by Sims (cf. op. cit.). For its order we have

$$\#ON = 2^9.3^4.5.7^3.11.19.31$$

$$= 460815505920. \tag{256}$$

It is probably fair to say that the non-monstrous sporadic groups have been little more than curiosities to non-specialists in finite group theory up until the appearance of O'Nan moonshine. It is pleasing to witness such a group involve itself in one of the most important open problems in number theory.

Another connection between moonshine and arithmetic geometry has been demonstrated by Beneish in [248]. In this case the group in question is $M_{24}$—the star of Mathieu moonshine Sect. 24—and the arithmetic application is to counting points on elliptic curves defined over finite fields. A distinguishing feature of this connection is the recent appearance [249] of explicit vertex algebraic constructions.

For a third and final connection to arithmetic we mention the classical fact, due to Lehner [250, 251], that the Fourier coefficients (25) of the elliptic modular invariant (24) satisfy strong congruences modulo powers of particular primes. For example we have $c(p^a n) \equiv 0 \bmod p^a$ in the notation of (25) for $p \in \{2, 3, 5, 7, 11\}$ and any positive integers $a$ and $n$, and in fact similar but even stronger congruences (see op. cit.) for $p \in \{2, 3, 5\}$. Given the role (31) that the elliptic modular invariant plays in moonshine (cf. Conjecture 1, Theorems 8–9) it is natural to ask to what extent such congruences carry over the McKay–Thompson series $T_g$ (32) for $g \in \mathbb{M}$. This question is addressed rather thoroughly in [252]. It would be interesting to have counterparts of this work for the other instances of moonshine that have so far come to light.

# References

1. J.H. Silverman, *The Arithmetic of Elliptic Curves*. Graduate Texts in Mathematics, vol. 106, 2nd edn. (Springer, Dordrecht, 2009). https://doi.org/10.1007/978-0-387-09494-6
2. A.W. Knapp, *Elliptic Curves*. Mathematical Notes, vol. 40 (Princeton University Press, Princeton, 1992)
3. A. Wiles, The Birch and Swinnerton-Dyer conjecture, in *The Millennium Prize Problems* (Clay Mathematics Institute, Cambridge, 2006), pp. 31–41
4. L.J. Mordell, On the rational solutions of the indeterminate equations of the 3rd and 4th degrees. Proc. Camb. Phil. Soc. **21**, 179–192 (1922)
5. A. Weil, *L'arithmétique sur les courbes algébriques*. Numdam (1928)
6. M.F. Vignéras, *Arithmétique des algèbres de quaternions*. Lecture Notes in Mathematics, vol. 800 (Springer, Berlin, 1980)
7. M. Deuring, Die Typen der Multiplikatorenringe elliptischer Funktionenkörper. Abh. Math. Semin. Univ. Hambg. **14**(1), 197–272 (1941). https://doi.org/10.1007/BF02940746
8. A.P. Ogg, Automorphismes de courbes modulaires. In: *Séminaire Delange-PisotPoitou* (16e année: 1974/75). Théorie des nombres, Fasc. 1, Exp. No. 7 (Secrétariat Mathématique, Paris, 1975), p. 8
9. M. Aschbacher, The status of the classification of the finite simple groups. Not. Am. Math. Soc. **51**(7), 736–740 (2004)
10. R. Solomon, A brief history of the classification of the finite simple groups. Bull. Amer. Math. Soc. (N.S.) **38**(3), 315–352 (2001). https://doi.org/10.1090/S0273-0979-01-00909-0
11. Z. Janko, A new finite simple group with Abelian 2-Sylow subgroups. Proc. Nat. Acad. Sci. U. S. A. **53**, 657–658 (1965)
12. R.W. Carter, *Simple Groups of Lie Type*. Wiley Classics Library (John Wiley & Sons, Inc., New York, 1989). Reprint of the 1972 original, A Wiley-Interscience Publication
13. R. Steinberg, *Lectures on Chevalley Groups*. University Lecture Series, vol. 66 (American Mathematical Society, Providence, 2016). https://doi.org/10.1090/ulect/066. Notes prepared by John Faulkner and Robert Wilson, Revised and corrected edition of the 1968 original [ MR0466335], With a foreword by Robert R. Snapp
14. R.A. Wilson, *The Finite Simple Groups*. Graduate Texts in Mathematics, vol. 251 (Springer-Verlag London, Ltd., London, 2009). https://doi.org/10.1007/978-1-84800-988-2
15. R.L. Griess Jr., The structure of the "monster" simple group, in *Proceedings of the Conference on Finite Groups* (Academic Press, New York, 1976), pp. 113–118
16. R.L. Griess Jr., The friendly giant. Invent. Math. **69**(1), 1–102 (1982). https://doi.org/10.1007/BF01389186
17. V.M. Aricheta, Supersingular elliptic curves and moonshine. Symmetry Integr. Geom. Methods Appl. **15**, Paper No. 007, 17 (2019). https://doi.org/10.3842/SIGMA.2019.007
18. J.P. Serre, *A Course in Arithmetic* (Springer, New York, 1973). Translated from the French, Graduate Texts in Mathematics, No. 7
19. G. Shimura, *Introduction to the Arithmetic Theory of Automorphic Functions*. Publications of the Mathematical Society of Japan, No. 11 (Iwanami Shoten Publishers, Tokyo, 1971). Kanô Memorial Lectures, No. 1
20. P.A. Griffiths, Introduction to algebraic curves. Translations of Mathematical Monographs, vol. 76 (American Mathematical Society, Providence, 1989). Translated from the Chinese by Kuniko Weltin
21. J.H. Conway, S.P. Norton, Monstrous moonshine. Bull. London Math. Soc. **11**(3), 308–339 (1979). https://doi.org/10.1112/blms/11.3.308s
22. J.G. Thompson, Some numerology between the Fischer-Griess Monster and the elliptic modular function. Bull. London Math. Soc. **11**(3), 352–353 (1979)
23. J. Conway, R. Curtis, S. Norton, R. Parker, R. Wilson, With computational assistance from J. G. Thackray, *Atlas of Finite Groups. Maximal Subgroups and Ordinary Characters for Simple Groups* (Clarendon Press, Oxford, 1985)

24. J.G. Thompson, Finite groups and modular functions. Bull. London Math. Soc. **11**(3), 347–351 (1979)
25. J.F. Duncan, Arithmetic groups and the affine $E_8$ Dynkin diagram, in *Groups and Symmetries*. CRM Proceedings & Lecture Notes, vol. 47 (American Mathematical Society, Providence, 2009), pp. 135–163
26. R.E. Borcherds, Monstrous moonshine and monstrous Lie superalgebras. Invent. Math. **109**(2), 405–444 (1992). https://doi.org/10.1007/BF01232032
27. N.M. Katz, B. Mazur, *Arithmetic Moduli of Elliptic Curves*. Annals of Mathematics Studies, vol. 108 (Princeton University Press, Princeton, 1985). https://doi.org/10.1515/9781400881710
28. G. Veneziano, Construction of a crossing-simmetric, Regge-behaved amplitude for linearly rising trajectories. Nuovo Cimento A Serie **57**, 190–197 (1968). https://doi.org/10.1007/BF02824451
29. S. Fubini, G. Veneziano, Duality in operator formalism. Nuovo Cim. **A67**, 29–47 (1970). https://doi.org/10.1007/BF02728411
30. I.B. Frenkel, J. Lepowsky, A. Meurman, A natural representation of the Fischer-Griess Monster with the modular function $J$ as character. Proc. Nat. Acad. Sci. U. S. A. **81**(10), 3256–3260 (1984)
31. I.B. Frenkel, J. Lepowsky, A. Meurman, A moonshine module for the Monster, in *Vertex Operators in Mathematics and Physics*. Mathematical Sciences Research Institute Publications, vol. 3 (Springer, New York, 1985), pp. 231–273
32. I.B. Frenkel, J. Lepowsky, A. Meurman, Vertex operator algebras and the Monster. Pure and Applied Mathematics, vol. 134 (Academic Press Inc., Boston, 1988)
33. V.G. Kac, Simple irreducible graded Lie algebras of finite growth. Izv. Akad. Nauk SSSR Ser. Mat. **32**, 1323–1367 (1968)
34. I.L. Kantor, Graded Lie algebras. Trudy Sem. Vektor. Tenzor. Anal. **15**, 227–266 (1970)
35. R.V. Moody, A new class of Lie algebras. J. Algebra **10**, 211–230 (1968)
36. J. Lepowsky, R.L. Wilson, Construction of the affine Lie algebra $A_1^{(1)}$. Comm. Math. Phys. **62**(1), 43–53 (1978)
37. I.B. Frenkel, V.G. Kac, Basic representations of affine Lie algebras and dual resonance models. Invent. Math. **62**(1), 23–66 (1980/81)
38. G. Segal, Unitary representations of some infinite-dimensional groups. Comm. Math. Phys. **80**(3), 301–342 (1981)
39. W. Fulton, J. Harris, *Representation Theory*. Graduate Texts in Mathematics, vol. 129 (Springer-Verlag, New York, 1991). https://doi.org/10.1007/978-1-4612-0979-9. A first course, Readings in Mathematics
40. J.E. Humphreys, *Introduction to Lie Algebras and Representation Theory* (Springer, New York, 1972). Graduate Texts in Mathematics, vol. 9
41. J. Lepowsky, Introduction, in *Vertex Operators in Mathematics and Physics*. Mathematical Sciences Research Institute Publications, vol. 3 (Springer, New York, 1985), pp. 1–13. https://doi.org/10.1007/978-1-4613-9550-8_1
42. J.H. Conway, N.J.A. Sloane, Sphere packings, lattices and groups, in *Grundlehren der Mathematischen Wissenschaften [Fundamental Principless of Mathematical Sciences]*, vol. 290, 3rd edn. (Springer, New York, 1999). https://doi.org/10.1007/978-1-4757-6568-7. With additional contributions by E. Bannai, R.E. Borcherds, J. Leech, S.P. Norton, A.M. Odlyzko, R.A. Parker, L. Queen and B.B. Venkov
43. S. Mandelstam, Dual-resonance models. Phys. Rep. **13**, 259–353 (1974)
44. S. Mandelstam, Introduction to string models and vertex operators, in *Vertex Operators in Mathematics and Physics*. Mathematical Sciences Research Institute Publications, vol. 3 (Springer, New York, 1985), pp. 15–35. https://doi.org/10.1007/978-1-4613-9550-8_2
45. J.H. Schwarz, Dual-resonance theory. Phys. Rep. **8**, 269–335 (1973)
46. V.G. Kac, *Infinite-Dimensional Lie algebras*, 3rd edn. (Cambridge University Press, Cambridge, 1990). https://doi.org/10.1017/CBO9780511626234

47. J.H. Conway, L. Queen, Computing the character table of a Lie group, in *Finite Groups—Coming of Age*. Contemporary Mathematics, vol. 45 (American Mathematical Society, Providence, 1985), pp. 51–87. https://doi.org/10.1090/conm/045/822234

48. V.G. Kac, An elucidation of: "Infinite-dimensional algebras, Dedekind's $\eta$-function, classical Möbius function and the very strange formula". $E_8^{(1)}$ and the cube root of the modular invariant $j$. Adv. Math. **35**(3), 264–273 (1980). https://doi.org/10.1016/0001-8708(80)90052-3

49. L. Queen, Modular functions and finite simple groups, in *The Santa Cruz Conference on Finite Groups*. Proceedings of Symposia in Pure Mathematics, vol. 37 (American Mathematical Society, Providence, 1980), pp. 561–566

50. M.S. Virasoro, Subsidiary conditions and ghosts in dual resonance models. Phys. Rev. **D1**, 2933–2936 (1970). https://doi.org/10.1103/PhysRevD.1.2933

51. A.A. Belavin, A.M. Polyakov, A.B. Zamolodchikov, Infinite conformal symmetry in two-dimensional quantum field theory. Nucl. Phys. B **241**(2), 333–380 (1984). https://doi.org/10.1016/0550-3213(84)90052-X

52. J. Leech, Notes on sphere packings. Canad. J. Math. **19**, 251–267 (1967)

53. J.H. Conway, A characterisation of Leech's lattice. Invent. Math. **7**, 137–142 (1969)

54. J.H. Conway, A perfect group of order 8, 315, 553, 613, 086, 720, 000 and the sporadic simple groups. Proc. Nat. Acad. Sci. U. S. A. **61**, 398–400 (1968)

55. J.H. Conway, A group of order 8, 315, 553, 613, 086, 720, 000. Bull. London Math. Soc. **1**, 79–88 (1969)

56. J.H. Conway, Three lectures on exceptional groups, in *Finite Simple Groups (Proceedings of the International Conference, Oxford, 1969)* (Academic Press, London, 1971), pp. 215–247

57. J.H. Conway, A simple construction for the Fischer-Griess monster group. Invent. Math. **79**(3), 513–540 (1985). https://doi.org/10.1007/BF01388521.

58. J.H. Conway, A simple construction for the Fischer-Griess monster group. Invent. Math. **79**(3), 513–540 (1985)

59. R. Borcherds, Vertex algebras, Kac-Moody algebras, and the Monster. Proc. Natl. Acad. Sci. U. S. A. **83**(10), 3068–3071 (1986)

60. E. Frenkel, V. Kac, A. Radul, W. Wang, $\mathcal{W}_{1+\infty}$ and $\mathcal{W}(\mathfrak{gl}_N)$ with central charge $N$. Comm. Math. Phys. **170**(2), 337–357 (1995)

61. E. Frenkel, D. Ben-Zvi, *Vertex Algebras and Algebraic Curves*. Mathematical Surveys and Monographs, vol. 88, 2nd edn. (American Mathematical Society, Providence, 2004)

62. V. Kac, *Vertex Algebras for Beginners*. University Lecture Series, vol. 10, 2nd edn. (American Mathematical Society, Providence, 1998)

63. R.E. Borcherds, Quantum vertex algebras, in *Taniguchi Conference on Mathematics Nara '98*. Advanced Studies in Pure Mathematics, vol. 31 (Mathematical Society of Japan, Tokyo, 2001), pp. 51–74

64. M.M. Patnaik, Vertex algebras as twisted bialgebras: on a theorem of Borcherds, in *Communicating Mathematics*. Contemporary Mathematics, vol. 479 (American Mathematical Society, Providence, 2009), pp. 223–238. https://doi.org/10.1090/conm/479/09354

65. T. Creutzig, J.F.R. Duncan, W. Riedler, Self-dual vertex operator superalgebras and superconformal field theory. J. Phys. A **51**(3), 034001, 29 (2018). https://doi.org/10.1088/1751-8121/aa9af5

66. M.R. Gaberdiel, An introduction to conformal field theory. Rep. Prog. Phys. **63**, 607–667 (2000). https://doi.org/10.1088/0034-4885/63/4/203

67. C. Dong, Vertex algebras associated with even lattices. J. Algebra **161**(1), 245–265 (1993). https://doi.org/10.1006/jabr.1993.1217

68. C. Dong, R.L. Griess Jr., C.H. Lam, Uniqueness results for the moonshine vertex operator algebra. Amer. J. Math. **129**(2), 583–609 (2007). https://doi.org/10.1353/ajm.2007.0009

69. J. van Ekeren, Lattices, vertex algebras, and modular categories. J. Geom. Phys. **126**, 27–41 (2018). https://doi.org/10.1016/j.geomphys.2018.01.008

70. C.H. Lam, H. Shimakura, Classification of holomorphic framed vertex operator algebras of central charge 24. Am. J. Math. **137**(1), 111–137 (2015). https://doi.org/10.1353/ajm.2015.0001

71. C.H. Lam, H. Shimakura, Inertia groups and uniqueness of holomorphic vertex operator algebras (2018). e-prints. arXiv:1804.02521

72. C.H. Lam, H. Shimakura, 71 holomorphic vertex operator algebras of central charge 24. Bull. Inst. Math. Acad. Sin. (N.S.) **14**(1), 87–118 (2019)

73. C.H. Lam, H. Shimakura, Reverse orbifold construction and uniqueness of holomorphic vertex operator algebras. Trans. Amer. Math. Soc. **372**(10), 7001–7024 (2019). https://doi.org/10.1090/tran/7887

74. C.H. Lam, H. Yamauchi, The FLM conjecture and framed VOA, in *Vertex Operator Algebras and Related Areas*. Contemporary Mathematics, vol. 497 (American Mathematical Society, Providence, 2009), pp. 125–138. https://doi.org/10.1090/conm/497/09774

75. J. van Ekeren, S. Möller, N. Scheithauer, Construction and classification of holomorphic vertex operator algebras. Journal für die reine und angewandte Mathematik (Crelles Journal) (2017)

76. L. Dixon, D. Friedan, E. Martinec, S. Shenker, The conformal field theory of orbifolds. Nucl. Phys. B **282**(1), 13–73 (1987). https://doi.org/10.1016/0550-3213(87)90676-6

77. S. Hamidi, C. Vafa, Interactions on orbifolds. Nucl. Phys. B **279**(3–4), 465–513 (1987). https://doi.org/10.1016/0550-3213(87)90006-X

78. J.A. Harvey, Twisting the heterotic string. In: *Workshop on Unified String Theories* (World Scientific Publishing, Singapore, 1986), pp. 704–718

79. R. Dijkgraaf, V. Pasquier, P. Roche, Quasi Hopf algebras, group cohomology and orbifold models. Nucl. Phys. B **18B**, 60–72 (1991). https://doi.org/10.1016/0920-5632(91)90123-V. Recent advances in field theory (Annecy-le-Vieux, 1990)

80. R. Dijkgraaf, C. Vafa, E. Verlinde, H. Verlinde, The operator algebra of orbifold models. Comm. Math. Phys. **123**(3), 485–526 (1989)

81. J. Lepowsky, H. Li, *Introduction to Vertex Operator Algebras and Their Representations*. Progress in Mathematics, vol. 227 (Birkhäuser Boston Inc., Boston, 2004). https://doi.org/10.1007/978-0-8176-8186-9

82. T. Gannon, Much ado about Mathieu. Adv. Math. **301**, 322–358 (2016). https://doi.org/10.1016/j.aim.2016.06.014

83. R. Borcherds, Generalized Kac-Moody algebras. J. Algebra **115**(2), 501–512 (1988). https://doi.org/10.1016/0021-8693(88)90275-X

84. G.H. Hardy, E.M. Wright, *An Introduction to the Theory of Numbers*, 6th edn. (Oxford University Press, Oxford, 2008). Revised by D. R. Heath-Brown and J. H. Silverman, With a foreword by Andrew Wiles

85. Y. Zhu, Modular invariance of characters of vertex operator algebras. J. Am. Math. Soc. **9**(1), 237–302 (1996)

86. C. Dong, H. Li, G. Mason, Modular invariance of trace functions in orbifold theory and generalized Moonshine. Commun. Math. Phys. **214**, 1–56 (2000)

87. C. Dong, X. Lin, S.H. Ng, Congruence property in conformal field theory. Algebra Number Theory **9**(9), 2121–2166 (2015). https://doi.org/10.2140/ant.2015.9.2121

88. C. Dong, L. Ren, Congruence property in orbifold theory. Proc. Am. Math. Soc. **146**(2), 497–506 (2018). https://doi.org/10.1090/proc/13748

89. C. Dong, Z. Zhao, Modularity in orbifold theory for vertex operator superalgebras. Comm. Math. Phys. **260**(1), 227–256 (2005). https://doi.org/10.1007/s00220-005-1418-2

90. M. Miyamoto, Modular invariance of vertex operator algebras satisfying $C_2$-cofiniteness. Duke Math. J. **122**(1), 51–91 (2004). https://doi.org/10.1215/S0012-7094-04-12212-2

91. M.P. Tuite, Monstrous Moonshine from orbifolds. Comm. Math. Phys. **146**(2), 277–309 (1992)

92. M.P. Tuite, On the relationship between Monstrous Moonshine and the uniqueness of the Moonshine module. Comm. Math. Phys. **166**(3), 495–532 (1995)

93. M.P. Tuite, Monstrous Moonshine and the uniqueness of the Moonshine module, in *Low-Dimensional Topology and Quantum Field Theory*. NATO Advanced Science Institutes Series B: Physics, vol. 315 (Plenum, New York, 1993), pp. 289–296

94. M.P. Tuite, Monstrous Moonshine and orbifolds, in *Groups, Difference Sets, and the Monster*. Ohio State University Mathematics Research Institute Publication, vol. 4 (de Gruyter, Berlin, 1996), pp. 443–461

95. N.M. Paquette, D. Persson, R. Volpato, Monstrous BPS-algebras and the superstring origin of moonshine. Commun. Number Theory Phys. **10**(3), 433–526 (2016). https://doi.org/10.4310/CNTP.2016.v10.n3.a2

96. N.M. Paquette, D. Persson, R. Volpato, BPS algebras, genus zero and the heterotic Monster. J. Phys. A **50**(41), 414001, 17 (2017). https://doi.org/10.1088/1751-8121/aa8443

97. J.F.R. Duncan, I.B. Frenkel, Rademacher sums, moonshine and gravity. Commun. Number Theory Phys. **5**(4), 1–128 (2011)

98. S.P. Norton, More on moonshine, in *Computational Group Theory* (Academic Press, London, 1984), pp. 185–193

99. H. Rademacher, The Fourier Series and the functional equation of the absolute modular invariant $J(\tau)$. Am. J. Math. **61**(1), 237–248 (1939)

100. C. Lovelace, Pomeron form factors and dual Regge cuts. Phys. Lett. B **34**(6), 500–506 (1971). https://doi.org/https://doi.org/10.1016/0370-2693(71)90665-4

101. J.L. Gervais, B. Sakita, Field theory interpretation of supergauges in dual models. Nucl. Phys. **B34**, 632–639 (1971) [154(1971)]. https://doi.org/10.1016/0550-3213(71)90351-8

102. P. Ramond, Dual theory for free fermions. Phys. Rev. **D3**, 2415–2418 (1971). https://doi.org/10.1103/PhysRevD.3.2415

103. A. Neveu, J.H. Schwarz, Factorizable dual model of pions. Nucl. Phys. **B31**, 86–112 (1971). https://doi.org/10.1016/0550-3213(71)90448-2

104. A. Neveu, J.H. Schwarz, Quark model of dual pions. Phys. Rev. **D4**, 1109–1111 (1971). https://doi.org/10.1103/PhysRevD.4.1109

105. J.H. Schwarz, Physical states and pomeron poles in the dual pion model. Nucl. Phys. **B46**, 61–74 (1972). https://doi.org/10.1016/0550-3213(72)90201-5

106. J.F. Duncan, Super-Moonshine for Conway's largest sporadic group. Duke Math. J. **139**(2), 255–315 (2007)

107. K. Barron, $N = 1$ Neveu-Schwarz vertex operator superalgebras over Grassmann algebras and with odd formal variables, in *Representations and Quantizations (Shanghai, 1998)* (China Higher Education Press, Beijing, 2000), pp. 9–35

108. R. Heluani, V.G. Kac, Supersymmetric vertex algebras. Comm. Math. Phys. **271**(1), 103–178 (2007). https://doi.org/10.1007/s00220-006-0173-3

109. V. Kac, W. Wang, Vertex operator superalgebras and their representations, in *Mathematical Aspects of Conformal and Topological Field Theories and Quantum Groups*. Contemporary Mathematics, vol. 175 (American Mathematical Society, Providence, 1994), pp. 161–191. https://doi.org/10.1090/conm/175/01843.

110. K. Barron, A supergeometric interpretation of vertex operator superalgebras. Int. Math. Res. Not. (9), 409–430 (1996). https://doi.org/10.1155/S107379289600027X

111. K. Barron, The notion of $N = 1$ supergeometric vertex operator superalgebra and the isomorphism theorem. Commun. Contemp. Math. **5**(4), 481–567 (2003). https://doi.org/10.1142/S0219199703001051

112. R. Heluani, SUSY vertex algebras and supercurves. Comm. Math. Phys. **275**(3), 607–658 (2007). https://doi.org/10.1007/s00220-007-0325-0

113. A.J. Feingold, I.B. Frenkel, J.F.X. Ries, *Spinor Construction of Vertex Operator Algebras, Triality, and $E_8^{(1)}$*. Contemporary Mathematics, vol. 121 (American Mathematical Society, Providence, 1991). https://doi.org/10.1090/conm/121

114. J.F.R. Duncan, S. Mack-Crane, The moonshine module for Conway's group. Forum Math. Sigma **3**, e10, 52 (2015). https://doi.org/10.1017/fms.2015.7

115. E. Mathieu, Mémoire sur l'étude des fonctions de plusieurs quantités, sur la manière de les former et sur les substitutions qui les laissent invariables. Journal de Mathématiques Pures et Appliquées **6**, 241–323 (1861)
116. E. Mathieu, Sur la fonction cinq fois transitive de 24 quantités. Journal de Mathématiques Pures et Appliquées **18**, 25–46 (1873)
117. N.M. Katz, A. Rojas-León, P.H. Tiep, A rigid local system with monodromy group the big Conway group 2.Co_1 and two others with monodromy group the Suzuki group 6.Suz (2019). e-prints. arXiv:1901.03894
118. T. Nagell, Sur les propriétés arithmétiques des cubiques planes du premier genre. Acta Math. **52**(1), 93–126 (1929). https://doi.org/10.1007/BF02547402
119. R.J. Stroeker, B.M.M. de Weger, Solving elliptic Diophantine equations: the general cubic case. Acta Arith. **87**(4), 339–365 (1999). https://doi.org/10.4064/aa-87-4-339-365
120. S. Mukai, Finite groups of automorphisms of $K3$ surfaces and the Mathieu group. Invent. Math. **94**(1), 183–221 (1988). https://doi.org/10.1007/BF01394352
121. S. Kondo, Niemeier lattices, Mathieu groups, and finite groups of symplectic automorphisms of $K3$ surfaces. Duke Math. J. **92**(3), 593–603 (1998). https://doi.org/10.1215/S0012-7094-98-09217-1. With an appendix by Shigeru Mukai
122. Y.T. Siu, Every $K3$ surface is Kähler. Invent. Math. **73**(1), 139–150 (1983). https://doi.org/10.1007/BF01393829
123. W.P. Barth, K. Hulek, C.A.M. Peters, A. Van de Ven, Compact complex surfaces, *Ergebnisse der Mathematik und ihrer Grenzgebiete. 3. Folge. A Series of Modern Surveys in Mathematics [Results in Mathematics and Related Areas. 3rd Series. A Series of Modern Surveys in Mathematics]*, vol. 4, 2nd edn. (Springer, Berlin, 2004)
124. I.V. Dolgachev, J. Keum, Finite groups of symplectic automorphisms of $K3$ surfaces in positive characteristic. Ann. Math. (2) **169**(1), 269–313 (2009). https://doi.org/10.4007/annals.2009.169.269
125. I.V. Dolgachev, J. Keum, $K3$ surfaces with a symplectic automorphism of order 11. J. Eur. Math. Soc. (JEMS) **11**(4), 799–818 (2009). https://doi.org/10.4171/JEMS/167
126. S. Kondo, Maximal subgroups of the Mathieu group $M_{23}$ and symplectic automorphisms of supersingular $K3$ surfaces. Int. Math. Res. Not. Art. ID 71517, 9 (2006). https://doi.org/10.1155/IMRN/2006/71517
127. S. Mukai, H. Ohashi, Finite groups of automorphisms of Enriques surfaces and the Mathieu group $M_{12}$ (2014). e-prints. arXiv:1410.7535
128. A. Beauville, Géométrie des surfaces $K3$: modules et périodes. Société Mathématique de France, Paris (1985). Papers from the seminar held in Palaiseau, October 1981–January 1982, Astérisque No. 126 (1985)
129. D. Huybrechts, *Lectures on K3 Surfaces*. Cambridge Studies in Advanced Mathematics, vol. 158 (Cambridge University Press, Cambridge, 2016). https://doi.org/10.1017/CBO9781316594193
130. S. Kondo, A survey of finite groups of symplectic automorphisms of $K3$ surfaces. J. Phys. A **51**(5), 053003 (2018). https://doi.org/10.1088/1751-8121/aa9f7e
131. P.S. Aspinwall, $K3$ surfaces and string duality, in *Fields, Strings and Duality* (World Scientific Publishing, River Edge, 1997), pp. 421–540
132. T. Eguchi, H. Ooguri, Y. Tachikawa, Notes on the K3 Surface and the Mathieu group $M_{24}$. Exper. Math. **20**, 91–96 (2011)
133. L.A. Borisov, A. Libgober, Elliptic genera of toric varieties and applications to mirror symmetry. Invent. Math. **140**(2), 453–485 (2000). https://doi.org/10.1007/s002220000058
134. V. Gritsenko, Elliptic genus of Calabi-Yau manifolds and Jacobi and Siegel modular forms. Algebra i Analiz **11**(5), 100–125 (1999)
135. M.C.N. Cheng, J.F.R. Duncan, Optimal Mock Jacobi Theta Functions (2016). ArXiv e-prints
136. M.C.N. Cheng, J.F.R. Duncan, J.A. Harvey, Weight one Jacobi forms and umbral moonshine. J. Phys. A **51**(10), 104002, 37 (2018). https://doi.org/10.1088/1751-8121/aaa819
137. M. Eichler, D. Zagier, *The Theory of Jacobi Forms* (Birkhäuser, Basel, 1985)

138. A. Dabholkar, S. Murthy, D. Zagier, Quantum Black Holes, Wall Crossing, and Mock Modular Forms (2012). e-prints arXiv:1208.4074
139. M.C.N. Cheng, J.F.R. Duncan, Meromorphic Jacobi forms of half-integral index and umbral moonshine modules. Commun. Math. Phys. **370**(3), 759–780 (2019). https://doi.org/10.1007/s00220-019-03540-2
140. A.J. Feingold, I.B. Frenkel, A hyperbolic Kac-Moody algebra and the theory of Siegel modular forms of genus 2. J. Math. Ann. **263**, 87–144 (1983)
141. G. Höhn, Komplexe elliptische Geschlechter und $S\hat{1}$-aequivariante Kobordismustheorie (Complex elliptic genera and $S\hat{1}$-equivariant cobordism theory) (2004). arXiv Mathematics e-prints math/0405232
142. I.M. Krichever, Generalized elliptic genera and Baker-Akhiezer functions. Mat. Zametki **47**(2), 34–45, 158 (1990). https://doi.org/10.1007/BF01156822
143. V. Gritsenko, *Complex Vector Bundles and Jacobi Forms*, vol. 1103 (Kyoto University, Kyoto, 1999), pp. 71–85. Automorphic forms and $L$-functions
144. P.S. Landweber (ed.), *Elliptic Curves and Modular Forms in Algebraic Topology*. Lecture Notes in Mathematics, vol. 1326 (Springer, Berlin, 1988). https://doi.org/10.1007/BFb0078035
145. S. Ochanine, Sur les genres multiplicatifs définis par des intégrales elliptiques. Topology **26**(2), 143–151 (1987). https://doi.org/10.1016/0040-9383(87)90055-3
146. S. Ochanine, What is. . . an elliptic genus? Not. Am. Math. Soc. **56**(6), 720–721 (2009)
147. T. Eguchi, K. Hikami, Enriques moonshine. J. Phys. A **46**(31), 312001, 11 (2013). https://doi.org/10.1088/1751-8113/46/31/312001
148. S. Govindarajan, S. Samanta, Two moonshines for $L_2(11)$ but none for $M_{12}$. Nucl. Phys. B **939**, 566–598 (2019). https://doi.org/10.1016/j.nuclphysb.2019.01.004
149. P. Appell, Sur les fonctions doublement périodiques de troisième espèce. Ann. Sci. École Norm. Sup. (3) **1**, 135–164 (1884)
150. M. Lerch, Bemerkungen zur Theorie der elliptischen Funktionen. Jahrbuch über die Fortschritte der Mathematik **24**, 442–445 (1892)
151. M. Lerch, Poznámky k theorii funcí elliptických. Rozpravy České Akademie Císaře Františka Josefa pro vědy, slovesnost a umění v praze **24**, 465–480 (1892)
152. D. Zagier, Ramanujan's mock theta functions and their applications (after Zwegers and Bringmann–Ono). Astérisque (326), Exp. No. 986, vii–viii, 143–164 (2007). Séminaire Bourbaki, vol. 2007/2008
153. S. Zwegers, Mock Theta Functions (2008). e-prints. arXiv:0807.4834
154. S. Zwegers, Multivariable Appell functions and nonholomorphic Jacobi forms. Res. Math. Sci. **6**(1), Paper No. 16, 15 (2019). https://doi.org/10.1007/s40687-019-0178-0
155. T. Creutzig, G. Höhn, Mathieu moonshine and the geometry of K3 surfaces. Commun. Number Theory Phys. **8**(2), 295–328 (2014)
156. M.C.N. Cheng, $K3$ surfaces, $N = 4$ dyons and the Mathieu group $M_{24}$. Commun. Number Theory Phys. **4**(4), 623–657 (2010). https://doi.org/10.4310/CNTP.2010.v4.n4.a2
157. A. Sen, Black holes and the spectrum of half-BPS states in $N = 4$ supersymmetric string theory. Adv. Theor. Math. Phys. **9**(4), 527–558 (2005)
158. M.R. Gaberdiel, S. Hohenegger, R. Volpato, Mathieu twining characters for $K3$. J. High Energy Phys. (9), 058, 20 (2010). https://doi.org/10.1007/JHEP09(2010)058
159. M.R. Gaberdiel, S. Hohenegger, R. Volpato, Mathieu Moonshine in the elliptic genus of $K3$. J. High Energy Phys. (10), 062, 24 (2010). https://doi.org/10.1007/JHEP10(2010)062
160. T. Eguchi, K. Hikami, Note on twisted elliptic genus of $K3$ surface. Phys. Lett. B **694**(4–5), 446–455 (2011). https://doi.org/10.1016/j.physletb.2010.10.017
161. M. Krauel, G. Mason, Vertex operator algebras and weak Jacobi forms. Int. J. Math. **23**(6), 1250024, 10 (2012). https://doi.org/10.1142/S0129167X11007677
162. R. Beals, R. Wong, Special functions, *Cambridge Studies in Advanced Mathematics*, vol. 126 (Cambridge University Press, Cambridge, 2010). https://doi.org/10.1017/CBO9780511762543. A graduate text

163. F.W.J. Olver, D.W. Lozier, R.F. Boisvert, C.W. Clark (eds.), *NIST Handbook of Mathematical Functions*. U.S. Department of Commerce, National Institute of Standards and Technology, Washington, DC (Cambridge University Press, Cambridge, 2010). With 1 CD-ROM (Windows, Macintosh and UNIX)

164. G.N. Watson, *A Treatise on the Theory of Bessel Functions*. Cambridge Mathematical Library (Cambridge University Press, Cambridge, 1995). Reprint of the second (1944) edition

165. M.C.N. Cheng, J.F.R. Duncan, On Rademacher sums, the largest Mathieu group and the holographic modularity of moonshine. Commun. Number Theory Phys. **6**(3), 697–758 (2012). https://doi.org/10.4310/CNTP.2012.v6.n3.a4

166. M.C.N. Cheng, J.F.R. Duncan, Rademacher sums and Rademacher series, in *Conformal Field Theory, Automorphic Forms and Related Topics*. Contributions in Mathematical and Computational Sciences, vol. 8 (Springer, Heidelberg, 2014), pp. 143–182

167. W. Duke, Almost a century of answering the question: what is a mock theta function? Not. Am. Math. Soc. **61**(11), 1314–1320 (2014). https://doi.org/10.1090/noti1185

168. A. Folsom, What is . . . a mock modular form? Notices Amer. Math. Soc. **57**(11), 1441–1443 (2010)

169. K. Ono, Unearthing the visions of a master: harmonic Maass forms and number theory, in *Current Developments in Mathematics, 2008* (International Press, Somerville, 2009), pp. 347–454

170. B.C. Berndt, R.A. Rankin, Ramanujan, *History of Mathematics*, vol. 9 (American Mathematical Society, Providence/London Mathematical Society, London, 1995). Letters and commentary

171. J. Bruinier, K. Ono, Heegner divisors, *L*-functions and harmonic weak Maass forms. Ann. Math. (2) **172**(3), 2135–2181 (2010). https://doi.org/10.4007/annals.2010.172.2135

172. E. Witten, Elliptic genera and quantum field theory. Comm. Math. Phys. **109**(4), 525–536 (1987)

173. T. Kawai, Y. Yamada, S.K. Yang, Elliptic genera and $N = 2$ superconformal field theory. Nucl. Phys. B **414**(1–2), 191–212 (1994). https://doi.org/10.1016/0550-3213(94)90428-6

174. M. Ademollo, L. Brink, A. D'Adda, R. D'Auria, E. Napolitano, S. Sciuto, E. del Giudice, P. di Vecchia, S. Ferrara, F. Gliozzi, R. Musto, R. Pettorino, Supersymmetric strings and colour confinement. Phys. Lett. B **62**(1), 105–110 (1976). https://doi.org/10.1016/0370-2693(76)90061-7

175. T. Eguchi, A. Taormina, Unitary representations of the $N = 4$ superconformal algebra. Phys. Lett. B **196**(1), 75–81 (1987). https://doi.org/10.1016/0370-2693(87)91679-0

176. T. Eguchi, A. Taormina, Character formulas for the $N = 4$ superconformal algebra. Phys. Lett. B **200**(3), 315–322 (1988). https://doi.org/10.1016/0370-2693(88)90778-2

177. T. Eguchi, K. Hikami, Superconformal algebras and mock theta functions. II. Rademacher expansion for $K3$ surface. Commun. Number Theory Phys. **3**(3), 531–554 (2009). https://doi.org/10.4310/CNTP.2009.v3.n3.a4

178. T. Eguchi, H. Ooguri, A. Taormina, S.K. Yang, Superconformal algebras and string compactification on manifolds with SU($n$) holonomy. Nucl. Phys. B **315**(1), 193–221 (1989). https://doi.org/10.1016/0550-3213(89)90454-9

179. H. Ooguri, Superconformal symmetry and geometry of Ricci-flat Kähler manifolds. Int. J. Mod. Phys. A **4**(17), 4303–4324 (1989). https://doi.org/10.1142/S0217751X89001801

180. M.R. Gaberdiel, S. Hohenegger, R. Volpato, Symmetries of K3 sigma models. Commun. Number Theory Phys. **6**(1), 1–50 (2012). https://doi.org/10.4310/CNTP.2012.v6.n1.a1

181. M.R. Gaberdiel, R. Volpato, Mathieu moonshine and orbifold K3s, in *(Conformal Field Theory, Automorphic Forms and Related Topics)*. Contributions in Mathematical and Computational Sciences, vol. 8 (Springer, Heidelberg, 2014), pp. 109–141

182. J.F.R. Duncan, S. Mack-Crane, Derived equivalences of K3 surfaces and twined elliptic genera. Res. Math. Sci. **3**, Art. 1, 47 (2016). https://doi.org/10.1186/s40687-015-0050-9

183. M.C.N. Cheng, F. Ferrari, S.M. Harrison, N.M. Paquette, Landau-Ginzburg orbifolds and symmetries of K3 CFTs. J. High Energy Phys. **2017**, Article number: 46 (2017). https://doi.org/10.1007/JHEP01(2017)046

184. M.C.N. Cheng, S.M. Harrison, R. Volpato, M. Zimet, K3 string theory, lattices and moonshine. Res. Math. Sci. **5**(3), Paper No. 32, 45 (2018). https://doi.org/10.1007/s40687-018-0150-4

185. M. Gell-Mann, M. Lévy, The axial vector current in beta decay. Nuovo Cimento (10) **16**, 705–726 (1960)

186. S.V. Ketov, *Quantum Non-linear Sigma-Models*. Texts and Monographs in Physics (Springer, Berlin, 2000). https://doi.org/10.1007/978-3-662-04192-5. From quantum field theory to supersymmetry, conformal field theory, black holes and strings

187. W. Nahm, K. Wendland, A Hiker's guide to $K3$ – aspects of $N = (4, 4)$ superconformal field theory with central charge $c = 6$. Commun. Math. Phys. **216**, 85–138 (2001)

188. K. Wendland, Snapshots of conformal field theory, in *Mathematical Aspects of Quantum Field Theories*. Mathematical Physics Studies (Springer, Cham, 2015), pp. 89–129

189. K. Wendland, K3 en route from geometry to conformal field theory, in *Geometric, Algebraic and Topological Methods for Quantum Field Theory* (World Scientific Publishing, Hackensack, 2017), pp. 75–110

190. K. Barron, Axiomatic aspects of $N = 2$ vertex superalgebras with odd formal variables. Comm. Algebra **38**(4), 1199–1268 (2010). https://doi.org/10.1080/00927870902828900

191. K. Barron, The moduli space of $N = 2$ super-Riemann spheres with tubes. Commun. Contemp. Math. **9**(6), 857–940 (2007). https://doi.org/10.1142/S0219199707002666

192. D. Ben-Zvi, R. Heluani, M. Szczesny, Supersymmetry of the chiral de Rham complex. Compos. Math. **144**(2), 503–521 (2008). https://doi.org/10.1112/S0010437X07003223

193. R. Heluani, Supersymmetry of the chiral de Rham complex. II. Commuting sectors. Int. Math. Res. Not. IMRN (6), 953–987 (2009)

194. R. Heluani, J. van Ekeren, Characters of topological $N = 2$ vertex algebras are Jacobi forms on the moduli space of elliptic supercurves. Adv. Math. **302**, 551–627 (2016). https://doi.org/10.1016/j.aim.2016.05.018

195. J. van Ekeren, Superconformal vertex algebras and Jacobi forms, in *Perspectives in Lie Theory*. Springer INdAM Series, vol. 19 (Springer, Cham, 2017), pp. 315–330

196. N. Hitchin, Hyper-Kähler manifolds. Astérisque, tome 206, Exp. No. 748, 137–166 (1992). Séminaire Bourbaki, vol. 1991/92

197. D. Huybrechts, Generalized Calabi-Yau structures, $K3$ surfaces, and $B$-fields. Int. J. Math. **16**(1), 13–36 (2005). https://doi.org/10.1142/S0129167X05002734

198. D. Huybrechts, On derived categories of K3 surfaces, symplectic automorphisms and the Conway group, in *Development of moduli theory—Kyoto 2013*. Advanced Studies in Pure Mathematics, vol. 69 (Mathematics Society, Tokyo, 2016), pp. 387–405. https://doi.org/10.2969/aspm/06910387

199. L. Queen, Modular functions arising from some finite groups. Math. Comp. **37**(156), 547–580 (1981). https://doi.org/10.2307/2007446

200. G. Mason, Finite groups and modular functions, in *The Arcata Conference on Representations of Finite Groups* (Arcata, Calif., 1986). Proceedings of Symposia in Pure Mathematics, vol. 47 (American Mathematical Society, Providence, 1987), pp. 181–210. With an appendix by S. P. Norton

201. S. Norton, From Moonshine to the Monster, in *Proceedings on Moonshine and Related Topics* (Montréal, QC, 1999). CRM Proceedings & Lecture Notes, vol. 30 (American Mathematical Society, Providence, 2001), pp. 163–171

202. G. Höhn, Generalized Moonshine for the Baby Monster, in *Workshop and Conference on Infinite Dimensional Lie Theory and Its Applications*, Toronto (2003)

203. G. Höhn, *Selbstduale Vertexoperatorsuperalgebren und das Babymonster*. Bonner Mathematische Schriften [Bonn Mathematical Publications], vol. 286 (Universität Bonn, Mathematisches Institut, Bonn, 1996). Dissertation, Rheinische Friedrich-Wilhelms-Universität Bonn, Bonn, 1995

204. H. Yamauchi, 2A-orbifold construction and the baby-monster vertex operator superalgebra. J. Algebra **284**(2), 645–668 (2005). https://doi.org/10.1016/j.jalgebra.2004.09.039

205. S. Carnahan, Generalized Moonshine IV: Monstrous Lie algebras (2012). e-prints. arXiv:1208.6254
206. S. Carnahan, Generalized moonshine I: genus-zero functions. Algebra Number Theory 4(6), 649–679 (2010). https://doi.org/10.2140/ant.2010.4.649
207. S. Carnahan, Generalized moonshine, II: Borcherds products. Duke Math. J. 161(5), 893–950 (2012). https://doi.org/10.1215/00127094-1548416
208. S. Carnahan, Fricke Lie algebras and the genus zero property in Moonshine. J. Phys. A 50(40), 404002, 21 (2017). https://doi.org/10.1088/1751-8121/aa781d
209. S. Carnahan, 51 constructions of the Moonshine module. Commun. Number Theory Phys. 12(2), 305–334 (2018). https://doi.org/10.4310/cntp.2018.v12.n2.a3
210. M.C.N. Cheng, J.F.R. Duncan, J.A. Harvey, Umbral moonshine. Commun. Number Theory Phys. 8(2), 101–242 (2014). https://doi.org/10.4310/CNTP.2014.v8.n2.a1
211. M.C.N. Cheng, J.F.R. Duncan, J.A. Harvey, Umbral moonshine and the Niemeier lattices. Res. Math. Sci. 1(3), 1–81 (2014)
212. J.F.R. Duncan, M.J. Griffin, K. Ono, Proof of the umbral moonshine conjecture. Res. Math. Sci. 2(26) (2015)
213. K. Ono, L. Rolen, S. Trebat-Leder, Classical and umbral moonshine: connections and p-adic properties. J. Ramanujan Math. Soc. 30(2), 135–159 (2015)
214. V. Anagiannis, M.C.N. Cheng, S.M. Harrison, $K3$ elliptic genus and an umbral moonshine module. Comm. Math. Phys. 366(2), 647–680 (2019). https://doi.org/10.1007/s00220-019-03314-w
215. J. Duncan, J. Harvey, The umbral moonshine module for the unique unimodular Niemeier root system. Algebra Number Theory 11(3), 505–535 (2017). https://doi.org/10.2140/ant.2017.11.505
216. J.F.R. Duncan, A. O'Desky, Super vertex algebras, meromorphic Jacobi forms and umbral moonshine. J. Algebra 515, 389–407 (2018). https://doi.org/10.1016/j.jalgebra.2018.08.017
217. M.R. Gaberdiel, D. Persson, H. Ronellenfitsch, R. Volpato, Generalized Mathieu Moonshine. Commun. Number Theory Phys. 7(1), 145–223 (2013). https://doi.org/10.4310/CNTP.2013.v7.n1.a5
218. M.R. Gaberdiel, D. Persson, R. Volpato, Generalised moonshine and holomorphic orbifolds, in String-Math 2012. Proceedings of Symposia in Pure Mathematics, vol. 90 (American Mathematical Society, Providence, 2015), pp. 73–86. https://doi.org/10.1090/pspum/090/01520
219. M.C.N. Cheng, P. de Lange, D.P.Z. Whalen, Generalised umbral moonshine. Symmetry Integ. Geom. Methods Appl. 15, Paper No. 014, 27 (2019). https://doi.org/10.3842/SIGMA.2019.014
220. A. Taormina, K. Wendland, The Conway Moonshine Module is a Reflected K3 Theory (2017). e-prints. arXiv:1704.03813
221. M.C.N. Cheng, J.F.R. Duncan, S.M. Harrison, S. Kachru, Equivariant K3 invariants. Commun. Number Theory Phys. 11(1), 41–72 (2017). https://doi.org/10.4310/CNTP.2017.v11.n1.a2
222. J. Bryan, G. Oberdieck, CHL Calabi-Yau threefolds: Curve counting, Mathieu moonshine and Siegel modular forms (2018). e-prints. arXiv:1811.06102
223. A. Taormina, K. Wendland, The overarching finite symmetry group of Kummer surfaces in the Mathieu group $M_{24}$. J. High Energy Phys. (8), 125 (2013)
224. A. Taormina, K. Wendland, Symmetry-surfing the moduli space of Kummer K3s, in String-Math 2012. Proceedings of Symposia in Pure Mathematics, vol. 90 (American Mathematical Society, Providence, 2015), pp. 129–153. https://doi.org/10.1090/pspum/090/01522
225. A. Taormina, K. Wendland, A twist in the $M_{24}$ Moonshine story. Confluentes Math. 7(1), 83–113 (2015). https://doi.org/10.5802/cml.19
226. A. Taormina, K. Wendland, Not doomed to fail. J. High Energy Phys. (9), 062 (2018). https://doi.org/10.1007/jhep09(2018)062
227. K. Wendland, Hodge-elliptic genera and how they govern K3 theories. Comm. Math. Phys. 368(1), 187–221 (2019). https://doi.org/10.1007/s00220-019-03425-4

228. M.R. Gaberdiel, C.A. Keller, H. Paul, Mathieu moonshine and symmetry surfing. J. Phys. A **50**(47), 474002, 29 (2017). https://doi.org/10.1088/1751-8121/aa915f
229. M.C.N. Cheng, S. Harrison, Umbral moonshine and $K3$ surfaces. Comm. Math. Phys. **339**(1), 221–261 (2015). https://doi.org/10.1007/s00220-015-2398-5
230. J.A. Harvey, B.C. Rayhaun, Traces of singular moduli and moonshine for the Thompson group. Commun. Number Theory Phys. **10**(1), 23–62 (2016). https://doi.org/10.4310/CNTP.2016.v10.n1.a2
231. J.G. Thompson, A conjugacy theorem for $E_8$. J. Algebra **38**(2), 525–530 (1976). https://doi.org/10.1016/0021-8693(76)90235-0
232. P.E. Smith, A simple subgroup of $M?$ and $E_8(3)$. Bull. London Math. Soc. **8**(2), 161–165 (1976). https://doi.org/10.1112/blms/8.2.161
233. M.J. Griffin, M.H. Mertens, A proof of the Thompson moonshine conjecture. Res. Math. Sci. **3**(3), 36 (2016). https://doi.org/10.1186/s40687-016-0084-7
234. R.L. Griess Jr., C.H. Lam, Groups of Lie type, vertex algebras, and modular moonshine. Electron. Res. Announc. Math. Sci. **21**, 167–176 (2014). https://doi.org/10.3934/era.2014.21.167
235. R.L. Griess Jr., C.H. Lam, Groups of Lie type, vertex algebras, and modular moonshine. Int. Math. Res. Not. IMRN (21), 10716–10755 (2015). https://doi.org/10.1093/imrn/rnv003
236. R.E. Borcherds, Automorphic forms on $O_{s+2,2}(\mathbf{R})$ and infinite products. Invent. Math. **120**(1), 161–213 (1995). https://doi.org/10.1007/BF01241126
237. J.F.R. Duncan, J.A. Harvey, B.C. Rayhaun, Thompson moonshine revisited. In preparation
238. R.E. Borcherds, Automorphic forms with singularities on Grassmannians. Invent. Math. **132**(3), 491–562 (1998). https://doi.org/10.1007/s002220050232
239. E. Hofmann, Liftings and Borcherds products, in *L-Functions and Automorphic Forms*. Contributions in Mathematical and Computational Sciences, vol. 10 (Springer, Cham, 2017), pp. 333–366
240. J.H. Bruinier, *Borcherds Products on O(2, l) and Chern Classes of Heegner Divisors*. Lecture Notes in Mathematics, vol. 1780 (Springer, Berlin, 2002). https://doi.org/10.1007/b83278
241. J.F.R. Duncan, M.H. Mertens, K. Ono, O'Nan moonshine and arithmetic (2017). e-prints. arXiv:1702.03516
242. J.F.R. Duncan, M.H. Mertens, K. Ono, Pariah moonshine. Nat. Commun. **8**, 670 (2017). https://doi.org/10.1038/s41467-017-00660-y
243. J.F.R. Duncan, From the Monster to Thompson to O'Nan (2019). e-prints. arXiv:1909.09684
244. D. Goldfeld, Gauss's class number problem for imaginary quadratic fields. Bull. Amer. Math. Soc. (N.S.) **13**(1), 23–37 (1985). https://doi.org/10.1090/S0273-0979-1985-15352-2
245. D. Goldfeld, The Gauss class number problem for imaginary quadratic fields, in *Heegner Points and Rankin L-Series*. Mathematical Sciences Research Institute Publications, vol. 49 (Cambridge University Press, Cambridge, 2004), pp. 25–36. https://doi.org/10.1017/CBO9780511756375.004
246. D. Zagier, L-series of elliptic curves, the Birch-Swinnerton-Dyer conjecture, and the class number problem of Gauss. Notices Amer. Math. Soc. **31**(7), 739–743 (1984)
247. M.E. O'Nan, Some evidence for the existence of a new simple group. Proc. London Math. Soc. (3) **32**(3), 421–479 (1976). https://doi.org/10.1112/plms/s3-32.3.421
248. L. Beneish, Quasimodular moonshine and arithmetic connections. Trans. Am. Math. Soc. **372**(12), 8793–8813 (2019). https://doi.org/10.1090/tran/7874
249. L. Beneish, Module constructions for certain subgroups of the largest Mathieu group (2019). e-prints. arXiv:1912.04373
250. J. Lehner, Divisibility properties of the Fourier coefficients of the modular invariant $j(\tau)$. Am. J. Math. **71**, 136–148 (1949). https://doi.org/10.2307/2372101
251. J. Lehner, Further congruence properties of the Fourier coefficients of the modular invariant $j(\tau)$. Am. J. Math. **71**, 373–386 (1949). https://doi.org/10.2307/2372252
252. R.C. Chen, S. Marks, M. Tyler, $p$-adic properties of Hauptmoduln with applications to moonshine. Symmetry Integ. Geom. Methods Appl. **15**, Paper No. 033, 35 (2019). https://doi.org/10.3842/SIGMA.2019.033

# Modified Elliptic Genus

## Valery Gritsenko

## Contents

1   Automorphic Correction of Elliptic Genus ................................................... 88
   1.1   Witten Genus .......................................................................... 89
   1.2   Elliptic Genus of Calabi–Yau Manifolds ............................................. 90
   1.3   Complex Curves, Surfaces and Threefolds ........................................... 95
   1.4   When $\chi(M, E; \tau, z)$ is Holomorphic? ........................................ 96
2   The Graded Algebra $J_{0,*/2}$ Over $\mathbb{Z}$ ........................................ 98
3   Some Applications ........................................................................ 107
   3.1   Calabi–Yau Fivefolds and the Euler Number Modulo 24 ............................ 107
   3.2   Exact Formulae for MWG for Dimension $d = 6, 7, 9$ and Rank $r < d$ ............. 109
   3.3   Relations Between $\hat{A}$-Genera ................................................ 112
   3.4   Vector Bundle of Rank 2 ........................................................... 115
4   $A_2^{(2)}$-Genus ........................................................................ 117
References ................................................................................. 118

**Abstract** This mini course is an additional part to my semester course on the theory of Jacobi modular forms given at the mathematical department of NRU HSE in Moscow (see Gritsenko Jacobi modular forms: 30 ans après; COURSERA (12 lectures and seminars), 2017–2019). This additional part contains some applications of Jacobi modular forms to the theory of elliptic genera and Witten genus. The subject of this course is related to my old talk given in Japan (see Gritsenko (Proc Symp "Automorphic forms and L-functions" **1103**:71–85, 1999)). Our approach based on special representations of Jacobi theta-series founded in Gritsenko (Proc Symp "Automorphic forms and L-functions" **1103**:71–85, 1999) (see Lemma 1.5 below). In the next future we hope to present new automorphic invariants of holomorphic varieties at the automorphic seminar of the International laboratories

V. Gritsenko (✉)
University of Lille, Lille, France

International Laboratory of Mirror Symmetry and Automorphic Forms, NRU HSE, Moscow, Russia
e-mail: valery.gritsenko@univ-lille.fr

© Springer Nature Switzerland AG 2020
V. A. Gritsenko, V. P. Spiridonov (eds.), *Partition Functions and Automorphic Forms*, Moscow Lectures 5,
https://doi.org/10.1007/978-3-030-42400-8_2

of mirror symmetry and automorphic forms of NRU HSE. So this mini course will be also an introduction to this new theory.

The elliptic genus of a Calabi–Yau manifold is an automorphic form in two variables with integral Fourier coefficients. More exactly this is a weak Jacobi form of weight zero. For arbitrary holomorphic vector bundle over a complex manifold we define an automorphic correction of its elliptic genus which is a meromorphic Jacobi form. We calculate this Jacobi form explicitly in the case of manifolds of small dimensions. Using arithmetic properties of the graded ring of integral Jacobi forms we obtain some geometrical applications.

# 1 Automorphic Correction of Elliptic Genus

Let $M$ be an almost complex compact manifold $M$ of (complex) dimension $d$ and let $E$ be a complex vector bundle over $M$. For two formal variables $q = \exp(2\pi i \tau)$ and $y = \exp(2\pi i z)$ where $\tau \in \mathbb{H}_1$ is in the upper half-plane and $z \in \mathbb{C}$ one defines a formal power series $\mathbf{E}_{q,y} \in K(M)[[q, y^{\pm 1}]]$

$$\mathbf{E}_{q,y} = \bigotimes_{n \geq 0} \bigwedge_{-q^n y^{-1}} E^* \otimes \bigotimes_{n \geq 1} \bigwedge_{-q^n y} E \otimes \bigotimes_{n=1}^{\infty} S_{q^n} T_M^* \otimes \bigotimes_{n=1}^{\infty} S_{q^n} T_M \qquad (1.1)$$

where $T_M$ denotes the holomorphic tangent bundle of $M$ and

$$\bigwedge_x E = \sum_{k \geq 0} (\wedge^k E) x^k, \qquad S_x E = \sum_{k \geq 0} (S^k E) x^k$$

are formal power series with exterior powers and symmetric powers of a bundle as coefficients. We propose the following

**Definition 1.1** *Modified Witten genus* $\chi(M, E; \tau, z)$ of a complex vector bundle $E$ of rank $r$ over a compact (almost) complex manifold $M$ of dimension $d$ is defined as follows

$$\chi(M, E; \tau, z) = q^{(r-d)/12} y^{r/2} \int_M \exp\left(\frac{1}{2}(c_1(E) - c_1(T_M))\right)$$

$$\exp\left((p_1(E) - p_1(T_M)) \cdot G_2(\tau)\right) \exp\left(-\frac{c_1(E)}{2\pi i} \frac{\vartheta_z}{\vartheta}(\tau, z)\right) \mathrm{ch}(\mathbf{E}_{q,y}) \, \mathrm{td}(T_M)$$

where $c_1(E)$ and $p_1(E)$ are the first Chern and Pontryagin class of $E$, td is the Todd class, $\mathrm{ch}(\mathbf{E}_{q,y})$ is the Chern character which we apply to each coefficient of the formal power series and the integral $\int_M$ denotes the evaluation of the top degree differential form on the fundamental cycle of the manifold.

In the definition we use Jacobi theta-series of level two $\vartheta(\tau, z) = -i\vartheta_{11}(\tau, z)$

$$\vartheta(\tau, z) = \sum_{n \equiv 1 \bmod 2} (-1)^{\frac{n-1}{2}} q^{\frac{n^2}{8}} y^{\frac{n}{2}} = -q^{1/8} y^{-1/2} \prod_{n \geq 1} (1 - q^{n-1}y)(1 - q^n y^{-1})(1 - q^n),$$

$\vartheta_z(\tau, z) = \dfrac{\partial \vartheta}{\partial z}(\tau, z)$ and $G_2(\tau) = -\frac{1}{24} + \sum_{n=1}^{\infty} \sigma_1(n) q^n$ is a quasi-modular Eisenstein series of weight 2 where $\sigma_1(n)$ is the sum of all positive divisors of $n$.

*Remark 1.2 (Differentiable Manifolds)* In this paper we shall mainly apply Definition 1.1 for the case of complex manifolds but one can give the same definition for differential manifolds. The natural set up of Definition 1.1 is the following. Let $M$ be a compact even-dimensional Spin$^c$-manifold without boundary and $E$ a Hermitian $C^\infty$-vector bundle over $M$. Then one can define Todd class for the manifold $M$ using $\hat{A}$-genus:

$$\mathrm{td}(T_M) := e^{c_1/2}\hat{A}(T_M), \quad \text{where } \hat{A}(T_M) = \prod_i \frac{x_i/2}{\sinh(x_i/2)}.$$

(This definition depends on the choice of the spinor bundle.) Then

$$\mathrm{ind}(D_E^+) = \int_M \mathrm{ch}(E)\mathrm{td}(T_M)$$

is the index of a twisted Dirac operator and we can give Definition 1.1 in this case.

## 1.1 Witten Genus

As a limit case of the definition above one obtains the Witten genus (see [26–28], [19]). Let assume that $M$ admits a spin structure (i.e. the second Witney-Stiefel class $w_2(M)$ is zero or $c_1(T_M) = 0 \mod 2$) and $p_1(M) = 0$. Let $E = M \times \mathbb{C}^r$ be the trivial vector bundle of rank $r$ over $M$. Then $\mathrm{ch}(\bigwedge_x E) = (1 + x)^r$ and

$$q^{r/12} y^{r/2} \mathrm{ch}\left( \bigotimes_{n=0}^{\infty} \bigwedge_{-y^{-1}q^n} E^* \otimes \bigotimes_{n=1}^{\infty} \bigwedge_{-yq^n} E \right) = \left( \frac{\vartheta(\tau, z)}{\eta(\tau)} \right)^r.$$

Thus

$$q^{d/12} \chi(M, M \times \mathbb{C}^r; \tau, z) = \frac{\vartheta(\tau, z)^r}{\eta(\tau)^r} \int_M \prod_{i=1}^{d} \frac{x_i/2}{\sinh(x_i/2)} \prod_{n=1}^{\infty} \frac{1}{(1 - q^n e^{x_i})(1 - q^n e^{-x_i})}$$

$$= \hat{A}\left( M, \bigotimes_{n=1}^{\infty} S_{q^n}(T_M \oplus T_M^*) \right) \frac{\vartheta(\tau, z)^r}{\eta(\tau)^r} = \text{Witten genus } (M) \frac{\vartheta(\tau, z)^r}{\eta(\tau)^{r+2d}}.$$

If we take the trivial vector bundle of rank 0, then

$$\chi(M, 0; \tau, z) = \chi(M; \tau) = \frac{\text{Witten genus}(M)}{\eta(\tau)^{2d}}.$$

This is a modular form in $\tau$ with respect to $SL_2(\mathbb{Z})$.

This calculation shows that Definition 1.1 is non-stable. In some questions one can use a stable variant of the series $\mathbf{E}_{q,y}$

$$\mathbf{E}_{q,y}^{st} = \bigotimes_{n=0}^{\infty} \bigwedge\nolimits_{-y^{-1}q^n}(E^* - r) \otimes \bigotimes_{n=1}^{\infty} \bigwedge\nolimits_{-yq^n}(E - r) \otimes \bigotimes_{n=1}^{\infty} S_{q^n}(TM^* - d) \otimes \bigotimes_{n=1}^{\infty} S_{q^n}(TM - d)$$

where $r = \text{rank } E$ and $d = \dim M$. Then, after evident changing of $q$- and $y$-factors before the integral in Definition 1.1, we obtain

$$\chi^{st}(M, M \times \mathbb{C}^r; \tau, z) = \text{Witten genus}(M; \tau).$$

In this paper we shall use the non-stable variant of Definition 1.1.

## 1.2   Elliptic Genus of Calabi–Yau Manifolds

This case is of some interest in physics. Let $E = T_M$ and $c_1(T_M) = 0$. Then there are no correction terms in Definition 1.1. Thus the elliptic genus is, up to the factor $y^{d/2}$, the Euler–Poicaré characteristic of the element $\mathbf{E}_{q,y}$. This function is *the elliptic genus* of the Calabi–Yau manifold $M$ or the genus one partition function of the supersymmetric $(2, 2)$ sigma model whose target space is $M$ (see [21]).

$$\chi(M, T_M; \tau, z) = \text{Elliptic genus}(M; \tau, z) = y^{d/2} \int_M \text{ch}(\mathbf{E}_{q,y}) \, \text{td}(T_M).$$

We remark that in this case all Fourier coefficients of the elliptic genus are integral according to the Riemann–Roch–Hirzebruch formula. They are equal to the index of a Dirac operator twisted with a corresponding vector bundle coefficient of the formal power series $\mathbf{E}_{q,y}$.

It is known that the elliptic genus of a Calabi–Yau manifold is a modular form in variables $\tau$ and $z$, i.e., it is a weak Jacobi form of weight 0 and index $d/2$. If $c_1(T_M) \neq 0$, *then the elliptic genus of $M$ defined above is not a modular form in $\tau$ and $z$*. We add three correction factors in Definition 1.1 in order to obtain a function with a good behaviour with respect to the modular transformations in two variables $\tau$ and $z$. If $E = T_M$ and $c_1(T_M) \neq 0$, then the integral of Definition 1.1 contains the only correction term

$$\exp\left(-\frac{c_1(T_M)}{2\pi i} \frac{\vartheta_z}{\vartheta}(\tau, z)\right).$$

Thus the elliptic genus of $M$ (as a function in two variables) is equal to the zeroth term in a sum of $d + 1$ summands of the modified genus. These summands correspond to all powers of the first Chern class of $M$

$$\chi(M, T_M; \tau, z) = \text{Elliptic genus}\,(M; \tau, z) + \sum_{n=1}^{d} \left( \int_M c_1(M)^n(\ldots) \right).$$

The elliptic genus is not an automorphic form in two variables but the modified elliptic genus is. The main result of this section is

**Theorem 1.3** *Let $E$ be a complex (holomorphic) vector bundle of rank $r$ over a compact complex (Spin$^c$) manifold $M$ of dimension $d$ (real dimension $2n$). Let $\chi(M, E; \tau, z)$ be the modified Witten genus. Then*

$$\chi(M, E; \tau, z) \left( \frac{\vartheta(\tau, z)}{\eta(\tau)} \right)^{d-r}$$

*is a weak Jacobi form of weight 0 and index $d/2$. In particular, $\chi(M, E; \tau, z)$ is a weak Jacobi form if $\operatorname{rank}(E) \geq \dim(M)$.*

First we recall the definition of Jacobi forms of the type we need in this paper. Let $t \geq 0$ and $k$ be integral or half-integral. Let $v$ be a character of finite order (or a multiplier system for half-integral $k$) of $SL_2(\mathbb{Z})$. A holomorphic function $\phi(\tau, z)$ on $\mathbb{H}_1 \times \mathbb{C}$ is called a *weak Jacobi form of weight $k$ and index $t$ with character $v$* if it satisfies the functional equations

$$\phi\left( \frac{a\tau + b}{c\tau + d}, \frac{z}{c\tau + d} \right) = v(\gamma)(c\tau + d)^k\, e^{2\pi it \frac{cz^2}{c\tau + d}}\, \phi(\tau, z) \quad (\gamma = \left( \begin{smallmatrix} a & b \\ c & d \end{smallmatrix} \right) \in SL_2(\mathbb{Z}))$$

(1.3a)

and

$$\phi(\tau, z + \lambda\tau + \mu) = (-1)^{2t(\lambda+\mu)}\, e^{-2\pi it(\lambda^2\tau + 2\lambda z)}\phi(\tau, z) \quad (\lambda, \mu \in \mathbb{Z}) \quad (1.3b)$$

and $\phi(\tau, z)$ has the Fourier expansion of the type

$$\phi(\tau, z) = \sum_{\substack{n \geq 0\; l \in t + \mathbb{Z}}} f(n, l)\, q^n y^l.$$

We denote the space of all week Jacobi forms of weight $k$, index $t$ and character (or multiplier system) $v$ by $J_{k,t}(v)$. The space $J_{k,t}(v)$ is finite dimensional (see [5]). The only difference with [5] is that we admit Jacobi forms of half-integral index. One of the main examples of a weak Jacobi form of half-integral weight with trivial

$SL_2$-character is the quotient of the theta-series $\theta(\tau, z)$ by the cube of the Dedekind $\eta$-function

$$\phi_{-1,1/2}(\tau, z) = \vartheta(\tau, z)/\eta(\tau)^3 = (r^{1/2} - r^{-1/2}) + q(\ldots) \in J_{-1,\frac{1}{2}}.$$

*Proof of Theorem 1.3* To prove the theorem we represent $\chi(M, E; \tau, z)$ in terms of the theta-series. Let $c(E)$ be the total Chern class of the vector bundle $E$

$$c(E) = \sum_{i=0}^{r} c_i(E) = \prod_{i=1}^{r}(1 + x_i)$$

where $x_i = 2\pi i \xi_i$ $(1 \leq i \leq r)$ are the formal Chern roots of $E$. We denote by $x'_j = 2\pi i \zeta_j$ $(1 \leq j \leq d)$ the Chern roots of $T_M$. We recall that

$$\mathrm{ch}\left(\bigwedge_t E\right) = \prod_{i=1}^{r}(1 + te^{x_i}), \qquad \mathrm{ch}\left(S_t E\right) = \prod_{i=1}^{r} \frac{1}{1 - te^{x_i}}.$$

According to the last formulae we have

$$\mathrm{ch}\left(\mathbf{E}_{q,y}\right)\mathrm{td}\left(T_M\right) = \prod_{n=1}^{\infty} \prod_{j=1}^{d} \prod_{i=1}^{r} \frac{(1 - q^{n-1}y^{-1}e^{-x_i})(1 - q^n ye^{x_i})}{(1 - q^{n-1}e^{-x'_j})(1 - q^n e^{x'_j})} x'_j.$$

Therefore

$$q^{(r-d)/12} y^{r/2} \exp\left(\frac{1}{2}(c_1(E) - c_1(T_M))\right)\mathrm{ch}\left(\mathbf{E}_{q,y}\right)\mathrm{td}\left(T_M\right) =$$

$$(-1)^{r-d} \prod_{i=1}^{r} \frac{\vartheta(\tau, -z - \xi_i)}{\eta(\tau)} \prod_{j=1}^{d} \frac{\eta(\tau)}{\vartheta(\tau, -\zeta_j)}(2\pi i \zeta_j) \qquad (1.4)$$

where

$$\eta(\tau) = q^{1/24} \prod_{n \geq 1}(1 - q^n) = \sum_{n \in \mathbb{N}} \left(\frac{12}{n}\right) q^{n^2/24}$$

is the Dedekind $\eta$-function which is the $SL_2(\mathbb{Z})$-cusp form of weight $1/2$ with multiplier system $v_\eta$ of order 24. To put the last expression under the integral we obtain the following formula for the modified elliptic genus

$$
\chi(M, E; \tau, z) = \int_M \prod_{i=1}^{r} \exp\left(-4\pi^2 G_2(\tau)\xi_i^2 - \frac{\vartheta_z}{\vartheta}(\tau, z)\xi_i\right) \frac{\vartheta(\tau, z + \xi_i)}{\eta(\tau)} \times
$$

$$
\prod_{j=1}^{d} \exp\left(4\pi^2 G_2(\tau)\zeta_i^2\right) \frac{\eta(\tau)}{\vartheta(\tau, \zeta_j)} (2\pi i \zeta_j).
$$

$$(1.5)$$

We shall calculate the top differential form under the integral using Lemmas 1.4 and 1.5 above.

**Lemma 1.4** *Let $\phi \in J_{k,m}^{mer}(v)$ and $\psi \in J_{k',t}^{mer}(v')$ be Jacobi forms. (We also admit here Jacobi forms meromorphic in $z$.) Then the functions*

$$
\Phi_\phi(\tau, z) = \exp\left(-8\pi^2 m \, G_2(\tau) z^2\right) \phi(\tau, z)
$$

*and*

$$
\Phi_{\phi,\psi}(\tau, z, \xi) = \exp\left(-8\pi^2 m \, G_2(\tau)\xi^2 - \frac{m\psi_z(\tau, z)}{t\phi(\tau, z)}\xi\right) \phi(\tau, z + \xi),
$$

*where $\psi_z = \dfrac{\partial \psi}{\partial z}$, transforms like a modular form in $\tau$ and a Jacobi form of index $m$ in $\tau$ and $z$ respectively. More exactly they satisfy the equations*

$$
\Phi_\phi(\frac{a\tau + b}{c\tau + d}, \frac{z}{c\tau + d}) = v(\gamma)(c\tau + d)^k \Phi(\tau, z)
$$

$$
\Phi_{\phi,\psi}(\frac{a\tau + b}{c\tau + d}, \frac{z}{c\tau + d}, \frac{\xi}{c\tau + d}) = v(\gamma)(c\tau + d)^k e^{2\pi i m \frac{cz^2}{c\tau + d}} \Phi_{\phi,\psi}(\tau, z, \xi)
$$

$$
\Phi_{\phi,\psi}(\tau, z + \lambda\tau + \mu, \xi) = (-1)^{(\lambda + v)} e^{-2\pi i m(\lambda^2\tau + 2\lambda z)} \Phi_\phi(\tau, z, \xi)
$$

*for any $\gamma = \left(\begin{smallmatrix} a & b \\ c & d \end{smallmatrix}\right) \in SL_2(\mathbb{Z})$ and $\lambda, \mu \in \mathbb{Z}$.*

**Proof** From the definition of Jacobi forms follows that

$$
\frac{\phi_z}{\phi}(\gamma < \tau >, \frac{z}{c\tau + d}) = (c\tau + d)\frac{\phi_z}{\phi}(\tau, z) + 4\pi i t c z
$$

$$
\frac{\phi_z}{\phi}(\tau, z + \lambda\tau + \mu) = \frac{\phi_z}{\phi}(\tau, z) - 4\pi i t\lambda.
$$

We recall that $G_2(\tau)$ is a quasi-modular form of weight 2, i.e. it satisfies

$$G_2(\gamma < \tau >) = (c\tau + d)^2 G_2(\tau) - \frac{c(c\tau + d)}{4\pi i}.$$

From this Lemma 1.4 follows.

According to Lemma 1.4 the coefficients $f_n(\tau)$ and $\phi_n(\tau, z)$ of the Taylor expansion of $\Phi_\phi$ and $\Phi_{\phi, \psi}$

$$\Phi_\phi(\tau, z) = \sum_{n \in \mathbb{Z}} f_n(\tau) z^n, \qquad \Phi_{\phi, \psi}(\tau, z, \xi) = \sum_{n \in \mathbb{Z}} \phi_n(\tau, z) \xi^n. \qquad (1.6)$$

are transformed like a $SL_2(\mathbb{Z})$-modular form of weight $k + n$ and like a Jacobi form of weight $k + n$ and index $m$ respectively. One should take only a finite number of terms in the Taylor expansions in order to find the cohomology classes from $H^{2d}(M; \mathbb{C})$ under the integral (1.5). The corresponding coefficients are transformed like Jacobi form of weight zero and index $r/2$ with character $v_\eta^{2(r-d)}$ of $SL_2(\mathbb{Z})$. It gives us a type of the modular behaviour of the modified Witten genus.

To formulate the next lemma we need to recall the definition of the Weierstrass $\wp$-function

$$\wp(\tau, z) = z^{-2} + \sum_{\substack{\omega \in \mathbb{Z}\tau + \mathbb{Z} \\ \omega \neq 0}} \left( (z + \omega)^{-2} - \omega^{-2} \right) \in J_{2,0}^{mer}$$

which is a meromorphic Jacobi form of weight 2 and index 0 with pole of order 2 along $z \in \mathbb{Z}\tau + \mathbb{Z}$.

**Lemma 1.5** *We have*

$$\exp\left( -4\pi^2 G_2(\tau) \xi^2 - \frac{\vartheta_z}{\vartheta}(\tau, z) \xi \right) \frac{\vartheta(\tau, z + \xi)}{\eta(\tau)} = \exp\left( -\sum_{n \geq 2} \wp^{(n-2)}(\tau, z) \frac{\xi^n}{n!} \right)$$

*where* $\wp^{(n)}(\tau, z) = \dfrac{\partial^n}{\partial z^n} \wp(\tau, z)$.

**Proof** Jacobi form $\phi_{-1, \frac{1}{2}}$ (see (1.5)) has the following exponential representation as a Weierstrass $\sigma$-function (see, for example, Appendix I in [19])

$$\phi_{-1, \frac{1}{2}}(\tau, z) = \frac{\vartheta(\tau, z)}{\eta(\tau)^3} = (2\pi i z) \exp\left( \sum_{k \geq 1} \frac{2}{(2k)!} G_{2k}(\tau)(2\pi i z)^{2k} \right) \qquad (1.7)$$

where $G_{2k}(\tau) = -B_{2k}/4k + \sum_{n=1}^{\infty} \sigma_{2k-1}(n) q^n$ is the Eisenstein series of weight $2k$. (For each $\tau \in \mathbb{H}_1$ the product is normally convergent in $z \in \mathbb{C}$.) Since one can obtain the Weierstrass $\wp$-function as the second derivative of the Jacobi theta-series

$\frac{\partial^2}{\partial z^2} \log \vartheta(\tau, z) = -\wp(\tau, z) + 8\pi^2 G_2(\tau)$, the identity (1.7) implies that

$$\wp^{(n-2)}(\tau, z) = \frac{(-1)^n (n-1)!}{z^n} + 2 \sum_{k \geq 2, \, 2k \geq n} (2\pi i z)^{2k} G_{2k}(\tau) \frac{z^{(2k-n)}}{(2k-n)!}.$$

After that the formula of the lemma follows by direct calculation.

Now we can finish the proof of Theorem 1.3. According to the formula of Lemma 1.4 the Chern roots $x_i$ $(1 \leq i \leq r)$ of the vector bundle $E$ and the Chern roots $x'_j$ $(1 \leq j \leq d)$ of the manifold $M$ can be spitted under the integral in (1.6), i.e.,

$$\chi(M, E; \tau, z) = \frac{\vartheta^r}{\eta^{r+2d}} \int_M P(E; \tau, z) \cdot W(M; \tau). \tag{1.8}$$

The first factor depends only on the vector bundle $E$

$$P(E; \tau, z) = \exp\left(-\sum_{n \geq 2} \frac{\wp^{(n-2)}(\tau, z)}{(2\pi i)^n n!} \left(\sum_{i=1}^r x_i^n\right)\right).$$

The second factor is the Witten factor

$$W(M; \tau) = \exp\left(2 \sum_{k \geq 2} \frac{G_{2k}(\tau)}{(2k)!} \left(\sum_{j=1}^d x_j'^{2k}\right)\right)$$

which determines the elliptic genus of the manifold $M$ as a function in one variable $\tau$ (see Sect. 1.1). The derivation of order $(n-2)$ of the Weierstrass $\wp$-function is a meromorphic Jacobi form of weight $n$ and index 0 with pole of order $n$ along $z = 0$. Thus the coefficient of a monomial in $x_i$, $x'_j$ of the total degree $d$ in (1.8) is a meromorphic Jacobi form of weight 0 and index $r/2$ with pole of order not bigger than $(d-r)$. Thus the product $\vartheta(\tau, z)^{d-r} \chi(M, E, \tau, z)$ is holomorphic on $\mathbb{H}_1 \times \mathbb{C}$. This is weak Jacobi form since its Fourier expansion does not contain negative powers of $q$. Theorem 1.3 is proved.

## 1.3  Complex Curves, Surfaces and Threefolds

From (1.8) proved above, it follows that for an arbitrary complex vector bundle over a complex curve $M_1$

$$\chi(M_1, E_r; \tau, z) \equiv 0.$$

Let us calculate $\chi(M, E; \tau, z)$ for surfaces and three-folds. Using (1.8) we have

$$\chi(M_2, E_r; \tau, z) = \frac{1}{2(2\pi i)^2}\left(\sum_{i=1}^{r} x_i^2\right)[M]\frac{\vartheta^r}{\eta^{r+4}}\wp = \frac{1}{24}(c_1(E)^2 - 2c_2(E))\,\phi_{0,1}\left(\frac{\vartheta}{\eta}\right)^{r-2}$$

where

$$\phi_{0,1}(\tau, z) = -\frac{3}{\pi^2}\frac{\wp(\tau, z)\vartheta(\tau, z)^2}{\eta(\tau)^6} = (y + 10 + y^{-1}) + q(10y^{\pm 2} - 88y^{\pm 1} - 132) + \dots$$

$$(1.9)$$

is the unique up to a constant weak Jacobi form of weight 0 and index 1. Therefore $\chi(M_2, T_M; \tau, z) = -\frac{1}{8}\text{sign}(M_2)\phi_{0,1}(\tau, z)$. In particular, $\phi_{0,1}(\tau, z)$ *is the elliptic genus of an Enriques surface.*

The same formulae (1.8) gives us the following result for arbitrary vector bundle over a complex three-fold

$$\chi(M_3, E_r; \tau, z) = -\frac{1}{(2\pi i)^3}\text{ch}_3(E)\frac{\vartheta(\tau, z)^r\wp_z(\tau, z)}{\eta(\tau)^{r+6}}$$

Compare poles we obtain that this is a product of a weak Jacobi form $\phi_{0,\frac{3}{2}}$ of index $3/2$ by the theta-series to the power $(r - 3)$, where

$$\phi_{0,\frac{3}{2}}(\tau, z) = \frac{\vartheta(\tau, 2z)}{\vartheta(\tau, z)} = -\frac{\vartheta(\tau, z)^3\wp_z(\tau, z)}{(2\pi i)^3\eta(\tau)^9} \in J_{0,\frac{3}{2}} \tag{1.10}$$

Thus

$$\chi(M_3, E_r; \tau, z) = \frac{1}{6}(c_1(E)^3 - 3c_1(E)c_2(E) + 3c_3(E))\phi_{0,\frac{3}{2}}(\tau, z)\phi_{0,\frac{1}{2}}(\tau, z)^{r-3}.$$

## 1.4 When $\chi(M, E; \tau, z)$ is Holomorphic?

Theorem 1.3 implies that the modified Witten genus can have a pole only at $y = 1$. It is holomorphic in $\tau$ and $z$ if $\text{rank}(E) \geq \dim_{\mathbb{C}}(M)$. It is also clear from the representation (1.5) that $\chi(M, E; \tau, z)$ is holomorphic for arbitrary $d$ and $r$ if $c_1(E) = 0$ over $\mathbb{R}$.

The examples above show us that the function $\chi(M, E; \tau, z)$ can have a pole at $y = 1$ if rank of the vector bundle is smaller than the dimension of the manifold. Let us calculate the $q^0$-term of the function $q^{(r-d)/12}\chi(M_d, E_r; \tau, z)$ if $r < d$. From the exponential formula (1.7) for $\vartheta$ follows that

$$(2\pi i)^{-1}\frac{\vartheta_z}{\vartheta}(\tau, z) = \left(\frac{1}{2} - \frac{1}{1 - y}\right) - \sum_{n=1}^{\infty}\sum_{k=1}^{\infty}\frac{2}{(2k - 1)!}\sigma_{2k-1}(n)(2\pi i z)^{2k-1}q^n.$$

Thus the $q^0$-term is equal to

$$y^{\frac{r}{2}} \int_M e^{(p_1(E)-p_1(T_M))G_2(\tau)} e^{\frac{c_1(E)}{1-y}} \operatorname{ch}\Big( \bigwedge_{-y^{-1}} E^* \Big) \hat{A}(T_M).$$

Writing down the polynomial $\bigwedge_{-y^{-1}} E^*$ in powers of $(y-1)$

$$\bigwedge_{-y^{-1}} E^* = y^{-r} \sum_{l=0}^{r} (y-1)^{r-l} \sum_{i_1 < \cdots < i_l} (1 - e^{-x_{i_1}}) \ldots (1 - e^{-x_{i_l}})$$

$$= y^{-r} \sum_{l=0}^{r} (y-1)^{r-l} \sum_{p=0}^{l} C_{r-p}^{l-p} \operatorname{ch}(\wedge^p E^*)$$

we obtain that the maximal negative power of $(y-1)$ comes into $q^0$-term with coefficient

$$(y-1)^{-(d-r)} \sum_{l=0}^{r} \frac{(-1)^{d-l}}{(d-l)!} \int_M c_1(E)^{d-l} c_l(E). \tag{1.11}$$

In particular this coefficient does not depend on the $G_2(\tau)$-exponential factor. From the Taylor expansion (1.6) follows that if a weak Jacobi form $\phi$ vanishes for $z = 0$, then is vanishes with order at least four. Thus if $d - 4 \leq r < d$ and (1.11) vanishes, then the weak Jacobi form $\chi(M_d, E_r; \tau, z)(\vartheta/\eta)^{d-r}$ vanishes. Therefore the Jacobi form $\chi(M_d, E_r; \tau, z)$ is holomorphic.

We recall the definition of the twisted $\hat{A}$-genus:

$$\hat{A}(M, W) = \int_M \operatorname{ch}(W \otimes K_M^{1/2}) \operatorname{td}(T_M) = \int_M \operatorname{ch}(W) \hat{A}(T_M). \tag{1.12}$$

Let us assume that $\chi(M_d, E_r; \tau, z)$ is holomorphic and let $p_1(M) = p_1(E)$. Then using the formula for $\vartheta_z/\vartheta$ and the second expression for $\wedge_{-y^{-1}} E^*$ above we can calculate the first Fourier coefficients of the modified Witten genus

$$\chi(M_d, E_r; \tau, z) = q^{\frac{r-d}{12}} y^{-\frac{r}{2}} \Big( y^r \int_M \hat{A}(T_M) + y^{r-1} \int_M (-\operatorname{ch}(E^*) - c_1(E)) \hat{A}(T_M)$$

$$+ y^{r-2} \int_M \Big( -c_1(E) + \frac{c_1(E)^2}{2} + c_1(E) \operatorname{ch}(E^*) + \operatorname{ch}(\wedge^2 E^*) \Big) \hat{A}(T_M) + \ldots \Big).$$

In particular, if $d$ is even, then the first two Fourier coefficients are equal to $\hat{A}$-genera $\hat{A}(M)$ and $\hat{A}(M, E)$:

$$\chi(M_{2k}, E_r; \tau, z) = q^{\frac{r-2k}{12}} \big( y^{\frac{r}{2}} \hat{A}(M) - y^{\frac{r}{2}-1} \hat{A}(M, E_r) + \ldots \big) \quad \text{(for arbitrary } c_1(E))$$

$$\tag{1.13}$$

If $c_1(E) = 0$ over $\mathbb{R}$ then

$$q^{\frac{d-r}{12}} \chi(M_d, E_r; \tau, z) \equiv \sum_{m=0}^{r} (-1)^m y^{\frac{r}{2}-m} \hat{A}(M, \wedge^m E^*) \mod q. \tag{1.14}$$

## 2  The Graded Algebra $J_{0,*/2}$ Over $\mathbb{Z}$

If $c_1(E) = 0$, $p_1(E) = p_1(M)$ and $c_1(M) \equiv 0 \mod 2$ in $H^2(M, \mathbb{Z})$, then Fourier coefficients of the Jacobi form $\chi(M, E; \tau, z)$ are integral numbers. They coincide with indices of twisted Dirac operators (see Remark 1.2). In particular the elliptic genus of a Calabi–Yau manifold is a weak Jacobi form of weight 0 with integral Fourier coefficients. In this section we study the structure of the graded ring of all weak Jacobi forms of weight zero and integral or half-integral with integral Fourier coefficients.

Firstly we show how are related weak Jacobi forms of integral and half-integral indices.

**Lemma 2.1**  *Let m be integral, then*

$$J^w_{2k,m+\frac{1}{2}} = \phi_{0,\frac{3}{2}} \cdot J^w_{2k,m-1}, \qquad J^w_{2k+1,m+\frac{1}{2}} = \phi_{-1,\frac{1}{2}} \cdot J^w_{2k+2,m}.$$

*where $\phi_{0,\frac{3}{2}}(\tau, z)$ and $\phi_{-1,\frac{1}{2}}$ were defined in (1.7) and (1.10).*

**Proof**  This is follows from the fact that $\mathrm{div}(\phi_{2k,m+\frac{1}{2}}) \supset \{z \equiv \frac{1}{2}, \frac{\tau}{2}, \frac{\tau+1}{2}$ $\mod \mathbb{Z}\tau + \mathbb{Z}\}$ and $\mathrm{div}(\phi_{2k+1,m+\frac{1}{2}}) \supset \{\mathbb{Z}\tau + \mathbb{Z}\}$.

Let us denote by $J^{\mathbb{Z}}_{0,t}$ the $\mathbb{Z}$-module of all weak Jacobi forms of weight 0 and index $t$ with integral Fourier coefficients. We consider the rings

$$J^{\mathbb{Z}}_{0,*} = \bigoplus_{m \in \mathbb{Z}_{\geq 0}} J^{\mathbb{Z}}_{0,m}, \qquad J^{\mathbb{Z}}_{0,*/2} = \bigoplus_{t \in \frac{1}{2}\mathbb{Z}_{\geq 0}} J^{\mathbb{Z}}_{0,t}.$$

A weak Jacobi form does not contain negative powers of $q$ in its Fourier expansion. We denote by

$$J^{\mathbb{Z}}_{0,*}(q) = \{\phi \in J^{\mathbb{Z}}_{0,*} \mid \phi(\tau, z) = \sum_{n \geq 1, l \in \mathbb{Z}} a(n, l) q^n y^l\}$$

the ideal of all weak Jacobi forms without $q^0$-term.

**Lemma 2.2** *The ideal $J_{0,*}^{\mathbb{Z}}(q)$ is principal. It is generated by a weak Jacobi form of weight 0 and index 6*

$$\xi_{0,6}(\tau, z) = \Delta(\tau)\phi_{-1,\frac{1}{2}}(\tau, z)^{12} = \frac{\vartheta(\tau, z)^{12}}{\eta(\tau)^{12}} = q(r^{\frac{1}{2}} - r^{-\frac{1}{2}})^{12} + q^2(\ldots).$$

*Proof* Let $\phi_{0,m} \in J_{0,*}^{\mathbb{Z}}(q)$. According to Lemma 2.1 we can assume that $m$ is integral. Let us consider the Taylor expansion of type (1.6) of $\phi$

$$\exp(-8\pi^2 m G_2(\tau)z^2)\phi(\tau, z) = \sum_{n \in \mathbb{Z}_{\geq 0}} f_{2n}(\tau)z^{2n}$$

where $f_k(\tau)$ is a $SL_2(\mathbb{Z})$-modular form of weight $k$. The $f_{2n}(\tau)$ is a cusp form since $\phi(\tau, z)$ does not contain a $q^0$-term. Thus we have $n \geq 6$ in the last summation and $\phi(\tau, z)$ has zero of order at least 12 along $z = 0$.

**Corollary 2.3** *A weak Jacobi form of weight 0 is uniquely defined by its $q^0$-term if its index is less than 6 or equal to $\frac{13}{2}$.*

We shall see that the graded ring $J_{0,*}^{\mathbb{Z}}$ is generated by four Jacobi form with $q^0$-terms of type $y + c + y^{-1}$. These are the Jacobi forms $\phi_{0,1}$ and $\phi_{0,3}$ (see (1.9)–(1.10))

$$\phi_{0,1}(\tau, z) = (y + 10 + y^{-1}) + q(\ldots),$$

$$\phi_{0,3}(\tau, z) = \left(\frac{\vartheta(\tau, 2z)}{\vartheta(\tau, z)}\right)^2 = (y^{\pm 1} + 2) + 2q(-y^{\pm 3} - y^{\pm 2} + y^{\pm 1} + 2) + q^2(\ldots)$$

$$(2.1a)$$

which are related to the elliptic genus of Calabi–Yau manifolds of dimension 2 and 3 and Jacobi forms

$$\phi_{0,2}(\tau, z) = \frac{\vartheta(\tau, 2z)^5 + \vartheta(\tau, 4z)\vartheta(\tau, z)^4}{\vartheta(\tau, z)^3\vartheta(\tau, 2z)\vartheta(\tau, 3z)} = (y^{\pm 1} + 4) + q(y^{\pm 3} - 8y^{\pm 2} - y^{\pm 1} + 16) + \ldots,$$

$$\phi_{0,4}(\tau, z) = \frac{\vartheta(\tau, 3z)}{\vartheta(\tau, z)} = (y^{\pm 1} + 1) - q(y^{\pm 4} + y^{\pm 3} - y^{\pm 1} - 2) + q^2(\ldots)$$

$$(2.1b)$$

which were used in [13] for construction of Borcherds products. (We note that the coefficients of any Jacobi form of even weight and integral index satisfy $a(n, l) = a(n, -l)$.) One can write down the first three functions as symmetric polynomials in the quotients $\vartheta_{ab}(\tau, z)/\vartheta_{ab}(\tau, 0)$ the Jacobi theta-series of level 2. (We use here the notation for these functions from [16].) To check the next formulae one should compare only $q^0$-terms of corresponding Jacobi form. Let put

$$\xi_{00} = \frac{\vartheta_{00}(\tau, z)}{\vartheta_{00}(\tau, 0)}, \quad \xi_{10} = \frac{\vartheta_{10}(\tau, z)}{\vartheta_{10}(\tau, 0)}, \quad \xi_{01} = \frac{\vartheta_{01}(\tau, z)}{\vartheta_{01}(\tau, 0)}.$$

Then we have

$$\phi_{0,1}(\tau, z) = 4(\xi_{00}^2 + \xi_{10}^2 + \xi_{01}^2), \qquad \phi_{0,\frac{3}{2}}(\tau, z) = 4\xi_{00}\xi_{10}\xi_{01}$$

$$\phi_{0,2}(\tau, z) = 2\big((\xi_{00}\xi_{10})^2 + (\xi_{00}\xi_{01})^2 + (\xi_{10}\xi_{01})^2\big). \tag{2.1c}$$

The basis weak Jacobi forms $\phi_{0,1}$, $\phi_{0,3/2}$, $\phi_{0,2}$ appears also in physics ([3], [4], [6], [20] and many others).

**Theorem 2.4 (see [7])**

1. *The graded rings $J_{0,*/2}^{\mathbb{Z}}$ and $J_{0,*}^{\mathbb{Z}}$ of all weak Jacobi forms of weight $0$ with integral coefficients are finite generated:*

$$J_{0,*/2}^{\mathbb{Z}} = \mathbb{Z}[\phi_{0,1}, \phi_{0,\frac{3}{2}}, \phi_{0,2}, \phi_{0,4}], \quad and \quad J_{0,*}^{\mathbb{Z}} = \mathbb{Z}[\phi_{0,1}, \phi_{0,3}, \phi_{0,2}, \phi_{0,4}].$$

*The Jacobi forms $\phi_{0,1}$, $\phi_{0,\frac{3}{2}}$, $\phi_{0,2}$ are algebraic independent and*

$$4\phi_{0,4} = \phi_{0,1}\phi_{0,\frac{3}{2}}^2 - \phi_{0,2}^2. \tag{2.2}$$

2. *Let $m$ be a positive integer. The module*

$$J_{0,m}^{\mathbb{Z}}/J_{0,m}^{\mathbb{Z}}(q) = \mathbb{Z}[\psi_{0,m}^{(1)}, \dots, \psi_{0,m}^{(m)}]$$

*is a free $\mathbb{Z}$-module of rank $m$. We can chose a basis in it with the following $q^0$-terms*

$$[\psi_{0,m}^{(n)}(\tau, z)]_{q^0} = y^n - n^2 y + (2n^2 - 2) - n^2 y^{-1} + y^{-n} \qquad (2 \le n \le m),$$

$$[\psi_{0,m}^{(1)}]_{q^0} = \frac{1}{(12, m)}\big(my + (12 - 2t) + my^{-1}\big)$$

*where $(12, m)$ is the greatest common divisor of $12$ and $m$*

**Proof** The torsion relation $4\phi_4 = \phi_1\phi_3 - \phi_2^2$ easily follows from Corollary 2.3. The form $\xi_6$ generating the principle ideal $J_{0,*}^{\mathbb{Z}}(q)$ is also a polynom in $\phi_1$, $\phi_{3/2}$ and $\phi_2$. We have

$$\xi_6 = -\phi_1^2\phi_4 + 9\phi_1\phi_2\phi_3 - 8\phi_2^3 - 27\phi_3^2 \tag{2.3}$$

(To prove it one needs to check that the $q$-constant term of the write hand side is zero and to calculate one coefficient with the first power $q$.)

Let us prove that $\phi_1$, $\phi_{3/2}$ and $\phi_2$ are algebraically independent. For this we consider its values at $z = \frac{1}{2}$. First we have

$$\phi_{3/2}(\tau, \frac{1}{2}) \equiv 0, \quad \phi_4(\tau, \frac{1}{2}) \equiv -1, \quad \phi_2(\tau, \frac{1}{2}) \equiv 2.$$

(The first two identities follow from definition and the third one is a corollary of (2.2).) The restriction of

$$\alpha(\tau) := \phi_1(\tau, \frac{1}{2}) = 8 + 2^8 q + 2^{11} q^2 + 11 \cdot 2^{10} q^3 + 3 \cdot 2^{14} q^4 + \dots. \tag{2.4}$$

From (2.3) it follows that

$$\alpha(\tau)^2 - 64 = 2^{12} \frac{\Delta(2\tau)}{\Delta(\tau)}. \tag{2.5}$$

The last function is, up to factor $2^{12}$, the "Hauptmodul" for the congruence subgroup $\Gamma_0(2)$. Since only one from three Jacobi forms of different indices gives us non-constant function for $z = 1/2$ they are algebraically independent.

To prove that they generate the full graduate ring we construct a basis with properties indicated in the second point of the theorem. Let us consider a weak Jacobi form of weight 0 and integral index $m$

$$\phi(\tau, z) = \sum_{n \geq 0, l \in \mathbb{Z}} a(n, l) q^n y^l.$$

It is known that *the norm* $4mn - l^2$ of the index of the Fourier coefficient $a(n, l)$ of a weak Jacobi form is bounded from bellow. More exactly we have $4mn - l^2 > -m^2$ (see [5]). Thus the $q^0$-term $[\phi]_{q^0} = \sum_{-m \leq l \leq m} a(0, l) y^l$ is a polynom in $y^{\pm 1}$ of degree not bigger than $m$. The coefficient of this polynomial satisfy an important relation as it follows from the next lemma proved in [13].

**Lemma 2.5** *For arbitrary nearly holomorphic Jacobi form* $\phi_{0,m} = \sum_{n,l} a(n, l) q^n r^l$ *of weight 0 and integral index $m$ (i.e. $\Delta(\tau)^N \phi_{0,m}$ is weak holomorphic for some $N$) the following identity is valid*

$$m \sum_l a(0, l) - 24m \sum_{n<0,l} \sigma_1(n) a(n, l) - 6 \sum_l l^2 a(0, l) = 0.$$

*Proof of the Lemma (See [13, §2.])* We use the differential operator

$$L_k = 8\pi i \frac{\partial}{\partial \tau} - \frac{\partial^2}{\partial z^2} - \left(\frac{2k-1}{z}\right) \frac{\partial}{\partial z}$$

defined in [5, §3]. If the Jacobi form $\phi_{12,t}(\tau, z) = \Delta(\tau)\phi_{0,t}(\tau, z)$ is weak holomorphic (i.e. when $f(n, l) \neq 0$ implies $n \geq -1$), then $t \sum_l f(0, l) - 24tf(-1, l) - 6 \sum_l l^2 f(0, l)$ is the constant term of $(\mathcal{D}_2 \phi_{12,t})(\tau)$ (see [5, Theorem 3.1]) which is a cusp form of weight 14 for $SL_2(\mathbb{Z})$. Thus it is equal to zero. For arbitrary $\phi_{0,t}$ let us consider

$$f_2(\tau) = \Delta(\tau)^{-1} L_{12}(\Delta(\tau)\phi_{0,t}(\tau, z))|_{z=0}$$

which is a nearly holomorphic $SL_2(\mathbb{Z})$-form of weight 2. Let $\phi_{0,t}(\tau, z) = \sum_{\nu \geq 0} \chi_\nu(\tau) z^\nu$ is a Taylor expansion around $z = 0$ where $\chi_\nu(\tau) = \frac{(2\pi i)^\nu}{\nu!} \sum_n \left( \sum_l l^\nu f(n, l) \right) q^n$. Thus

$$f_2(\tau) = 8\pi i \left( \frac{\Delta'(\tau)}{\Delta(\tau)} - \chi_0'(\tau) \right) - 48\chi_2(\tau)$$

where $(2\pi i)^{-1} \frac{\Delta'(\tau)}{\Delta(\tau)} = E_2(\tau) = 1 - 24 \sum_{n \geq 1} \sigma_1(n) q^n$. It shows that the sum in the right hand side of the identity of the lemma is (up to the constant $(4\pi i)^2$) the constant term of $f_2(\tau)$. But the constant term of any nearly holomorphic modular form of weight 2 is equal to zero. The lemma is proved.

For a weak Jacobi form the relation of the last lemma is reduced to

$$m \sum_l a(0, l) = 6 \sum_l l^2 a(0, l).$$

This implies that the polynom $[\phi]_{q^0}(y^\pm)$ can not be a constant, and if it is "linear", i.e. $[\phi]_{q^0} = ay + b + ay^{-1}$, then $a$ is divisible by $\frac{m}{(m, 12)}$. It follows that $m$ Jacobi forms of type described at the statement 1 of Theorem 2.4 form a basis of the module $J_{0,m}^{\mathbb{Z}} / J_{0,m}^{\mathbb{Z}}(q)$.

Let us construct a basis with this property by induction. For $m = 1, 2, 3, 4, 6, 8, 12$ we set $\psi_m^{(1)} = \phi_m$ where $\phi_1, \ldots, \phi_4$ were defined above and we put

$$\phi_6(\tau, z) = \phi_2 \phi_4 - \phi_3^2 = (y + y^{-1}) + q(\ldots),$$

$$\phi_8(\tau, z) = \phi_2 \phi_6 - \phi_4^2 = (2y - 1 + 2y^{-1}) + q(\ldots),$$

$$\phi_{12}(\tau, z) = \phi_4 \phi_8 - 2\phi_6^2 = (y - 1 + y^{-1}) + q(\ldots).$$

Now we can define weak Jacobi forms $\psi_m^{(1)}$ using the following procedure. Let $m \geq 5$. We set

$$\psi_{m,I} = \widetilde{\psi}_{m-4} \phi_4 + \widetilde{\psi}_{m-2} \phi_2 - 2\widetilde{\psi}_{m-3} \phi_3 = my^{\pm 1} + (12 - 2m) + q(\ldots)$$

where $\tilde{\psi}_m := (12, m)\psi_m^{(1)}$. We can take it as $\psi_m^{(1)}$ if $m$ is coprime with 12. If $m \equiv 0$ mod 2 and $m \geq 6$ we put $\psi_{m,II} = \frac{1}{2}\psi_{m,I}$ which has integral Fourier coefficients. If $m \equiv 0 \mod 3$ and $m \geq 9$ we define

$$\psi_{m,III} = \frac{2}{3}\tilde{\psi}_{m-3}\phi_3 + \frac{1}{3}\tilde{\psi}_{m-6}\phi_6 - \tilde{\psi}_{m-4}\phi_4 = \frac{m}{3}y^{\pm 1} + (4 - \frac{2m}{3}) + q(\ldots).$$

For $m \equiv 0 \mod 4$ and $m > 12$ we take

$$\psi_{m,IV} = \frac{1}{4}[\tilde{\psi}_{m-12}\phi_{12} + \tilde{\psi}_{m-4}\phi_4 - \tilde{\psi}_{m-8}\phi_8] = \frac{m}{4}y^{\pm 1} + (3 - \frac{m}{2}) + q(\ldots).$$

In the remain cases $m \equiv 0 \mod 6$ and $m \equiv 0 \mod 12$ $(m > 12)$ we put

$$\psi_{m,VI} = \frac{1}{2}\psi_{m,III} = \frac{m}{6}y^{\pm 1} + (2 - \frac{m}{3}) + q(\ldots),$$

$$\psi_{m,XII} = \frac{2}{3}\tilde{\psi}_{m-3}\phi_3 - \frac{1}{2}\tilde{\psi}_{m-4}\phi_4 - \frac{1}{6}\tilde{\psi}_{m-6}\phi_6 + \frac{1}{12}\tilde{\psi}_{m-12}\phi_{12} = \frac{m}{12}y^{\pm 1} + \frac{6-m}{6} + q(\ldots).$$

Thus we finished the construction of functions $\psi_m^{(1)}$ if we put $\psi_m^{(1)} := \psi_{m,D}$ for $(m, 12) = D$.

Next we construct $\psi_m^{(n)}$ for $1 < n \leq m$. Due to existence of the function $\xi_6$ such a basis is not unique. We shall construct functions $\psi_m^{(n)}$ in such a way that they will have a good restriction to $z = \frac{1}{2}$. For $n = 2$ we put

$$\psi_m^{(2)} = \tilde{\psi}_{m-3}\phi_3 - \tilde{\psi}_{m-4}\phi_4 - \tilde{\psi}_m \qquad (m \geq 5)$$

and

$$\psi_2^{(2)} = \phi_1^2 - 24\phi_2, \quad \psi_3^{(2)} = \phi_1\phi_2 - 18\phi_3, \quad \psi_4^{(2)} = \phi_1\phi_3 - 16\phi_4.$$

For $m = 3$ we put

$$\psi_m^{(3)} = \psi_{m-3}^{(2)}\phi_3 + 2\psi_m^{(2)} \quad (m \geq 5) \quad \text{and} \quad \psi_m^{(3)} = \psi_{m-1}^{(2)}\phi_1 - 6\psi_m^{(2)} \quad (m = 3 \text{ or } 4).$$

For arbitrary $n \geq 4$ we set

$$\psi_m^{(n)} = \psi_{m-3}^{(n-1)}\phi_3 - 2\psi_m^{(n-1)} - \psi_m^{(n-2)} + (n-1)^2\psi_m^{(2)} \qquad (\text{if } m - 2 \geq n)$$

and

$$\psi_m^{(n)} = \psi_{m-1}^{(n-1)}\phi_1 - 10\psi_m^{(n-1)} - \psi_m^{(n-2)} + (n-1)^2\psi_m^{(2)} \qquad (\text{if } n = m \text{ or } m - 1).$$

The functions $\psi_m^{(n)}$ constructed above have $q$-constant term indicated in 2. Moreover as functions in $\tau$ for $z = \frac{1}{2}$ they are equal to a constant or a constant times $\alpha(\tau)$ since $\phi_3(\tau, \frac{1}{2}) \equiv 0$.

**Corollary 2.6** *For the function constructed above we have*

$$\psi_m^{(1)}(\tau, \frac{1}{2}) = \begin{cases} \frac{12-4m}{(12,m)} & m \equiv 0 \mod 2 \\ \frac{3-m}{2(3,m)} \alpha(\tau), & m \equiv 1 \mod 2. \end{cases}$$

*and if $n \leq m - 2$ we have*

$$\psi_m^{(n)}(\tau, \frac{1}{2}) = \begin{cases} 4n^2 \text{ if } n \text{ is even or } 4n^2 - 4 \text{ if } n \text{ is odd} & m \equiv 0 \mod 2 \\ \frac{n^2}{2}\alpha(\tau) \text{ if } n \text{ is even or } \frac{n^2-1}{2}\alpha(\tau) \text{ if } n \text{ is odd} & m \equiv 1 \mod 2. \end{cases}$$

Now we can finish the proof of Theorem 2.4. Each function of the basis $\psi_m^{(n)}$ of the module $J_{0,m}^{\mathbb{Z}}/J_{0,m}^{\mathbb{Z}}(q)$ constructed above is a polynom in the basic Jacobi forms $\phi_1, \ldots, \phi_4$. Using Lemma 2.2 we see that the ring $J_{0,*}^{\mathbb{Z}}$ is generated by four forms $\phi_1, \ldots, \phi_4$. The corresponding statement for the ring $J_{0,*/2}^{\mathbb{Z}}$ follows now from the first identity of Lemma 2.1. Theorem 2.4 is proved.

Theorems 1.3 and 2.4 give us a description of the type of functions $\chi(M_d, E_r, \tau, z)$ and their $q^0$-terms if $r \geq d$. If $r < d$ the modified Witten genus is not a weak Jacobi form. Let us assume that $r < d$ and that $\chi(M_d, E_r, \tau, z)$ is holomorphic in $z$. Then according the Definition 1.1 and Theorem 1.3

$$\eta(\tau)^{2d-2r}\chi(M_d, E_r, \tau, z) \in J_{d-r,r/2}^w.$$

Thus to obtain a description of MEG for $r < d$ which will be similar to the case $r \geq d$ considered above we need a result about generators of the bigraded ring of all week Jacobi forms with integral coefficients

$$J_{*,*/2}^{w,\mathbb{Z}} = \bigoplus_{k \in \mathbb{Z}, m \in \mathbb{Z}} J_{k,m/2}^{w,\mathbb{Z}}.$$

The main difference between the $\mathbb{Z}$-modules $J_{0,m/2}^{w,\mathbb{Z}}$ and $J_{k,m/2}^{w,\mathbb{Z}}$, which is rather important for the subject, is the fact that for a positive weight there are Jacobi forms whose $q^0$-term is equal to 1. (According Theorem 2.4 or Lemma 2.5 this is impossible in the case of weight 0.)

If $m = 0$, one can construct such a modular form in $\tau$ as a product of the Eisenstein series $E_k(\tau) = 1 - \frac{4k}{B_{2k}} \sum_{n=1}^{\infty} \sigma_{k-1}(n)q^n$ of weight 4 and 6. For $m > 1$ one can define (see [5]) the series of Eisenstein–Jacobi

$$E_{k,m}(\tau, z) = \frac{1}{2} \sum_{\substack{c,d \in \mathbb{Z} \\ (c,d)=1}} \sum_{l \in \mathbb{Z}} (c\tau+d)^{-k} \exp\left(2\pi i\left(l^2\frac{a\tau+b}{c\tau+d} + 2l\frac{z}{c\tau+d} - \frac{cz^2}{c\tau+d}\right)\right).$$

It is known that $E_{k,m} = \sigma_{k-1}(m)^{-1} E_{k,1}|_k T_-(m)$, where $T_-(m)$ is a Hecke operator and $\sigma_{k-1}(m) = \sum_{d|m} d^{k-1}$. More exactly, if

$$E_{k,1}(\tau, z) = 1 + \sum_{n>0, l\in\mathbb{Z}} a(n, l)q^n y^l,$$

one has

$$E_{k,m}(\tau, z) = 1 + \sigma_{k-1}(m)^{-1} \sum_{n>0, l\in\mathbb{Z}} \sum_{d|(n,l,m)} d^{k-1} a(\frac{nm}{d^2}, \frac{l}{d})q^n y^l.$$

In general, the Fourier coefficients of $E_{k,m}$ are not integers. As in the theory of the Eisenstein series for $SL_2(\mathbb{Z})$ there is a formula for the Eisenstein-Jacobi series $E_{4,1}$, $E_{4,2}$ and $E_{4,3}$ in terms of the theta-series for the unimodular lattice

$$\mathbb{E}_8 = \{ \ell \in \mathbb{Z}^8 \cup \mathbb{Z}^8 + (\frac{1}{2}, \ldots, \frac{1}{2}) \mid \sum_{i=1}^{8} l_i \in 2\mathbb{Z} \}.$$

For instance, the following formulae hold

$$E_{4,1}(\tau, z) = \sum_{\ell\in\mathbb{E}_8} q^{\frac{1}{2}(l_1^2+\cdots+l_8^2)} y^{\frac{1}{2}(l_1+\cdots+l_8)} =$$

$$1 + q(y^{\pm 2} + 56y^{\pm 1} + 126) + q^2(126y^{\pm 2} + 576y^{\pm 1} + 756) + \ldots.$$

There are similar formulae for $E_{4,2}$ and $E_{4,3}$ (see [5, §7] for details). It follows that the Fourier coefficients of these Eisenstein–Jacobi series are integers.

Using a weak Jacobi form

$$\phi_{-2,1}(\tau, z) = \phi_{-1,\frac{1}{2}}(\tau, z)^2 = \frac{\vartheta(\tau, z)^2}{\eta(\tau)^6}.$$

one can find some relations between Eisenstein–Jacobi series and the generates of the ring of Jacobi forms of weigh 0. Using the same arguments like in the proof of Lemma 2.2 one gets

$$E_4\phi_{0,1} - E_6\phi_{-2,1} = 12E_{4,1} \qquad E_{4,1}\phi_{0,1} - E_{6,1}\phi_{-2,1} = 12E_{4,2}$$

$$E_6\phi_{0,1} - E_4^2\phi_{-2,1} = 12E_{6,1} \qquad E_{6,1}\phi_{0,1} - E_4(\tau)E_{4,1}\phi_{-2,1} = 12E_{6,2}$$

and

$$E_{4,1}\phi_{0,2} - E_4\phi_{0,3} = 2E_{4,3} \qquad E_{4,2}\phi_{0,1} - E_{4,1}\phi_{0,2} = 6E_{4,3}.$$

The series $E_{4,1}$ and $E_{6,1}$ satisfy also the following identity

$$E_6(\tau)E_{4,1}(\tau, z) - E_4(\tau)E_{6,1}(\tau, z) = 144\phi_{10,1}$$

where $\phi_{10,1} = \Delta(\tau)\phi_{-2,1} = \eta(\tau)^{18}\vartheta(\tau, z)^2$ is the unique (up to a constant) Jacobi cusp form of weight 10 and index 1. Using this formula and above formulae for $E_{6,1}$ and $E_{6,2}$ one can see that the Eisenstein–Jacobi series $E_{6,1}$ and $E_{6,2}$ also have integral Fourier coefficients. Moreover one can see that

$$E_{4,1}(\tau, z) \equiv E_{6,1}(\tau, z) \mod 24, \qquad E_{4,2} \equiv E_{4,1}^2 \mod 12$$

and

$$\phi_{0,1}(\tau, z) - \phi_{-2,1}(\tau, z) \equiv 12(1 + \sum_{n \geq 1} q^{n^2} y^{\pm 2n}) \mod 24.$$

For the pair $(k, m) = (6, 3)$ the Jacobi form

$$E'_{6,3} = \frac{1}{2}(E_{6,1}\phi_{0,2} - E_6\phi_{0,3}) \tag{2.6}$$

has $q^0$-term equal to 1. This is not a series of Eisenstein since one can prove that

$$E'_{6,3} = E_{6,3} + \frac{22}{61}\Delta\phi_{-2,1}^3.$$

Now we construct integral Jacobi forms starting with 1. For $m = 0, 1, 2, 3$ and even $2k \geq 4$ we can take the form $\tilde{E}_{2k,m} = E_4^a E_6^\varepsilon E_{k',m}$ ($\varepsilon = 0$ or 1) which has $q^0$-term equal to 1. Then using the forms $\tilde{\psi}_{0,m}^1$ constructed in Theorem 2.4 we can determine

$$\tilde{E}_{2k}(\tau)\tilde{\psi}_{0,m}^1 - \tilde{E}_{2k,1}\tilde{\psi}_{0,m-1}^1 = (y - 2 + y^{-1}) + q(\ldots).$$

Combining the last form with similar forms $\tilde{E}_{2k,m_0}\tilde{\psi}_{0,m-m_0}^1$ ($m_0 = 0, 1, 2, 3$) one can find a weak Jacobi forms

$$\tilde{E}_{2k,m} = 1 + q(\ldots) \qquad (2k \geq 4, m \geq 0) \tag{2.7}$$

with integral Fourier coefficients. We note that the form $\tilde{E}_{2k,m}$ *is constructed as a polynomial over* $\mathbb{Z}$ *in the Eisenstein series* $E_4$ *et* $E_6$, *the Eisenstein–Jacobi series* $E_{4,1}, E_{4,2}, E_{4,3}, E_{6,1}, E_{6,2}$ *and the Jacobi forms* $E'_{6,3}, \phi_{-2,1}, \phi_{0,1}, \ldots, \phi_{0,4}$. It gives us the following

**Theorem 2.7 (See [8])**

1. *The bigraded ring of all weak Jacobi forms with integral Fourier coefficients has the following generators over* $\mathbb{Z}$

$$J_{*,*/2}^{w,\mathbb{Z}} = \mathbb{Z}[E_4(\tau), E_6(\tau), \Delta(\tau), E_{4,1}, E_{4,2}, E_{4,3}, E_{6,1}, E_{6,2}, E'_{6,3},$$

$$\phi_{0,1}, \phi_{0,2}, \phi_{0,\frac{3}{2}}, \phi_{0,4}, \phi_{-1,\frac{1}{2}}]$$

where $E_{k,m}(\tau)$ are the Eisenstein–Jacobi series, and $E'_{6,3}$ is a Jacobi form defined in (2.6).

2. *Over the ring* $\mathbb{Z}[12^{-1}]$ *we have*

$$J_{*,*/2}^{w,\mathbb{Z}[12^{-1}]} = \mathbb{Z}[12^{-1}]\left[E_4(\tau), E_6(\tau), \phi_{0,1}, \phi_{0,\frac{3}{2}}, \phi_{-1,\frac{1}{2}}\right].$$

**Proof**

1. Using the Jacobi form $\widetilde{E}_{2k,m}$ constructed above and the forms $\psi_{0,m}^{(n)}$ from Theorem 2.4 we can reduce any $\phi \in J_{2k,m}^{w,\mathbb{Z}}$ $(2k \geq 4)$ to a form whose $q^0$-term vanishes. Then dividing this form by $\Delta(\tau)$, we get a weak Jacobi form of a smaller weight. If the weight $2k < 4$ and $2k \neq 0$, then using the same arguments as in the proof of Lemma 2.2 we have

$$J_{2k,m}^{w,\mathbb{Z}} = \phi_{-2,1}^{2-k} J_{4,m+k-2}^{w,\mathbb{Z}}.$$

If the weight is odd we have

$$J_{2k+1,m}^{w,\mathbb{Z}} = \phi_{-1,2} J_{2k+2,m-2}^{\mathbb{Z},f} \qquad \text{where} \quad \phi_{-1,2} = \phi_{-1,\frac{1}{2}}\phi_{0,\frac{3}{2}} = \eta(\tau)^{-3}\vartheta(\tau, 2z).$$

2. To prove the result over $\mathbb{Z}[12^{-1}]$ it is enough to take in mind that

$$24\phi_{0,2} = \phi_{0,1}^2 - E_4\phi_{-2,1}^2$$

$$3 \cdot 144\phi_{0,3} = \phi_{0,1}^3 - 3E_4\phi_{0,1}\phi_{-2,1}^2 + 2E_6\phi_{-2,1}^3$$

and to use the identities for the Eisenstein–Jacobi series mentioned above.

See more details in [17].

# 3   Some Applications

## 3.1   Calabi–Yau Fivefolds and the Euler Number Modulo 24

Let $M_d$ be an almost complex compact manifold if complex dimension $d$ such that $c_1(M_d) = 0$ over $\mathbb{R}$. According to the Riemann–Roch–Hirzebruch formula and Theorem 1.3 for a manifold with trivial the first Chern class $\chi(M, T_M; \tau, z)$ is a weak Jacobi form with integral Fourier coefficients. Its $q^0$-term is equal to the $y$-genus of Hirzebruch of the manifold $M$. More exactly

$$\chi(M, T_M; \tau, z) = y^{d/2} \sum_{p=0}^{d} (-1)^p y^{-p} \chi^p(M) + q(\ldots) \qquad (3.1)$$

where $\chi^p(M) = \chi(M, \wedge^p T_M^*) = (\mathrm{ch}(\wedge^p T_M^*)\mathrm{td}(T_M))[M]$.

The value of any weak Jacobi form from $J_{0,*/2}$ at $z = 0$ is a holomorphic automorphic function of weight 0 for $SL_2(\mathbb{Z})$. Thus it is a constant. According (3.1) the value of the elliptic genus is the Euler number of $M$

$$\chi(M, T_M; \tau, 0) = \sum_{p=0}^{d} (-1)^p \chi^p = e(M) \qquad (c_1(M) = 0 \text{ in } H^2(M, \mathbb{R})).$$

The space $J_{0,\frac{5}{2}}$ of weak Jacobi forms of weight 0 and index $3/2$ has dimension one. Its generator is the form

$$\phi_{0,\frac{5}{2}}(\tau, z) = \phi_{0,\frac{3}{2}}(\tau, z)\phi_{0,1}(\tau, z) = -\frac{3}{\pi^2}\vartheta(\tau, 2z)\vartheta(\tau, z)\wp(\tau, z)\eta(\tau)^{-6}$$

$$= \left[ (y^{\pm\frac{3}{2}} + 11y^{\pm\frac{1}{2}}) + q(-y^{\pm\frac{7}{2}} + \ldots) + q^2(-11y^{\pm\frac{9}{2}} + \ldots) \right]$$

($y^{\pm m}$ means that we have two summands with $y^m$ and $y^{-m}$ respectively). Thus

$$\chi(M_5, T_{M_5}; \tau, z) = \frac{e(M_5)}{24} \cdot \phi_{0,\frac{5}{2}}(\tau, z) = \frac{e(M_5)}{24}(y^{\pm\frac{3}{2}} + 11y^{\pm\frac{1}{2}} + +q(\ldots)).$$

This proves that *the Euler number of any Calabi–Yau 5-fold is divisible by 24 and its cohomological invariants $\chi^P(CY_5)$ satisfy the relations*

$$\chi^1(CY_5) = -\frac{1}{24}e(CY_5), \qquad \chi^2(CY_5) = \frac{11}{24}e(CY_5).$$

In particular, for the Hodge numbers of an arbitrary strict Calabi–Yau five-fold the following equality holds

$$11(h^{1,1} - h^{1,4}) = h^{2,2} - h^{2,3} + 10(h^{2,1} - h^{3,1}).$$

For an arbitrary complex vector bundle $E_r$ of rank $r$ over an almost complex compact manifold $M_5$ of complex dimension 5 the formula (1.8) gives us

$$\chi(M_5, E_r; \tau, z) = \frac{1}{720}\phi_{0,5/2}(\tau, z)\int_{M_5}(c_1^5 - 5c_1^3c_2 + 15c_1^2c_3 - 30c_1c_4 + 30c_5)(E_r).$$

If $0 < r < 5$ the modified Witten genus has a pole of order $(5 - r)$ along $z = 0$ if it is not identically equal to 0.

If $r \geq 5$, $c_1(E_r) = 0$ and $p_1(E_r) = p_1(M_5)$ then the fact that $J_{0,5/2}$ is generated by $\phi_{0,5/2}$ implies

$$\chi(M_5, E_r; \tau, z) = A(M_5, E_r)\phi_{0,\frac{1}{2}}(\tau, z)\phi_{0,\frac{5}{2}}(\tau, z).$$

We obtain

$$\sum_{m=1}^{r-1}(-1)^{m+1}y^{\frac{r}{2}-m}\hat{A}(M_5,\wedge^m E_r) = A(M_5,E_r)(y^{\frac{1}{2}}-y^{-\frac{1}{2}})^{r-5}(y^{\pm\frac{3}{2}}+11y^{\pm\frac{1}{2}})$$

In particular ($r \geq 5$, $c_1(E_r) = 0$ and $p_1(E_r) = p_1(M_5)$),

$$A(M_5,\wedge^2 E_r) = (r-16)A(M_5,E_r), \quad A(M_5,\wedge^3 E_r) = \frac{(r-6)(r-27)}{2}A(M_5,E_r),$$

$$A(M_5,\wedge^4 E_r) = \frac{(r-8)(r^2-43r+192)}{6}A(M_5,E_r).$$

The result about the Euler number of a Calabi–Yau five-folds obtained above is a particular case of the following general fact

**Proposition 3.1** *Let $M_d$ be an almost complex manifold of complex dimension d such that $c_1(M) = 0$ in $H^2(M,\mathbb{R})$. Then the Euler number $e(M_d)$ satisfies the following congruence modulo 24*

$$d \cdot e(M_d) \equiv 0 \mod 24.$$

*If $c_1(M) = 0$ in $H^2(M,\mathbb{Z})$, then we have a more strong congruence*

$$e(M) \equiv 0 \mod 16 \qquad if \; d \equiv 2 \mod 8.$$

For $d = 3$ it was proved by F. Hirzebruch in 1960. For a hyper-Kähler compact manifold it was proved by S. Salamon in [24]. For a Calabi–Yau four-fold it was proved in [25] where it was shown that the number $\frac{e(M)}{24}$ is obstruction to cancelling the tadpole. We obtain this result for an arbitrary almost complex manifold as a corollary of a more general result about $\hat{A}$-genera (see Proposition 3.2 bellow.)

*Remark* After my talk on elliptic genera at a seminar of MPI in Bonn in April 1997 Professor F. Hirzebruch informed me that the result of Proposition 3.1 was known for him (non-published). Using some natural examples he also proved that this properties of divisibility of the Euler number modulo 24 is strict (see [18]).

## 3.2 Exact Formulae for MWG for Dimension d = 6, 7, 9 and Rank r < d

The $J_{0,1}$, $J_{0,3/2}$, $J_{0,5/2}$ are the only spaces of type $J_{0,*/2}$ having dimension 1. Nevertheless there are some other cases except manifolds of dimension 2, 3 and 5 when the modified Witten genus is uniquely defined up to a constant. *Let us consider a vector bundle $E_r$ over $M_d$ with $r < d$ such that $\chi(M_d,E_r;\tau,z)$ is holomorphic.*

(For example, one can consider $E_r$ with $c_1(E) = 0$.) According to Theorem 1.3 we have

$$\chi(M_d, E_r; \tau, z) \in \eta(\tau)^{2r-2d} J^w_{d-r, \frac{r}{2}} \qquad (r \le d). \qquad (3.2)$$

It is easy to find the list of possible $(d, r)$ when $\dim J_{d-r, \frac{r}{2}} = 1$. For such $(d, r)$ the MWG is defined uniquely up to a constant factor:

$r = 2, \ d = 4; \quad r = 3, \ d = 4, \ 7, \ 9, \ 11, \ 13, \ 17; \quad r = 4, \ d = 7, \ 9, \ 11, \ 13, \ 17;$

$r = 5, \ d = 7; \quad r = 6, \ d = 7.$

If $d = 4$, $1 \le r < 4$ and $\chi(M_4, E_r; \tau, z)$ is holomorphic, then this form coincides with the MEG of the trivial vector bundle of the same rank over $M_4$

$$\chi(M_4, E_r; \tau, z) = \hat{A}(M_4) j(\tau)^{1/3} \left( \frac{\vartheta(\tau, z)}{\eta(\tau)} \right)^r$$

where $j(\tau) = E_4(\tau)^3 / \Delta_{12}(\tau)$. Note that the Witten genus of a complex manifold of an odd dimension is always equal to 0. Using the results about Jacobi modular forms we obtain the following formulae for $(d, r)$ from the list above (recall that we assume that $c_1(E_r) = 0$ and $p_1(E_r) = p_1(M_d)$)

$$\chi(M_{2k+1}, E_3; \tau, z) = \hat{A}(M_{2k+1}, E_3) E_{2k-2}(\tau) \frac{\vartheta(\tau, 2z)}{\vartheta(\tau, z)} \eta(\tau)^{-4k+4} \qquad k = 3, 4, 5, 6, 8$$

$$\chi(M_{2k+1}, E_4; \tau, z) = \hat{A}(M_{2k+1}, E_4) E_{2k-2}(\tau) \vartheta(\tau, 2z) \eta(\tau)^{-4k+3} \qquad k = 3, 4, 5, 6, 8$$

$$\chi(M_7, E_5; \tau, z) = \hat{A}(M_7, E_5) j(\tau)^{1/3} \vartheta(\tau, z) \vartheta(\tau, 2z) \eta(\tau)^{-2}$$

$$\chi(M_7, E_6; \tau, z) = \hat{A}(M_7, E_6) j(\tau)^{1/3} \vartheta(\tau, z)^2 \vartheta(\tau, 2z) \eta(\tau)^{-3}$$

Thus if $d = 7$ then the modified Witten genus $\chi(M_7, E_r; \tau, z)$ for all $r < 7$ is defined uniquely up to a constant if it is holomorphic. The last identity gives us the following relation between $\hat{A}$-genus

$$\hat{A}(M_7, \wedge^2 E_6) = 2\hat{A}(M_7, E_6).$$

There are some number of interesting cases when holomorphic form $\chi(M, E; \tau, z)$ is defined by two constants. Let us analyse the case of complex dimension 6 which is related to the $\mathbb{Z}$-module $J^{w, \mathbb{Z}}_{4, 1}$ generated by the forms $E_6(\tau) \phi_{-2, 1}(\tau, z)$ and $E_4(\tau, z)$. We again assume that $p_1(E_r) = p_1(M_6)$ and that $\chi(M_6, E_r; \tau, z)$ is holomorphic. For $1 \le r < 6$ the following formula holds

$$\chi(M_6, E_r; \tau, z) = \left[ \hat{A}(M_6) E_6(\tau) \phi_{-2, 1}(\tau, z) + (r\hat{A}(M_6) - \hat{A}(M_6, E_r)) E_{4, 1}(\tau, z) \right] \times$$

$$\vartheta(\tau, z)^{r-2} \eta(\tau)^{-r-6} \qquad (3.3)$$

In particular, if additionally assume $c_1(E_r) = 0$ then the following identity follow

$$\hat{A}(M_6, \wedge^2 E_4) = 2\hat{A}(M_6, E_4) - 2\hat{A}(M_6)$$

$$\hat{A}(M_6, \wedge^2 E_5) = 3\hat{A}(M_6, E_5) - 5\hat{A}(M_6).$$

If $\hat{A}(M_6) = 0$ (we recall that $\hat{A}(M) = 0$ for any connected spin manifold with a nontrivial $S^1$ action, see [2]), then the (stable) modified Witten genus of $M_6$ can be equal to the holomorphic Eisenstein–Jacobi series $E_{4,1}$. If $\hat{A}(M_6) = 0$, then

$$\chi(M_6, E_2; \tau, z) = -\hat{A}(M_6, E_2) E_{4,1}(\tau, z) \eta(\tau)^{-8}.$$

We can give some other examples of vector bundles of this 'Eisenstein type'. If $\hat{A}(M_8) = \hat{A}(M_8, E_4) = 0$, then

$$\chi(M_8, E_4; \tau, z) = -\hat{A}(M_8, E_4) E_{4,2}(\tau, z) \eta(\tau)^{-8}.$$

If $\hat{A}(M_{10}) = \hat{A}(M_{10}, E_6) = \hat{A}(M_{10}, \wedge^2 E_6) = 0$, then

$$\chi(M_{10}, E_6; \tau, z) = -\hat{A}(M_{10}, \wedge^3 E_6) E_{4,3}(\tau, z) \eta(\tau)^{-8}.$$

(In the last two example we also assume that $c_1(E_r) = 0$.)
    There are the formulae similar to (3.3) for

$$r = 2, \ d = 8, \ 10; \quad r = 3, \ d = 8, \ 10; \quad r = 5, \ 6, \ d = 9, \ 11, \ 13; \quad r = 7, \ 8, \ d = 9.$$

Let us consider the case of dimension $d = 9$ and $5 \le r < 9$ when $c_1(E) = 0$ and $p_1(E_r) = p_1(M_9)$. Then the MWG is given by the formula

$$\chi(M_9, E_r; \tau, z) = \vartheta(\tau, 2z)\vartheta(\tau, z)^{r-6}\eta(\tau)^{-r-3} \times$$

$$\left[\hat{A}(M_9, E_r) E_6(\tau)\phi_{-2,1}(\tau, z) + \left((r-4)\hat{A}(M_9, E_r) - \hat{A}(M_9, \wedge^2 E_r)\right) E_{4,1}(\tau, z)\right].$$

In particular, we obtain

$$\hat{A}(M_9, \wedge^3 E_7) = \hat{A}(M_9, \wedge^2 E_7) - \hat{A}(M_9, E_7),$$

$$\hat{A}(M_9, \wedge^3 E_8) = 2\hat{A}(M_9, \wedge^2 E_8) - \hat{A}(M_9, E_8).$$

## 3.3 Relations Between $\hat{A}$-Genera

In Sects. 3.1 and 3.2 we analyse the $q^{min}$-part of the Fourier expansion of MEG. Let us calculate the next term in its Fourier expansion, i.e. coefficients corresponding to the power $q^{min+1}$. We assume bellow that $c_1(E) = 0$ over $\mathbb{R}$ and $p_1(M) = p_1(E)$. Under these conditions

$$\chi(M_d, E_r; \tau, z) = q^{(r-d)/12} y^{r/2} \int_M \mathrm{ch}(\mathbf{E}_{q,y}) \, \mathrm{ch}(K_M^{1/2}) \mathrm{td}(T_M)$$

where

$$\mathbf{E}_{q,y} = \sum_{m=0}^{r} (-1)^m y^{-m} \wedge^m E^* +$$

$$q \sum_{n=-1}^{r+1} (-1)^n y^{-n} \left( \wedge^{n-1} E^* \otimes E^* \oplus \wedge^n E^* \otimes (T_M \oplus T_M^*) \oplus \wedge^{n+1} E^* \otimes E \right) + q^2(\dots).$$

Thus we have

$$q^{\frac{d-r}{12}} \chi(M_d, E_r; \tau, z) = q^{\frac{r-d}{12}} \hat{A}_y(M, E) + q^1 \sum_{n=-1}^{r+1} (-1)^n y^{\frac{r}{2}-n} \alpha_n(M_d, E_r) + \dots$$

$$(3.4)$$

where we put

$$\hat{A}_y(M, E) = \sum_{m=0}^{r} (-1)^m y^{\frac{r}{2}-m} \hat{A}(M, \wedge^m E^*)$$

and

$$\alpha_n(M_d, E_r) = \hat{A}(M, \wedge^{n-1} E^* \otimes E^*) + \hat{A}(M, \wedge^n E^* \otimes (T_M \oplus T_M^*)) + \hat{A}(M, \wedge^{n+1} E^* \otimes E).$$

(We assume that $\wedge^{-2} E = \wedge^{-1} E = 0$.)

**Proposition 3.2** Let $M = M_d$, $E = E_r$, $p_1(M) = p_1(E)$ and $c_1(E) = 0$.

1. If $r = d$, then we have the following identity

$$\frac{d}{2} \sum_{m=0}^{d} (-1)^m \hat{A}(M, \wedge^m E^*) = 6 \sum_{m=0}^{r} (-1)^m (\frac{r}{2} - m)^2 \hat{A}(M, \wedge^m E^*). \quad (3.5a)$$

2. If $r > d$, then the polynom $\hat{A}_y(M, E)$ has zero of order at least $(r - d)$ at $y = 1$.

3. If $0 < d - r < 4$, then $\hat{A}_y(M, E)$ has zero of order at least $4 - d + r$ at $y = 1$.
4. Let $0 < d - r \le 12$. We put $v = \frac{1}{2}(14 - d + r)$, if $d - r$ is even, and $v = \frac{1}{2}(13 - d + r)$ if $d - r$ is odd. Then the following two relations hold

$$\sum_{k=0}^{v} \frac{(-24)^k}{r^k(2k)!(v-k)!}\left[(24k - 20v - 4)\sum_{m=0}^{r}(-1)^m(\frac{r}{2} - m)^{2k}\hat{A}(M, \wedge^m E^*)+ \right.$$

$$\left. \sum_{n=-1}^{r+1}(-1)^n(\frac{r}{2} - n)^{2k}\alpha_n(M, E)\right] = 0$$

*if $r - d$ is even and*

$$\sum_{k=0}^{v} \frac{(-24)^k}{r^k(2k+1)!(v-k)!}\left[(24k - 20v - 2)\sum_{m=0}^{r}(\frac{r}{2} - m)^{2k+1}\hat{A}(M, \wedge^m E^*)+ \right.$$

$$\left. \sum_{n=-1}^{r+1}(\frac{r}{2} - n)^{2k+1}\alpha_n(M, E)\right] = 0$$

*if $r - d$ is odd.*

*Example 3.3* Let $d = 14$, $r = 2$, $p_1(E) = p_1(M)$ and $c_1(E) = 0$. Then the constraint of 4 has a form

$$68\hat{A}(M, E) - 58\hat{A}(M) - 10\hat{A}(M, S^2E) = 20\hat{A}(M, T_M) + 2\hat{A}(M, T_M \otimes E).$$

**Proof** The first and the second statements are direct corollaries of Theorem 1.3 and Lemma 2.5. If $0 < d - r < 4$ and $c_1(E) = 0$, then $\chi(M, E, \tau, z)$ is holomorphic and $(\vartheta(\tau, z)\eta(\tau)^{-1})^{d-r}\chi(M, E, \tau, z)$ is a weak Jacobi form of weight 0 with zero along $z = 0$. Thus it has zero of order least 4. It proves 3.

To prove the forth statement we can use the same method as in Lemma 2.5. Let us consider

$$\tilde{\chi}(\tau, z) = \eta(\tau)^{2d-2r-24}\chi(M, E, \tau, z) \in J^{nh}_{d-r-12, r/2}.$$

This is a nearly holomorphic Jacobi form of non-positive weight $d - r - 12$. The minimal possible power of $q$ in its Fourier expansion is $-1$. The coefficient of the Taylor expansion of $exp(-4\pi^2 r G_2(\tau)z^2)\tilde{\chi}(\tau, z)$ corresponding to $z^{14-v}$ is a modular form in $\tau$ of weight 2. If $d - r$ is even, this modular form is equal to

$$\sum_{k=0}^{v}(-4\pi^2 r G_2(\tau))^{\frac{14-v}{2}-k}\frac{1}{(2k)!(v-k)!}\frac{\partial^{2k}\tilde{\chi}}{\partial z^{2k}}(-4\pi^2 r G_2(\tau))^{v-k} \in M_2^{qh}(SL_2(\mathbb{Z})).$$

To prove the relations one needs to calculate the constant term of its Fourier expansion which is zero. The Fourier expansion of $\tilde{\chi}(\tau, z)$ starts with $q^{-1}$. Thus to find a constant term of the modular form above we can replace $G_2(\tau)^n$ with its linear term. After some calculation one obtains the constrains written in 4.

Let us consider the case when $M$ is almost complex of even complex dimension $d$, $E = T_M$ and $c_1(T_M) = 0$ over $\mathbb{R}$ (for example, if $M$ is a Calabi–Yau manifold), then the first relation of the last proposition can be written as

$$24 \sum_{p=0}^{d} (-1)^p \chi^p(M)(\frac{d}{2} - p)^2 = d \cdot e(M). \tag{3.5b}$$

This identity was rediscovered several times in the mathematical and physical papers: in [22] it was obtained using the second derivative of the Hirzebruch $y$-genus (see also [24]); in physics it was found as a corollary of the sum rule for the charges in $N = 2$ super-conformal field theories (see [1]). Lemma 2.5 provides an automorphic proof of (3.5a) and (3.5b). As we have seen above this approach gives us a more general statement since we can apply Lemma 2.5 in the case of a vector bundle of rank $r < d$ or $r > d$ to get a constraint for its cohomological invariants. If $d > r$ and $d - r \equiv 0 \mod 12$, then Lemma 2.5 gives us a relation immediately. If $d - r \not\equiv 0 \mod 12$, then we can use the same arguments as above. In general, the coefficients of high $q$-powers of $\mathbf{E}_{q,y}$ will be involved but in any case there is a relation between coefficients of MWG.

The statement 2 of Proposition 3.2 is a generalisation of the statement 1 in the case of $d \geq r$.

*Example 3.4 Vector bundle over a manifold of complex dimension 4.* Let us consider the case $d = 4$, $r \geq 4$. Let $p_1(E_4) = p_1(M_4)$. We do not assume no addition condition on $c_1(E_4)$ in this example. According to Theorem 1.3 and (1.12) ($\chi(M_4, E_r; \tau, z)$ is holomorphic in $z$ is $r \geq d$) we have the following formula for MWG

$$\chi(M_4, E_r; \tau, z) = \left[\hat{A}(M_4)\psi_{0,2}^{(2)}(\tau, z) + \left(r\hat{A}(M_4) - \hat{A}(M_4, E_r)\right)\phi_{0,2}(\tau, z)\right]\left(\frac{\vartheta(\tau, z)}{\eta(\tau)}\right)^{r-4} \tag{3.6}$$

where

$$\psi_{0,2}^{(2)}(\tau, z) = \phi_{0,1}^2(\tau, z) - 24\phi_{0,2}(\tau, z) = y^{\pm 2} - 4y^{\pm 1} + 6 + q(\ldots).$$

Thus one can determines all $\hat{A}(M_4, \wedge^m E_4)$ using only two $\hat{A}$-genera $\hat{A}(M_4)$ and $\hat{A}(M_4, E_4)$. In particular, one has

$$\hat{A}(M_4, \wedge^2 E_r) = (r - 8)\hat{A}(M_4, E_r) + \frac{r^2 - 7r + 56}{2}\hat{A}(M_4).$$

The Jacobi form $\phi_{0,2}(\tau, z)$ is one of the generators of the graded ring $J_{0,*}$ and the formula for the MWG above shows us that

$$\phi_{0,2}(\tau, z) = \chi(M_4, T_M; \tau, z) \qquad \text{if } \hat{A}(M) = 0, \ \hat{A}(M_4, T_M) = -1.$$

We recall that $\hat{A}(M) = 0$ for any connected spin manifold with a nontrivial $S^1$ action. If $A(M) = 0$ then the condition $\hat{A}(M_4, T_M) = -1$ is equivalent to the condition on the Euler number $e(M_4) = -6$ or on the signature $\text{sign}(M_4) = -2$ of the manifold.

## 3.4   Vector Bundle of Rank 2

In (3.3) we calculated the MEG of vector bundles of rank 2 over manifolds of dimension $d = 6$. For $d = 8, 10$ we have a formula similar to (3.3)

$$\chi(M_d, E_2) = \left[ \hat{A}(M_d) E_d(\tau) \phi_{-2,1} + (2\hat{A}(M_d) - \hat{A}(M_d, E_2)) E_{d-2,1} \right] \eta(\tau)^{-8} \quad (d = 6, 8, 10)$$

where we assume that $p_1(E_2) = p_1(M_d)$ and that the MWG is holomorphic. (The first Chern class $c_1(E)$ is arbitrary.)

Let us consider the case $d = 12$ and $d = 14$ in which we add to $p_1(E_2) = p_1(M_d)$ the condition that $c_1(E_2) = 0$. The $\mathbb{Z}$-module $J_{d-2,1}^{w,\mathbb{Z}}$ is of rank 3. According to Theorem 2.7 one can find a basis in the module of integral Jacobi forms and one finds that

$$\chi(M_{12}, E_2; \tau, z) \in \mathbb{Q} < E_4^3(\tau)\phi_{-2,1}, \ E_6(\tau)E_{4,1}, \ \Delta(\tau)\phi_{-2,1} > \eta(\tau)^{-20}$$

$$\chi(M_{14}, E_2; \tau, z) \in \mathbb{Q} < E_{14}(\tau)\phi_{-2,1}, \ E_8(\tau)E_{4,1}, \ \Delta(\tau)\phi_{0,1} > \Delta(\tau)^{-1}.$$

For $d = 14$ we obtain the following formula

$$\chi(M, E; \tau, z) = \hat{A}(M)E_{14}(\tau)\phi_{-2,1}\Delta(\tau)^{-1} +$$

$$\left(2\hat{A}(M) - \hat{A}(M, E)\right)E_8(\tau)E_{4,1}\Delta(\tau)^{-1} + c\phi_{0,1} \quad (d = 14, \ r = 2) \tag{3.7}$$

where

$$c = 56\hat{A}(M, E) - 119\hat{A}(M) + 2\hat{A}(M, S^2E) + 2\hat{A}(M, T_M).$$

The following particular examples are specially interesting. Let $\hat{A}(M) = 0$. Then

$$\chi(M, E; \tau, z) = -\hat{A}(M, E)\phi_{0,1}^{(-1)} + \hat{A}(M, S^2E) + 2\hat{A}(M, T_M)\phi_{0,1}$$

where the Jacobi form

$$\phi_{0,1}^{(-1)}(\tau, z) = q^{-1} + 70 + y^{\pm 2} + q(\ldots)$$

was used in [11] for construction of the Igusa modular forms $\Delta_{35}$ of weight 35 with respect to $Sp_4(\mathbb{Z})$. The Fourier coefficients of $\phi_{0,1}^{(-1)}$ and of $\phi_{0,1}$ are multiplicities of special Lorentzian Kac–Moody Lie algebras of Borcherds type (see [12]). If we suppose that

$$\hat{A}(M) = 0, \; \hat{A}(M, E) = -2, \; \hat{A}(M, S^2 E) + 2\hat{A}(M, T_M) = 0 \qquad (3.8)$$

(according to Example 3.3 the last relation in (3.8) is equivalent to $\hat{A}(M, T_M \otimes E) = -68$.) Then *the second quantised elliptic genus of the $(0, 2)$-symmetric non-linear sigma-model related to the vector bundle $E_2 \to M_{14}$ which satisfies (3.8) is equal, up to a simple factor, to the $-2$-power of the Igusa modular form of weight 35 and it is related to the automorphic correction of the simplest hyperbolic Kac–Moody algebra.* (See [11] and [23] for more details).

As we mentioned in the introduction the elliptic genus of a Calabi–Yau manifold is a partition function of the corresponding supersymmetric non-linear sigma-model. The following corollary follows immediately from Theorem 1.3.

**Corollary 3.5** *If two MEG are equal $\chi(M_d, E_r; \tau, z) = \chi(M_{d'}, E_{r'}; \tau, z)$ then $r = r'$ and $d \equiv d' \mod 12$.*

This situation can occur. Let us consider (3.7) in which we assume that

$$d = 14, \; r = 2 \;\; \hat{A}(M) = 0, \; \hat{A}(M, E) = 0, \; \hat{A}(M, S^2 E) + 2\hat{A}(M, T_M) = 2. \qquad (3.9)$$

(The last relation in (3.9) is equivalent to $\hat{A}(M, T_M \otimes E) = -10$.) Then the MWG of a such rank 2 vector bundle over a manifold of dimension 14 is equal to the elliptic genus of a $K3$ surface

$$\chi(M_{14}, E_2; \tau, z) = \chi(K3, T_{K3}; \tau, z).$$

Another example of this type one can construct in dimension 12. Let us assume that

$$d = 12, \; r = 2, \;\; \hat{A}(M) = 0, \; \hat{A}(M, E) = 0. \qquad (3.10)$$

For such a vector bundle we have then

$$\chi(M_{12}, E) = A(M, T_M \otimes E) \left( \frac{\vartheta(\tau, z)}{\eta(\tau)} \right)^2.$$

One can consider the Jacobi form $\vartheta(\tau, z)^2 \eta(\tau)^{-2}$ as the MWG of the trivial vector bundle of rank 2 over a point. Thus the partition function of $(0, 2)$-symmetric non-linear sigma model constructed by the vector bundle $E_2 \to M_{12}$ satisfying (3.10) is of this form.

# 4 $A_2^{(2)}$-Genus

The formal power series $\mathbf{E}_{q,y}$ over $K(M)$ (see (1.1)) is a geometric analog of the Jacobi theta-series $\vartheta(\tau, z)$ which is the denominator function of the affine algebra $A_1^{(1)}$. A similar construction we can propose for an arbitrary affine Lie algebra using some constructions from [9, 14, 15]. In this section we consider the case of $A_2^{(2)}$ in way proposed in [8].

Let us define

$$\mathbf{E}_{q,y}^{(2)} = \bigwedge\nolimits_{y^{-1}} E^* \otimes \bigotimes_{n \geq 1} \bigwedge\nolimits_{-q^n y^{-2}} \Psi_2(E^*) \otimes \bigotimes_{n \geq 1} \bigwedge\nolimits_{-q^n y^2} \Psi_2(E)$$

$$\otimes \bigotimes_{n \geq 1} S_{q^n y^{-1}} E^* \otimes \bigotimes_{n \geq 1} S_{q^n y} E \otimes \bigotimes_{n=1}^{\infty} S_{q^n}(T_M \oplus T_M^*)$$

where $\Psi_2(E)$ is the second Adams operation on vector bundle $e$. We remind that

$$\mathrm{ch}(\Psi_2(E)) = \mathrm{ch}(E \otimes E) - \mathrm{ch}(\wedge^2 E) = \sum_{i=1}^{r} e^{2x_i}.$$

For the algebra $A_2^{(2)}$ we can give the following definition (compare with Definition 1.1).

**Definition 4.1 ($A_2^{(2)}$-Genus)** $\alpha(M, E; \tau, z)$ of a complex vector bundle $E$ of rank $r$ over a compact (almost) complex manifold $M$ of dimension $d$ is defined as follows

$$\chi(M, E; \tau, z) = q^{-d/12} y^{r/2} \int_M \exp\left(\frac{1}{2}(c_1(E) - c_1(T_M))\right)$$

$$\exp\left((3p_1(E) - p_1(T_M)) \cdot G_2(\tau)\right) \exp\left(-\frac{c_1(E)}{2\pi i} \frac{\partial}{\partial z} \log(\phi_{0,3/2}(\tau, z))\right) \mathrm{ch}(\mathbf{E}_{q,y}^{(2)}) \, \mathrm{td}(T_M).$$

where $c_1(E)$ and $p_1(E)$ are the first Chern and Pontryagin class of $E$. td is the Todd class, $\mathrm{ch}(\mathbf{E}_{q,y})$ is the Chern character which we apply to each coefficient of the formal power series and the integral $\int_M$ denotes the evaluation of the top degree differential form on the fundamental cycle of the manifold.

The series $\mathbf{E}_{q,y}$ was a geometric variant of the Jacobi triple product and $\mathbf{E}_{q,y}^{(2)}$ is a geometric analog of the quintuple product

$$\vartheta_{3/2}(\tau, z) = \sum_{n \in \mathbb{Z}} \left( \frac{12}{n} \right) q^{n^2/24} r^{n/2}$$

$$= q^{\frac{1}{24}} r^{-\frac{1}{2}} \prod_{n \geq 1} (1 + q^{n-1} r)(1 + q^n r^{-1})(1 - q^{2n-1} r^2)(1 - q^{2n-1} r^{-2})(1 - q^n)$$

where

$$\left( \frac{12}{n} \right) = \begin{cases} 1 & \text{if } n \equiv \pm 1 \bmod 12 \\ -1 & \text{if } n \equiv \pm 5 \bmod 12 \\ 0 & \text{if } (n, 12) \neq 1. \end{cases}$$

We note that

$$\vartheta_{3/2}(\tau, z) = \frac{\eta(\tau) \vartheta(\tau, 2z)}{\vartheta(\tau, z)} \in J_{\frac{1}{2}, \frac{3}{2}}(v_\eta).$$

**Theorem 4.2** *Let E be a complex (holomorphic) vector bundle of rank r over a compact complex manifold M of dimension d. Let $\alpha(M, E; \tau, z)$ be the $A_2^{(2)}$-genus. Then*

$$\eta(\tau)^{d+r} \vartheta_{3/2}(\tau, z)^{d-r} \alpha(M, E; \tau, z) \in J_{d, \frac{3}{2}d}^w$$

*is a weak Jacobi form of weight d and index* $3d/2$.

***Proof*** The proof of the theorem is similar to the proof of Theorem 1.3.

# References

1. O. Aharony, S. Yankielowicz, A.N. Schellekens, Charge sum rules in $N = 2$ theories. Nucl. Phys. **B418**, 157 (1994)
2. M.F. Atiyah, F. Hirzebruch, *Spin Manifolds and Group Actions, Essays in Topology and Related Subjects* (Springer-Verlag, Berlin, 1970) pp 18–28
3. R. Dijkgraaf, E. Verlinde, H. Verlinde, Counting dyons in $N = 4$ string theory. Nucl. Phys. **B484**, 543–561 (1997)
4. R. Dijkgraaf, G. Moore, E. Verlinde, H. Verlinde, Elliptic genera of symmetric products and second quantized strings. Commun. Math. Phys. **185**, 197–209 (1997)
5. M. Eichler, D. Zagier, *The Theory of Jacobi Forms*. Progress in Mathematics, vol. 55 (Birkhäuser, Basel, 1985)
6. T. Eguchi, H. Ooguri, A. Taormina, S.-K. Yang, Superconformal algebras and string compactification on manifolds with $SU(N)$ holonomy. Nucl. Phys. **B315**, 193 (1989)

7. V. Gritsenko, Elliptic genus of Calabi–Yau manifolds and Jacobi and Siegel modular forms. Algebra i Analyz **11**, 100–25 (1999); English transl. in St. Petersburg Math. J. **11**, 781–804 (2000)

8. V. Gritsenko, Complex vector bundles and Jacobi forms. Proc. Symp. "Automorphic forms and L-functions" **1103**, 71–85 (1999)

9. V. Gritsenko, Reflective modular forms and applications. Russian Math. Surv. **73**(5), 797–864 (2018)

10. V. Gritsenko, Jacobi modular forms: 30 ans après; COURSERA (12 lectures and seminars) (2017–2019)

11. V.A. Gritsenko, V.V. Nikulin, The Igusa modular forms and "the simplest" Lorentzian Kac–Moody algebras. Matem. Sbornik **187**, 1601–1643 (1996)

12. V.A. Gritsenko, V.V. Nikulin, Siegel automorphic form correction of some Lorentzian Kac–Moody Lie algebras. Amer. J. Math. **119**, 181–224 (1997)

13. V.A. Gritsenko, V.V. Nikulin, Automorphic forms and Lorentzian Kac-Moody algebras. Part II. Inter. J. Math. **9**(2), 201–275 (1998)

14. V.A. Gritsenko, V.V. Nikulin, Lorentzian Kac-Moody algebras with Weyl groups of 2-reflections. Proc. Lond. Math. Soc. **116**(3), 485–533 (2018)

15. V. Gritsenko, N.-P. Skoruppa, D. Zagier, Theta blocks (2019). arXiv:1907.00188

16. V. Gritsenko, H. Wang, Powers of Jacobi triple product, Cohens numbers and the Ramanujan $\Delta$-function. Eur. J. Math. **4**(2), 561–584 (2018)

17. V. Gritsenko, H. Wang, Graded rings of integral Jacobi forms. J. of Num. Theory (2020). https://dx.doi.org/10.1016/j.jnt.2020.03.006

18. F. Hirzebruch, Letter to V. Gritsenko from 11 August 1997. St. Petersburg Math. J. **11**, 805–806 (2000)

19. F. Hirzebruch, T. Berger, R. Jung, *Manifolds and Modular Forms* (Kluwer Academic Publishers, Dordrecht, 1992)

20. T. Kawai, K. Mohri, Geometry of (0, 2) Landau–Ginzburg orbifolds. Nucl. Phys. **B425**, 191–216 (1994)

21. T. Kawai, Y. Yamada, S.-K. Yang, Elliptic genera and N=2 superconformal field theory. Nucl. Phys. **B414**, 191–212 (1994)

22. A.S. Libgober, J.W. Wood, Uniqueness of the complex structure of Kähler manifolds of certain homotopy types. J. Differ. Geom. **32**, 139–154 (1990)

23. G. Moore, String duality, automorphic forms and generalized Kac–Moody algbceras. Nucl. Phys. Proc. Suppl. **67**, 56–67 (1998)

24. S.M. Salamon, On the cohomology of Kähler and hyper-Kähler manifolds. Topology **35**, 137–155 (1996)

25. S. Sethi, C. Vafa, E. Witten, Constraints on low-dimensional string compactications. Nucl. Phys **B480**, 213–224 (1996)

26. E. Witten, Elliptic genera and quantum field theory. Commun. Math. Phys. **109**, 525 (1987)

27. E. Witten, The index of the dirac operator in loop space, in *Elliptic Curves and Modular Forms in Algebraic Topology* (Springer, Berlin, 1988), pp. 161–181

28. D. Zagier, Note on the Landweber–Strong elliptic genus, in *Elliptic Curves and Modular Forms in Algebraic Topology* (Springer, Berlin, 1988), pp. 216–224

# Superconformal Indices and Instanton Partition Functions

**Seok Kim**

## Contents

1   Introduction..................................................................... 121
2   Instanton Partition Functions: Basics and Calculus ........................ 123
3   Instantons in 5d QFTs........................................................ 132
   3.1   5d Instantons for 5d SCFTs .......................................... 133
   3.2   5d Instantons for 6d SCFTs on $S^1$ ............................... 138
4   Instanton Strings in 6d QFTs ................................................ 145
   4.1   Strings of 6d $\mathcal{N} = (2, 0)$ Theories ...................... 150
   4.2   Strings of 6d $\mathcal{N} = (1, 0)$ Theories ...................... 152
5   Superconformal Observables .................................................. 156
   5.1   6d $(2, 0)$ Index and its Physics ................................... 157
   5.2   5d Index and its Physics ............................................ 170
6   Conclusion..................................................................... 172
References ........................................................................ 173

**Abstract** We review recent advances in superconformal field theories in 5 and 6 spacetime dimensions based on the instanton partition functions. After reviewing the basics of the instanton partition functions, we use them to study 5d SCFTs, and also 6d SCFTs compactified on a circle. Various non-perturbative aspects of these systms are discussed. We also discuss the related partition functions called superconformal indices.

## 1   Introduction

Quantum field theory (QFT) is an essential tool of modern physics, from particle physics to condensed matter physics and optics, among many. Its progresses

S. Kim (✉)
Department of Physics and Astronomy, Center for Theoretical Physics, Seoul National University, Seoul, South Korea
e-mail: skim@phya.snu.ac.kr

© Springer Nature Switzerland AG 2020
V. A. Gritsenko, V. P. Spiridonov (eds.), *Partition Functions and Automorphic Forms*, Moscow Lectures 5,
https://doi.org/10.1007/978-3-030-42400-8_3

triggered major advances in physics, while its conceptual/technical limitations often set the current edges of our understandings of Nature.

Traditional methods of QFT are often based on Lagrangian, either at weak or strong coupling. However, inspired by string theory from around 20 years ago, it has been gradually realized that some QFT's do not seem to admit obvious Lagrangian descriptions, despite with an abstract argument for their existence. Such examples have been most dramatically found in spacetime dimension $d > 4$, in which consistent interacting quantum field theories without physical pathologies have been unknown.

In $d = 6$, maximal superconformal field theories were indirectly discovered in [1]. Generalizations to 6d SCFTs with less SUSY are found, e.g. in [2–5]. A typical aspect is that the SCFTs have interacting tensionless strings as their light degrees of freedom. One does not know how to describe them, or whether they can be described at all, by local Lagrangian QFTs. In $d = 5$, the first examples of SCFTs engineered from string theory are [6]. Some immediate generalizations of these constructions can be found in [7, 8] (see also [5]). These are mostly found to be strong coupling limits of non-renormalizable 5d Yang–Mills theories, at which point infinitely many species of particles become massless. Again, it is unknown how to describe such systems using local Lagrangian QFTs. So in both 6d and 5d, the main method of 'constructing' such SCFTs was via string theory. In particular, in recent years, some efforts have been made to fully classify possible 6d $\mathcal{N} = (1, 0)$ SCFTs, where the main tools are 6d effective field theory in the tensor branch and F-theory engineering on singular ellitic Calabi–Yau threefold $(CY_3)$: see [9, 10] and references thereof. More recently, there have been similar studies on 5d $\mathcal{N} = 1$ SCFTs by using 5d effective field theory in the Coulomb branch and also considering M-theory on singular $CY_3$ [11, 12].

Studies of such QFTs have also crucially relied on string theory engineering, lacking proper QFT set-ups. Recent studies enabled us to study various QFT partition functions, combining string theory set-ups, intuitions from effective field theory, and SUSY techniques. In many partition functions, important roles are played by the so-called instantons partition functions [13] of non-renormalizable gauge theories in $d = 5, 6$. In this paper, we shall review the last partition functions, and how they can be applied to better understand aspects of SCFTs in $d > 4$.

Historically, the instanton partition functions of [13] were first studied to derive the Seiberg–Witten theory of 4d $\mathcal{N} = 2$ gauge theories in the Coulomb branch. This is a supersymmetric partition function on $\mathbb{R}^4$ with the so-called Omega deformation, acquiring non-perturbative contributions from multi-instantons localized at the origin of $\mathbb{R}^4$. It has been known from then that considering a 5d uplift of this problem is helpful, both conceptually and technically. In the latter set-up, the 5d gauge theory is put on $\mathbb{R}^4 \times S^1$ with Omega deformation on $\mathbb{R}^4$, and one is interested in the 4d effective field theory in the Coulomb branch. Since one also integrates out the massive KK modes on $S^1$, one obtains a prepotential containing the KK energy scale. The full partition function on $\mathbb{R}^4 \times S^1$ has alternative interpretation, rather than a tool of computing the quantum prepotential. Interpreting $S^1$ as the Euclidean 'thermal' circle, one can interpret this BPS partition function as a Witten index which counts

certain BPS states. From this perspective, this provides important spectral information on 5d gauge theories themselves, often related to 5d SCFTs with massive
deformations and Coulomb VEV. Further connections to topological strings [14, 15]
triggered the developments of the so-called topological vertex formalisms [16].

In this framework, the partition functions of 5d gauge theories will reduce (in a
sector with given instanton number) to the quantum mechanical path integral over
the instanton moduli space. This is a quantum mechanical non-linear sigmar model
(NLSM). This quantum mechanical system is well known to be incomplete, due to
the small instanton singularity. In our 5d gauge theories, it is most natural to regard
this incompleteness as reflecting the non-renormalizablity of 5d gauge theories. The
5d gauge theories we discuss are believed to have UV completions given by 5d
SCFTs, often from string theory arguments. Often, one can use the very string theory
set-up to provide UV completions of the instanton quantum mechanics. This will
lead us to the ADHM-like quantum mechanics, which are also often called gauged
linear sigma model (GLSM).

In Sect. 2, we shall start our discussions by presenting these ADHM quantum
mechanics for instanton solitons in 5d, and review how they can be used to
compute the partition functions of [13]. Various 'prescriptions' adopted in [13] were
'derived' only recently [17]. These rigorous derivations were partly motivated after
encountering tricky examples of 5d SCFTs and gauge theories in the recent years, in
which simple prescriptions were hard to find. Many technical and conceptual issues
will be discussed in this section.

In Sect. 3, we shall use these partition functions to explore the physics of 5d
SCFTs in the Coulomb branch. We discuss UV enhanced symmetries of the 5d
gauge theory partition functions for 5d SCFTs. We also discuss the 5d gauge
theory partition functions for 6d SCFTs compactified on an extra circle. The non-
perturbative emergence of the KK circle by instantons will be discussed.

In Sect. 4, we shall study the partition functions of instanton strings, or more
generally self-dual strings, in 6d SCFTs. 2d GLSM approaches and other comple-
mentary approaches have been used to explore the physics of 6d $\mathcal{N} = (2, 0)$ and
$\mathcal{N} = (1, 0)$ SCFTs in the so-called tensor branch.

In Sect. 5, some superconformal observables such as the superconformal index
will be discussed. In many of these partition functions, it turns out that the instanton
partition functions play important roles. We shall discuss several interesting physics
of the 5d and 6d SCFTs using these partition functions.

## 2 Instanton Partition Functions: Basics and Calculus

We start by discussing the 4d $\mathcal{N} = 2$ gauge theories, and afterwards discuss their
5d/6d uplifts. The system preserves 8 Hermtian supercharges, which we call

$$Q_\alpha^A \ , \ \overline{Q}_{\dot{\alpha}}^A \ . \tag{2.1}$$

$A = 1, 2$ is the doublet index of $SU(2)_R$ R-symmetry, $\alpha = 1, 2$, $\dot{\alpha} = 1, 2$ are for $SU(2)_l \times SU(2)_r \sim SO(3, 1)$ Lorentz symmetry. In the Euclidean QFT, the last symmetry is replaced by $SO(4)$.

$\mathcal{N} = 2$ gauge theories are determined by specifying the gauge group $G$ and matter contents, the representation **R** under $G$. The vector multiplet is in the adjoint representation of $G$, consisting of: gauge field $A_\mu$, complex scalar $\Phi$, and fermions. Matters are given by the hypermultiplet, in the representation **R** of $G$. It consists of: two complex scalars $q_A = (q, \tilde{q}^\dagger)$, and fermions. The hypermultiplet scalar is in the doublet of $SU(2)_R$. In terms of these fields, the supersymmetric Lagrangian density is given by

$$g_{YM}^2 \mathcal{L} = -\frac{1}{4}\text{tr}(F_{\mu\nu}F^{\mu\nu}) - \frac{1}{2}\text{tr}(D_\mu\Phi D^\mu\Phi) + \frac{1}{4}[\Phi, \Phi^\dagger]^2$$
$$-|D_\mu q_A|^2 - |(\Phi)_{\mathbf{R}}q_A|^2 \qquad (2.2)$$
$$-\frac{1}{2}\sum_{i=1}^{3}(q_A(\tau^i)^A_B q^{\dagger B})^2 + (\text{terms involving fermions}) .$$

In this Lagrangian, there exists a continuous parameter $g_{YM}$, the gauge coupling. An extra possible term is the topological theta term, of the form $\frac{\theta}{2\pi}\int \text{tr}(F \wedge F)$.

The classical $\mathcal{N} = 2$ gauge theory can be extended to 5 and 6 dimensions, to gauge theories preserving 8 Hermitian SUSY. In 5d, one obtains $\mathcal{N} = 1$ gauge theories. The vector multiplet consists of: $A_\mu$ with $\mu = 0, \cdots, 4$, a real scalar $\Phi$, and fermions. In 6d, one obtains $\mathcal{N} = (1, 0)$ chiral gauge theories. The vector multiplet consists of $A_\mu$ with $\mu = 0, \cdots, 5$, and a left-chiral gaugino. Hypermultiplets assume similar forms as in 4d. Note that in 6d, the fermions in the hypermultiplet are right-chiral. The 5d SUSY Lagrangian can be optionally supplemented by the 5d Chern–Simons term, if $G = SU(N)$, with quantized CS coefficients. These terms will be discussed later when it becomes necessary. The classical 6d gauge theories are often subject to quantum consistency conditions, coming from local and globally gauge anomalies. These issues will be addressed in Sect. 4.

The 4d and 5d gauge theories have moduli space of vacua coming from the vacuum expectation values (VEV) of $\Phi$. We call this 'Coulomb branch.' The 6d gauge theory does not have Coulomb branch of vacua, since the vector multiplet does not contain a scalar. This branch will be one of the main subjects in this paper. In this paper, we shall often be interested in the Coulomb branches of gauge theories compactified to 4d. Namely, we consider 5d gauge theories on $\mathbb{R}^4 \times S^1$, and 6d gauge theories on $\mathbb{R}^4 \times T^2$. The 4d Coulomb branch is locally given by $\mathbb{C}^r$, where $r$ is the rank of the gauge group $G$, and each factor of $\mathbb{C}$ comes from the VEV $v = \langle\Phi\rangle$ of the complex scalar $\Phi$. The 5d theories on $\mathbb{R}^4 \times S^1$ has Coulomb branches from the VEV $v = \langle\Phi + iA_4\rangle$, where $A_4$ is the Wilson line on spatial $S^1$. The local form of the moduli space is $(\mathbb{R} \times S^1)^r$, i.e. product of cylinders. In 6d, the expectation value is given to two Wilson lines $v\langle A_4 + iA_5\rangle$ on $T^2$. $v$ lives on a torus, so that the moduli space locally takes the form of $(T^2)^r$.

At generic point in the Coulomb branch, gauge symmetry $G$ is broken to $U(1)^r$. Massless fields in the Coulomb branch are given by $r$ Abelian vector multiplets. Integrating out the massive fields, such as massive W-bosons and matters, and also the Kaluza–Klein (KK) fields along the extra dimension $S^1$, $T^2$, one obtains a nontrivial effective action of these massless fields. The effective action is governed by a holomorphic function $\mathcal{F}(v)$ of the Coulomb VEVs $v$, called prepotential. It was explained in [18, 19] that, for certain class of theories, the form of prepotential can be inferred by: knowing the weak-coupling behaviors of $\mathcal{F}(v)$ at large $v$, some simple assumptions on the singularity structures of $\mathcal{F}$, and holomorphic of $\mathcal{F}$.

A microscopic derivation of the results of [18, 19] is given in [13, 20]. One idea is to first consider the partition function of QFT in the Omega-deformed $\mathbb{R}^4$. This deformation is a sort of curved space deformation of $\mathbb{R}^4$, whose explanation we postpone till below. It will be much easier to first explain it as chemical potentials on $\mathbb{R}^4 \times S^1$. The partition function $Z(v, q, \epsilon_{1,2})$ depends on the Coulomb VEV, the gauge coupling $g_{YM}$ by $q \equiv \exp\left[-\frac{4\pi^2}{g_{YM}^2}\right]$, and the Omega deformations $\epsilon_{1,2}$. Decomposing the prepotential $\mathcal{F}$ by the classical one $\mathcal{F}_{cl}$, perturbative correction $\mathcal{F}_{pert}$, and the instanton correction $\mathcal{F}_{inst}$, the nontrivial part is $\mathcal{F}_{inst}$. It has been found in [13] that this part is related to the Omega deformed partition function as

$$Z \stackrel{\epsilon_{1,2}\to 0}{\longrightarrow} \exp\left[-\frac{\mathcal{F}_{inst}(v, q)}{\epsilon_1 \epsilon_2} + \text{(less divergent terms in } \epsilon_{1,2})\right]. \qquad (2.3)$$

So once one can microscopically compute $Z$, it could be used to address $\mathcal{F}_{inst}$, and hopefully derive the results of [18, 19].

Another idea of [13], to explicitly compute $Z$, was the following. Note that the instanton partition function admits an expansion

$$Z(v, q, \epsilon_{1,2}) = \sum_{k=0}^{\infty} q^k Z_k(v, \epsilon_{1,2}), \qquad (2.4)$$

where $Z_0 \equiv 1$ by definition. Nekrasov [13] and Nekrasov and Shadchin [21] compute the coefficient $Z_k$ in the $k$ instanton sector, when the gauge group $G$ of the gauge theory is classical, i.e. $G = A_{N-1} = SU(N)$, $B_N = SO(2N+1)$, $C_N = Sp(N)$, $D_N = SO(2N)$. Namely, $Z_k$ is given by a matrix integral, where the matrices are given by the ADHM data for $k$ instantons in the gauge theory. The ADHM matrix models will be introduced shortly.

At this moment, it is useful [13] to consider a 5d uplift of $Z_k$. Then, regarding the $S^1$ as the Euclidean time direction, each $Z_k$ acquires the interpretation of a Witten index over the $k$ instanton Hilbert space. The instantons which contribute to this index has a classical soliton representation. They solve the following PDE,

$$F_{\mu\nu} = \star_4 F_{\mu\nu} \qquad (2.5)$$

on spatial $\mathbb{R}^4$, where $\mu, \nu = 1, \cdots, 4$, and is independent of time. In particular, our interest is on the solutions with finite energy density ($\sim$ particle mass), which is attained only when

$$k = \frac{1}{8\pi^2} \int_{\mathbb{R}^4} \text{tr}(F \wedge F) \tag{2.6}$$

is quantized to integers. For self-dual instantons obeying (2.5), $k$ should be positive. The index can be evaluated as the Witten index for the ADHM quantum mechanics of $k$ instantons. The index is defined as follows:

$$\text{Tr}_k \left[ (-1)^F e^{-\epsilon_1(J_1 + J_R)} e^{-\epsilon_2(J_2 + J_R)} e^{-m \cdot F} \right], \tag{2.7}$$

where $J_{1,2}$ are two $U(1)^2 \subset SO(4)$ angular momenta on $\mathbb{R}^4$, and $J_R$ is the Cartan of $SU(2)_R$ R-symmetry of the 5d SYM, inherited to the instanton quantum mechanics. $F$ collectively denote other flavor symmetries of the system, whose chemical potentials are collectively shown as $m$. $\epsilon_1$ and $\epsilon_2$ appearing here implicitly defines the Omega deformation of $\mathbb{R}^4$.

Equation (2.7) can be computed from the following ADHM quantum mechanical system for instantons. The quantum mechanics preserves 4 Hermitian supersymmetry. In the sense that its classical Lagrangian and SUSY variation can be reduced from 2d gauge theories, we call it $\mathcal{N} = (0, 4)$ supersymmetry. See [22, 23] for more detailed explanations on the notion and notations. Let us denote by $G_N$ the possible classical 5d gauge groups $SU(N)$, $SO(N)$, $Sp(N)$. In the instanton quantum mechanics, $G_N$ appears as a gauge symmetry. The quantum mechanics has its own gauge symmetry, which we call $\hat{G}_k$. In the three cases for $G_N$, $\hat{G}_k$ are given by $U(k)$, $Sp(k)$, $O(k)$, respectively. Then, the following 1d gauge fields are introduced, to account for zero modes in the instanton background incurred by 5d vector multiplet fields:

$$\text{1d hyper } a_{\alpha\dot{\beta}}, \Psi_\alpha^A : \textbf{adj} \in U(k) \,;\, \textbf{anti} \in Sp(k) \,;\, \textbf{sym} \in O(k) \tag{2.8}$$

$$\text{1d hyper } q_{\dot{\alpha}}, \psi^A : (\mathbf{k}, \overline{\mathbf{N}}) \in (U(k), SU(N)); \; (\mathbf{2k}, \mathbf{N}) \in (Sp(k), SO(N));$$

$$(\mathbf{k}, \mathbf{2N}) \in (O(k), Sp(N)),$$

where **anti**, **sym** denote rank 2 antisymmetric/symmetric representations. The bosonic part of these fields is called the 'ADHM data.' They are originally introduced in [24, 25] as the matrices which describe the instanton moduli space after imposing certain algebraic constraints. In our context, the moduli space will define the target space for the nonlinear sigma model, which arises as the moduli space approximation describing the low energy excitations of the solitons. In the context of our 1d gauge theory, the algebraic constraint will arise as the vanishing condition of the bosonic potentials for $a_{\alpha\dot{\beta}}, q_{\dot{\alpha}}$. In the context of $\mathcal{N} = (0, 4)$ gauge theory, the potentials will arise from the D-term potentials, which we describe below.

To complete the discussions of the ADHM quantum mechanics, one should introduce the 1d vector multiplets, in the adjoint representation of $\hat{G}_k$. It consists of a worldline vector potential $A_t$, a real scalar $\varphi$, and the gauginos $\bar{\lambda}^A_{\dot{\alpha}}$. Its coupling to the matter fields, and further interactions, are explained in [22, 23]. The Lagrangian of this system is given by

$$
L_{QM} = \frac{1}{g^2_{QM}} \text{tr} \left[ \frac{1}{2}(D_t\varphi)^2 + \frac{1}{2}(D_t a_m)^2 + D_t q_{\dot{\alpha}} D_t \bar{q}^{\dot{\alpha}} + \frac{1}{2}[\varphi, a_m]^2 \right.
$$
$$
- (\varphi \bar{q}^{\dot{\alpha}} - \bar{q}^{\dot{\alpha}} v)(q_{\dot{\alpha}} \varphi - v q_{\dot{\alpha}})
$$
$$
\left. - \frac{1}{2} D_I D_I + (\text{terms involving fermions}) \right]  \qquad (2.9)
$$

where $I = 1, 2, 3$ is for the triplet of $SU(2)_r$, and $a_m \sim a_{\alpha\dot{\beta}}(\bar{\sigma}_m)^{\dot{\beta}\alpha}$ is introduced with $m = 1, \cdots, 4$ being the vector index of $\mathbb{R}^4$. Here, the D-term fields are given by

$$
D_I = q_{\dot{\alpha}}(\tau_I)^{\dot{\alpha}}_{\dot{\beta}} \bar{q}^{\dot{\beta}} + (\tau_I)^{\dot{\alpha}}_{\dot{\beta}}[a_{\alpha\dot{\alpha}}, \bar{a}^{\alpha\dot{\beta}}] .  \qquad (2.10)
$$

If one has extra 5d hypermultiplets, they contribute extra zero mode fields to the 1d ADHM GLSM. See, for instance, [21] for the details. The 5d hypermultiplets primarily contribute to 1d Fermi multiplets, while GLSM UV completions (see below for details) demand extra twisted hypermultiplets in the GLSM. The specific examples of extra zero modes caused by 5d hypermultiplets will be explained later, when we discuss examples in the next sections.

Before explaining how to compute $Z_k$ from this quantum mechanics, we digress for a while to explain the physical sense in which the above GLSM describes the instantons' low energy moduli space dynamics. The main point is that the UV GLSM comes with a super-renormalizable gauge coupling, where $g^2_{QM}$ has dimension of mass$^3$. So in UV, all the degrees of freedom, represented as 1d hypermultiplets, Fermi multiplets, vector multiplets represent their degrees or freedom as visible in the Lagrangian. However, the IR strong coupling effects give distinct fates to their natures at low energy. First of all, the ADHM hypermultiplet fields $a_{\alpha\dot{\beta}}$, $q_{\dot{\alpha}}$ are not all low energy degrees of freedom in the non-linear sigma model (NLSM): some of them are subject to be lifted by constraints $D = 0$, where the D-term fields are given by (2.10). This is an example of some UV 1d fields being integrated out in IR NLSM. On the other hand, there are other types of degrees of freedom, which are absent in IR NLSM but are caused during making a UV completion to the GLSM. One universal possible example of this sort is the fields in the 1d vector mutiplet. The worldline vector potential $A_t$ is non-dynamical, just playing the role of algebraic auxiliary field for the Gauss law constraint for $\hat{G}_k$ gauge symmetry. However, other fields like $\varphi$ and $\bar{\lambda}$ could potentially serves as extra degrees of freedom. The IR fate of these fields depend on specific models. In

some examples, after the coarse-graining of the 1d RG flow, they may be integrated
out and not affect the IR physics. In other case, they may be present as the light
degrees of freedom in IR but decoupled from the other physical degrees of freedom
describing instantons. In still another case, these fields may fail to decouple in any
obvious way from the ADHM degrees of freedom, making the use of the UV GLSM
in question. There are some bottom-up and top-down criteria for judging which of
the three cases above a specific model belongs to. Top-down assessments can often
be obtained from D-brane considerations, engineering the 5d/6d SCFTs from 5-
brane webs [26]. A closely related IR assessments can be found, e.g. in [17, 27].
Extra 1d twisted hypermultiplets from 5d hypermultiplets also may contain such
decoupled degrees of freedom: see, e.g. [17, 28]. In this review, we shall mostly
discuss examples of 5d/6d SCFTs which are free of such extra decoupled degrees
of freedom in their instanton calculus. However, we shall explain later when such
subtleties arise.

With these possible subtleties in mind, we now explain how to compute $Z_k$ from
the ADHM quantum mechanics. The index for the ADHM quantum mechanics is
compute from in [17]. The result is given by a contour integral formula, where the
number of integral variables is given by the rank of $\hat{G}_k$. They come from the Cartans
of the constant values of $\phi \equiv \varphi + i A_\tau$, where $\tau$ is the imaginary time coordinate in
the Euclidean quantum mechanics. Let us label by $I = 1, \cdots, \hat{r}$ the Cartans of $\hat{G}$
of ranke $\hat{r}$. (Note that $\hat{r} = k$ for $\hat{G} = U(k)$, $Sp(k)$, and $\hat{r} = \lfloor k/2 \rfloor$ for $\hat{G} = O(k)$.)
The contour integral takes the following form:

$$Z_k = \frac{1}{|W(\hat{G})|} \oint \left[ \prod_{I=1}^{\hat{r}} \frac{d\phi_I}{2\pi i} \right] Z_{\text{vec}}(\phi, v, \epsilon_{1,2}) Z_{\text{hyper}}(\phi, v, \epsilon_{1,2}, m) , \qquad (2.11)$$

where $Z_{\text{vec}}$ and $Z_{\text{hyper}}$ denote the contributions from 5d vector and hypermultiplets,
respectively. The forms of this integrands and the choice of contour will be
explained now. From the 5d vector multiplet in the adjoint representation of $G$,
the 1d ADHM fields contribute to the integrand of the following form:

$$Z_{\text{vec}} = \frac{\prod_{\alpha \in \text{root}(\hat{G})} 2 \sinh\left(\frac{\alpha(\phi)}{2}\right) \cdot \prod_{\alpha \in \text{adj}(\hat{G})} 2 \sinh\left(\frac{\alpha(\phi)+2\epsilon_+}{2}\right)}{\left( \prod_{\hat{\rho} \in \text{fund}(\hat{G})} \prod_{\rho \in \text{fund}(G)} 2 \sinh\left(\frac{\hat{\rho}(\phi)-\rho(\alpha)+\epsilon_+}{2}\right) \right)}$$
$$\times \prod_{\rho \in R(\hat{G})} 2 \sinh\left(\frac{\rho(\phi)+\epsilon_1}{2}\right) \cdot 2 \sinh\left(\frac{\rho(\phi)+\epsilon_2}{2}\right) \bigg) \qquad (2.12)$$

where $\epsilon_+ \equiv \frac{\epsilon_1+\epsilon_2}{2}$. For the $Z_{\text{hper}}$ from 5d hypermultiplets, we refer the reader to the
results summarized in [17, 28]. We shall explain specific examples of $Z_{\text{hyper}}$ later,
when necessary.

Now we explain the choice of the integral contour for $\phi_I$. The result of the
contour integral for (2.11) is given by a specific residue sum for $Z_{1\text{-loop}} \equiv$
$Z_{\text{vec}} Z_{\text{hyper}}$, called the Jeffrey-Kirwan (JK) residue (see [29] and references theirn).
To explain this in the simplest possible manner, we focus on a pole of $Z_{1\text{-loop}}$ near

$\phi = \phi_*$. Making a Laurent expansion of $Z_{1\text{-loop}}$ around $\phi = \phi_*$, the possibly nonzero residues are obtained only from the 'simple pole' parts, which are linear combinations of the functions of the form

$$\frac{1}{Q_{j_1}(\phi - \phi_*) \cdots Q_{j_r}(\phi - \phi_*)} . \tag{2.13}$$

$Q_{j_1}, \cdots, Q_{j_r}$ are chosen in $\mathbf{Q}(\phi_*)$, which are the charge vectors appearing in the sinh factors appearing in the denominator of $Z_{1\text{-loop}}$. See [17] for how to make a Laurent expansions of $Z_{1\text{-loop}}$ and to extract out the terms of the form (2.13). The so-called Jeffrey-Kirwan residue JK-Res, that is relevant for writing down our index, also refers to the auxiliary vector $\eta$ in the space of charges $\mathbf{Q}$. JK-Res is defined by Benini et al. [29]

$$\text{JK-Res}(\mathbf{Q}_*, \eta) \frac{d\phi_1 \wedge \cdots \wedge d\phi_r}{Q_{j_1}(\phi) \cdots Q_{j_r}(\phi)} = \begin{cases} \left|\det(Q_{j_1}, \cdots, Q_{j_r})\right|^{-1} & \text{if } \eta \in \text{Cone}(Q_{j_1}, \cdots, Q_{j_r}) \\ 0 & \text{otherwise} \end{cases} . \tag{2.14}$$

'Cone' denotes the cone spanned by the $r$ independent vectors. Namely, $\eta \in \text{Cone}(Q_1, \cdots, Q_r)$ if $\eta = \sum_{i=1}^r a_i Q_i$ with positive coefficients $a_i$. Although this definition apparently looks over-determining JK-Res as a linear functional, it is known to be consistent: see [29] and references therein. See [17] for more explanations of these structures, in the context of the instanton quantum mechanics. The index $Z_k$ is then given by Benini et al. [29]

$$Z = \frac{1}{|W|} \sum_{\phi_*} \text{JK-Res}(\mathbf{Q}(\phi_*), \eta) \, Z_{1\text{-loop}}(\phi, \epsilon_+, z) . \tag{2.15}$$

Note that the result may in principle depend on the choice of $\eta$. However, it was argued in [17] that this dependence is absent for $G = Sp(N), SO(N)$. For $G = U(N)$, it may sometimes appear that the choice of $\eta$ yields different results. However, it was shown (also illustrated with examples) that this difference only has to do with the extra decoupled degrees of freedom in the UV GLSM, so it only demands the separation out of these extra contributions from the index. See [17] for more details. In the rest of this paper, we shall deal with simple examples in which $Z_k$ are manifestly independent of the choice of the auxiliary vector $\eta$.

What we have presented so far is a 'modern formulation' (and generalization) of the core results derived in [13, 20, 21], where the latter derivations are valid for some classes of QFT models. More concretely, [13, 20] derived the 4d and 5d instanton partition functions with gauge group $G = SU(N)$ (or $U(N)$), optionally with hypermultiplets in the fundamental representation of the adjoint representations. In [21], the 4d and 5d partition functions are derived for $G = SO(N)$ and $Sp(N)$. The derivations of [21] are valid also after adding hypermultiplet matters in the fundamental representation. However, after adding hypermultiplet matters in the higher rank representations than the fundamentals, i.e. rank 2 or higher, it was sometimes unclear how to choose the integral contour for $\phi_I$, and then how to interpret $Z_k$ as a partition function for the 5d QFT. Also, just with fundamental hypermultiplets, it has been observed [30] that the naive contour integral formulae (e.g. trying to apply the results of [21, 28]) if the number of 5d hypermultiplets is too many (but still within the range in which one believes them to describe 5d SCFTs). These issues were discussed in [17] in detail, both physically and computationally. The elaboration on the computational side is the results summarized above. Some conceptual issues will be addressed in Sect. 3 with examples.

For certain gauge theories, the index $Z_k$ can be expressed in a more concrete form than the above contour integral formula. We would like to present such simpler formulae for the cases with $G = SU(N)$ or $U(N)$, thus with $\hat{G} = U(k)$. With these groups, the nonzero residues of JK-Res can sometimes be classified very neatly in terms of the colored Young diagrams [13, 31, 32]. The formulae we shall present apply to $SU(N)$ or $U(N)$ theories with hypermultiplet matters in either fundamental representations or adjoint representations. (For similar formulae with more 'exotic' gauge groups and matters, see e.g. [17] for more details.) Similar Young diagram sums for the index often appear from the topological vertex approaches, and some examples of this sort will be presented in Sect. 4. These simplified formulae hold when the nontrivial pole locations $\phi_*$ are classified. This is possible for $G = SU(N)$ when the nonzero residues come only from the contour integrand $Z_{\text{vec}}$ from 5d vector multiplets. Inserting the roots $\alpha$ and weight vectors $\rho$, $\hat{\rho}$ for $SU(N)$, $U(k)$, one can show [13, 31, 32] that $\phi_*$ is labeled by all possible $N$-colored Young diagrams with total box number $k$. By the latter, one means a collection of $N$ Young diagrams $Y = (Y_1, Y_2, \cdots, Y_N)$, whose total box number is constrained as $|Y| \equiv \sum_{i=1}^{N} |Y_i| = k$. (Some $Y_i$'s may be void, i.e. with no boxes.) Let us label each box in $Y_i$ by $s = (m, n)$, when the box is on the $m$'th row and on the $n$'th column of the Young diagram $Y_i$. $i = 1, \cdots, N$ and $s = (m, n)$ will label all the $k$ boxes of $Y$. The classification of the pole location $\phi = \phi_*$ is given by specifying its $k$ components, $\phi_I$, $I = 1, \cdots, k$. It was found that the label $I$ can be replaced by $i, s = (m, n)$, i.e. by boxes of $Y$. Each box represents a component $\phi(s)$, which is given by Nekrasov [13], Flume and Poghossian [31], Bruzzo et al. [32]

$$\phi(s) = v_i - \epsilon_+ - (n - 1)\epsilon_1 - (m - 1)\epsilon_2 . \tag{2.16}$$

The notation used here follows [27]. The residue sum formula for $Z_k$ is now given by the sum over $Y$. Had one been considering 5d pure $\mathcal{N} = 1$ super-Yang–Mills, i.e. with only vector multiplet, the residue sum formula becomes [13, 31, 32]

$$Z_k = \sum_{Y; |Y|=k} \prod_{i=1}^{N} \prod_{s \in Y_i} \frac{(-1)^{Nk}}{2 \sinh \frac{E_{ij}(s)}{2} \cdot 2 \sinh \frac{E_{ij}(s) - 2\epsilon_+}{2}} \tag{2.17}$$

where

$$E_{ij}(s) \equiv v_i - v_j - \epsilon_1 h_i(s) + \epsilon_2(v_j(s) + 1) \, . \tag{2.18}$$

Here, $h_i(s)$ denotes the horizontal distance from the box $s$ to the right end of $Y_i$, and $v_j(s)$ denotes the vertical distance from $s$ to the bottom end of $Y_j$. See, e.g. [27, 33] for more explanations and examples. When there are hypermultplets so that there is an extra factor $Z_{\text{hyper}}(\phi, v, \epsilon_{1,2}, m)$ in the contour integrand, there can be two possibilities. Firstly, if $Z_{\text{hyper}}$ may contribute to nontrivial poles in the JK-Res, meaning that $Z_{\text{hyper}}$ diverges at a $\phi_*$. In this case, the simple Young diagram formula general breaks down. Secondly, if $Z_{\text{hyper}}$ does not affect the determination the pole location $\phi_*$, one continues to have the poles classified by $Y$. In this case, the index simply becomes

$$Z_k = \sum_{Y; |Y|=k} \prod_{i=1}^{N} \prod_{s \in Y_i} \frac{(-1)^{Nk}}{2 \sinh \frac{E_{ij}(s)}{2} \cdot 2 \sinh \frac{E_{ij}(s) - 2\epsilon_+}{2}} \cdot Z_{\text{hyper}}(\phi(s), v, \epsilon_{1,2}, m) \, .$$

$$\tag{2.19}$$

This formula holds, e.g. for many hypermultiplets in fundamental representation, in which case only 1d Fermi multiplet contributes to $Z_{\text{hyper}}$ so that there cannot be any poles caused by $Z_{\text{hyper}}$ [28]. More interestingly, if there is a single hypermultiplet in the adjoint representation, $Z_{\text{hyper}}$ contains poles in general [13, 28]. However, by carefully selecting the poles to be kept in the JK-Res sum, one can show that no poles from $Z_{\text{hyper}}$ are to be kept [17]. This fact was noticed earlier in the literature, either as a working prescription [13, 32] or from an alternative supersymmetric localization calculus known as the 'Higgs branch localization' [33]. Finally, there are more exotic examples in which this formula holds [27], applicable to $SO(7)$ instantons with hypermultiplets in **8** or $G_2$ instantons with hypermultiplets in **7**, etc.

Finally in this section, we comment on the instantons strings of 6d supersymmetric Yang–Mills theories. The 6d Yang–Mills theory has its inverse coupling $1/g_{\text{YM}}^2$ at (mass)$^2$ order, and is of course non-renormalizable by itself. For this massive theory to make sense in 6d SCFT, we have to couple the 6d vector supermultiplet to a 6d self-dual tensor multiplet. (See Sect. 4 for more explanations.) One can give a vacuum expectation value to the real scalar in the tensor multiplet, which sets the value of this inverse YM coupling. So the Yang–Mills theory coupled to the (Abelian) tensor multiplet provides a tensor branch effective action of 6d SCFTs, if the gauge group and extra hypermultiplet matter contents are properly chosen. One also find a quantum consistency requirement for such super-Yang–Mills theories, coming from the local and global anomaly cancelations. Again, more details will be explained in Sect. 4, only with specific examples.

With this understanding, one considers the localized soliton configurations satisfying (2.5) on $\mathbb{R}^4$ slice of $\mathbb{R}^{5,1}$. The solutions being independent of the longitudinal $\mathbb{R}^{1,1}$, they represent stringy configurations of the 6d SCFTs in the tensor branch. After compactifying a space longitudinal to the string to $S^1$, one can study the BPS states of the wrapped strings, also with momentum on $S^1$ as well as other nonzero charges. From the viewpoint of the 2d QFT living on the strings, this is an index of the Euclidean QFT on $T^2$, called elliptic genus. For certain instanton strings, the ADHM-like 2d gauge theory descriptions are available, which can be used to compute the elliptic genus $Z_n(\tau, v, \epsilon_{1,2}, m)$ at definite string winding number $n$. $n$ is a parameter analogous to $k$ in the Witten index for instanton particles. All the definition of the index is given in the same manner as $Z_k$ above, except that one now has an extra chemical potential $\tau$ conjugate to the worldsheet momentum $P$. They are reflected in the trace definition of the index by the factor of

$$e^{2\pi i \tau P} . \tag{2.20}$$

The formula for the elliptic genera for GLSM are available in [29]. It takes a very similar form as the index $Z_k$ presented in this section, with the following simple replacement. Firstly, all $2 \sinh \frac{z}{2}$ functions in the integrand $Z_{\text{vec}} Z_{\text{hyper}}$ are replaced as

$$2 \sinh \frac{z}{2} \rightarrow \frac{i\theta_1(\tau, \frac{z}{2\pi i})}{\eta(\tau)} , \tag{2.21}$$

where $z$ is a linear combination of chemical potentials except $\tau$. The chemical potentials $\phi, v, \epsilon_{1,2}, m$ now live on tori, rather than cylinders. In particular, the contour integral variable $\phi$ also lives on a torus. The contour integral is given again by a particular residue sum, which is precisely the same Jeffrey-Kirwan residues [29] that we explained in the 1d context.

So far, all the formal aspects of $Z_n$ look more or less similar as $Z_k$ for instanton particles in 5d. In Sect. 4, we shall explain surprising subtleties and new aspects for instanton strings, or more generally 6d self-dual strings in the tensor branch.

# 3   Instantons in 5d QFTs

In this section, we illustrate how to use the results of the previous section for 5d instanton particles. Namely, we shall first use our 5d gauge theory calculus to study 5d QFTs. In this context, instantons realize non-perturbative particles which become massless in the 5d SCFT limit. Then, we shall explain how to use 5d instanton calculus to study 6d SCFTs compactified on a circle. In this context, the 5d instanton particles become the KK momentum modes of the compactified $S^1$.

## 3.1 5d Instantons for 5d SCFTs

There are many 5d SCFTs which can be engineered from branes or geometry. In this section, we shall only consider specific examples. The class of 5d SCFTs discussed in this section i the one originally found in [6]. As we shall review now, these SCFTs have one dimensional Coulomb branch, and are related by mass deformations to 5d $\mathcal{N} = 1$ gauge theories with $SU(2)$ gauge group and $N_f \leq 7$ fundamental hypermultiplets. The gauge theory has $N_f + 1$ massive parameters, preserving $\mathcal{N} = 1$ supersymmetry. $N_f$ of them $m_a$ are the masses for the $N_f$ quark fields. This is associated with the $U(1)^{N_f} \subset SO(2N_f)$ flavor symmetries which rotate the quarks. The last massive parameter is the inverse-gauge coupling $1/g_{YM}^2$, which has dimension of mass in 5d. This is also associated with a global symmetry of the system, coming from the following topologically conserved current:

$$j_\mu \sim \star_5 \text{tr}(F \wedge F)_\mu . \tag{3.1}$$

One finds a $U(1)$ symmetry associated with $j_\mu$. The conserved quantity for $j_\mu$ is nothing but the instanton number, carried by the instanton soliton that we studied in Sect. 2. In fact, it is a general feature of 5d SCFTs that their relevant deformations are always associated with global symmetries [34]. To summarize, the gauge theory has $N_f + 1$ massive parameters associated with the $SO(2N_f) \times U(1)$ global symmetries. Turning off all these mass parameters, one expects to find a 5d SCFT [6]. In particular, since turning off $1/g_{YM}^2$ is a strong coupling limit of the 5d gauge theory, the SCFT can be regarded as sitting at the strong coupling limit of gauge theory. Since we have no control over this non-renormalizable gauge theory at strong coupling, this expectation has to be supported by using string theory embedding. For instance, [6] engineered this system by $N$ D4-branes probing an O8$^-$ plane and $N_f$ D8-branes, and took the strong coupling and low energy decoupling limits.

As mentioned in the introduction, we do not have a known Lagrangian description of this 5d SCFT. In various aspects, the CFT limit $m_a \to 0$, $1/g_{YM}^2 \to 0$ exhibits features different from standard gauge theories. One aspect is that the distinction between the solitonic particles and elementary particles become irrelevant. Namely, with mass deformations set at $m_i \ll \frac{1}{g_{YM}^2}$, the elementary 'quark' fields have masses at $m_i$, while the instanton soliton has much bigger mass $\frac{4\pi^2}{g_{YM}^2}$. Treating all these mass parameters at similar order and taking them to be small, one finds all of these particles being light at the same order. Here, note that the independent particle species involving solitons are much richer than those involving elementary quark fields only. For instance, this can be studied very rigorously in the BPS sector and Coulomb branch, by studying the instanton partition function of Sect. 2. This is because infinite tower of bound states are created if instantons are involved as constituents. Therefore, in the 5d SCFT limit $m_a, 1/g_{YM}^2 \to 0$, infinitely many species of particles are becoming massless. If any Lagrangian description exists, one challenge is to naturally exhibit this feature after generic relevant deformations.

One characteristic feature of the strong coupling fixed point is a symmetry enhancement. Namely, at general values of $m_a$, $1/g_{YM}^2$, one only finds $U(1)^{N_f+1}$ symmetry. When $m_a = 0$, still at $1/g_{YM}^2 \neq 0$, this enhances to $SO(2N_f) \times U(1)$, which is of course readily visible in the gauge theory. As one further sends the last mass parameter zero $1/g_{YM}^2 \to 0$, string theory predicts the following strong coupling enhancement of global symmetry [6]:

$$SO(2N_f) \times U(1) \rightarrow E_{N_f+1}. \tag{3.2}$$

For instance, this can be seen by dualizing the type I$'$ brane engineering of [6] (which uses D4-D8-D8) to the heterotic string theory. Although this symmetry is perturbatively visible in the heterotic side, this is a non-perturbative symmetry enhancement from the type I$'$ viewpoint. Instanton particles are expected to play crucial roles in making this enhancement possible. One goal of this subsection is to explain how one can test and further study such non-perturbative symmetries from the instanton partition functions.

Before getting into the detailed studies using instanton partition functions, let us summarize the patterns of enhancement in some detail, for each $N_f$. They are given as follows:

$$N_f = 0 : E_1 = SU(2) \supset U(1)_I, \tag{3.3}$$

$$N_f = 1 : E_2 = SU(2) \times SU(2) \supset SO(2) \times U(1)_I,$$

$$N_f = 2 : E_3 = SU(3) \times SU(2) \supset SO(4) \times U(1) \text{ (where } SU(3) \supset SU(2) \times U(1)_I),$$

$$\mathbf{8} \to \mathbf{3}_0 + \mathbf{1}_0 + \mathbf{2}_1 + \mathbf{2}_{-1}$$

$$N_f = 3 : E_4 = SU(5) \supset SO(6) \times U(1)_I, \quad \mathbf{24} \to \mathbf{15}_0 + \mathbf{1}_0 + \mathbf{4}_1 + \bar{\mathbf{4}}_{-1}$$

$$N_f = 4 : E_5 = SO(10) \supset SO(8) \times U(1)_I, \quad \mathbf{45} \to \mathbf{28}_0 + \mathbf{1}_0 + (\mathbf{8}_s)_{-1} + (\mathbf{8}_c)_1$$

$$N_f = 5 : E_6 \to SO(10) \times U(1)_I, \quad \mathbf{78} \to \mathbf{45}_0 + \mathbf{1}_0 + \mathbf{16}_{-1} + \overline{\mathbf{16}}_1$$

$$N_f = 6 : E_7 \supset SO(12) \times U(1)_I, \quad \mathbf{133} \to \mathbf{66}_0 + \mathbf{1}_0 + \mathbf{32}_1 + \mathbf{32}_{-1} + \mathbf{1}_2 + \mathbf{1}_{-2}$$

$$N_f = 7 : E_8 \supset SO(14) \times U(1)_I, \quad \mathbf{248} \to \mathbf{91}_0 + \mathbf{1}_0 + \mathbf{64}_1 + \overline{\mathbf{64}}_{-1} + \mathbf{14}_2 + \mathbf{14}_{-2}.$$

The branching rules for the adjoint representations of $E_{N_f}$ is shown, where the subscripts are the $U(1)_I$ instanton charges of the generators. Note that the extra conserved currents than $SO(2N_f) \times U(1)_I$ all carry nonzero instanton charges. This implies that the extra currents are realized only at strong coupling, when instantons become light.

In order to see these non-perturbative symmetries from the instanton partition functions in a simple manner, it is helpful to make an alternative expansion of $Z(q, v, \epsilon_{1,2}, m)$. In Sect. 2, one computed the coefficients of the expansion $Z =$

$\sum_{k=0}^{\infty} Z_k q^k$ using ADHM quantum mechanics. To see the symmetry enhancement, it is crucial to keep $q$ and $e^{-m_a}$'s on equal footing, as they are expected to combine and form the fugacities of $E_{N_f+1}$. Here, note that going to Coulomb branch with nonzero $v$ does not break the $E_{N_f+1}$ symmetry. So one makes a double expansion of $Z$ in $e^{-v}$ and $q$, focus on each order in the $e^{-v}$ expansion, and then collect various terms at different $q$ order. The coefficients at given order in $e^{-v}$ is expected to combine into characters of $E_{N_f+1}$. Such rearrangements of terms will be made after we compute $Z_k$ till higher enough order to see nontrivial structures.

Now we turn to the computational side, for the coefficients $Z_k$, as there turn out to be subtle but interesting physics. The $SU(2)$ instanton calculus with $N_f \leq 7$ fundamental hypermultiplets can be approached using three alternative approaches. One distinctive aspect is that $SU(2)$ can be regarded as a special case of either $SU(N)$ at $N = 2$, or $Sp(N)$ at $N = 1$ since $Sp(1) \sim SU(2)$. Such distinction is irrelevant in 5d SYM, at least at the perturbative level in which global structure of the gauge group is irrelevant. For instance, if one studies non-local defect operators, the distinction might matter. In our problem, studying non-perturbative instantons, the distinction matters for a slightly different reason. This is because the ADHM quantum mechanics does depend on whether one views this instantons as a special case of $SU(N)$ instantons or $Sp(N)$ instantons. The two ADHM quantum mechanics models are different, to start with, e.g. having $U(k)$ and $O(k)$ worldline gauge symmetry. This basically has to do with the fact that the ADHM quantum mechanics is a UV uplift of the non-linear sigma model on instanton moduli space. The latter has some UV incompleteness coming from small instanton singularities, which are resolved by GLSM uplifts. Of course, we are computing BPS partition functions which are invariant under the 1d RG flow, so one may expect that the two GLSMs may naturally yield same results. If this holds, this will be a sort of 1d IR duality. It is quite fair to expect such dualities, *supposing that* both descriptions make good sense in UV. It will turn out that there are three possible 1d UV uplifts of the NLSMs, and some of them become sick as $N_f$ increases beyond certain critical values. We first briefly summarize the three descriptions, together with their string theory engineerings, explaining when some of them go pathological.

The first description is obtained by regarding $SU(2)$ as a special case of $SU(N)$, and use $U(k)$ ADHM quantum mechanics for the $k$ instanton dynamics. With $N_f$ fundamental hypermultiplets in 5d, this can be engineered from 5-branes webs in the IIB string theory for $N_f \leq 4$. This is realized by first having segments of two parallel horizontal D5-branes, suspended between two vertical NS5-branes to realize $SU(2)$ gauge groups. This is supplemented by $N_f \leq 4$ semi-infinite horizontal D5-branes. Some examples are shown, e.g. in Fig. 4 of [17]. For $N_f > 5$, the two NS5-branes should cross each other at some point, failing to provide the desired 5d gauge theory as the decoupled 5d QFT system in IR (at least naively). Even for $N_f > 5$, there are ways of studying the corresponding 5d SCFT from similar 5-brane web setting, e.g. by introducing 7-branes as endpoints of 5-brane webs and making some deformations to the s-called Tao diagrams [35]. In this setting, one may use the topological vertex method to compute the instanton partition function, in seme sense. However, for $N_f > 5$, the ADHM quantum mechanics formulation fails to provide good descriptions. In 1d GLSM, $N_f$ fundamental hypermultiplets in 5d induces $N_f$ Fermi multiplet fields that are fundamental in $U(k)$. Consider the Coulomb branch moduli space in 1d, spanned by the scalar $\varphi$ in 1d vector multiplet. For simplicity, let us set $k = 1$ so that $\varphi$ is a real number. One can show (see e.g. [17]) that a 1-loop effective potential is induced by integrating out the 1d hypermultiplets and Fermi multplet, which is given by

$$V_{\text{eff}}(\varphi) \propto +(4 - N_f)|\varphi| . \tag{3.4}$$

Namely, the 1-loop effect induces a confining potential for $\varphi$ at $N_f < 4$, and a runaway potential to $|\varphi| \to \infty$ for $N_f > 4$. Here, note that $\varphi$ is part of the UV extra degrees of freedom, as one makes a 1d UV uplift of the NLSM on instanton moduli space. The above confining potential for $N_f < 4$ implies that quantum effects lift these light degrees of freedom. In 5-brane web picture, such as Fig. 4 of [17], instanton is a horiontal segment of D1-brane suspended between NS5-branes. $\varphi$ is the vertical position of such D1-brane. The linear confining potential means that, if the D1-brane tries to move away from the center of the web, to $|\varphi| \to \infty$, the length of the string grows. This is a technically good aspect, in that the UV uplift has no extra degrees of freedom. At $N_f = 4$, the 1-loop potential vanishes and the classical flat direction is unlifted. This corresponds to the right figure of Fig. 4 of [17]. So the 1d UV uplift does have extra light degrees of freedom. However, it can still be argued that such extra degrees of freedom decouples in IR with the physical ones on the instanton solitons of 5d QFT. It has been discussed in detail in section 3.4 of [17] how to separated them in our $Z$. However, for $N_f > 4$, $\varphi$ experiences a runaway potential to large $\varphi$. This corresponds to the fact that two NS5-branes cross each other. Then one does not know whether the extra UV d.o.f. from $\varphi$ decouple from the 5d QFT d.o.f. or not. In other words, the ADHM quantum mechanics becomes useless at $N_f > 4$, and perhaps is a wrong UV uplift. At $N_f < 4$, the calculus of $Z_k$ is quite straightforward. One uses the general results presented in Sect. 2, with

$$Z_{\text{hyper}} = \prod_{I=1}^{k} \prod_{a=1}^{N_f} 2 \sinh \frac{\phi_I + m_a}{2} \tag{3.5}$$

for the 1d Fermi multiplet fields. The Young diagram formula also holds.

There is another brane description which provides the good ADHM quantum mechanics till $N_f \leq 6$. This description views the gauge group as $Sp(1)$, and uses $O(k)$ gauged quantum mechanics for the ADHM description. In the 5-brane web setting, one has configurations similar to the Fig. 4 of [17], but with an extra O7$^-$ plane and 4 D7-branes inserted. See the last two paragraphs of section 3.4.1 of [17]. In this setting, 4 D7-branes provide four quarks in the fundamental representation.

In addition, one can attach up to two extra semi-infinite D5-branes (not double-counting the mirror images of O7). Thus one can add up to $N_f \leq 6$ fundamental hypermultiplets in this brane setting. This description has been used in [30] till $N_f \leq 5$ to study the $E_{N_f+1}$ symmetry enhancements. At $N_f = 6$, there is an extra issue of extra Coulomb branch states which decouples from 5d QFT. This effect was taking into account in [17], after which one sees the expected $E_7$ enhancement from this model. However, the case with $N_f = 7$ cannot be treated well in this setting, for the same reason as given in the previous paragraph. The contour integral formula for $Z_k$ is a bit more cumbersome than those of the previous paragraph, partly because $O(k)$ holonomies on $S^1$ consist of two parts, in $O(k)_+$ and $O(k)_-$. We refer to [17] for the details.

Finally, going back to the description of [19] using D4-D8-O8 system, the instanton quantum mechanics is obtained by putting $k$ extra D0-branes in this brane background. Considering generic $N \geq 1$ D4-branes to start with, the 5d QFT is an $Sp(N)$ gauge theory with $N_f$ fundamental and 1 rank 2 antisymmetric hypermultiplets. As before, $N_f$ fundamental hypermultiplets in 5d induces $N_f$ $O(k)$ fundamental Fermi multiplets in 1d. The 5d antisymmetric hypermultiplet induces complicated 1d fields. See appendix A of [17] for a summary. In eqn. (A.2) there, the underlined fields on first, second, third lines are 1d d.o.f. induced from the $Sp(N)$ antisymmetric hypermultiplet in 5d. The details will not be needed here. The point is that, such underlined 1d degrees of freedom exists even at $N = 1$. This might be a puzzle at first sight, since in 5d, $Sp(1)$ antisymmetric representation means neutral. So if one trusts the perturbative intuition from 5d SYM, this should not yield any effect in the 1d system. However, this is the UN incompleteness of the 5d Yang–Mills becomes manifest. One should really rely on the string theory engineering of the ADHM construction, which demands extra d.o.f. induced from '$Sp(1)$ antisymmetric hypermultiplet' in 5d. So one should include these fields, and carry out a careful index calculus as explained in our Sect. 2. One finally has to carefully factor out the extra UV d.o.f. as in section 3.4 of [17]. This way, one can indeed study the instanton partition function for all $N \leq 7$, seeing $E_{N_f+1}$ symmetry from the index.

To see the enhancement from the calculation, one has to multiply the perturbative partition function $Z_{\mathrm{pert}}$ to the instanton partition function $Z$ that we have been discussing, for an obvious reason. The perturbative partition function in the Coulomb branch (with $v > 0$) is given by

$$Z_{\mathrm{pert}} = PE\left[-\frac{2\cosh\epsilon_+}{2\sinh\frac{\epsilon_{1,2}}{2}}e^{-2v} + \frac{\chi_{\mathbf{v}}^{SO(2N_f)}(m)}{2\sinh\frac{\epsilon_{1,2}}{2}}e^{-v}\right], \qquad (3.6)$$

where $\chi_{\mathbf{v}} = \sum_{a=1}^{N_f}(e^{m_a} + e^{-m_a})$ is the $SO(2N_f)$ character for the vector representation. We simply show a few leading terms in the series expansion of

$Z_{\text{pert}}Z$, for sample theories at $N_f = 0$ and $N_f = 5$ [36]:

$$N_f = 0: 1 + \frac{t+q}{(1-t)(1-q)}\chi_2^{E_1}(q)A^2 + \left[\frac{(q^2+t^2)(q+t+q^2+t^2+qt(1+q+t))}{tq(1-t^2)(1-q^2)}\right.$$

$$\left. + \frac{(q+t+q^2+t^2+qt(1+q+t))}{(1-t)(1-t^2)(1-q)(1-q^2)}\chi_3^{E_1}(q)\right]A^4 + \mathcal{O}(A^6),$$ (3.7)

$$N_f = 5: 1 - \frac{q^{1/2}t^{1/2}}{(1-t)(1-q)}\chi_{27}^{E_6}A + \left[\frac{q+t}{(1-t)(1-q)}\chi_{27}^{E_6}\right.$$ (3.8)

$$\left. + \frac{qt}{(1-t)(1-q)}\chi_{351}^{E_6} + \frac{qt(q+t)}{(1-t)(1-t^2)(1-q)(1-q^2)}(\chi_{27}^{E_6})^2\right]A^2 + \cdots$$

where $(t, q) \equiv (e^{-\epsilon_1}, e^{\epsilon_2})$. $A$'s are basically the exponential of $SU(2)$ Coulomb VEV, which have to be suitably shifted by $E_{N_f+1}$ masses [36]. For instance, at $N_f = 0$, $A^4 = e^{-4v}q$. See [36] for the definition of $A$ at $N_f = 5$. These results show the enhanced $E_{N_f+1}$ symmetry, in that the index is rearranged into their characters, obeying the Weyl symmetry of $E_{N_f+1}$. At $N_f = 0$, this is simply the $q \to q^{-1}$ symmetry with $A$ fixed. At $N_f = 5$, the symmetry enhancement $SO(10) \times U(1)_I \to E_6$ is confirmed by showing the representations of $SO(10) \times U(1)_I$ combine well according to the following branching rules:

$$\mathbf{27} \to \mathbf{1}_{-4} + \mathbf{10}_2 + \mathbf{16}_{-1}$$ (3.9)

$$\mathbf{351} \to \mathbf{10}_2 + \overline{\mathbf{16}}_5 + \mathbf{16}_{-1} + \mathbf{45}_{-4} + \mathbf{120}_2 + \mathbf{144}_{-1}.$$

## 3.2   5d Instantons for 6d SCFTs on $S^1$

We now study instantons of 5d gauge theories, which describe 6d SCFTs compactified on $S^1$. 5d instanton particles here play the role of Kaluza–Klein momentum modes of the 6d QFT along $S^1$. For simplicity, in this subsection we shall focus on the 6d $(2, 0)$ SCFT of $A_{N-1}$ type. However, there are various generalizations in the literature, which we summarize here first. 6d $\mathcal{N} = (2, 0)$ theories of $D_N$ type have been studied using 5d descriptions in [37], using instantons of 5d maximal super-Yang–Mills with $SO(2N)$ gauge group. Similar techniques can be used to study 6d $\mathcal{N} = (1, 0)$ SCFTs on $S^1$. Perhaps the simplest such theory is the E-string theory, for M5-branes probing an M9-brane which hosts $E_8$ symmetry [38]. This is related by an $S^1$ compactification with $E_8$ holonomy to 5d $Sp(N)$ gauge theories with $N_f = 8$ fundamental and one antisymmetric hypermultiplets. Its instanton partition functions have been studied in [17, 35, 39], among others. Recently, many more 5d descriptions are engineered for 6d $(1, 0)$ theories compactified on $S^1$, either from geometric considerations or type IIB brane webs [11, 12, 40–44].

Now we focus on the maximal superconformal field theory in 6d of $A_{N-1}$ type. In M-theory, this system is engineered by taking $N$ coincident M5-branes, in the flat M-theory background. In the low energy limit, the system contains a 6d SCFT on M5-branes' worldvolume which is decoupled from the bulk. In the Coulomb branch, we take $N$ M5-branes separated along one of the five transverse directions of $\mathbb{R}^5$. There are self-dual strings provided by open M2-branes, suspended between separated M5-branes along this direction, also wrapping $\mathbb{R}^{1,1} \subset \mathbb{R}^{5,1}$ of the 5-brane worldvolume.

We are interested in the index of the circle compactified self-dual strings. The index is invariant under the change of continuous parameters of the theory, as well as other background parameters. So we can take the circle radius to be very small, and use the type IIA string theory description for the computation. M5-branes reduce to multiplet D4-branes, on which one finds the 5d maximal supersymmetric Yang–Mills theory with $U(N)$ gauge group. Let us denote by $v_i$ (with $i = 1, \cdots, N$) the $N$ scalar VEVs, or positions of $N$ M5-branes along $\mathbb{R} \subset \mathbb{R}^5$. These are related to the 5d VEVs by a multiplication of $R$. Let $n_i$ denote the number of self-dual strings ending on a given M5-branes, with orientations taken into account. If the strings have $k$ units of Kaluza–Klein momentum bound to them, one obtains in the small $R$ limit a system of $k$ D0-branes bound to fundamental strings with charges $n_i$ ending on the $i$'th $N$ D4-branes. In particular, the energy of the compactified self-dual strings is bounded as

$$E \geq \frac{k}{R} + v_i n_i \ . \tag{3.10}$$

In the regime with very small $R$, where we plan to compute the index, we can use the effective description with fixed $k$, as the particles with large rest mass $\sim R^{-1}$ become non-relativistic. So the quantum mechanics of $k$ D0-branes bound to $N$ D4-branes would capture the index $Z_k(v, \epsilon_{1,2}, m)$. The quantum numbers $n_i$ will be realized as $U(N)$ Noether charges of this ADHM quantum mechanics. This is simply the decoupling limit of the $k$ D0-branes bound to D4-branes, and could also be regarded as the discrete lightcone quantization (DLCQ) of M5-branes at given $k$ [45].

The quantum mechanics of $k$ D0-branes on $N$ D4-branes (in the Coulomb phase) preserves 8 SUSY, since the D0-D4 system preserves $\frac{1}{4}$ of the type IIA SUSY. The system has $SO(4) \sim SU(2)_l \times SU(2)_r$ rotation symmetry on D4 worldvolume transverse to D0, and $SO(5)$ rotation transverse to the D4's. When D4's are displaced along one of the five directions of $\mathbb{R}^5$, with VEV $v = \text{diag}(v_1, \cdots, v_N)$, $SO(5)$ is broken to $SO(4)' \sim SU(2)_L \times SU(2)_R$. We denote by $\alpha, \dot{\alpha}, a, \dot{a}$ the doublet indices of the four $SU(2)$'s, respectively, in the order presented above. $\dot{a}$ is equivalent to $A$ doublet index in the previous section, for $SU(2)_R$ in the $\mathcal{N} = (0, 4)$ SUSY. The 8 supercharges can be written by $Q_{\dot{\alpha}}^a$, $Q_{\dot{\alpha}}^{\dot{a}}$ which satisfy reality conditions. The 1d degrees of freedom are:

$$\text{D0-D0 strings}: U(k) \text{ adjoint } A_0, \quad (\varphi^{1,2,3,4} \sim \varphi_{a\dot{a}}, \varphi^5), \quad \lambda_\alpha^a, \quad \lambda_{\dot{\alpha}}^{\dot{a}}$$

$$U(k) \text{ adjoint } a_m \sim a_{\alpha\dot{\alpha}}, \quad \lambda_\alpha^a, \quad \lambda_\alpha^{\dot{a}}$$

$$\text{D0-D4 strings}: U(k) \times U(N) \text{ bi-fundamental } q_{\dot{\alpha}}, \quad \psi^a, \quad \psi^{\dot{a}} \qquad (3.11)$$

with $m = 1, \cdots, 4$. The D4-D4 strings move along $\mathbb{R}^4$ transverse to the D0's, and decouple at low energy. (These will be perturbative 5d SYM degrees.) This system can be formally obtained by a dimensional reduction of a 2 dimensional $\mathcal{N} = (4, 4)$ SUSY gauge theory, in which $Q_\alpha^a$ and $Q_{\dot{\alpha}}^{\dot{a}}$ respectively define 4 left-moving and right-moving supercharges. The first line of the above field content is called the $(4, 4)$ vector multiplet. The second and third line separately form a hypermultiplet. The action of this system is very standard, and could be found e.g. in [33], whose notations we followed here.

The index $Z_{k,}(v, \epsilon_{1,2}, m)$ is defined in this quantum mechanics by

$$Z_{k,\text{inst}}(v^I, \epsilon_{1,2}, m) = \text{Tr}\left[(-1)^F e^{-v \cdot n} e^{-2\epsilon_+(J_r + J_R)} e^{-2\epsilon_- J_l} e^{-m J_L}\right], \qquad (3.12)$$

where $n = (n_1, \cdots, n_N)$ denotes the $U(1)^N \subset U(N)$ charge, $\epsilon_\pm \equiv \frac{\epsilon_1 \pm \epsilon_2}{2}$, and $J_l, J_r, J_L, J_R$ are the Cartans of $SU(2)_l, SU(2)_r, SU(2)_L, SU(2)_R$, respectively. The trace is over the Hilbert space of the $k$ instanton quantum mechanics. Note that the measure in the trace commutes with two supercharges $Q = Q_{\dot{1}}^{\dot{1}}, Q^\dagger = -Q_{\dot{2}}^{\dot{2}}$ among $Q_{\dot{\alpha}}^{\dot{a}}$, as we explained at the beginning of this section.

The contour integral, or residue sum, expression for this index was obtained by Nekrasov [13]. It can be understood from the results summarized in our Sect. 2, for a 5d vector multiplet and an adjoint hypermultiplet. The contour integrand is given by $Z_{1\text{-loop}} \equiv Z_{\text{vec}}(\phi, v, \epsilon_{1,2}) Z_{\text{hyper}}(\phi, v, \epsilon_{1,2}, m)$, where

$$Z_{1\text{-loop}} = \frac{\prod_{I \neq J} 2 \sinh \frac{\phi_{IJ}}{2} \cdot \prod_{I,J=1}^k 2 \sinh \frac{\phi_{IJ}+2\epsilon_+}{2}}{\prod_{I,J=1}^k 2 \sinh \frac{\phi_{IJ}+\epsilon_1}{2} \cdot 2 \sinh \frac{\phi_{IJ}+\epsilon_2}{2}} \cdot \prod_{I,J=1}^k \frac{2 \sinh \frac{\phi_{IJ}\pm m-\epsilon_-}{2}}{2 \sinh \frac{\phi_{IJ}\pm m-\epsilon_+}{2}}$$

$$\cdot \prod_{I=1}^k \prod_{i=1}^N \frac{2 \sinh \frac{m \pm (\phi_I - v_i)}{2}}{2 \sinh \frac{\epsilon_+ \pm (\phi_I - v_i)}{2}}, \qquad (3.13)$$

where $\phi_{IJ} \equiv \phi_I - \phi_J$, and the sinh expressions with $\pm$ in the arguments mean multiplying the sinh factors with all possible signs. The factor on the second line comes from the integral over the fundamental hypermultiplet $q_{\dot{\alpha}}, \psi^a, \psi^{\dot{a}}$, and the second factor on the right hand side of the first line comes from the adjoint hypermultiplet $a_m, \lambda_\alpha^a, \lambda_{\dot{\alpha}}^{\dot{a}}$. Finally, the first factor on the first line comes from the vector multiplet nonzero modes $A_\tau, \varphi^I, \lambda_{\dot{\alpha}}^a, \lambda_{\dot{\alpha}}^{\dot{a}}$. As reviewed in Sect. 2, the associated residue sum is given by Flume and Poghossian [31], Bruzzo et al. [32], and Kim [33]

$$Z_{k,\text{inst}}(v, \epsilon_{1,2}, m) = \sum_{|Y|=k} \prod_{i,j=1}^{N} \prod_{s \in Y_i} \frac{\sinh \frac{E_{ij}+m-\epsilon_+}{2} \sinh \frac{E_{ij}-m-\epsilon_+}{2}}{\sinh \frac{E_{ij}}{2} \sinh \frac{E_{ij}-2\epsilon_+}{2}} \qquad (3.14)$$

with $E_{ij} = v_i - v_j - \epsilon_1 h_i(s) + \epsilon_2(v_j(s) + 1)$.

The result (3.14) is useful to understand various aspects of the $(2, 0)$ theory in Coulomb phase, and its self-dual strings compactified on a circle [33]. It is also useful to understand the conformal phase (with zero Coulomb VEV) of the theory. An early finding of this sort was that (3.14) could be used to study the index of the DLCQ $(2, 0)$ theory, which is the 6d CFT compactified on a light-like circle. Namely, one takes (3.14) and suitably integrates over the Coulomb VEV $v$ with Haar measure inserted, to extract out the gauge invariant spectrum [33]. More recently, and this will be reviewed in our Sect. 5, (3.14) was used as the building block of more sophisticated CFT observable, the superconformal index on $S^5 \times S^1$. Again several factors of the form (3.14) are multiplied (with other factors that we shall call the 'classical measure,' see Sect. 3), and we suitably integrate over the Coulomb VEV parameter $v$.

Now we explain what kind of interesting 6d physics one can study with this instanton partition function. This will basically be a review of [33]. A closely related study is made in [46], from the 2d QFT living on the 6d self-dual string worldsheet. The 2d QFT approach will be explained in Sect. 4. Among many, there are three major findings made in [33], using the instanton partition function: (1) clarifying the Kaluza–Klein (KK) mode structures of single M5-brane theory from 5d viewpoint, (2) similar studies for multiple M5-branes, (3) exploration of the so-called nonrelativistic superconformal index for the DLCQ M5-brane theory. In this section, we shall explain (1), and also (2) briefly. For (3), we refer the readers to [33].

Having computed the partition function $Z_k$ for a given quantum mechanics, we then consider their generating function $Z = \sum_k Z_k q^k$. We complete it by multiplying the perturbative partition function to $Z$, which in the case of 5d maximal super-Yang–Mills theory is given by Kim [33]

$$Z_{\text{pert}} = PE \left[ \frac{\sinh \frac{m \pm \epsilon_+}{2}}{\sinh \frac{\epsilon_{1,2}}{2}} \chi_{\mathbf{adj}}^{U(N)}(v) \right]. \qquad (3.15)$$

The partition function $Z_{\text{pert}}Z$ is that of 5d maximal SYM on $\mathbb{R}^4 \times S^1$, where $S^1$ is the temporal circle. Since the instanto number $k$ provides the KK momenta, we expect this to be a 6d partition function on $\mathbb{R}^4 \times T^2$. The second circle of $T^2$ is the spatial one, expected to be 'emergent' after one sums over $k$. Whether the partition function behaves properly as the 6d partition function is the essence of the questions (1) and (2) above.

The question is answered in detail in [33] for $N = 1$. This question might (apparently) look uninteresting, in that the expected 6d system is the free QFT on $S^1$, made of self-dual tensor multiplet. So rather than exploring novel aspects of interacting 6d QFTs, this is a test on the approaches based on 5d instanton particles. However, since the 5d approach is closed related to the type IIA approach to the M-theory on a circle (with extra D4-branes), this is sort to testing the old and famous idea of the M-theory deconstructed by IIA D0-branes.

At any rate, if the instanton partition function is well organized to see the 6d QFT physics, one expects that the single particle index exhibits same degeneracy at all KK momentum level $k$. This is because the expected 6d system consists of free fields, whose KK decomposition will yield same massive field contents at every $k$. Indeed, this was shown in [33], with precisely the expected charge contents for BPS states. To see this, one has to better understand the structures of the residue sums given in terms of Young diagrams. For instance, after some rearrangements, it was shown that [33]

$$Z_1 = \frac{\sinh \frac{m+\epsilon_-}{2} \sinh \frac{m-\epsilon_-}{2}}{\sinh \frac{\epsilon_1}{2} \sinh \frac{\epsilon_2}{2}} \tag{3.16}$$

$$Z_2 = \frac{Z_1^2 + Z_1(2\epsilon_{1,2}, 2m)}{2} + Z_1$$

$$Z_3 = \frac{Z_1^3 + 3Z_1 \cdot Z_1(2\epsilon_{1,2}, 2m) + 2Z_1(3\epsilon_{1,2}, 3m)}{6} + Z_1^2 + Z_1$$

and so on. The meaning of these formulae is as follows. The first two terms on the right hand side of $Z_2$ provide the partition function of two identical particles, where each particle's spectrum is described by $Z_1$. In particular, the two particles are unbound to each other. Had there not been the last term of $Z_2$, this would have been the partition function of non-interacting two identical particles. The last term $Z_1$ is thus interpreted as the single particle partition function, made of the bound states of the two identical particles. Since these two particles are mutually BPS and thus do not exchange forces, the bound states are all threshold bounds, having zero binding energy. Therefore, we indeed find that the single particle partition functions at both $k = 1, 2$ are given by $Z_1$. With these understood, let us then consider $Z_3$. The first three terms are again 3-particle partition function of identical three particles, unbound to others. The next term $Z_1^2$ is a 2-particle partition function, one $Z_1$ coming from a particle at $k = 1$ and another coming from a 2-body bound state at $k = 2$ that we have just identified. Indeed, since we have completely identified the particle contents till $k \leq 2$, this term should exist as the two particles do not feel mutual forces. Then, the last term $Z_1$ is interpreted as coming from 3-body threshold bound states. Again, the 3-body bound state has the same spectrum as the single particle states at $k = 1, 2$, described by the same partition function $Z_1$.

So these formulae illustrate that, up to $k \leq 3$, the single particle spectrum is indeed consistent with the expectation from the 6d free QFT. One can make this study more systematic, by trying to rearrange the infinite sum $Z = \sum_k Z_k q^k$ and see this structure generally. In fact, the Young diagram sum for $N = 1$ has been explicitly carried out in [47]. The result is given by

$$Z(q, \epsilon_{1,2}, m) = PE\left[Z_1(\epsilon_{1,2}, m)\frac{q}{1-q}\right] \equiv \exp\left[\sum_{n=1}^{\infty}\frac{1}{n}Z_1(n\epsilon_{1,2}, nm)\frac{q^n}{1-q^n}\right].$$
(3.17)

[Note that there is no $v$ dependence at $N = 1$ because it correspond to decoupled $U(1)$.] Expanding this in $q$, one indeed finds coefficients of the forms (3.16). The expression inside $PE$ is the index over single particles, given by

$$Z_1(q + q^2 + q^3 + \cdots).$$
(3.18)

This implies that the single particle spectrum of instantons are indeed the same at all $k$, consistent with the 6d free QFT picture. A more detailed study of the form of $Z_1$ reveals that this is indeed the index over single massive tensor multiplet in 5d, which originates from the 6d self-dual tensor multiplet [33].

The partition functions at $N > 1$ are more involved, since the 6d QFTs are interacting. Especially, compactifying an interacting QFT on $S^1$, one cannot predict that the spectrum is same at all $k$'s.[1] However, studies are made in [33] (see also [46] and our Sect. 4.1) by expanding $Z_{\text{pert}}Z$ in the Coulomb VEV's $v$. We first explain the precise meaning of this expansion, and the significance of such studies from the 6d viewpoint.

We are in the Coulomb branch of the 5d Yang–Mills, or equivalently in the tensor branch of the 6d QFT (whose meaning will be explained in Sect. 4). This corresponds to separating $N$ D4-branes (or $N$ M5-branes) along one transverse direction. We introduce $N$ Coulomb VEV's, $v_1, \cdots, v_N$, labeling their positions. Using the Weyl symmetry of $U(N)$, we can order them as $v_1 > v_2 > \cdots > v_N$. Then one can expand $Z_{\mathbb{R}^4 \times T^2} \equiv Z_{\text{pert}}Z$ in positive powers of the $N - 1$ fugacities $e^{-(v_i-v_{i+1})}$ ($< 1$). They are the fugacities of the $U(1)^{N-1} \subset SU(N)$ electric charges, carried by the W-bosons. They come from open strings suspended between D4-branes. The expansion takes the form of

$$Z_{\mathbb{R}^4 \times T^2}(q, v, \epsilon_{1,2}, m) = \sum_{n_1, \cdots, n_{N-1}=0}^{\infty} e^{-\sum_{i=1}^{N-1} n_i(v_i - v_{i+1})} Z_{\{n\}}(q, \epsilon_{1,2}, m).$$
(3.19)

---

[1]There appears to be confusions on this point in some literature, expecting unique threshold KK bounds even at $N > 1$. We find that there are no physical grounds to expect so. One the other hand, in the analogous problem of D0-brane threshold bounds in the IIA string theory, one *does* expect unique threshold bounds at all $k$'s. This is because the expected 11d theory is supergravity, which is free at low energy.

The coefficients $Z_{\{n\}}$ now has the following interpretation. Note that the open strings with charges $\{n\}$ uplift in M-theory to open M2-branes suspended between M5-branes, providing the self-dual strings on the M5-brane worldvolume. The strings wrap the circle $S^1$. The charges $\{n\}$ thus uplift to the winding numbers of these strings. In the partition function, their worldsheets wrap $T^2$. [This is the reason why we gave the name $Z_{\mathbb{R}^4 \times T^2}$.] So $Z_{\{n\|}$ is interpreted as the elliptic genus of the 2d QFT living on these strings, at definite winding number. Defining $q \equiv e^{2\pi i \tau}$, $\tau$ is the complex structure of $T^2$, since its conjugate instanton number uplifts to the momentum on $S^1$.

We expect very strong properties of the elliptic genera, $Z_n(\tau, \epsilon_{1,2}, m)$, coming from the modular invariance. More precisely, this is a partition function of mass-deformed 2d QFT's, so one expects a specific modular anomaly. (See, for instance, [29].) So although we do not have very specific expectation on what $Z_n$ should be, unlike the case with $N = 1$, we have definite expectations on how $Z_n$ should behave under the modular transformations. Testing this is basically asking whether the 5d instanton sum properly exhibits the structure expected from emergent $S^1$ geometry. This will be addressed in Sect. 4.1 in detail. Here we focus on a special sector in which we do have concrete expectations on the functional form of $Z_n$.

The specific sector we would like to study is the single string sector, $n = 1$, of the QFT at $N = 2$. This correspond to the sector with one W-boson in 5d. This sector can also be embedded into single string sectors of $U(N)$ corresponding to the simple roots of $SU(N)$. Consider a 6d self-dual strings consisting of single open M2-brane, suspended between two separated M5-branes. One also allows momentum to flow on the string. If one compactifies this system on $S^1$ with string wrapped on it, one gets back to our 5d description with a W-boson, with various numbers of instantons bound to it. On the other hand, compactifying the 6d system on $S^1$ with the string transverse to it, one obtains a 5d gauge theory with one monopole string. One can further compactify it on another $S^1$, where the monopole string wraps this circle. In the latter description, one can rely on the description of moduli space dynamics, for a single $SU(2)$ monopole string. This is given by a non-linear sigma model on $\mathbb{R}^3 \times S^1$, which is the moduli space of this monopole. The $\mathbb{R}^3$ factor comes from the position of this string on uncompactified spatial $\mathbb{R}^3$. $S^1$ factor comes from the unbroken $U(1) \subset SU(2)$ gauge orientation. In the M5-brane picture, this circle should be originating from the position of the self-dual string along the transverse $S^1$, which we compactified transverse to the string. The NLSM has $\mathcal{N} = (4, 4)$ supersymmetry on the worldsheet.

The problem of monopole string is in general not identical to the problem of self-dual string, in that we compactified an extra direction on $S^1$. However, at the level of NLSM, it is easy to go back to the latter problem by decompactifying the $S^1$ in the target space. Therefore, one obtains an $\mathcal{N} = (4, 4)$ non-linear sigma model on $\mathbb{R}^4$, as the worldsheet description of single self-dual string at $N = 2$. This is a free QFT. [In fact it is free already as a QFT on the monopole string, having a locally flat target space $\mathbb{R}^3 \times S^1$.] So one can easily compute the elliptic genus of this free

QFT, whose result is [33]

$$Z_1 = \frac{\theta_1(\frac{m+\epsilon_+}{2\pi i})\theta_1(\frac{m-\epsilon_+}{2\pi i})}{\theta_1(\frac{\epsilon_1}{2\pi i})\theta_1(\frac{\epsilon_2}{2\pi i})} = \frac{\sinh \frac{m+\epsilon_+}{2} \sinh \frac{m-\epsilon_+}{2}}{\sinh \frac{\epsilon_1}{2} \sinh \frac{\epsilon_2}{2}} \prod_{n=1}^{\infty} \frac{(1-q^n e^{\pm m \pm \epsilon_+})}{(1-q^n e^{\pm \epsilon_1})(1-q^n e^{\pm \epsilon_2})} .$$

(3.20)

Here, the factors with $\pm$ signs inside the product are all multiplied. In [33], the 5d partition function $Z_{\text{pert}}Z$ was expanded in $e^{-(v_1-v_2)}$, and the $n = 1$ order terms are collect up to a very high order (i.e. till $k = 10$), which completely agrees with (3.20). This establishes a nontrivial support to the fact that the 5d partition function indeed sees the right structure of the 6d QFT (i.e. the 2d QFT on the strings). More systematic studies of this sort are explained in Sect. 4.1, using the results of [46].

# 4  Instanton Strings in 6d QFTs

6d Yang–Mills theories are non-renormalizable. It should be understood as effective field theories of consistent QFTs. We consider 6d $\mathcal{N} = (1, 0)$ supersymmetric Yang–Mills theories which arise as the so-called tensor branch effective action of 6d SCFTs. Such systems should always have Abelian self-dual tensor multiplets, coupling to the vector multiplets. Let us first explain the structures of these Yang–Mills theories in detail, as well as related systems which have only tensor multiplets without vector multiplets.

For simplicity, we only consider the cases with one tensor multiplet, and a vector multiplet whose gauge group $G$ is simple. The tensor multiplet consists of a 2-form $B_{\mu\nu}$ (whose 3-form flux satisfies a self-duality condition), a real scalar $\Phi$, and fermions. When the 6d theory has a gauge symmetry $G$, its vector multiplet consists of the gauge field $A_\mu$ and fermions. There may be hypermultiplet matters in the representation $\mathbf{R}$ of $G$. In the tensor branch, this system admits an effective field theory description in which the VEV $\langle \Phi \rangle > 0$ sets the inverse gauge coupling $\frac{1}{g_{YM}^2}$ of the 6d Yang–Mills theory. The bosonic part of the tensor/vector multiplet action is given by

$$S_{\text{vector+tensor}}^{\text{bos}} = \int \left[ \frac{1}{2} d\Phi \wedge \star d\Phi + \frac{1}{2} H \wedge \star H \right] + \sqrt{c} \int [-\Phi \text{tr}(F \wedge \star F) + B \wedge \text{tr}(F \wedge F)]$$

(4.1)

with certain $c > 0$ that depends on the theory, where

$$H \equiv dB + \sqrt{c} \, \text{tr}\left( AdA - \frac{2i}{3}A^3 \right).$$

(4.2)

This action should be understood as providing the equation of motion by varying the action with $B_{\mu\nu}$, after which the self-duality constraint $H = \star H$ is imposed by hand. The hypermultiplet part of the action (coupling to the vector multiplet) is standard, which we do not explain here.

The equation of motion for $B$ is given by

$$d \star H \, (= dH) = \sqrt{c} \, \mathrm{tr}(F \wedge F) \,. \tag{4.3}$$

Self-dual string solutions should have their tensions proportional to $\langle \Phi \rangle$, and source nonzero $H = \star H$. The configurations with nonzero $\mathrm{tr}(F \wedge F)$, namely the instanton string solitons, provide such sources. They are extended along $\mathbb{R}^{1,1} \subset \mathbb{R}^{5,1}$ and satisfy

$$F = \pm \star_4 F \tag{4.4}$$

on the transverse $\mathbb{R}^4$, where the instanton number $k = \frac{1}{8\pi^2} \int_{\mathbb{R}^4} \mathrm{tr}(F \wedge F)$ is quantized. BPS self-dual strings further satisfy $H = \mp \star_4 d\Phi$. Here, the upper/lower signs correspond to $k > 0$ and $k < 0$, respectively. We shall consider self-dual instantons with $k > 0$.

Let us explain various consistency conditions for the 6d SCFTs, from the Yang–Mills description. We first discuss the gauge anomaly cancelation. Gauge anomalies come both from tree and 1-loop levels, which should cancel each other via the Green-Schwarz mechanism [48, 49]. Under the gauge transformation $\delta A_\mu = D_\mu \epsilon$ and certain $\delta B_{\mu\nu}$ to be determined, one finds

$$\delta H = d\delta B + 2\sqrt{c} \, \mathrm{tr}[D\epsilon \wedge F] - \sqrt{c} \, d\mathrm{tr}[A \wedge D\epsilon] = d\delta B + \sqrt{c} \, d\,[2\mathrm{tr}(\epsilon F) - \mathrm{tr}(A \wedge D\epsilon)] \,. \tag{4.5}$$

We want the 3-form field strength $H$ to be physical. $\delta B$ can be chosen to achieve $\delta H = 0$, i.e. $\delta B = -\sqrt{c} \, \mathrm{tr}[2\epsilon F - A \wedge D\epsilon]$. Note that, had the vector fields been Abelian, one would have obtained $\delta B_{\mu\nu} = -\sqrt{c} \, \epsilon F_{\mu\nu}$. Then, the action suffers from the classical anomaly

$$\delta S = \delta \left[ \sqrt{c} \int B \wedge \mathrm{tr}(F \wedge F) \right] = -\sqrt{c} \int \delta B \wedge \mathrm{tr}(F \wedge F) \,. \tag{4.6}$$

This turns out to contribute a term proportional to $c \, \mathrm{tr}(F^2)^2$, to the anomaly polynomial 8-form. The 1-loop anomaly from the box diagrams, where fermions in the vector and hypermultiplets run through the loop, contributes terms proportional to $\mathrm{tr}_{\mathrm{adj}}(F^4)$ and $\mathrm{tr}_{\mathbf{R}}(F^4)$. For the net anomaly to cancel, the combination of the quartic Casimirs $\mathrm{tr}_{\mathrm{adj}}(F^4)$ and $\mathrm{tr}_{\mathbf{R}}(F^4)$ appearing at 1-loop should factorize to a square of quadratic Casimirs. The factorization condition of the quartic Casimir severely constrains possible gauge groups and matters. Firstly, the factorization can happen when $G$ does not have independent quartic Casimirs. Among simple groups, this is true for $G = SU(2), SU(3), G_2, F_4, E_6, E_7, E_8$. So for these $G$, one can introduce hypermultiplets in any representation $\mathbf{R}$, if it is compatible with $c > 0$. Exceptionally, $G = SO(8)$ may also yield anomaly-free systems. This is because $\text{tr}_{\text{adj}}(F^4)$ can be factorized as $\text{tr}(F^2)^2$ for $SO(8)$. So the 6d $SO(8)$ super-Yang–Mills without matters can be made anomaly free. If $\mathbf{R}$ is taken to be a suitable representation of $SO(8)$ whose quatic Casimir factorizes, such hypermultiplets can also be introduced. An example is $\mathbf{R} = n(\mathbf{8}_v \oplus \mathbf{8}_s \oplus \mathbf{8}_c)$. There are other possible choices of $G$ and $\mathbf{R}$, such as $G = SU(N)$ and $N_f = 2N$ fundamental hypermultiplets.

The global anomalies [50] further constrain possible $\mathbf{R}$. For instance, among the gauge groups mentioned in the previous paragraph, the cases with $G = SU(2)$, $G_2$ suffer from global anomalies without matters. So if one restricts oneself to theories with no hypermultiplets, $G = SU(2)$ and $G_2$ are forbidden. More generally, the numbers of some simple hypermultiplets are constrained as [50]

$$
\begin{aligned}
SU(2) &: n_{\mathbf{2}} = 4, 10, \cdots && \in 4 + 6\mathbb{Z}_{\geq 0} \\
SU(3) &: n_{\mathbf{3}} = 0, 6, 12, \cdots && \in 6\mathbb{Z}_{\geq 0} \\
G_2 &: n_{\mathbf{7}} = 1, 4, 7, \cdots && \in 1 + 3\mathbb{Z}_{\geq 0} \,,
\end{aligned}
\tag{4.7}
$$

where $\mathbb{Z}_{\geq 0} = \{0, 1, 2, 3, \cdots\}$, to avoid global anomalies.

Going back to the ordinary gauge anomaly, suppose that the 1-loop anomaly factorizes. Then, the coefficient $c$ appearing in the classical action has to be tuned to have the classical anomaly to cancel the factorized 1-loop anomaly. One has to make sure that the value of $c$ that ensures anomaly cancelation is positive. If this condition is not met, one fails to have a conformal field theory decoupled from gravity. Here, we note that the signs in front of the 1-loop anomaly contributions are different for vector multiplet and hypermultiplets. Anomaly from vector multiplet tends to increase $c$, while those from hypermultiplets tend to decrease $c$. So to have $c > 0$, $\mathbf{R}$ should not be too big [51]. For instance, in the list (4.7), $c > 0$ is satisfied at $n_{\mathbf{2}} = 4, 10$ for $SU(2)$, at $n_{\mathbf{6}} = 0, 6, 12$ for $SU(3)$, and at $n_{\mathbf{7}} = 1, 4, 7$ for $G_2$.

We are going to consider the instanton strings' partition functions in the 6d Yang–Mills theories of the sort explained so far, compactified on $S^1$. The full partition function, summed over the instanton strings' winding numbers on $S^1$, would be a partition function on $\mathbb{R}^4 \times T^2$. Here, since the tensor VEV $v = \langle \Phi \rangle$ sets the effective 6d coupling constant $1/g_{\text{YM}}^2$, it also sets the tension of the instanton strings. Therefore, $v$ is regarded as the chemical potential coupling to the instanton number ($\sim$ wining number of instanton strings).

Before proceeding, we also note that the notion of self-dual strings is more general than just instanton strings. This is because the former also has meaning in the 6d SCFTs which do not contain 6d vector multiplets. Typical examples are 6d $\mathcal{N} = (2, 0)$ SCFTs, or the E-string theories, whose tensor branch effective actions contain free tensor multiplets only. Even in these cases, there are strings which source the tensor field $H$ by delta functions supported on the string worldsheet:

$$
d \star H = dH \sim \delta^{(4)} \,.
\tag{4.8}
$$

Note that the right hand side is sourced by a singular function, rather than a smooth source $\sim \mathrm{tr}(F \wedge F)$, due to the lack of smooth gauge theory solitons. However, the quantum partition functions of these strings can be treated in a uniform manner.

In particular, we shall be interested in the Witten index which counts the BPS degeneracies of these strings wrapping the circle. Namely, the 6d CFT is put on $\mathbb{R}^{4,1} \times S^1$, and there are $r$ real scalar VEVs $v^I$ ($I = 1, \cdots, r$). The index is defined by

$$Z_{\{n_I\}}(\tau, \epsilon_{1,2}, m) = \mathrm{Tr}\left[(-1)^F q^{\frac{H'+P}{2}} e^{-\epsilon_1(J_1+J_R)-\epsilon_2(J_2+J_R)} e^{-m\cdot\mathcal{F}}\right], \qquad (4.9)$$

where $H'$ is the energy over the string rest mass $Rv^I n_I$, $q \equiv e^{2\pi i\tau}$, $J_1 \equiv J_l + J_r$, $J_2 \equiv J_r - J_l$, and $\mathcal{F}$ collectively denotes all the other conserved global charges which commute with the supercharges. The charges appearing inside the trace is chosen so that they commute with the two supercharges $Q_{\dot{1}}^1$, $Q_{\dot{2}}^2$, among $Q_{\dot{\alpha}}^A$. From the 5d or 6d superalgebra, the most general states preserving these two supercharges will be the $\frac{1}{2}$-BPS states preserving all $Q_{\dot{\alpha}}^A$. So with this index we are counting the states in the $\frac{1}{2}$-BPS multiplet, with a further refinement given by $J_r + J_R$ (which does not commute with all four $Q_{\dot{\alpha}}^A$). We also define the partition function $Z(v^I, \tau, \epsilon_{1,2}, m)$ by summing over the winding numbers of the self-dual strings,

$$Z(v^I, \tau, \epsilon_{1,2}, m) = \sum_{n_1, \cdots, n_r=0}^{\infty} e^{-v^I n_I} Z_{n_I}(\tau, \epsilon_{1,2}, m), \qquad (4.10)$$

where $Z_{n_I=0} \equiv 1$. Here, we introduce the chemical potentials $v^I$ conjugate to the winding numbers $n_I$.

$Z(v^I, \tau, \epsilon_{1,2}, m)$ is computed in various ways. Currently, in nontrivial theories, it is only computable in series expansions. One series expansion takes the form of (4.10), and the coefficients $Z_{n_I}(\tau, \epsilon_{1,2}, m)$ are computed from the elliptic genera of suitable 2 dimensional supersymmetric quantum field theories living on the worldsheets of these strings. A different kind of series expansion can be made in $q = e^{2\pi i\tau}$, when $q \ll 1$:

$$Z(v^I, \tau, \epsilon_{1,2}, m) = \sum_{k=0}^{\infty} q^k Z_k(v^I, \epsilon_{1,2}, m). \qquad (4.11)$$

The momentum charge $k$ on $S^1$ is given a weight $q^k$. These Kaluza–Klein momentum states are regarded as massive particles in 5d. $Z_k(v^I, \epsilon_{1,2}, m)$ can be computed from the quantum mechanics of the 'instanton solitons' of 5 dimensional gauge theory, if one has a 5d weakly coupled SYM description at small radius. This is what we have explained in Sect. 3.2. Here we note that the terminology 'instantons' in Sect. 3.2 and this section should not be confused. In Sect. 3.2, they are

5d particles providing KK modes along the 6d circle $S^1$. In this section, 'instantons' will denote instanton strings with respect to the 6d gauge group.

We first explain the general ideas of computing the coefficients $Z_{n_I}(q, \epsilon_{1,2}, m)$, before studying examples. The computation essentially relies on the string theory completion of the 6d SCFT, and suitable decoupling limits when the contribution of some charges to the BPS mass become large.

As the 6d SCFT lacks intrinsic definition, we rely on its string theory or M-theory engineering. In all such constructions, one engineers suitable string/M-theory backgrounds, and takes suitable low energy decoupling limits in which the 6 dimensional states decouple from the bulk states (e.g. 10/11 dimensional gravity, stringy states, so on). After this limit, certain 6 dimensional decoupled sector of 6d SCFT exists. Furthermore, we are interested in the 1+1 dimensional strings in the Coulomb phase, with nonzero VEV for the 6d scalar $v$ whose mass dimension is 2. The tension of the self-dual strings is proportional to $v$. We compactify these strings on a circle with radius $R$, and consider the states with windings $n_I$ and momentum $k$. As long as $\frac{1}{R} \ll vR$, the winding energy cost $\sim vR$ is much bigger than the worldsheet momenta's energy cost $\frac{1}{R}$, so that one can rely on Born-Oppenheimer approximation. Namely, we study the sector with fixed windings $n_I$, focussing on the worldsheet dynamics, with the guarantee that different sectors of the 6d theory at different winding numbers $n_I$ decouple in this regime. (This is like nonrelativistic approximation when the rest mass is much larger than the kinetic energy, in which case one can fix the particle number.) In this regime $R^{-1} \ll vR$, the characteristic wavelength $\frac{1}{R}$ of our interest on the worldsheet should be much smaller than $v^{\frac{1}{2}}$. $Z_{n_I}(q, \epsilon_{1,2}, m)$ can be computed in this regime by studying the 2d QFT living on the strings' worldsheet at fixed $n_I$. Although the computation is done in the regime $R^{-1} \ll v^{\frac{1}{2}}$, one may trust the index beyond this regime since it is independent of the continuous parameters $R, v$.

We generally expect the 2d QFT to be an interacting conformal field theory. The computation of the observables is generally very difficult with strongly interacting QFT. Here, the crucial step is to engineer a 2d gauge theory which is weakly coupled in UV, and flows to the desired interacting CFT in the IR. The construction of the UV gauge theory will often be easy with brane construction engineering of the 6d SCFT and the associated self-dual strings. Such UV gauge theories are constructed for the self-dual strings of a few interesting 6d CFTs, such as 'M-strings' [46], 'E-strings' [39], and some others [27, 52–54]. The UV gauge theories for many interesting self-dual strings are still unknown at the moment (and they are not guaranteed to exist at all). With a weakly-coupled UV gauge theory which flows to the desired CFT, the elliptic genus can in principle be easily computed from the UV theory [29], as the elliptic genus is independent of the continuous coupling parameters of the theory.

Before proceeding with concrete examples, we also comment that the quantity $Z(v^I, \tau, \epsilon_{1,2}, m)$ can often be computed from topological string amplitudes on suitable Calabi–Yau threefolds. This happens when the 6d SCFTs are engineered from F-theory on singular elliptic Calabi–Yau threefolds [2, 3, 5]. Changing the moduli of $CY_3$ in a way that specific 2-cycles shrink to zero volume, one obtains a 6 dimensional singularity which supports decoupled degrees of freedom at low energy, defining 6d SCFTs. One important ingredient of these theories is D3-branes wrapping these collapsing 2-cycles, which yield self-dual strings that become tensionless in the singular limit. Therefore, the volume moduli of these 2-cycles are the Coulomb branch VEVs $v^I$ in the 6d tensor supermultiplets.

So in this setting, we consider the F-theory on $\mathbb{R}^{4,1} \times S^1 \times CY_3$ in the Coulomb phase. We wrap D3-branes along $S^1$ times the 2-cycles in $CY_3$. This system can be T-dualized on $S^1$ to the dual circle $\tilde{S}^1$ of the type IIA theory. The D3-branes map to D2-branes transverse to $\tilde{S}^1$. Consider the regime with large $S^1$, or equivalently small $\tilde{S}^1$, and make an M-theory uplift on an extra circle $S_M^1$. Then $\tilde{S}^1$ and $S_M^1$ combine to a torus and fiber the 4d base of the original $CY_3$ we started from, meaning that we get M-theory on the same $CY_3$. The self-dual string winding numbers over the 2-cycles maps to the M2-brane winding numbers on the same cycles. The momentum on $S^1$ maps to M2-brane winding number on the $T^2$ fiber. So the counting of the self-dual string states maps to counting the wrapped M2-branes on $CY_3$ in M-theory. The last BPS spectrum is computed by the topological string partition function on $CY_3$ [14, 15]. In particular, consider an expansion of $Z_{\mathbb{R}^4 \times T^2}$ in the rotation parameters $\epsilon_1, \epsilon_2$ given by

$$
Z_{\mathbb{R}^4 \times T^2}(v^I, q, \epsilon_{1,2}, m) = \exp \left[ \sum_{n \geq 0, g \geq 0} (\epsilon_1 + \epsilon_2)^n (\epsilon_1 \epsilon_2)^{g-1} F^{(n,g)}(v^I, q, m) \right].
$$
(4.12)

The coefficients of the expansion $F^{(n,g)}(v^I, q, m) = \sum_{n_I, k, f} e^{-v^I n_I} q^k e^{-m \cdot f}$ $F^{(n,g)}_{n_I, k, f}$ are computed by the topological string amplitudes on $CY_3$. The series in (4.12) is the genus expansion of refined topological string. So from this viewpoint, the elliptic genus we study in this section is the all genus sum of the topological string amplitudes. A few low genus expansions are known for many interesting 6d self-dual strings. This provides an alternative method of computing some data of the full elliptic genus when neither 2d nor 1d gauge theories are known. For instance, see [52] for the results 6d strings engineered by F-theory on Hirzebruch surfaces, where many such strings do not have known gauge theory descriptions (yet).

## 4.1 Strings of 6d $\mathcal{N} = (2, 0)$ Theories

The self-dual strings of 6d $\mathcal{N} = (2, 0)$ theory of $A_{N-1}$ type can be engineered as follows. The 6d QFT's are engineered from $N$ parallel M5-branes. The strings are engineered by open M2-brane suspended between M5-branes. We would like to make a type IIA string theory reduction by compactifying this system transverse to both M2- and M5-branes. Then, this reduces to open D2-branes suspended between separated NS5-branes. On the $1 + 1$ dimensional worldvolume of the strings, one finds an $\mathcal{N} = (4, 4)$ gauge theory. This turns out to be quite inconvenient description to describe the self-dual strings, which is supposed to arise after the RG flow of the 2d gauge theory [46]. This is because of the following reason. The transverse space to the NS5-branes is $\mathbb{R}^4$. If the NS5-branes are separated along a line, the remaining transverse space if $\mathbb{R}^3$, exhibiting $SO(3)$ internal symmetry. Combined with the M-theory circle, the transverse space to the separated M5-branes is $\mathbb{R}^3 \times S^1$. In the strong coupling limit, this decompactifies to $\mathbb{R}^4$, with $SO(4)$ symmetry. It turns out that explicitly seeing both Cartans of $SO(4)$ is crucial for studying the elliptic genera of these strings. However, the 2d QFT from the IIA description summarized above does not see both Cartans, in the UV gauge theory.

In [46, 55], alternative UV gauge theories are engineered on the worldsheet of these strings. This is engineered by using the same configuration of D2-NS5-branes, but putting an extra D6-brane which is transverse to the $\mathbb{R}^3$. This system has one more $U(1)$ symmetry than the one without the D6-brane, which uplifts in the M-theory limit to the missing $SO(4)$ Cartan [46, 55]. For simplicity, we explain the 2d QFT only for the case with $A_1$, i.e. two M5-branes at $N = 2$. The 2d fields contents are first presented (in a more generalized version) in [55]:

$$(A_\mu, \lambda_{-\dot{\alpha}A}) \;:\; U(n) \text{ vector multiplet}$$

$$(a_{\alpha\dot{\beta}}, \Psi_{+\alpha A}) \;:\; U(n) \text{ adjoint hypermultiplet}$$

$$(q_{\dot{\alpha}}, \psi_{+A}) \;:\; U(n) \times SU(1) \text{ bifundamental hypermultiplet}$$

$$(\eta_-) \;:\; U(n) \times U(2) \text{ bifundamental Fermi multiplet}. \qquad (4.13)$$

The first and second lines come from the reduction of 3d fields on $n$ D2-branes on a segment, subject to the boundary conditions on two NS5-branes. In particular, the scalars $a_{\alpha\dot{\beta}}$ parametrize the positions of the strings on $\mathbb{R}^4$ worldvolume of the NS5-branes. The third line comes from the open strings connecting D2- and D6-branes on the segment between two NS5-branes. Here, $SU(1)$ formally means the gauge symmetry on single D6-branes, which is of course void. The final Fermi multiplets come from open strings connecting D2's and one of the two D6's across an NS5-brane. The two Fermi multiplets transform under $U(2)$, but the overall $U(1)$ of $U(2)$ suffers from mixed anomaly with $U(n)$ gauge symmetry, so is not a symmetry. The remaining $SU(2)$ combines with the $SU(2)_R$ symmetry rotating $A = 1, 2$ to form the enhanced $SO(4)$, as asserted above. (for $N > 2$, the last $SU(2)$ rotating Fermi multiplets get reduced to $U(1)$.) For $N > 2$, one more generally finds a $U(n_1) \times \cdots \times U(n_{N-1})$ quiver gauge theory in 2d with $\mathcal{N} = (0, 4)$ SUSY. See [46, 55] for the details.

The elliptic genera of these self-dual strings $Z_{\{n\}}(\tau, \epsilon_{1,2}, m)$ can be computed straightforwardly from these 2d gauge theories, using the formula of [29]. For instance, at $N = 2$, one finds

$$Z_n = (-1)^n \sum_{Y; |Y| = n} \prod_{s \in Y} \frac{\theta(m + \phi(s))\theta(m - \phi(s))}{\theta(E(s))\theta(E(s) - 2\epsilon_+)} \qquad (4.14)$$

with

$$E(s) = -\epsilon_1 h(s) + \epsilon_2(v(s) + 1) \ , \quad \theta(z) \equiv \frac{i\theta_1(\tau|\frac{z}{2\pi i})}{\eta(\tau)} \ . \tag{4.15}$$

At $n = 1$, this includes the result shown in Sect. 3.2 for the monopole strings (derived from free 2d NLSM). These elliptic genera explicitly satisfies the required modular anomalies un the S-duality $\tau \to -\frac{1}{\tau}$ [29].

These are the partition functions carrying self-dual string charges. The full partition function on $\mathbb{R}^4 \times T^2$ is the grand partition function of $Z_n$'s, multiplied by the $U(1)^2$ index for the two Abelian M5-branes. This is explained in Sect. 3.1,

$$Z_{U(1)} \equiv PE \left[ \frac{\sinh \frac{m \pm \epsilon_+}{2}}{\sinh \frac{\epsilon_{1,2}}{2}} \right] \ . \tag{4.16}$$

One finds

$$Z_{\mathbb{R}^4 \times T^2} = Z_{U(1)}^2 \sum_{n=0}^{\infty} e^{-n(v_1 - v_2)} Z_n \ . \tag{4.17}$$

In [46], this was shown to be agreeing with the 5d instanton partition function of Sect. 3.2, by making double expansions in $q$ and $e^{-(v_1 - v_2)}$ to high orders. Similar comparisons are made for the QFT's with $N > 2$.

The 2d QFT's on these strings, especially their modular anomalies, can be used to study many interesting properties of M5-branes. For instance, in [56], the modular anomalies of $Z_{n_1, n_2, \cdots, n_{N-1}}$ was used to show that the free energy of $\log Z_{\mathbb{R}^4 \times T^2}$ scales like $N^3$ in a Cardy-like asymptotic limit. See [56] for more details.

## 4.2 Strings of 6d $\mathcal{N} = (1, 0)$ Theories

The 6d $\mathcal{N} = (1, 0)$ SCFTs can be divided into two classes: those which have gauge symmetry (6d vector multiplet), and those which do not. In the former case, the self-dual strings have instanton string representations. In the latter case, they do not have smooth soliton representations. In both cases, the analysis of their quantum partition functions are somewhat similar. Therefore, we discuss both of them in this subsection.

Perhaps the most well explored self-dual strings in the $(1, 0)$ SCFTs are the so-called E-strings [38, 57]. E-strings are the self-dual strings suspended between M5-branes and the M9-plane. M9-plane is a Horava-Witten wall of the heterotic M-theory [58]. An M9-plane supports $E_8$ gauge symmetry, i.e. it hosts 10d super-Yang–Mills theory with $E_8$ gauge group. Therefore, the 6d SCFT on M5-branes probing M9 sees $E_8$ as a global symmetry. This symmetry is also seen by the 2d strings. The strings are again provided by the open M2-branes suspended between M5-M9. In this paper, we consider the simplest E-string theory containing one M5-brane, with rank 1 tensor branch. 2d description of their strings is found find [39]. For the E-strings in higher rank E-string theories, we refer the readers to [23].

To engineer the 2d gauge theories on E-strings, one compactifies the M-theory on a small circle to get a IIA description. The M-theory is put on $\mathbb{R}^{8,1} \times S^1 \times \mathbb{R}^+$. The M9-plane is located at the tip of the half line $\mathbb{R}^+ = \mathbb{R}/\mathbb{Z}_2$. To get a weakly coupled IIA description, one has to turn on a holonomy of $E_8$ so that the symmetry is broken to $SO(16)$ (see, e.g. [39] for a summary). Then upon such reduction, an M9-plane reduces to an O8$^-$ plane and 8 D8-branes on top of it. It is the $S^1$ holonomy on finite M-theory circle $S^1$ which breaks $E_8 \to SO(16)$. In this setting, E-strings are realized as $n$ D2-branes suspended between NS5-O8. The 2d $\mathcal{N} = (0, 4)$ gauge theory has $O(n)$ gauge symmetry, due to the orientifold projection, with the following field contents:

$$(A_\mu, \lambda_{-\dot{\alpha}A}) \quad : \quad O(n) \text{ vector multiplet}$$

$$(a_{\alpha\dot{\beta}}, \Psi_{+\alpha A}) \quad : \quad O(n) \text{ symmetric hypermultiplet}$$

$$\Xi_l \quad : \quad O(n) \times SO(16) \text{ Fermi multiplet}, \tag{4.18}$$

with $l = 1, \cdots, 16$. $\Xi_l$ are left-moving Majorana fermions. Its elliptic genera can be computed by slightly generalizing the calculus of [29], which was studied in [39].

In other cases with rank 1 tensor branches, one finds nontrivial 6d gauge symmetries. So one expects that the 2d gauge theories on the worldsheets of self-dual strings to be the ADHM-type models. Such gauge theory descriptions are naturally expected only when the 6d gauge group $G$ is classical, since we do not expect D-brane descriptions for exceptional gauge theories. For instance, let us consider the 'simplest' 6d theories at rank 1 tensor branch, with non-Higgsable gauge symmetries. These theories are sometimes called 'minimal 6d SCFTs' [52]. These theories are shown in Table 1. The cases with $n = 1, 2$ are the E-string theory, and $\mathcal{N} = (2, 0)$ theory of $A_1$ type, respectively. In other cases, at $n \geq 3$, all but the cases with $n = 3, 4$ have exceptional gauge groups. The ADHM-like descriptions for these exceptional instanton strings are indeed unknown (to date). In the rest of this subsection, we discuss the case with $n = 4$, and then with $n = 3$.

The case with $n = 4$ was discussed in [52]. The 6d SCFT can be engineered by type IIA branes, placing a segment of O6$^-$ and 4 D6-branes between two NS5-branes. The self-dual strings are segments of D2-branes suspended between the two

**Table 1** Symmetries/matters of minimal SCFTs

| $n$ | 1 | 2 | 3 | 4 | 5 | 6 | 7 | 8 | 12 |
|---|---|---|---|---|---|---|---|---|---|
| Gauge symmetry | – | – | $SU(3)$ | $SO(8)$ | $F_4$ | $E_6$ | $E_7$ | $E_7$ | $E_8$ |
| Global symmetry | $E_8$ | $SO(5)_R$ | – | – | – | – | – | – | – |
| Matters | | | – | – | – | – | $\frac{1}{2}\mathbf{56}$ | – | – |

NS5-branes. The fields contents are given by Haghighat et al. [52]

$$(A_\mu, \lambda_{-\dot\alpha A}) \; : \; Sp(n) \text{ vector multiplet}$$

$$(a_{\alpha\dot\beta}, \Psi_{+\alpha A}) \; : \; Sp(n) \text{ antisymmetric hypermultiplet}$$

$$(q_{\dot\alpha}, \psi_{+A}) \; : \; Sp(n) \times SO(8) \text{ bifundamental hypermultiplet}. \qquad (4.19)$$

This is nothing but a 2d uplift of the ADHM quantum mechanics explained in Sect. 2, for $G = SO(8)$ and $\hat{G} = Sp(n)$. Although the 2d uplift may naively sound straightforward, it should satisfy very strong quantum consistency conditions. This is because the 2d system is chiral. Note that with $\mathcal{N} = (0, 4)$ supersymmetry, the vector multiplet contains left-moving fermions while hypermultiplets contain right-moving fermions. Their contributions to the $Sp(n)$ gauge anomaly should cancel, for this QFT to make sense. It is shown that this system is free of the $Sp(n)$ anomaly [52, 54]. This can be anticipated since this QFT is derived from a consistent D-brane configuration. The elliptic genera of these instantons strings can be computed from these 2d QFTs, as studied in [52].

Now let us turn to the case with $n = 3$. The 6d Yang–Mills theory has $SU(3)$ gauge group, so it naively appears that its instanton strings can be described by a 2d ADHM-like gauge theories. Somewhat curiously, this expectation turns out to be wrong. This is because there are no purely D-brane construction for this 6d system, in which $SU(3)$ is realized on a stack of three D-branes. Namely, say in F-theory, this is realized by wrapping various nontrivial 7-branes on a 2-cycle, so that mutually non-local $(p, q)$ string junctions provide the $SU(3)$ W-bosons. For instance, see [59] for detailed illustrations. So although $SU(3)$ is mathematically a classical group, the construction of this gauge theory follows precisely the same pattern as the constructions of exceptional gauge theories in F-theory. Therefore, the 2d version of the $SU(3)$ ADHM gauge theories have no D-brane derivations.

Incidently, such 2d gauge theories suffer from gauge anomalies. On $k$ instanton strings, the naive $\mathcal{N} = (0, 4)$ ADHM theory has the following field contents:

$$(A_\mu, \lambda_{-\dot\alpha A}) : U(k) \text{ vector multiplet}$$

$$(q_{\dot\alpha}, \psi_{+A}) : \text{hypermultiplets in } (\mathbf{k}, \bar{\mathbf{3}})$$

$$(a_{\alpha\dot\beta}, \Psi_{+\alpha A}) : \text{hypermultiplets in } (\mathbf{adj}, \mathbf{1}). \qquad (4.20)$$

This gauge theory has anomalous $U(k)$ gauge symmetry [54], defining an inconsistent system. We note here that, although there are no D-brane derivations, [54] found a remedy of the sickness of this 2d ADHM quiver, by adding more fields and interactions, which does describe $SU(3)$ instanton strings of the 6d SCFT. To this end, it will be useful to use the $\mathcal{N} = (0, 2)$ off-shell formalism, following [22]. A vector multiplet decomposes into a $(0, 2)$ vector multiplet and an adjoint Fermi multiplet. A hypermultiplet decomposes into a pair of chiral multiplets in conjugate representations. Thus we obtain

$$(A_\mu, \lambda_{-\dot{\alpha}A}) \rightarrow (A_\mu, \lambda_0, D) + (\lambda, G_\lambda)_{(R,J)=(1,-1)}$$

$$(q_{\dot{\alpha}}, \psi_{+A}) \rightarrow (q, \psi_+)_{(R,J)=(0,\frac{1}{2})} + (\tilde{q}, \tilde{\psi}_+)_{(R,J)=(0,\frac{1}{2})}$$

$$(a_{\alpha\dot{\beta}}, \Psi_{+\alpha A}) \rightarrow (a, \Psi)_{(R,J,J_l)=(0,\frac{1}{2},\frac{1}{2})} + (\tilde{a}, \tilde{\Psi})_{(R,J,J_l)=(0,\frac{1}{2},-\frac{1}{2})}. \quad (4.21)$$

Here, the $(0, 2)$ R-charge is given by $R = 2J_R$, where $J_R$ is the Cartan of $SU(2)_R$, so that $R[Q] = -1$ for $Q \equiv Q^{\dot{1}2} \sim Q_{\dot{2}1}$. $J \equiv J_r + J_R$ and $J_l$ are treated as flavor symmetries in the $(0, 2)$ setting, where $J_l$ and $J_r$ are the Cartans of $SU(2)_l$ and $SU(2)_r$, respectively.

Now we present the modified 2d quiver for the $SU(3)$ instanton strings. We first explain the fields. To the above ADHM fields, we add the following $\mathcal{N} = (0, 2)$ supermultiplets:

$$(\phi, \chi) : \text{chiral multiplet in } (\overline{\mathbf{k}}, \overline{\mathbf{3}})$$

$$(b, \xi) + (\tilde{b}, \tilde{\xi}) : \text{two chiral multiplet in } (\overline{\textbf{anti}}, \mathbf{1})$$

$$(\hat{\lambda}, \hat{G}) : \text{complex Fermi multiplet in } (\textbf{sym}, \mathbf{1})$$

$$(\check{\lambda}, \check{G}) : \text{complex Fermi multiplet in } (\textbf{sym}, \mathbf{1})$$

$$(\zeta, G_\zeta) : \text{complex Fermi multiplet in } (\overline{\mathbf{k}}, \mathbf{1}). \quad (4.22)$$

Here, '**anti**' and '**sym**' denote rank 2 antisymmetric and symmetric representations of $U(k)$, respectively. [The actual field contents are slightly more complicated than the above, for subtle reasons which we shall not need to discuss in this paper. See [54] for the details.] With added fields, the UV theory only has $\mathcal{N} = (0, 2)$ supersymmetry. With these fields, one can show that the $U(k)$ gauge anomaly completely cancels [54], thus enabling this system to be consistent quantum mechanically.

It turns out that to realize the correct moduli space of $SU(3)$ instantons, one has to turn on interactions which only preserve $\mathcal{N} = (0, 1)$ supersymmetry. Therefore, if this system describes the $SU(3)$ instanton strings of the 6d $\mathcal{N} = (1, 0)$ SCFT, we expect a supersymmetry enhancement in IR of the 2d QFT.

The elliptic genera of these 2d QFT's we studied in [54]. Very marvelously, these elliptic genera successfully explains various BPS invariants computed from the topological string methods [52], whose physical setting was explained at the beginning of this section.

This is quite a surprising finding, in that the partition functions for instantons which are not engineered from D-branes are computed from ADHM-like gauge theories. Reducing the above complicated quiver to 1d, one obtains a much more complicated quantum mechanical system than the $SU(3)$ ADHM construction. The two quantum mechanical indices were shown to completely agree with each other

in 1d [54]. So this is a very novel, alternative, ADHM-like formalism to compute
Nekrasov's instanton partition function.

This idea can be slightly generalized to provide ADHM-like gauge theory
descriptions on the instanton worldvolume for certain exceptional gauge groups and
matters. Namely, in 6d, the non-Higgsable $SU(3)$ gauge theory is related to the
following Higgsable 6d gauge theories:

$$(SU(3)) \leftarrow (G_2, n_7 = 1) \leftarrow (SO(7), n_7 = 0, n_8 = 2) \leftarrow (SO(8), n_{8_v} = n_{8_s} = n_{8_c} = 1)$$

$$\leftarrow \begin{cases} (SO(N), n_N = N - 7, n_S = \frac{16}{d_S})_{N=9,\cdots,12} \\ \\ (F_4, n_{26} = 2) \leftarrow (E_6, n_{27} = 3) \leftarrow (E_7, n_{\frac{1}{2}56} = 5) \leftarrow (E_8, n_{\text{inst}} = 9) \end{cases} \tag{4.23}$$

The arrows $\leftarrow$ denote Higgsings by giving VEV's to charged hypermultiplets.
The pair $(G, n_{\mathbf{R}})$ denote the gauge group and the number of hypermultiplets in
representation $\mathbf{R}$, respectively. By slightly generalizing the alternative ADHM-like
quiver above for $SU(3)$ instantons, [27] found ADHM-like gauge theory descriptions for $G_2$ instantons with certain numbers $n_7$ of hypermultiplets in $\mathbf{7}$. Similarly,
[27] found ADHM-like descriptions for $SO(7)$ instantons with hypermultiplets in
$\mathbf{8}$. These ADHM-like quivers turned out to be useful to study more general strings of
non-Higgsable SCFTs containing $G_2$ gauge groups or $SO(7)$ matters in the spinor
representation $\mathbf{8}$.

## 5 Superconformal Observables

In this section, we discuss some partition functions of the 5d/6dsuperconformal field
theories on compact Euclidean spaces. Many of these observables are expressed in
terms of the Coulomb/tensor branch partition functions, i.e. the partition functions
of instanton particles or strings that we discussed so far. Especially, in this paper,
we shall discuss the so-called superconformal indices, which are BPS partition
functions on $S^5 \times S^1$ or $S^4 \times S^1$.

We first discuss the superconformal index of 6d SCFTs on $S^5 \times S^1$. It uses the
Coulomb branch partition function $Z_{\mathbb{R}^4 \times T^2}(\tau, v, \epsilon_1, \epsilon_2, m_0)$ that we explained in
Sects. 3 and 4, either in terms of the instanton particles' partition function in 5d
SYM, or the self-dual strings' partition function of 6d SCFTs in the tensor branch.
The expression is given by Kim and Kim [60], Lockhart and Vafa [61], Kim et
al. [62]

$$Z_{S^5 \times S^1}(\beta, m, \omega_i) = \frac{e^{-S_{\text{bkgd}}}}{|W(G_r)|} \int_{-\infty}^{\infty} \left[ \prod_{I=1}^{r} d\phi_I \right] e^{-S_0(\phi, \beta, \omega_i)}$$

$$\times Z_{\mathbb{R}^4 \times T^2} \left( \frac{2\pi i}{\beta \omega_1}, \frac{\phi}{\omega_1}, \frac{2\pi i \omega_{21}}{\omega_1}, \frac{2\pi i \omega_{31}}{\omega_1}, 2\pi i \left( \frac{m}{\omega_1} + \frac{3}{2} \right) \right)$$

$$\cdot Z_{\mathbb{R}^4 \times T^2}\left(\frac{2\pi i}{\beta \omega_2}, \frac{\phi}{\omega_2}, \frac{2\pi i \omega_{32}}{\omega_2}, \frac{2\pi i \omega_{12}}{\omega_2}, 2\pi i \left(\frac{m}{\omega_2} + \frac{3}{2}\right)\right)$$

$$\times Z_{\mathbb{R}^4 \times T^2}\left(\frac{2\pi i}{\beta \omega_3}, \frac{\phi}{\omega_3}, \frac{2\pi i \omega_{13}}{\omega_3}, \frac{2\pi i \omega_{23}}{\omega_3}, 2\pi i \left(\frac{m}{\omega_3} + \frac{3}{2}\right)\right),$$

$$S_0 = \frac{2\pi^2 \text{tr}(\phi^2)}{\beta \omega_1 \omega_2 \omega_3}, \tag{5.1}$$

where the arguments in $Z_{\mathbb{R}^4 \times T^2}$ are listed in the order of $Z_{\mathbb{R}^4 \times T^2}(\tau, v, \epsilon_1, \epsilon_2, m)$. Here, $\omega_{ij} \equiv \omega_i - \omega_j$. $G_r$ is the gauge group of the low energy 5d SYM that one obtains by reducing the 6d SCFT, and $W(G_r)$ is the Weyl group of $G_r$. More abstractly, in the 6d CFT, $W(G_r)$ acquires meaning as the Weyl group acting on the tensor branch as $\mathbb{R}^r / W(G_r)$, where $\phi_I$ parametrizes the tensor branch $\mathbb{R}^r$. This expression has been proposed with two different motivations. Lockhart and Vafa [61] argued for such an expression from the topological string theory. On the other hand, [60, 62] arrived at this answer from the partition functions of 5d super-Yang–Mills on $S^5$. In Sect. 5.1, we shall discuss the physics of the 6d $\mathcal{N} = (2, 0)$ theories from this partition function.

Similarly, the superconformal index of 5d SCFTs on $S^4 \times S^1$ has been explored in [30]. The expression is given in terms of the Coulomb branch partition function on $\mathbb{R}^4 \times S^1$ as follows:

$$Z_{S^4 \times S^1}(\epsilon_{1,2}, a, \{m\}) = \oint [d\alpha] Z_{\mathbb{R}^4 \times S^1}(i\alpha, \epsilon_{1,2}, q, m_a) Z_{\mathbb{R}^4 \times S^1}(-i\alpha, \epsilon_{1,2}, q^{-1}, -m_a), \tag{5.2}$$

where $\pm i\alpha$ are inserted into the place of the Coulomb VEV $v$.

In the Sect. 5.1, we shall study the physics of (5.1) for the (2, 0) theory, in various cases in which (5.1) can be handled more concretely. In Sect. 5.2, we shall study some physics of the 5d SCFTs using the superconformal index.

## 5.1 6d (2, 0) Index and its Physics

The index (5.1) has been studied in more detail for the (2, 0) SCFTs, especially in the $A_{N-1}$ case in which the 5d instanton counting has been best understood. The technical issue concerning the expression (5.1) is that it is given in a 'weak coupling' expansion form, taking the form of series expansion in $\beta$ and $e^{-\frac{4\pi^2}{\beta \omega_i}}$ when $\beta \ll 1$ after $\phi$ integral. However, the index structure of $Z_{S^5 \times S^1}$ will be best visible in the regime $\beta \gg 1$, as a series expansion in $e^{-\beta}$. At the moment, this re-expansion in the regime $\beta \gg 1$ has been achieved only in two special cases. One is the 6d Abelian (2, 0) index with all fugacities turned on, and another is the non-Abelian index with all but one fugacities tuned to special values. In this subsection, we shall explain these two.

**Abelian** $(2, 0)$ **Index** We should first explain what is the virtue of studying the Abelian theory, as the 6d theory is free. In fact the superconformal index of the free 6d $(2, 0)$ tensor multiplet is computed in [63]. In our convention, it is given by

$$Z_{S^5 \times S^1}(\beta, m, \omega_i) = e^{-\beta \epsilon_0} \exp\left[\sum_{n=1}^{\infty} \frac{1}{n} f(n\beta, nm, n\omega_i)\right], \tag{5.3}$$

$$f(\beta, m, \omega_i) = \frac{\left(e^{-\frac{3\beta}{2}}(e^{\beta m} + e^{-\beta m}) \atop \begin{array}{c} -(e^{-\beta(\omega_1+\omega_2)} + e^{-\beta(\omega_2+\omega_3)} + e^{-\beta(\omega_3+\omega_1)}) + e^{-3\beta} \end{array}\right)}{(1 - e^{-\beta\omega_1})(1 - e^{-\beta\omega_2})(1 - e^{-\beta\omega_3})}.$$

$\epsilon_0$ is the 'zero point energy' factor, which is in general regularization scheme dependent. We shall explain this factor in more detail below in this subsection. Since we know this (trivial) index concretely, one might wonder what is the virtue of getting it from (5.1). The first reason is simply to check that (5.1) correctly provides the well known results. The second reason is to emphasize the precise meaning of the formula (5.1). Equation (5.3) is given in the form of a series expansion in $e^{-\beta} \ll 1$. By expanding $f(\beta, m, \omega_i)$ in a series in $e^{-\beta}$, one would obtain an infinite product expression for $Z_{S^5 \times S^1}$ for the Abelian theory:

$$q^{\epsilon_0} \prod_{n_1=0}^{\infty} \prod_{n_1=0}^{\infty} \prod_{n_3=0}^{\infty} \frac{\left(\begin{array}{c}(1 - q^{2+n_1+n_2+n_3}\zeta_1^{n_1-1}\zeta_2^{n_2}\zeta_3^{n_3})(1 - q^{2+n_1+n_2+n_3}\zeta_1^{n_1}\zeta_2^{n_2-1}\zeta_3^{n_3}) \\ \times(1 - q^{2+n_1+n_2+n_3}\zeta_1^{n_1}\zeta_2^{n_2}\zeta_3^{n_3-1})\end{array}\right)}{\left(\begin{array}{c}(1 - yq^{\frac{3}{2}+n_1+n_2+n_3}\zeta_1^{n_1}\zeta_2^{n_2}\zeta_3^{n_3})(1 - y^{-1}q^{\frac{3}{2}+n_1+n_2+n_3}\zeta_1^{n_1}\zeta_2^{n_2}\zeta_3^{n_3}) \\ \times(1 - q^{3+n_1+n_2+n_3}\zeta_1^{n_1}\zeta_2^{n_2}\zeta_3^{n_3})\end{array}\right)},$$

where $q = e^{-\beta}$, $\zeta_i = e^{-\beta a_i}$ (with $\omega_i \equiv 1 + a_i$), $y = e^{-\beta m}$. This is well defined for small enough $q$.

Now to see if this index is reproduced from (5.1), we should sum over all the $q$ series appearing in the $Z_{\mathbb{R}^4 \times T^2}$ factors, and make a 'strong coupling' expansion to compare with the known 6d index. Alternatively, one can make a 'weakly coupled' expansion of (5.3) and confirm that we obtain (5.1). Using suitable contour integral expression for $\log Z_{S^5 \times S^1}$ obtained from (5.3) [62], one can make an expansion of (5.3) which is given by

$$Z_{S^5 \times S^1} = \left[\frac{\beta\omega_1\omega_2\omega_3}{2\pi}\right]^{\frac{1}{2}} e^{-\beta\epsilon_0 - \frac{\beta}{24}\left(1 + \frac{2a_1a_2a_3 + (1-a_1a_2-a_2a_3-a_3a_1)(\frac{1}{4}-m^2) + (\frac{1}{4}-m^2)^2}{\omega_1\omega_2\omega_3}\right)}$$

$$\times e^{\frac{\pi^2(\omega_1^2+\omega_2^2+\omega_3^2-2\omega_1\omega_2-2\omega_2\omega_3-2\omega_3\omega_1+4m^2)}{24\beta\omega_1\omega_2\omega_3}}$$

$$\cdot \, Z_{\text{pert}} \left( \frac{2\pi i \omega_{21}}{\omega_1}, \frac{2\pi i \omega_{31}}{\omega_1}, 2\pi i \left( \frac{m}{\omega_1} + \frac{3}{2} \right) \right)$$

$$\times \, Z_{\text{inst}} \left( \frac{2\pi i}{\beta \omega_1}, \frac{2\pi i \omega_{21}}{\omega_1}, \frac{2\pi i \omega_{31}}{\omega_1}, 2\pi i \left( \frac{m}{\omega_1} + \frac{3}{2} \right) \right)$$

$$\cdot \left( 1, 2, 3 \to 2, 3, 1 \right) \left( 1, 2, 3 \to 3, 1, 2 \right) , \tag{5.4}$$

where the last two factors the repetitions of the second line with the $1, 2, 3$ subscripts of $\omega_i$ permuted, $Z_{\text{pert}}$ is the perturbative $U(1)$ maximal SYM partition function on the $\Omega$ deformed $\mathbb{R}^4 \times S^1$, and $Z_{\text{inst}}$ is the 'instanton' part of the $U(1)$ maximal SYM on $\mathbb{R}^4 \times S^1$ [13, 20, 47]

$$Z_{\text{inst}}(\tau, \epsilon_1, \epsilon_2, m_0) = \exp \left[ \sum_{n=1}^{\infty} \frac{1}{n} \frac{\sinh \frac{n(m_0 + \epsilon_-)}{2} \sinh \frac{n(m_0 - \epsilon_-)}{2}}{\sinh \frac{n \epsilon_1}{2} \sinh \frac{n \epsilon_2}{2}} \frac{e^{2\pi i n \tau}}{1 - e^{2\pi i n \tau}} \right] , \tag{5.5}$$

which is identical to (3.14) when we expand (5.5) in $e^{2\pi i \tau}$. This is the instanton part of the partition function of 5d $U(1)$ maximal super-Yang–Mills theory, as explained in Sect. 3.2. So this part is consistent with (5.1). Note that neither $Z_{\text{pert}}$, $Z_{\text{inst}}$ depends on the $U(1)$ Coulomb VEV $\phi$. So the first factor $[\frac{\beta \omega_1 \omega_2 \omega_3}{2\pi}]^{\frac{1}{2}}$ can be replaced by

$$\left( \frac{\beta \omega_1 \omega_2 \omega_3}{2\pi} \right)^{\frac{1}{2}} = \int_{-\infty}^{\infty} d\phi \exp \left( - \frac{2\pi^2 \phi^2}{\beta \omega_1 \omega_2 \omega_3} \right) , \tag{5.6}$$

which is the Gaussian integral in (5.1) with the measure $e^{S_0}$. So the known index (5.3) would be completely agreeing with (5.1) if we identify $Z_{\mathbb{R}^4 \times T^2} = Z_{\text{pert}} Z_{\text{inst}}$ and if we take

$$\epsilon_0 = -\frac{1}{24} \left[ 1 + \frac{2a_1 a_2 a_3 + (1 - a_1 a_2 - a_2 a_3 - a_3 a_1)(\frac{1}{4} - m^2) + (\frac{1}{4} - m^2)^2}{\omega_1 \omega_2 \omega_3} \right] ,$$

$$S_{\text{bkgd}} = -\frac{\pi^2 (\omega_1^2 + \omega_2^2 + \omega_3^2 - 2\omega_1 \omega_2 - 2\omega_2 \omega_3 - 2\omega_3 \omega_1 + 4m^2)}{24 \beta \omega_1 \omega_2 \omega_3} . \tag{5.7}$$

We shall explain these identifications about $\epsilon_0$ and $S_{\text{bkgd}}$, in turn.

Firstly, $\epsilon_0$ is the vacuum 'Casimir energy' which we shall explain later in this subsection. For now, we simply regard (5.1) as giving a specific value of the vacuum energy at $\beta \gg 1$ expansion. Secondly, $S_{\text{bkgd}}$ couples the parameters $g_{YM}^2$, $m$ of the theory to the background parameters of $S^5$, such as $r_5$, $\omega_i$. In particular, it takes the form of the leading free energy in the 'high temperature' regime $\beta \ll 1$. From the analysis of one lower dimension on $S^5$, this data cannot be determined in a self-contained way, and should be given as an input. One can think about it in two different viewpoints. One may first regard $S_{\text{bkgd}}$ as our ignorance, but demand that we tune it so that the strong coupling expansion of (5.1) becomes an index. It is an extremely nontrivial request that tuning $S_{\text{bkgd}}$ in negative powers of $\beta$ yields an index at $\beta \gg 1$. As the above results in the Abelian theory shows, it completely fixes $S_{\text{bkgd}}$ if one can freely do both weak and strong coupling expansions. Furthermore, the general structures of the high temperature asymptotics of the 6d SCFT index was proposed in [64]. They only considered the angular momentum chemical potentials $\omega_i$, with $m = 0$, and completely fixed the $\beta$, $\omega_i$ dependence apart from a few central charge coefficients. Their proposal is consistent with (5.7) at $m = 0$.

So with these understood, we have confirmed that the partition function (5.1) at strong coupling $\beta \gg 1$ indeed reproduces the 6d Abelian $(2, 0)$ index.

**Unrefined Non-Abelian** $(2, 0)$ **Indices** Now we turn to more interesting non-Abelian indices. Again we shall restrict our interest to the $(2, 0)$ theory here, as we shall crucially use simplifications coming from extra SUSY when some chemical potentials are tuned. Namely, consider the following tuning of the $U(1) \subset SO(5)_R$ chemical potential for $\frac{R_1 - R_2}{2}$:

$$m = \frac{1}{2} - a_3 . \tag{5.8}$$

The index can then be written as (see [62] for the notations)

$$Z_{S^5 \times S^1}(\beta, \tfrac{1}{2} - a_3, a_i) = \text{Tr}\left[(-1)^F e^{-\beta(E-R_1)} e^{-\beta a_1(j_1 - j_3 - \frac{R_1 - R_2}{2})} e^{-\beta(j_2 - j_3 - \frac{R_1 - R_2}{2})}\right] . \tag{5.9}$$

Let us denote by $R_1$, $R_2$ the two Cartans of $SO(5)_R$ R-symmetry of the 6d $(2, 0)$ theory, and $J_1$, $J_2$, $J_3$ by the three angular momenta on $S^5$. The Poincare supercharges $Q^{R_1, R_2}_{J_1, J_2, J_3}$ and the conformal supercharges $S^{R_1, R_2}_{J_1, J_2, J_3}$ all carry the charges $\pm \equiv \pm\frac{1}{2}$ for $R_1$, $R_2$, $J_1$, $J_2$, $J_3$. The index above counts BPS states annihilated by the supercharges $Q \equiv Q^{++}_{---}$, $S \equiv S^{--}_{+++}$. At (5.8), two extra supercharges $Q^{+-}_{++-}$, $S^{-+}_{--+}$ commute with it. So the index exhibits more cancellations of bosons/fermions paired by the extra supercharges, which will make the index simpler. The SYM on $S^5$ will also preserve more SUSY. We will show shortly that the Eq. (5.1) can be exactly computed at this point. We also note that further tunings

$$m = \frac{1}{2}, \quad a_1 = a_2 = a_3 = 0 \tag{5.10}$$

will leave only one chemical potential $\beta$, in which case the measure of the index

$$Z_{S^5 \times S^1}(\beta, \tfrac{1}{2}, 0) = \text{Tr}\left[(-1)^F e^{-\beta(E-R_1)}\right] \tag{5.11}$$

commutes with 16 of the 32 supercharges of the $(2, 0)$ theory. Namely, the following 8 complex supercharges $Q^{+\pm}_{\pm\pm\pm}$ (with the $\pm$ subscripts satisfying $\pm\pm\pm = -$) and their conjugates $S^{-\pm}_{\pm\pm\pm}$ (with subscripts satisfying $\pm\pm\pm = +$) commute with $e^{-\beta(E-R_1)}$. The presence of 16 SUSY will have special implication on the index, especially concerning the 'zero point energy' of the vacuum which is captured by the index. Also, one would naturally expect that the circle reduction of the 6d theory at (5.10) will yield a maximal SYM which actually preserves 16 supercharges. This is indeed the case [60].

To understand the simplification at the level of the formula (5.1), we first study how the $\Omega$ background parameters and the mass parameters simplify. The effective $\Omega$ parameters $\epsilon_1$, $\epsilon_2$ and the mass $m_0$ in the three $Z_{\mathbb{R}^4 \times T^2}$ factors are given by

$$
\frac{1}{2\pi i}(\epsilon_1, \epsilon_2, m_0) = \left(\frac{\omega_2 - \omega_1}{\omega_1}, \frac{\omega_3 - \omega_1}{\omega_1}, \frac{m}{\omega_1} + \frac{3}{2}\right) \sim \left(\frac{\omega_2 - \omega_1}{\omega_1}, \frac{\omega_3 - \omega_1}{\omega_1}, \frac{m}{\omega_1} - \frac{1}{2}\right) : \text{1st}
$$

$$
\left(\frac{\omega_3 - \omega_2}{\omega_2}, \frac{\omega_1 - \omega_2}{\omega_2}, \frac{m}{\omega_2} - \frac{1}{2}\right) : \text{2nd}
$$

$$
\left(\frac{\omega_1 - \omega_3}{\omega_3}, \frac{\omega_2 - \omega_3}{\omega_3}, \frac{m}{\omega_3} - \frac{1}{2}\right) : \text{3rd} . \tag{5.12}
$$

We use $m_0$ for the effective mass parameter on $\mathbb{R}^4 \times T^2$, to avoid confusions with the actual mass parameter of 5d SYM on $S^5$, or the chemical potential $\beta m$ of the 6d index. Also, we used the fact that all parameters $\epsilon_1$, $\epsilon_2$, $m_0$ are periodic variables in $2\pi i$ shifts. So at (5.8), one finds

$$
\frac{1}{2\pi i}(\epsilon_1, \epsilon_2, m_0) = \left(\frac{\omega_2 - \omega_1}{\omega_1}, \frac{\omega_3 - \omega_1}{\omega_1}, \frac{\omega_2 - \omega_3}{2\omega_1}\right) : \text{1st}
$$

$$
\left(\frac{\omega_3 - \omega_2}{\omega_2}, \frac{\omega_1 - \omega_2}{\omega_2}, \frac{\omega_1 - \omega_3}{2\omega_2}\right) : \text{2nd}
$$

$$
\left(\frac{\omega_1 - \omega_3}{\omega_3}, \frac{\omega_2 - \omega_3}{\omega_3}, \frac{\omega_1 + \omega_2 - 2\omega_3}{2\omega_3}\right) : \text{3rd} . \tag{5.13}
$$

Defining $\epsilon_\pm \equiv \frac{\epsilon_1 \pm \epsilon_2}{2}$, we find that these effective parameters satisfy $m_0 = \epsilon_-$ in the first factor, $m_0 = -\epsilon_-$ in the second factor, and $m_0 = \epsilon_+$ in the third factor.

Let us explain that the partition function $Z_{\mathbb{R}^4 \times T^2}$ simplifies in all the three factors. Note that $Z_{\mathbb{R}^4 \times T^2}(q, v, \epsilon_{1,2}, m_0)$ takes the following form:

$$
Z_{\mathbb{R}^4 \times T^2} = Z_{\text{pert}}(v, \epsilon_{1,2}, m_0) Z_{\text{inst}}(q, v, \epsilon_{1,2}, m_0) , \tag{5.14}
$$

$$
Z_{\text{pert}} = \widetilde{PE} \left[ \frac{1}{2} \frac{\sin\frac{m_0 + \epsilon_+}{2} \sin\frac{m_0 - \epsilon_+}{2}}{\sin\frac{\epsilon_1}{2} \sin\frac{\epsilon_2}{2}} \chi_{\text{adj}}(v) \right] ,
$$

where $Z_{inst}$ is given in Sect. 2, and

$$\chi_{adj}(v) = \sum_{\alpha \in adj(G)} e^{\alpha(v)} . \tag{5.15}$$

$\widetilde{PE}$ is defined by expanding the function in $\widetilde{PE}[\cdots]$ in $e^{-\epsilon_{1,2}}$, $e^{-m_0}$, $e^{-\alpha(v_i)}$, and imposing

$$\widetilde{PE}[ne^{-x}] = \left[2\sinh\frac{x}{2}\right]^n = \left[\frac{e^{-\frac{x}{2}}}{1-e^{-x}}\right]^n , \quad \widetilde{PE}[f+g] = \widetilde{PE}[f]\widetilde{PE}[g] . \tag{5.16}$$

As for $Z_{inst}$, the $U(N)$ result is known well in the series expansion in $q$, as explained in Sects. 2 and 3. For $D_N$ cases, the $SO(2N)$ partition function has been computed rather recently in [17]. For $E_N$, almost nothing is known, although we shall say something about it below. Here we would like to emphasize the general structure of $Z_{pert}$ and $Z_{inst}$. Since $Z_{pert}$ and $Z_{inst}$ count BPS particles on $\mathbb{R}^{4,1}$ in the Coulomb phase, they carry universal prefactors from their center-of-mass supermultiplets. In particular, since perturbative particles and instantons preserve different 8 supercharges among the full 16 as massive vector and tensor multiplets, respectively, the prefactors appearing in the two parts are different. It is easy to check [33] that

$$Z_{pert} = \widetilde{PE}\left[I_+(\epsilon_{1,2}, m_0)(\cdots)\right] , \quad Z_{inst} = \widetilde{PE}\left[I_-(\epsilon_{1,2}, m_0)(\cdots)\right] . \tag{5.17}$$

$(\cdots)$ are the contributions from internal degrees of freedom of the BPS states, which are regular in $\epsilon_1 = \epsilon_2 = m_0 = 0$ the limit, and

$$I_\pm(\epsilon_{1,2}, m_0) \equiv \frac{\sin\frac{m_0+\epsilon_\pm}{2} \sin\frac{m_0-\epsilon_\pm}{2}}{\sin\frac{\epsilon_1}{2}\sin\frac{\epsilon_2}{2}} . \tag{5.18}$$

For $Z_{pert}$, this structure is already manifest in (5.14).

Firstly, at $m_0 = \pm\epsilon_-$, one finds from (5.18) and (5.17) that $I_-(\epsilon_{1,2}, \pm\epsilon_-) = 0$ and $Z_{inst} = 1$. Therefore, in the unrefined limit (5.8), one finds that the first and second $Z_{\mathbb{R}^4 \times T^2}$ factors in (5.1) reduce to the perturbative contributions at this point. Note also from (5.18) that $I_+(\epsilon_{1,2}, \pm\epsilon_-) = -1$. So applying this to (5.14), one obtains

$$Z_{\mathbb{R}^4 \times T^2} \rightarrow \widetilde{PE}\left[-\frac{1}{2}\chi_{adj}(v)\right] \tag{5.19}$$

at $m_0 = \pm\epsilon_-$. So applying this to the first and second factors of (5.1), one obtains

$$
Z^{(1)}_{\mathbb{R}^4 \times T^2} Z^{(2)}_{\mathbb{R}^4 \times T^2} \to \widetilde{PE}\left[-\frac{1}{2}(\chi_{\text{adj}}(v/\omega_1) + \chi_{\text{adj}}(v/\omega_2))\right] = \prod_{\alpha > 0} 2\sinh\frac{\alpha(v)}{\omega_1} \cdot 2\sinh\frac{\alpha(v)}{\omega_2} ,
$$
$$(5.20)$$

where the product is over positive roots of $G$, up to a possible overall sign on which we are not very careful. We then turn to $Z^{(3)}_{\mathbb{R}^4 \times T^2}$ at $m_0 = \epsilon_+$. From (5.17) or (5.14), it is obvious that $Z_{\text{pert}} = 1$ at $m_0 = \epsilon_+$, since $I_+ = 0$. So $Z_{\mathbb{R}^4 \times T^2}$ acquires $Z_{\text{inst}}$ contribution only at $m = \epsilon_+$. For $U(N)$, one can easily show from the $U(N)$ instanton partition function of Sects. 2 and 3 that

$$
Z_{\text{inst}}(\beta, \epsilon_{1,2}, m_0 = \pm\epsilon_+) = \frac{1}{\eta(\tau)^N} = e^{-\frac{\pi N i \tau}{12}} \prod_{n=1}^{\infty} \frac{1}{(1 - e^{2\pi n i \tau})^N} . 
$$
$$(5.21)$$

where $\tau = \frac{2\pi i}{\beta}$. Here, we have included the extra factor of

$$
e^{-S_{\text{bkgd}}} = e^{-\frac{\pi N i \tau}{12}} = e^{\frac{\pi^2 N}{6\beta}} ,
$$
$$(5.22)$$

which will be justified below. More generally, we shall present below nontrivial evidences that

$$
e^{-S_{\text{bkgd}}} Z_{\text{inst}}(\beta, \epsilon_{1,2}, m_0 = \pm\epsilon_+) = \frac{1}{\eta(\tau)^N}
$$
$$(5.23)$$

for all $U(N)$, $D_N = SO(2N)$, $E_N$ groups. This formula has been justified to some higher orders in $q$ for certain $D_N$ cases [37]. If one wishes to consider the interacting $A_{N-1}$ part only, instead of $U(N)$, one simply takes $Z_{\text{inst}} = \eta(\tau)^{-(N-1)}$ by dropping an overall $U(1)$ factor $Z^{U(1)}_{\text{inst}} = \eta(\tau)^{-1}$. So $e^{-S_{\text{bkgd}}} Z^{(3)}_{\mathbb{R}^4 \times T^2}$ simplifies to

$$
e^{-S_{\text{bkgd}}} Z^{(3)}_{\mathbb{R}^4 \times T^2} \to \frac{1}{\eta(\tau/\omega_3)^N} = \frac{1}{\eta(\frac{2\pi i}{\beta\omega_3})^N}
$$
$$(5.24)$$

in the limit (5.8), for all $A_N$, $D_N$, $E_N$ series.

With all the simplifications (5.20), (5.24), the partition function (5.1) on $S^5$ reduces to

$$
Z_{S^5 \times S^1}(\beta, m = \frac{1}{2} - a_3, a_i) = \frac{1}{\eta(\frac{2\pi i}{\beta\omega_3})^N} \cdot \frac{1}{W(G_N)} \int \prod_{I=1}^{N} d\phi_I e^{-\frac{2\pi^2 \text{tr}(\phi^2)}{\beta\omega_1\omega_2\omega_3}}
$$
$$
\times \prod_{\alpha > 0} 2\sinh\frac{\alpha(\phi)}{\omega_1} \cdot 2\sinh\frac{\alpha(\phi)}{\omega_2} .
$$
$$(5.25)$$

The integral over $\phi_I$ is simply a Gaussian integral. The result of the integral is

$$Z_{S^5 \times S^1}\left(\beta, m = \frac{1}{2} - a_3, a_i\right) = \left(\frac{\beta\omega_3}{2\pi}\right)^{\frac{N}{2}} \frac{1}{\eta(\frac{2\pi i}{\beta\omega_3})^N} e^{\frac{\beta c_2|G|}{24}\omega_3\left(\frac{\omega_1}{\omega_2} + \frac{\omega_2}{\omega_1}\right)}$$

$$\times \prod_{\alpha > 0} 2\sinh\left(\beta\omega_3 \frac{\alpha \cdot \rho}{2}\right), \tag{5.26}$$

where $\rho$ is the Weyl vector. Since $\eta(\tau)$ is a modular form, its expansion in the $\beta \gg 1$ regime is easy to understand. The result is given by

$$Z_{S^5 \times S^1}\left(\beta, \frac{1}{2} - a_3, a_i\right) = e^{\beta \frac{c_2|G|}{24} \frac{\omega_3}{\omega_1 \omega_2}(\omega_1 + \omega_2)^2} \prod_{\alpha > 0}(1 - e^{-\beta\omega_3 \alpha \cdot \rho}) \cdot \frac{1}{\eta(\frac{i\beta\omega_3}{2\pi})^N}$$
$$\tag{5.27}$$

$$= e^{\beta \frac{c_2|G|}{24} \frac{\omega_3}{\omega_1 \omega_2}(\omega_1 + \omega_2)^2 + \frac{N\beta\omega_3}{24}} \prod_{\alpha > 0}(1 - e^{-\beta\omega_3 \alpha \cdot \rho})$$

$$\cdot \prod_{n=1}^{\infty} \frac{1}{(1 - e^{-n\beta\omega_3})^N} .$$

After a little computation for the cases $G = U(N)$, $D_N$, $E_N$ (see [62] for the $A$ and $D$ cases), one obtains

$$Z_{S^5 \times S^1}\left(\beta, \frac{1}{2} - a_3, a_i\right) = e^{\beta \frac{c_2|G|}{24} \frac{\omega_3}{\omega_1 \omega_2}(\omega_1 + \omega_2)^2 + \frac{N\beta\omega_3}{24}} \prod_{s=0}^{\infty} \prod_{d = \deg[C(G)]} \frac{1}{1 - e^{-\beta\omega_3(d+s)}},$$
$$\tag{5.28}$$

where $d$ runs over the degrees of the possible Casimir operators $C(G)$ of the group $G$. More concretely, the degrees of the Casimir operators are

$$U(N) : 1, 2, \cdots, N \tag{5.29}$$

$$SO(2N) : 2, 4, \cdots, 2N - 2 \text{ and } N$$

$$E_6 : 2, 5, 6, 8, 9, 12$$

$$E_7 : 2, 6, 8, 10, 12, 14, 18$$

$$E_8 : 2, 8, 12, 14, 18, 20, 24, 30$$

for all ADE groups. We shall shortly give physical interpretations of these results.

Before proceeding to the interpretation of the result, we emphasize that the expression (5.1) obtained at $\beta \ll 1$ successfully becomes an index (or more generally, partition function which counts states) in the $\beta \gg 1$ regime, meaning that an expansion in $e^{-\beta} \ll 1$ has integer coefficients only. At this stage, we can justify the choice of $S_{\text{bkgd}} = -\frac{\pi^2 N}{6\beta}$. Just as in the Abelian case, we had to add this part by hand, as our supports on (5.1) came from one lower dimension in the high temperature regime. Namely, the leading singular behaviors of the free energy (coming in negative powers of $\beta$) have to be inputs in this approach. This input is all encoded in $S_{\text{bkgd}}$, in the form of the couplings of the parameters of the theory to the background gravity fields. As in the Abelian case, $S_{\text{bkgd}}$ in negative powers of $\beta$ is uniquely fixed by demanding the full quantity to be an index at $\beta \gg 1$. The structure of such couplings $S_{\text{bkgd}}$ has been explored in [64] in the case of $S^3 \times S^1$ using 3 dimensional supergravity, and similar studies are made on $S^5 \times S^1$. In particular, the absence of a term proportional to $\beta^{-3}$ in our $S_{\text{bkgd}}$ is consistent with what [64] proposes for the $(2, 0)$ theory.

Also note that the choice (5.23) for all ADE group is consistent with the requirement to have an index at $\beta \gg 1$, because the modular transformation of $\eta^{-N}$ at $\beta \gg 1$ regime exactly absorbs the factor $\left(\frac{\beta\omega_3}{2\pi}\right)^{\frac{N}{2}}$ in (5.26), which could have obstructed the index interpretation. As explained above, (5.23) was either derived or partially computed microscopically for the cases with AD. Below we shall provide more support of our choice (5.23) for E groups.

We now study the physics of (5.28). We first consider the 'spectrum' part of this index,

$$\prod_{s=0}^{\infty} \prod_{d=\deg[C(G)]} \frac{1}{1 - q^{d+s}} = PE\left[\frac{\sum_{d=\deg[C(G)]} q^d}{1 - q}\right], \qquad (5.30)$$

where we defined $q \equiv e^{-\beta\omega_3}$, and $PE$ here is defined in a more standard manner, $PE[f(x)] \equiv \exp\left[\sum_{n=1}^{\infty} \frac{1}{n} f(x^n)\right]$, without including the zero point energy factors. All coefficients of this index in $q$ expansion has positive coefficients, implying the possibility that this could actually be a partition function counting bosonic states/operators only. This is independently supported by other studies on the 6d $(2, 0)$ theory [65], which identified a closed 2d bosonic chiral subsector of the operator product expansions of local operators.

To give a more intuitive feelings on (5.30), we shall first explain an analogous situation in the 4d $\mathcal{N} = 4$ Yang–Mills theory with ADE gauge groups, in which (5.30) also emerges as the partition function of a class of BPS operators. In the 4d SYM, we are interested in gauge invariant BPS operators in the weakly coupled theory, consisting of one complex scalar $\Phi$ and one of the two holomorphic derivatives on $\mathbb{R}^4$, which we call $\partial$. The spectrum of these operators in the weakly coupled regime is worked out in [66]. In particular, we are interested in local operators $\mathcal{O}$ which are annihilated by a specific $Q$, $Q\mathcal{O} = 0$, with the equivalence relation $\mathcal{O} \sim \mathcal{O} + Q\lambda$, so we are interested in the cohomology elements. It was shown that the coholomology elements can be constructed using the $\Phi$ letters with $\partial$ derivatives only. The cohomology elements can be constructed by multiplying elements of the form

$$\partial^s f(\Phi), \tag{5.31}$$

where $f(\Phi)$ is any gauge invariant expression for the matrix $\Phi$, and then linearly superposing all possible operators constructed this way. So the question is to find the independent 'generators' taking the form of (5.31). Note that if $f$ satisfies $f(\Phi) = g(\Phi)h(\Phi)$ or $f(\Phi) = g(\Phi) + h(\Phi)$ with other gauge invariant expressions $g(\Phi), h(\Phi)$, then (5.31) is no longer an independent generator. With these considerations, if one takes $f(\Phi)$ to be all possible independent Casimir operators of the gauge group, then (5.31) forms the complete set of generators of the cohomology. The dimension of $\Phi$ is 1, so the dimension of the operator $f(\Phi)$ is the degree of the Casimir operator. So for instance, for ADE gauge groups, this leads to the scale dimension spectrum (5.29) of the $f(\Phi)$ appearing in the generator (5.31). For instance, the generators for $U(N)$ are $f(\Phi) = \mathrm{tr}(\Phi^n)$ with $n = 1, \cdots, N$. The generators for $SU(N)$ also takes the same form, with $n = 2, \cdots, N$. The generators for $SO(2N)$ are $f(\Phi) = \mathrm{tr}(\Phi^2), \cdots, \mathrm{tr}(\Phi^{2N-2})$ and $\mathrm{Pf}(\Phi) = \sqrt{\det\Phi}$. Thus the partition function for these generators, where the letters with scale dimension $\Delta$ are weighted by $q^\Delta$, is given by

$$z(q)_{\text{letter}} = \sum_{s=0}^{\infty} \sum_{d \in \deg[C(G)]} q^{d+s} = \frac{\sum_{d \in \deg[C(G)]} q^d}{1 - q}, \tag{5.32}$$

where $\Delta[\Phi] = \Delta[\partial] = 1$. Now the full set of the cohomology is obtained by forming the Fock space of the generators (5.31), and the partition function over this space is given by (5.30).

To summarize, from 4d maximal SYM, we obtained the same partition function as what we got for the 6d (2, 0) theory. This is not strange. For instance, had we been counting gauge invariant operators made of scalar $\Phi$ only without any derivatives, this would have given us the half-BPS states whose partition function is given by $PE[\sum_{d \in \deg[C(G)]} q^d]$. The half-BPS partition function is known to be universal in all maximal superconformal field theories, in 3,4,6 dimensions. There is also an explanation of this universality, by quantizing and counting the states of half-BPS giant gravitons in the AdS duals [67, 68]. Even after including one derivative, one can follow the D3-brane giant graviton counting of the partition function (5.30) in $AdS_5 \times S^5$ [66, 69], to quantize and count the M5-brane giant gravitons on $AdS_7 \times S^4$ to obtain the same partition function (at least for the A and D series). At this point, let us mention that the large $N$ limits of (5.30) for $U(N)$ and $SO(2N)$ completely agree with the supergravity indices on $AdS_7 \times S^4$ and $AdS_7 \times S^4/\mathbb{Z}_2$, respectively [60, 62]. It is also reassuring to find that the chiral algebra arguments of [65] naturally suggest the same partition function (5.30), for all ADE cases. So turning the logic around, the natural result (5.30) also supports our conjecture (5.23) on the instanton correction for the gauge groups E.

The partition function (5.30), with $d$ running over the degrees of the Casimir operators of $G = SU(N)$, $SO(2N)$, $E_N$, is known to be the vacuum character of the $W_G$ algebra. The appearance of the $W_G$ algebra in the superconformal index, and more generally in the chiral subsector of the 6d BPS operators, was asserted in [65] to be closely related to the appearance of the 2d Toda theories in the AGT correspondence [70, 71]. In fact the appearance of $W_G$ algebra and relation to the AGT relation are further supported recently, by considering the superconformal index with various defect operators [72]. Namely, for the $A_{N-1}$ theories, insertions of various dimension 2 and/or 4 defect operators to the unrefined index yielded the characters of various degenerate and semi-degenerate representations of $W_N$ algebra, and also the characters of the so-called $W_N^\rho$ representations when the dimension 4 operator is wrapped over the 2d plane where the chiral operators live. This appears to be very concrete supports of the predictions of the AGT correspondence [73] from the 6d index.

Finally, let us explain the prefactor of (5.28), which takes the form of $e^{-\beta(\epsilon_0)_{\text{SUSY}}}$ with

$$(\epsilon_0)_{\text{SUSY}} = -\frac{c_2|G|}{24}\frac{\omega_3}{\omega_1\omega_2}(\omega_1 + \omega_2)^2 - \frac{N\omega_3}{24}. \tag{5.33}$$

This formally takes the form of the 'vacuum energy' as it is conjugate to the chemical potential $\beta$ in the index. However, one needs to understand vacuum energies with care. As is obvious already in the free quantum field theory, vacuum energy is the summation of zero point energies of infinitely many harmonic oscillators, which is formally divergent. It is a quantity that has to be carefully defined and computed with regularization/renormalization. Since the regularization and renormalization are constrained by symmetry, it will be simplest to make the discussion with the special case (5.10) and (5.11) of our index. In this case, (5.33) simplifies to

$$(\epsilon_0)_{\text{SUSY}} = -\frac{c_2|G|}{6} - \frac{N}{24}. \tag{5.34}$$

More general cases are commented in [74].

Since $\beta$ is conjugate to $E - R_1$ in (5.11), the formal definition of $(\epsilon_0)_{\text{SUSY}}$ is given by the 'expectation value' of the charge $E - R_1$ for the vacuum on $S^5 \times \mathbb{R}$,

$$\langle E - R_1 \rangle = -\frac{\partial}{\partial\beta}\log Z_{S^5 \times S^1}\bigg|_{\beta\to\infty}. \tag{5.35}$$

This quantity has to be carefully defined. To concretely illustrate the subtleties, it will be illustrative to consider the free $(2, 0)$, consisting of an Abelian tensor multiplet. Then, (5.35) is given by the collection of the zero point values of $E - R_1$

for the free particle oscillators:

$$(\epsilon_0)_{\text{SUSY}} = \text{tr}\left[(-1)^F \frac{E - R_1}{2}\right] = \sum_{\text{bosonic modes}} \frac{E - R_1}{2} - \sum_{\text{fermionic modes}} \frac{E - R_1}{2}.$$

$$(5.36)$$

The trace is over the infinitely many free particle modes, and $E$, $R_1$ appearing in the sum are the values of $E$, $R_1$ carried by modes. This is similar to the ordinary Casimir energy defined by

$$\epsilon_0 \equiv \text{tr}\left[(-1)^F \frac{E}{2}\right] = \sum_{\text{bosonic modes}} \frac{E}{2} - \sum_{\text{fermionic modes}} \frac{E}{2}, \qquad (5.37)$$

which appears when one computes the partition function of a QFT on $S^n \times \mathbb{R}$ with inverse-temperature $\beta$ conjugate to the energy $E$ [75]. Both expressions are formal, and should be supplemented by a suitable regularization of the infinite sums. As in [75] for the latter quantity, one can use the charges carried by the summed-over states to provide regularizations. The charges that can be used in the regulator are constrained by the symmetries of the problem under considerations, which are different between (5.36) and (5.37). The latter is what is normally called the Casimir energy. Let us call $(\epsilon_0)_{\text{SUSY}}$ the supersymmetric Casimir energy [76].

For (5.37), the only charge that one can use to regularize the sum is energy $E$ [75]. This is because the symmetry of the path integral for the partition function on $S^n \times S^1$ includes all the internal symmetry of the theory, together with the rotation symmetry on $S^n$. Firstly, non-Abelian rotation symmetries forbid nonzero vacuum expectation values of angular momenta on $S^n$. Also, there are no sources which will give nonzero values for the internal charges: its expectation value is zero either if the internal symmetry is non-Abelian, or if there are sign flip symmetries of the Abelian internal symmetries. On the other hand, energy $E$ can be used in the regulator function. The remaining procedure of properly defining (5.37) is explained in [75]. One introduces a regulator function $f(E/\Lambda)$ with a UV cut-off $\Lambda$ (to be sent back to infinity at the final stage) which satisfies the properties $f(0) = 1$, $f(\infty) = 0$ and is sufficiently flat at $E/\Lambda = 0$: $f'(0) = 0$, $f''(0) = 0$, etc. The rigorous definition replacing (5.37) is given by

$$\text{tr}\left[(-1)^F \frac{E}{2} f(E/\Lambda)\right]. \qquad (5.38)$$

When energy level $E$ has an integer-spacing, $E = \frac{m}{r_5}$ with $m = 1, 2, 3, \cdots$, and the degeneracy for given $m$ is a polynomial of $m$ (as in [75]), one can show that this definition is the same as

$$\text{tr}\left[(-1)^F \frac{E}{2} e^{-\beta' E}\right] = -\frac{1}{2} \frac{d}{d\beta'} \text{tr}\left[(-1)^F e^{-\beta' E}\right], \qquad (5.39)$$

where small $\beta'$ is the regulator here. We shall use the latter regulator in our discussions.

On the other hand, the correct regularization of (5.36) is constrained by different symmetries. At $m = \pm\frac{1}{2}$, $a_i = 0$, the symmetry of the path integral is $SU(4|2)$ subgroup of $OSp(8^*|4)$, containing 16 supercharges. For instance, this is visible on the 5d SYM on $S^5$ [60], and is also manifest in (5.11). This subgroup is defined by elements of $OSp(8^*|4)$ which commute with $E - R_1$. So to respect the $SU(4|2)$ symmetry, only $E - R_1$ can be used to regularize the sum (5.36). So this sum can be regularized as $\mathrm{tr}\left[(-1)^F \frac{E-R_1}{2} f(\frac{E-R_1}{\Lambda})\right]$, or equivalently as

$$
\mathrm{tr}\left[(-1)^F \frac{E - R_1}{2} e^{-\beta'(E-R_1)}\right] = -\frac{1}{2}\frac{d}{d\beta'}\mathrm{tr}\left[(-1)^F e^{-\beta'(E-R_1)}\right]. \tag{5.40}
$$

The quantities

$$
f_{\mathrm{SUSY}}(\beta') = \mathrm{tr}\left[(-1)^F e^{-\beta'(E-R_1)}\right], \quad f(\beta') = \mathrm{tr}\left[(-1)^F e^{-\beta' E}\right] \tag{5.41}
$$

are computed in [63, 74] for the free $(2, 0)$ tensor multiplet, given by

$$
f(x) = \frac{5x^2(1 - x^2) - 16x^{\frac{5}{2}}(1 - x) + (10x^3 - 15x^4 + 6x^5 - x^6)}{(1 - x)^6},
$$

$$
f_{\mathrm{SUSY}}(x) = \frac{x}{1 - x}, \tag{5.42}
$$

where $x \equiv e^{-\frac{\beta'}{r_5}}$ with the $S^5$ radius $r_5$. From these expressions, one obtains

$$
-\frac{1}{2}\frac{d}{d\beta'} f(x) = \frac{5r_5}{16(\beta')^2} - \frac{25}{384r_5} + r_5^{-3}O(\beta')^2 \tag{5.43}
$$

and

$$
-\frac{1}{2}\frac{d}{d\beta'} f_{\mathrm{SUSY}}(x) = \frac{r_5}{2(\beta')^2} - \frac{1}{24r_5} + r_5^{-3}O(\beta')^2, \tag{5.44}
$$

as $\beta' \to 0$. As explained in [75], the first term $\frac{5r_5}{16(\beta')^2} \sim \frac{5}{16}r_5\Lambda^2$ of (5.43) should be canceled by a counterterm. This is because the vacuum value of $E$ has to be zero in the flat space limit $r_5 \to \infty$ from the conformal symmetry. A counterterm of the form $\Lambda^2 \int_{S^5 \times S^1} d^6x \sqrt{g}\, R^2$ or $(\beta')^{-2} \int_{S^5 \times S^1} d^6x \sqrt{g}\, R^2$ can cancel this divergence. Similarly, the first term of (5.44) has to be canceled by a counterterm of the same form. This is because the vacuum value of $E - R_1$ has to vanish in the flat space limit, required by the superconformal symmetry. After these subtractions and removing the regulator $\beta' \to 0$, one obtains

$$\epsilon_0 = -\frac{25}{384 r_5} , \quad (\epsilon_0)_{\text{SUSY}} = -\frac{1}{24 r_5} . \tag{5.45}$$

So although conceptually closely related, the two quantities are different observables. At least with the Abelian example above, we hope that we clearly illustrated the difference.

Considering that our $S^5$ partition function is constrained by $SU(4|2)$ SUSY in the path integral, it is very natural to expect that $(\epsilon_0)_{\text{SUSY}}$ of (5.34) is the supersymmetric Casimir energy. Note also that, $(\epsilon_0)_{SUSY} = -\frac{1}{24 r_5}$ computed above for the free 6d theory agrees with the zero point energy (5.34) computed from the $S^5$ partition function at $N = 1$, which concretely supports this natural expectation. (Note that $c_2|G| \equiv f^{abc} f^{abc} = 0$ for Abelian gauge group, and also that we absorbed the factor $\frac{1}{r_5}$ into $\beta$.) Even in the non-Abelian case, we think that (5.34) should be the supersymmetric Casimir energy $(\epsilon_0)_{\text{SUSY}}$, and not $\epsilon_0$.

The more conventional Casimir energy $\epsilon_0$ in the large $N$ limit has been computed in the $AdS_7 \times S^4$ gravity dual in the literature. The result is given by Awad and Johnson [77]

$$\epsilon_0 = -\frac{5 N^3}{24 r_5} , \tag{5.46}$$

while from the 5d maximal SYM with $U(N)$ gauge group, we obtain from (5.34)

$$(\epsilon_0)_{\text{SUSY}} = -\frac{N(N^2 - 1)}{6 r_5} \xrightarrow{N \to \infty} -\frac{N^3}{6 r_5} . \tag{5.47}$$

From the interpretation in our previous paragraph, we think it is likely that the disagreement of the two quantities is simply due to the fact that the gravity dual and the 5d SYM computed different observables. Assuming our interpretation, it will be interesting to study what kind of computation should be done in the gravity side to reproduce $(\epsilon_0)_{\text{SUSY}}$. We think the key is to keep SUSY manifest in the holographic renormalization computations, as this was what yielded two different Abelian observables (5.45).

## 5.2   5d Index and its Physics

In this subsection, we show some sample examples of the applications of the 5d superconformal index (5.2). We need some explanations of this formula. In Sects. 2 and 3, the partition function $Z_{\mathbb{R}^4 \times S^1}$ was understood as a series expansion in $q$, which is the fugacity conjugate to the instanton number $k > 0$. However, in 5.2, the integrand consists of two factors, one containing the instanton fugacity $q$ and another containing $q^{-1}$. Therefore, (5.2) cannot be understood as a series in $q$ or $q^{-1}$. Had we been understanding the instanton partition function $Z_{\mathbb{R}^4 \times S^1}$ in exact form, these comments would have been irrelevant. However, since we understan these partition functions only as series expansions at the moment, we should explain the way we should understand and use (5.2). The 5d superconformal index counts operators with positive scaling dimensions. They are therefore given by positive series expansions in $e^{-\epsilon_+}$, which is conjugate to the Cartan of the superconformal $SU(2)_R$ symmetry. Therefore, we should understand the two factors of $Z_{\mathbb{R}^4 \times S^1}$ in (5.2) as positive double series expansions in $t \equiv e^{-\epsilon_+}$ and $q^{\pm 1}$. Then, at a given positive order in $t$, one can show that the series in either $q^{\pm 1}$ are bounded to finite terms [30]. Therefore, one can regard (5.2) as a series in $t$, whose coefficients are invariant under the flip $q \to q^{-1}$. Since the last flip is often part of the enhanced Weyl symmetry of the UV global symmetries, the expansions explained in this paragraph naturally helps to make the index to exhibit UV global symmetries.

In [17, 30], the superconformal index on $S^4 \times S^1$ was explored to test enhanced UV global symmetries. Now in the modern understanding, we know that each Coulomb branch observable should exhibit UV global symmetry, as illustrated in Sect. 3.1. However, it is reassuring to see the enhancement in the superconformal index. For simplicity, we only show it for the 5d $SU(2)$ gauge theory at $N_f = 7$ fundamental hypermultiplets, expected to exhibit $E_8$ symmetry. The index (5.2) is given by

$$
\begin{aligned}
Z = {} & 1 + \chi^{E_8}_{248} t^2 + \chi_2(u) \left[ 1 + \chi^{E_8}_{248} \right] t^3 + \left[ 1 + \chi^{E_8}_{27000} + \chi_3(u) \left( 1 + \chi^{E_8}_{248} \right) \right] t^4 \\
& + \left[ \chi_2(u) \left( 1 + \chi^{E_8}_{248} + \chi^{E_8}_{27000} + \chi^{E_8}_{30380} \right) + \chi_4(u) \left( 1 + \chi^{E_8}_{248} \right) \right] t^5 \\
& + \left[ 2\chi^{E_8}_{248} + \chi^{E_8}_{30380} + \chi^{E_8}_{1763125} + \chi_3(u) \left( 2 + 2\chi^{E_8}_{133} + \chi^{E_8}_{3875} + 2\chi^{E_8}_{27000} + \chi^{E_8}_{30380} \right) \right. \\
& \left. + \chi_5(u) \left( 1 + \chi^{E_8}_{248} \right) \right] t^6 + \mathcal{O}\left( t^7 \right),
\end{aligned}
$$

$$(5.48)$$

where $u \equiv e^{-\epsilon_-}$. This index indeed exhibits $E_8$ enhancement. The relevant $E_8 \to SO(14) \times U(1)$ branching rules are

$$
\mathbf{248} = \mathbf{1}_0 + \mathbf{14}_2 + \mathbf{14}_{-2} + \mathbf{64}_{-1} + \overline{\mathbf{64}}_1 + \mathbf{91}_0,
$$

$$
\begin{aligned}
\mathbf{3875} = {} & \mathbf{1}_4 + \mathbf{1}_0 + \mathbf{1}_{-4} + \mathbf{14}_2 + \mathbf{14}_{-2} + \mathbf{64}_3 + \mathbf{64}_{-1} + \overline{\mathbf{64}}_1 + \overline{\mathbf{64}}_{-3} + \mathbf{91}_0 \\
& + \mathbf{104}_0 + \mathbf{364}_2 + \mathbf{364}_{-2} + \mathbf{832}_{-1} + \overline{\mathbf{832}}_1 + \mathbf{1001}_0,
\end{aligned}
$$

$$
\begin{aligned}
\mathbf{27000} = {} & 2 \times \mathbf{1}_0 + \mathbf{14}_2 + \mathbf{14}_{-2} + 2 \times \mathbf{64}_{-1} + 2 \times \overline{\mathbf{64}}_1 + 2 \times \mathbf{91}_0 \\
& + \mathbf{104}_4 + \mathbf{104}_0 + \mathbf{104}_{-4} + \mathbf{364}_2 + \mathbf{364}_{-2} \\
& + \mathbf{832}_3 + \mathbf{832}_{-1} + \overline{\mathbf{832}}_1 + \overline{\mathbf{832}}_{-3} + \mathbf{896}_2 + \mathbf{896}_{-2} + \mathbf{1001}_0 \\
& + \mathbf{1716}_{-2} + \overline{\mathbf{1716}}_2 + \mathbf{3003}_0 + \mathbf{3080}_0 + \mathbf{4928}_{-1} + \overline{\mathbf{4928}}_1,
\end{aligned}
$$

$$30380 = \mathbf{1}_0 + 2 \times \mathbf{14}_2 + 2 \times \mathbf{14}_{-2} + \mathbf{64}_3 + 2 \times \mathbf{64}_{-1} + 2 \times \overline{\mathbf{64}}_1 + \overline{\mathbf{64}}_{-3}$$
$$+ \mathbf{91}_4 + 3 \times \mathbf{91}_0 + \mathbf{91}_{-4} + \mathbf{104}_0 + \mathbf{364}_2 + \mathbf{364}_{-2}$$
$$+ \mathbf{832}_3 + 2 \times \mathbf{832}_{-1} + 2 \times \overline{\mathbf{832}}_1 + \overline{\mathbf{832}}_{-3} + \mathbf{896}_2 + \mathbf{896}_{-2}$$
$$+ \mathbf{1001}_0 + \mathbf{2002}_2 + \mathbf{2002}_{-2} + \mathbf{3003}_0 + \mathbf{4004}_0 + \mathbf{4928}_{-1} + \overline{\mathbf{4928}}_1,$$

$$1763125 = 2 \times \mathbf{1}_0 + 2 \times \mathbf{14}_2 + 2 \times \mathbf{14}_{-2} + 3 \times \mathbf{64}_{-1} + 3 \times \overline{\mathbf{64}}_1 + 3 \times \mathbf{91}_0$$
$$+ \mathbf{104}_4 + \mathbf{104}_0 + \mathbf{104}_{-4} + \mathbf{364}_2 + \mathbf{364}_{-2} + \mathbf{546}_6 + \mathbf{546}_2 + \mathbf{546}_{-2} + \mathbf{546}_{-6}$$
$$+ 2 \times \mathbf{832}_3 + 2 \times \mathbf{832}_{-1} + 2 \times \overline{\mathbf{832}}_1 + 2 \times \overline{\mathbf{832}}_{-3} + 2 \times \mathbf{896}_2 + 2 \times \mathbf{896}_{-2}$$
$$+ 2 \times \mathbf{1001}_0 + 2 \times \mathbf{1716}_{-2} + 2 \times \overline{\mathbf{1716}}_2 + \mathbf{2002}_2 + \mathbf{2002}_{-2}$$
$$+ 3 \times \mathbf{3003}_0 + 2 \times \mathbf{3080}_0 + \mathbf{4004}_4 + 2 \times \mathbf{4004}_0 + \mathbf{4004}_{-4}$$
$$+ 3 \times \mathbf{4928}_{-1} + 3 \times \overline{\mathbf{4928}}_1 + \mathbf{5625}_4 + \mathbf{5625}_0 + \mathbf{5625}_{-4}$$
$$+ \mathbf{5824}_3 + \mathbf{5824}_{-1} + \mathbf{5824}_{-5} + \overline{\mathbf{5824}}_5 + \overline{\mathbf{5824}}_1 + \overline{\mathbf{5824}}_{-3}$$
$$+ \mathbf{11648}_2 + \mathbf{11648}_{-2} + \mathbf{17472}_3 + \mathbf{17472}_{-1} + \overline{\mathbf{17472}}_1 + \overline{\mathbf{17472}}_{-3}$$
$$+ \mathbf{18200}_2 + \mathbf{18200}_{-2} + \mathbf{21021}_0 + \mathbf{21021}_{-4} + \overline{\mathbf{21021}}_4 + \overline{\mathbf{21021}}_0$$
$$+ \mathbf{24024}'_2 + \mathbf{24024}'_{-2} + \mathbf{27456}_3 + \overline{\mathbf{27456}}_{-3} + \mathbf{36608}_2 + \mathbf{36608}_{-2}$$
$$+ \mathbf{40768}_{-1} + \overline{\mathbf{40768}}_1 + \mathbf{45760}_3 + \mathbf{45760}_{-1} + \overline{\mathbf{45760}}_1 + \overline{\mathbf{45760}}_{-3}$$
$$+ \mathbf{58344}_0 + \mathbf{58968}_0 + \mathbf{64064}'_{-1} + \overline{\mathbf{64064}}'_1 + \mathbf{115830}_{-2} + \overline{\mathbf{115830}}_2$$
$$+ \mathbf{146432}_{-1} + \overline{\mathbf{146432}}_1 + \mathbf{200200}_0. \tag{5.49}$$

There are some more interesting recent applications of (5.2). Namely, some large $N$ 5d superconformal field theories are supposed to be dual to a gravity dual on $AdS_6$. More specifically, the 5d QFT on $N$ D4, $N_f$ D8 and an O8$^-$ is supposed to be dual to a massive IIA string theory in the background of warped $AdS_6 \times S^4/\mathbb{Z}_2$. The supersymmetric black holes in this background have been studied in [78, 79]. The index (5.2) of these 5d SCFTs can be used to microscopically count the entropies of these black holes [80].

# 6  Conclusion

In this review article, we explained the recent quantitative studies on quantum aspects of 5d and 6d superconformal field theories. We especially focussed on partition functions related to the instanton counting in the Coulomb or tensor branches.

We first reviewed the old and new computations of the instanton partition functions, calculated either via quantum mechanical Witten indices of 5d instanton particles, or 2d elliptic genera of 6d instanton strings. Although there are various approaches to compute them (e.g. topological strings, topological vertices), we focussed on the 1d and 2d gauge theory approaches to these partition functions. The Coulomb/tensor branch partition functions are used to study various interesting strong coupling physics (Sects. 3 and 4). We studied the strong coupling symmetry enhancements of 5d SCFTs, Kaluza–Klein deconstructions of 6d SCFTs on $S^1$ using 5d instantons, studies on the $N^3$ degrees of freedom on SCFTs for $N$ M5-branes, and explorations of exceptional instanton particles/strings. Finally, in Sect. 5, we reviewed the recent explorations on 5d and 6d superconformal indices on $S^4 \times S^1$ and $S^5 \times S^1$, respectively, which are expressed in terms of suitable instanton partition functions.

The review made in this paper deals with both technical advances in the calculus of supersymmetric partition functions, and also the new physics explored with them. As for the technical sides, Sect. 2 discussed the old and new progresses in the instanton partition function calculus. Also, Sect. 4.2 mostly explained technical advances in the 2d QFT methods of studying the 6d self-dual strings, especially concerning the calculus of their elliptic genera. Sections 3.1 and 5.2 mostly explained the applications of the 5d instanton particles' partition functions to explore the physics of 5d SCFTs. Sections 3.2, 4.1, and 5.1 explained the recent explorations of the physics of 6d $\mathcal{N} = (2, 0)$ SCFTs. However, some partition functions for higher dimensional QFTs were just computed rigorously, awaiting for physical applications. Especially, the 6d $\mathcal{N} = (1, 0)$ partition functions summarized in Sect. 4.2 should find more interesting physical applications.

As for very recent applications of the partition functions reviewed in this paper, we should first mention the studies of 6d index on $S^5 \times S^1$ made for the supersymmetric AdS$_7$ black holes in the large $N$ limit [81]. Also, the 5d SCFT indices on $S^4 \times S^1$ can be used to study the supersymmetric AdS$_6$ black holes in the large $N$ dual [80]. We are confident that more useful applications are to appear.

**Acknowledgements** This work is supported in part by the National Research Foundation of Korea (NRF) Grant 2018R1A2B6004914.

# References

1. E. Witten, Some comments on string dynamics (1995). [hep-th/9507121]
2. D.R. Morrison, C. Vafa, Compactifications of $F$-theory on Calabi-Yau threefolds. I. Nucl. Phys. B **473**, 74–92 (1996). https://doi.org/10.1016/0550-3213(96)00242-8. [hep-th/9602114]
3. D.R. Morrison, C. Vafa, Compactifications of $F$-theory on Calabi-Yau threefolds. II. Nucl. Phys. B **476**, 437–469 (1996). https://doi.org/10.1016/0550-3213(96)00369-0. [hep-th/9603161]
4. N. Seiberg, E. Witten, Comments on string dynamics in six-dimensions. Nucl. Phys. B **471**, 121–134 (1996). https://doi.org/10.1016/0550-3213(96)00189-7. [hep-th/9603003]

5. E. Witten, Phase transitions in $M$-theory and $F$-theory. Nucl. Phys. B **471**, 195–216 (1996). https://doi.org/10.1016/0550-3213(96)00212-X. [hep-th/9603150]

6. N. Seiberg, Five-dimensional SUSY field theories, nontrivial fixed points and string dynamics. Phys. Lett. B **388**, 753–760 (1996). https://doi.org/10.1016/S0370-2693(96)01215-4. [hep-th/9608111]

7. D.R. Morrison, N. Seiberg, Extremal transitions and five-dimensional supersymmetric field theories. Nucl. Phys. B **483**, 229–247 (1997). https://doi.org/10.1016/S0550-3213(96)00592-5. [hep-th/9609070]

8. K.A. Intriligator, D.R. Morrison, N. Seiberg, Five-dimensional supersymmetric gauge theories and degenerations of Calabi-Yau spaces. Nucl. Phys. B **497**, 56–100 (1997). https://doi.org/10.1016/S0550-3213(97)00279-4. [hep-th/9702198]

9. J.J. Heckman, D.R. Morrison, C. Vafa, On the classification of 6D SCFTs and generalized ADE orbifolds. J. High Energy Phys. **1405**, 028 (2014). https://doi.org/10.1007/JHEP05(2014)028. Erratum: JHEP **1506**, 017 (2015) doi.org/10.1007/JHEP06(2015)017. [arXiv:1312.5746 [hep-th]]

10. J.J. Heckman, D.R. Morrison, T. Rudelius, C. Vafa, Atomic classification of 6D SCFTs. Fortsch. Phys. **63**, 468–530 (2015). https://doi.org/10.1002/prop.201500024. [arXiv:1502.05405 [hep-th]]

11. P. Jefferson, H.C. Kim, C. Vafa, G. Zafrir, Towards classification of 5d SCFTs: single gauge node (2017). [arXiv:1705.05836 [hep-th]]

12. P. Jefferson, S. Katz, H.C. Kim, C. Vafa, On geometric classification of 5d SCFTs. J. High Energy Phys. **1804**, 103 (2018). https://doi.org/10.1007/JHEP04(2018)103. [arXiv:1801.04036 [hep-th]].

13. N.A. Nekrasov, Seiberg-Witten prepotential from instanton counting. Adv. Theor. Math. Phys. **7**(5), 831–864 (2003). https://doi.org/10.4310/ATMP.2003.v7.n5.a4. [hep-th/0206161]

14. R. Gopakumar, C. Vafa, M-theory and topological strings. I (1998). [hep-th/9809187]

15. R. Gopakumar, C. Vafa, M-theory and topological strings. II (1998). [hep-th/9812127]

16. M. Aganagic, A. Klemm, M. Marino, C. Vafa, The topological vertex. Commun. Math. Phys. **254**, 425–478 (2005). https://doi.org/10.1007/s00220-004-1162-z. [hep-th/0305132]

17. C. Hwang, J. Kim, S. Kim, J. Park, General instanton counting and 5d SCFT. J. High Energy Phys. **1507**, 063 (2015). https://doi.org/10.1007/JHEP07(2015)063. Addendum: JHEP **1604**, 094 (2016), doi:10.1007/JHEP04(2016)094, [arXiv:1406.6793 [hep-th]]

18. N. Seiberg, E. Witten, Electric - magnetic duality, monopole condensation, and confinement in $N = 2$ supersymmetric Yang-Mills theory. Nucl. Phys. B **426**, 19–52 (1994). https://doi.org/10.1016/0550-3213(94)90124-4. Erratum: [Nucl. Phys. B **430**, 485 (1994)], doi:10.1016/0550-3213(94)00449-8, [hep-th/9407087]

19. N. Seiberg, E. Witten, Monopoles, duality and chiral symmetry breaking in $N = 2$ supersymmetric QCD. Nucl. Phys. B **431**, 484–550 (1994). https://doi.org/10.1016/0550-3213(94)90214-3. [hep-th/9408099]

20. N. Nekrasov, A. Okounkov, Seiberg-Witten theory and random partitions. Prog. Math. **244**, 525–596 (2006). https://doi.org/10.1007/0-8176-4467-9_15. [hep-th/0306238]

21. N. Nekrasov, S. Shadchin, ABCD of instantons. Commun. Math. Phys. **252**, 359–391 (2004). https://doi.org/10.1007/s00220-004-1189-1. [hep-th/0404225]

22. D. Tong, The holographic dual of $AdS_3 \times S^3 \times S^3 \times S^1$. J. High Energy Phys. **1404**, 193 (2014). https://doi.org/10.1007/JHEP04(2014)193. [arXiv:1402.5135 [hep-th]]

23. J. Kim, S. Kim, K. Lee, Higgsing towards E-strings. arXiv:1510.03128 [hep-th]

24. M.F. Atiyah, N.J. Hitchin, V.G. Drinfeld, Y.I. Manin, Construction of instantons. Phys. Lett. A **65**, 185–187 (1978). https://doi.org/10.1016/0375-9601(78)90141-X

25. N.H. Christ, E.J. Weinberg, N.K. Stanton, General selfdual Yang-Mills solutions. Phys. Rev. D **18**, 2013–2025 (1978). https://doi.org/10.1103/PhysRevD.18.2013

26. O. Aharony, A. Hanany, B. Kol, Webs of $(p, q)$ 5-branes, five-dimensional field theories and grid diagrams. J. High Energy Phys. **9801**, 002 (1998). https://doi.org/10.1088/1126-6708/1998/01/002. [hep-th/9710116]

27. H.C. Kim, J.Kim, S. Kim, K.H. Lee, J. Park, 6d strings and exceptional instantons (2018). arXiv:1801.03579 [hep-th]
28. S. Shadchin, On certain aspects of string theory/gauge theory correspondence (2005). hep-th/0502180
29. F. Benini, R. Eager, K. Hori, Y. Tachikawa, Elliptic genera of 2d $\mathcal{N} = 2$ gauge theories. Commun. Math. Phys. **333**, 1241–1286 (2015). https://doi.org/10.1007/s00220-014-2210-y. [arXiv:1308.4896 [hep-th]]
30. H.C. Kim, S.S. Kim, K. Lee, 5-dim superconformal index with enhanced $E_n$ global symmetry. J. High Energy Phys. **1210**, 142 (2012). https://doi.org/10.1007/JHEP10(2012)142. [arXiv:1206.6781 [hep-th]]
31. R. Flume, R. Poghossian, An algorithm for the microscopic evaluation of the coefficients of the Seiberg-Witten prepotential. Int. J. Mod. Phys. A **18**, 2541–2563 (2003). https://doi.org/10.1142/S0217751X03013685. [hep-th/0208176]
32. U. Bruzzo, F. Fucito, J.F. Morales, A. Tanzini, Multiinstanton calculus and equivariant cohomology. J. High Energy Phys. **0305**, 054 (2003). https://doi.org/10.1088/1126-6708/2003/05/054. [hep-th/0211108]
33. H.C. Kim, S. Kim, E. Koh, K. Lee, S. Lee, On instantons as Kaluza-Klein modes of M5-branes. J. High Energy Phys. **1112**, 031 (2011). https://doi.org/10.1007/JHEP12(2011)031. [arXiv:1110.2175 [hep-th]].
34. C. Cordova, T.T. Dumitrescu, K. Intriligator, Deformations of superconformal theories. J. High Energy Phys. **1611**, 135 (2016). https://doi.org/10.1007/JHEP11(2016)135. [arXiv:1602.01217 [hep-th]]
35. S.S. Kim, M. Taki, F. Yagi, Tao probing the end of the world. Progress of Theoretical and Experimental Physics **2015**, 083B02 (2015). https://doi.org/10.1093/ptep/ptv108. [arXiv:1504.03672 [hep-th]]
36. V. Mitev, E. Pomoni, M. Taki, F. Yagi, Fiber-base duality and global symmetry enhancement. J. High Energy Phys. **1504**, 052 (2015). https://doi.org/10.1007/JHEP04(2015)052. [arXiv:1411.2450 [hep-th]]
37. Y. Hwang, J. Kim, S. Kim, M5-branes, orientifolds, and S-duality. J. High Energy Phys. **1612**, 148 (2016). https://doi.org/10.1007/JHEP12(2016)148. [arXiv:1607.08557 [hep-th]]
38. O.J. Ganor, A. Hanany, Small $E_8$ instantons and tensionless noncritical strings. Nucl. Phys. B **474**, 122–138 (1996). https://doi.org/10.1016/0550-3213(96)00243-X. [hep-th/9602120]
39. J. Kim, S. Kim, K. Lee, J. Park, C. Vafa, Elliptic genus of E-strings. J. High Energy Phys. **1709**, 098 (2017). https://doi.org/10.1007/JHEP09(2017)098. [arXiv:1411.2324 [hep-th]]
40. H. Hayashi, S.S. Kim, K. Lee, M. Taki, F. Yagi, A new 5d description of 6d D-type minimal conformal matter. J. High Energy Phys. **1508**, 097 (2015). https://doi.org/10.1007/JHEP08(2015)097. [arXiv:1505.04439 [hep-th]]
41. H. Hayashi, S.S. Kim, K. Lee, F. Yagi, 6d SCFTs, 5d dualities and tao web diagrams (2015). arXiv:1509.03300 [hep-th]
42. H. Hayashi, S.S. Kim, K. Lee, M. Taki, F. Yagi, More on 5d descriptions of 6d SCFTs. J. High Energy Phys. **1610**, 126 (2016). https://doi.org/10.1007/JHEP10(2016)126. [arXiv:1512.08239 [hep-th]]
43. H. Hayashi, S.S. Kim, K. Lee, F. Yagi, Equivalence of several descriptions for 6d SCFT. J. High Energy Phys. **1701**, 093 (2017). https://doi.org/10.1007/JHEP01(2017)093. [arXiv:1607.07786 [hep-th]]
44. H. Hayashi, K. Ohmori, 5d/6d DE instantons from trivalent gluing of web diagrams. J. High Energy Phys. **1706**, 078 (2017). https://doi.org/10.1007/JHEP06(2017)078. [arXiv:1702.07263 [hep-th]]
45. O. Aharony, M. Berkooz, S. Kachru, N. Seiberg, E. Silverstein, Matrix description of interacting theories in six-dimensions. Adv. Theor. Math. Phys. **1**, 148–157 (1997). https://doi.org/10.4310/ATMP.1997.v1.n1.a5. [hep-th/9707079]; O. Aharony, M. Berkooz, N. Seiberg, Light cone description of (2,0) superconformal theories in six-dimensions. Adv. Theor. Math. Phys. **2**, 119–153 (1998). https://doi.org/10.4310/ATMP.1998.v2.n1.a5. [hep-th/9712117]

46. B. Haghighat, A. Iqbal, C. Kozçaz, G. Lockhart, C. Vafa, M-strings. Commun. Math. Phys. **334**, 779–842 (2015). https://doi.org/10.1007/s00220-014-2139-1. [arXiv:1305.6322 [hep-th]]

47. A. Iqbal, C. Kozcaz, K. Shabbir, Refined topological vertex, cylindric partitions and the U(1) adjoint theory. Nucl. Phys. B **838**, 422–457 (2010). https://doi.org/10.1016/j.nuclphysb.2010. 06.010. [arXiv:0803.2260 [hep-th]]

48. M.B. Green, J.H. Schwarz, P.C. West, Anomaly free chiral theories in six-dimensions. Nucl. Phys. B **254**, 327–348 (1985). https://doi.org/10.1016/0550-3213(85)90222-6

49. A. Sagnotti, A note on the Green-Schwarz mechanism in open string theories. Phys. Lett. B **294**, 196–203 (1992). https://doi.org/10.1016/0370-2693(92)90682-T. [hep-th/9210127]

50. M. Bershadsky, C. Vafa, Global anomalies and geometric engineering of critical theories in six-dimensions (1997). hep-th/9703167

51. N. Seiberg, Nontrivial fixed points of the renormalization group in six-dimensions. Phys. Lett. B **390**, 169–171 (1997). https://doi.org/10.1016/S0370-2693(96)01424-4 [hep-th/9609161].

52. B. Haghighat, A. Klemm, G. Lockhart, C. Vafa, Strings of Minimal 6d SCFTs. Fortsch. Phys. **63**, 294–322 (2015). https://doi.org/10.1002/prop.201500014. [arXiv:1412.3152 [hep-th]]

53. A. Gadde, B. Haghighat, J. Kim, S. Kim, G. Lockhart, C. Vafa, 6d string chains. J. High Energy Phys. **1802**, 143 (2018). https://doi.org/10.1007/JHEP02(2018)143. [arXiv:1504.04614 [hep-th]]

54. H.C. Kim, S. Kim, J. Park, 6d strings from new chiral gauge theories (2016). arXiv:1608.03919 [hep-th]

55. B. Haghighat, C. Kozcaz, G. Lockhart, C. Vafa, Orbifolds of M-strings. Phys. Rev. D **89**, 046003 (2014). https://doi.org/10.1103/PhysRevD.89.046003. [arXiv:1310.1185 [hep-th]]

56. S. Kim, J. Nahmgoong, Asymptotic M5-brane entropy from S-duality. J. High Energy Phys. **1712**, 120 (2017). https://doi.org/10.1007/JHEP12(2017)120. [arXiv:1702.04058 [hep-th]]

57. A. Klemm, P. Mayr, C. Vafa, BPS states of exceptional noncritical strings. Nucl. Phys. Proc. Suppl. **58**, 177–194 (1997). https://doi.org/10.1016/S0920-5632(97)00422-2. [hep-th/9607139]

58. P. Horava, E. Witten, Heterotic and type I string dynamics from eleven-dimensions. Nucl. Phys. B **460**, 506–524 (1996). https://doi.org/10.1016/0550-3213(95)00621-4. [hep-th/9510209]

59. A. Grassi, J. Halverson, J.L. Shaneson, Non-Abelian Gauge Symmetry and the Higgs Mechanism in F-theory. Commun. Math. Phys. **336**, 1231–1257 (2015). https://doi.org/10. 1007/s00220-015-2313-0. [arXiv:1402.5962 [hep-th]]

60. H.C. Kim, S. Kim, M5-branes from gauge theories on the 5-sphere. J. High Energy Phys. **1305**, 144 (2013). [arXiv:1206.6339 [hep-th]]

61. G. Lockhart, C. Vafa, Superconformal partition functions and non-perturbative topological strings (2012). arXiv:1210.5909 [hep-th]

62. H.C. Kim, J. Kim, S. Kim, Instantons on the 5-sphere and M5-branes (2012). arXiv:1211.0144 [hep-th]

63. J. Bhattacharya, S. Bhattacharyya, S. Minwalla, S. Raju, Indices for superconformal field theories in 3,5 and 6 dimensions. J. High Energy Phys. **0802**, 064 (2008). https://doi.org/10. 1088/1126-6708/2008/02/064. [arXiv:0801.1435 [hep-th]]

64. L. Di Pietro, Z. Komargodski, Cardy formulae for SUSY theories in $d = 4$ and $d = 6$. J. High Energy Phys. **1412**, 031 (2014). https://doi.org/10.1007/JHEP12(2014)031. [arXiv:1407.6061 [hep-th]]

65. C. Beem, L. Rastelli, B.C. van Rees, $\mathcal{W}$ symmetry in six dimensions. J. High Energy Phys. **1505**, 017 (2015). https://doi.org/10.1007/JHEP05(2015)017. [arXiv:1404.1079 [hep-th]]

66. L. Grant, P.A. Grassi, S. Kim, S. Minwalla, Comments on 1/16 BPS quantum states and classical configurations. J. High Energy Phys. **0805**, 049 (2008). https://doi.org/10.1088/1126-6708/2008/05/049. [arXiv:0803.4183 [hep-th]]

67. G. Mandal, N.V. Suryanarayana, Counting 1/8-BPS dual-giants. J. High Energy Phys. **0703**, 031 (2007). https://doi.org/10.1088/1126-6708/2007/03/031. [hep-th/0606088]

68. S. Bhattacharyya, S. Minwalla, Supersymmetric states in M5/M2 CFTs. J. High Energy Phys. **0712**, 004 (2007). [hep-th/0702069 [HEP-TH]]

69. S.K. Ashok, N.V. Suryanarayana, Counting wobbling dual-giants. J. High Energy Phys. **0905**, 090 (2009). https://doi.org/10.1088/1126-6708/2009/05/090. [arXiv:0808.2042 [hep-th]]
70. L.F. Alday, D. Gaiotto, Y. Tachikawa, Liouville correlation functions from four-dimensional gauge theories. Lett. Math. Phys. **91**, 167–197 (2010). https://doi.org/10.1007/s11005-010-0369-5. [arXiv:0906.3219 [hep-th]]
71. N. Wyllard, $A_{N-1}$ conformal Toda field theory correlation functions from conformal $\mathcal{N} = 2SU(N)$ quiver gauge theories. J. High Energy Phys. **0911**, 002 (2009). https://doi.org/10.1088/1126-6708/2009/11/002. [arXiv:0907.2189 [hep-th]]
72. M. Bullimore, H.-C. Kim, The superconformal index of the (2,0) theory with defects (2014). [arXiv:1412.3872 [hep-th]]
73. L.F. Alday, D. Gaiotto, S. Gukov, Y. Tachikawa, H. Verlinde, Loop and surface operators in $\mathcal{N} = 2$ gauge theory and Liouville modular geometry. J. High Energy Phys. **1001**, 113 (2010). https://doi.org/0.1007/JHEP01(2010)113. [arXiv:0909.0945 [hep-th]]
74. H.C. Kim, S. Kim, S.S. Kim, K. Lee, The general M5-brane superconformal index (2013). [arXiv:1307.7660]
75. O. Aharony, J. Marsano, S. Minwalla, K. Papadodimas, M. Van Raamsdonk, The hagedorn/deconfinement phase transition in weakly coupled large N gauge theories. Adv. Theor. Math. Phys. **8**, 603–696 (2004). https://doi.org/10.4310/ATMP.2004.v8.n4.a1. [hep-th/0310285]
76. D. Cassani, D. Martelli, The gravity dual of supersymmetric gauge theories on a squashed $S^1 \times S^3$. J. High Energy Phys. **1408**, 044 (2014). https://doi.org/10.1007/JHEP08(2014)044. [arXiv:1402.2278 [hep-th]]
77. A.M. Awad, C.V. Johnson, Phys. Rev. D **63**, 124023 (2001). [hep-th/0008211]
78. D.D.K. Chow, Charged rotating black holes in six-dimensional gauged supergravity. Class. Quant. Grav. **27**, 065004 (2010). https://doi.org/10.1088/0264-9381/27/6/065004. [arXiv:0808.2728 [hep-th]]
79. S. Choi, C. Hwang, S. Kim, J. Nahmgoong, Entropy functions of BPS black holes in AdS$_4$ and AdS$_6$ (2018). arXiv:1811.02158 [hep-th]
80. S. Choi, S. Kim, Large AdS$_6$ black holes from CFT$_5$ (2019). arXiv:1904.01164 [hep-th]
81. S. Choi, J. Kim, S. Kim, J. Nahmgoong, Large AdS black holes from QFT (2018). arXiv:1810.12067 [hep-th]

# On Spinorial Representations of Involutory Subalgebras of Kac–Moody Algebras

**Axel Kleinschmidt, Hermann Nicolai, and Adriano Viganò**

## Contents

1 Introduction .................................................................... 180
2 Involutory Subalgebras of Kac–Moody Algebras .......................... 181
  2.1 Basic Definitions ......................................................... 181
  2.2 Maximal Compact Subalgebras .......................................... 184
3 Spin Representations ........................................................... 188
  3.1 The Simply-Laced Case ................................................. 190
  3.2 The Non-simply-laced Case ............................................ 198
4 Quotients ...................................................................... 203
  4.1 Spin-$\frac{1}{2}$ Quotients in $AE_d$ ........................................... 203
  4.2 Quotient Algebras for $K(E_{10})$ ...................................... 205
  4.3 General Remarks on the Quotients ...................................... 208
5 $K(E_{10})$ and Standard Model Fermions .................................... 211
References ....................................................................... 213

**Abstract** The representation theory of involutory (or 'maximal compact') subalgebras of infinite-dimensional Kac–Moody algebras is largely *terra incognita*, especially with regard to fermionic (double-valued) representations. Nevertheless, certain distinguished such representations feature prominently in proposals of possible symmetries underlying M theory, both at the classical and the quantum level.

Based on lectures given by H. Nicolai at the School *Partition Functions and Automorphic Forms*, BLTP JINR, Dubna, Russia, 28 January–2 February 2018.

A. Kleinschmidt (✉)
Max-Planck-Institut für Gravitationsphysik (Albert-Einstein-Institut), Potsdam, Germany

International Solvay Institutes, Brussels, Belgium
e-mail: axel.kleinschmidt@aei.mpg.de

H. Nicolai · A. Viganò
Max-Planck-Institut für Gravitationsphysik (Albert-Einstein-Institut), Potsdam, Germany

© Springer Nature Switzerland AG 2020
V. A. Gritsenko, V. P. Spiridonov (eds.), *Partition Functions
and Automorphic Forms*, Moscow Lectures 5,
https://doi.org/10.1007/978-3-030-42400-8_4

179

Here we summarise recent efforts to study *spinorial* representations systematically, most notably for the case of the hyperbolic Kac–Moody algebra $E_{10}$ where spinors of the involutory subalgebra $K(E_{10})$ are expected to play a role in describing algebraically the fermionic sector of $D = 11$ supergravity and M theory. Although these results remain very incomplete, they also point towards the beginning of a possible explanation of the fermion structure observed in the Standard Model of Particle Physics.

# 1 Introduction

Fermionic fields are central in all fundamental particle models as constituents of matter. They are characterised as being double-valued representations of the space-time symmetry group, notably the spin cover of the Lorentz group $\mathrm{Spin}(1, D - 1)$ in $D$ space-time dimensions. For instance, the Standard Model of particle physics contains 48 spin-$\frac{1}{2}$ particles, if one includes also right-handed neutrinos in the counting, and these transform also under the gauge symmetries $\mathrm{SU}(3)_c \times \mathrm{SU}(2)_w \times \mathrm{U}(1)_Y$. To date there is no satisfactory explanation of this precise set of spin-$\frac{1}{2}$ particles of the Standard Model of particle physics. Fermionic fields are also essential and inevitable in all supersymmetric theories, notably supergravity and superstring theory. Investigations of the bosonic sectors of these theories have led to conjectured gigantic symmetry groups governing the bosonic dynamics. The most well-known of these conjectures involve the infinite-dimensional Kac–Moody groups $E_{10}$ and $E_{11}$ [1–3]. Implicit in these conjectures is that there are also infinite-dimensional extensions of spin groups acting on the fermionic fields. At the Lie algebra level, these infinite-dimensional extensions are given by fixed points of an involution acting on the Kac–Moody algebra and are known by the name $K(E_{10})$ and $K(E_{11})$ and often called maximal compact subalgebras.[1]

In order to understand the fermions in the context of these conjectures it is therefore important to understand the fermionic (double-valued) representations of the maximal compact subalgebras $K(E_n)$. For $n \geq 9$, $K(E_n)$ does not belong to any standard class of Lie algebras, and there is no off-the-shelf representation theory available for finding the fermionic representations (whereas for finite-dimensional $E_n$ groups, i.e. $n \leq 8$, the maximal compact subalgebras *are* of standard type). By analysing the structure of maximal supergravity, some first spin-$\frac{1}{2}$ and spin-$\frac{3}{2}$ representations of $K(E_{10})$ were found in [4–7]. The fermionic representations found there also have the unusual property of unifying the distinct type IIA and type IIB spinors of $D = 10$ maximal supersymmetry [8, 9]. Moreover, as we shall also review in these proceedings, they hold the potential of explaining the fermionic content of the Standard Model of particle physics [10–12].

---

[1] Here and in the remainder of these notes, we abuse notation by using $E_n$ to denote both the Lie algebra and the associated group.

The purpose of this article is to summarise and systematise more recent advances on the representation theory of $K(E_{10})$ and similar (hyperbolic) Kac–Moody algebras [13–15]. Specifically, we shall explain:

- the construction of involutory subalgebras of Kac–Moody algebras;
- the basic conditions, called Berman relations, that any representation has to satisfy;
- the construction of spin-$\frac{1}{2}$ representations for simply-laced algebras;
- the extension up to spin-$\frac{7}{2}$ for the simply-laced $E_{10}$, $DE_{10}$ and the $AE_d$ family (for $d \geq 3$);
- the extension to higher spin for the non-simply-laced $AE_3$, $G_2^{++}$ and $BE_{10}$;
- the properties of these representations, in particular the fact that they are unfaithful and imply the existence of finite-dimensional quotients of the infinite-dimensional involutory algebras, as we will illustrate with several examples;
- the connection between $K(E_{10})$ and the fermions of the Standard Model.

Other work on fermionic representations includes [16–24]. Let us stress again that these results represent only the very first steps, and that many open problems remain which will probably require completely new insights and methods.

## 2 Involutory Subalgebras of Kac–Moody Algebras

We review some basic facts about Kac–Moody algebras and their maximal compact subalgebras, referring the reader to [25] for proofs and details.

### 2.1 Basic Definitions

An $n \times n$ matrix $A$ which satisfies

$$A_{ii} = 2 \quad \forall i = 1, \ldots, n, \tag{1a}$$

$$A_{ij} \in \mathbb{Z}_{\leq 0} \quad (i \neq j), \tag{1b}$$

$$A_{ij} = 0 \iff A_{ji} = 0, \tag{1c}$$

where $\mathbb{Z}_{\leq 0}$ denotes the non-positive integers, is called a *generalised Cartan matrix* (we will frequently refer to $A$ simply as the Cartan matrix). We always assume $\det A \neq 0$.

The *Kac–Moody algebra* $\mathfrak{g} \equiv \mathfrak{g}(A)$ is the Lie algebra generated by the $3n$ generators $\{e_i, f_i, h_i\}$ subject to the relations

$$[h_i, h_j] = 0, \tag{2a}$$

$$[h_i, e_j] = A_{ij} e_j, \tag{2b}$$

$$[h_i, f_j] = -A_{ij} f_j, \tag{2c}$$

$$[e_i, f_j] = \delta_{ij} h_j, \tag{2d}$$

$$\mathrm{ad}_{e_i}^{1-A_{ij}} e_j = 0, \tag{2e}$$

$$\mathrm{ad}_{f_i}^{1-A_{ij}} f_j = 0. \tag{2f}$$

Each triple $\{e_i, f_i, h_i\}$ generates a subalgebra isomorphic to $\mathfrak{sl}(2, \mathbb{C})$, as is evident from relations (2). This construction of the algebra in terms of generators and relations is generally referred to as the *Chevalley-Serre presentation*.

A generalised Cartan matrix $A$ is called *simply-laced* if the off-diagonal entries of $A$ are either 0 or $-1$. $A$ is called *symmetrizable* if there exists a diagonal matrix $D$ such that $DA$ is symmetric. We will say that $\mathfrak{g}$ is simply-laced (symmetrizable) if its generalised Cartan matrix $A$ is simply-laced (symmetrizable). Analogously, the algebra is *symmetric* if its Cartan matrix is. In the finite dimensional case an algebra is simply-laced if and only if it is symmetric, but this is not true in the general infinite-dimensional case. If the Cartan matrix $A$ is indefinite, the associated (infinite-dimensional) Kac–Moody algebra is called *indefinite*; it is called *hyperbolic* if the removal of any node in the Dynkin diagram leaves an algebra that is either of finite or affine type; with this extra requirement, the rank is bounded above by $r \leq 10$, with four such maximal rank algebras, $E_{10}$, $DE_{10}$, $BE_{10}$ and $CE_{10}$ [25]. Our interest here is mainly with hyperbolic algebras, as these appear to be the relevant ones for M theory.

The *Dynkin diagram* of a Kac–Moody algebra $\mathfrak{g}$ is a set of vertices and edges built with the following rules:

- for each $i = 1, \ldots, n$ there is an associated node in the diagram;
- if $|A_{ij}| = 1$ or $|A_{ji}| = 1$, the are $\max(|A_{ij}|, |A_{ji}|)$ lines between the nodes $i$ and $j$ and an arrow from $j$ to $i$ if $|A_{ij}| > |A_{ji}|$. If both $|A_{ij}| > 1$ and $|A_{ji}| > 1$, the line from $i$ to $j$ is decorated with the pair $(|A_{ij}|, |A_{ji}|)$.

Hence, one can read the Cartan matrix from the Dynkin diagram and vice versa.

Let $\mathfrak{h} := \mathrm{span}(h_1, \ldots, h_n)$ be the $n$-dimensional algebra of semi-simple elements, called the *Cartan subalgebra*. Furthermore, we let $\mathfrak{n}_+$ be the quotient of the free Lie algebra generated by $\{e_1, \ldots, e_n\}$ modulo the Serre relation (2e) and, similarly, $\mathfrak{n}_-$ is the quotient of the free Lie algebra generated by $\{f_1, \ldots, f_n\}$ modulo the Serre relation (2f). Then we have the triangular vector space decomposition

$$\mathfrak{g} = \mathfrak{n}_+ \oplus \mathfrak{h} \oplus \mathfrak{n}_-. \tag{3}$$

All three spaces are Lie subalgebras of $\mathfrak{g}$, but the sum is not direct as a sum of Lie algebras as for example $[\mathfrak{h}, \mathfrak{n}_\pm] \subset \mathfrak{n}_\pm$.

The number of the nested commutators which generate $\mathfrak{n}_+$ and $\mathfrak{n}_-$ is a priori infinite, but the Serre relations (2e) and (2f) impose non-trivial relations among them; this makes the algebra finite or infinite-dimensional, according to whether the generalised Cartan matrix is positive definite or not.

The adjoint action of the Cartan subalgebra $\mathfrak{h}$ on $\mathfrak{n}_+$ and $\mathfrak{n}_-$ is diagonal, which means that it is possible to decompose $\mathfrak{g}$ into eigenspaces

$$\mathfrak{g}_\alpha := \{x \in \mathfrak{g} \mid [h, x] = \alpha(h)x \quad \forall h \in \mathfrak{h}\}, \tag{4}$$

where $\alpha : \mathfrak{h} \to \mathbb{C}$ belongs to the dual space of the Cartan subalgebra. The eigenspaces $\mathfrak{g}_\alpha$ are called *root spaces* and the linear functionals $\alpha \in \mathfrak{h}^*$ *roots*. The *multiplicity* of a root $\alpha$ is defined as the dimension of the related root space $\mathfrak{g}_\alpha$. Relation (2b) implies that each $e_j$ is contained in an eigenspace $\mathfrak{g}_{\alpha_j}$, whose corresponding root $\alpha_j$ is called a *simple root*. Accordingly, $f_j$ belongs to $\mathfrak{g}_{-\alpha_j}$ with simple root $-\alpha_j$.

The lattice $Q := \bigoplus_{i=1}^n \mathbb{Z}\alpha_i$ provides a grading of $\mathfrak{g}$

$$[\mathfrak{g}_\alpha, \mathfrak{g}_\beta] \subseteq \mathfrak{g}_{\alpha+\beta}, \tag{5}$$

together with the root space decomposition

$$\mathfrak{g} = \bigoplus_{\alpha \in Q} \mathfrak{g}_\alpha, \; \mathfrak{g}_0 = \mathfrak{h}. \tag{6}$$

Any root of $\mathfrak{g}$ is a non-negative or non-positive integer linear combination of the simple roots $\alpha_i$.

Let us assume that $\mathfrak{g}$ is symmetrizable; this allows to define a bilinear form on the Cartan subalgebra $\mathfrak{h}$, which can be extended to an invariant bilinear form on the whole algebra $\mathfrak{g}$ obtaining the so-called *(generalised) Cartan–Killing form*. We are more interested in the scalar product $(\cdot \mid \cdot)$ among the roots, i.e. elements of $\mathfrak{h}^*$. For any pair of simple roots $\alpha_i, \alpha_j$ it is given by

$$(\alpha_i \mid \alpha_j) = B_{ij}, \tag{7}$$

where $B = DA$ is the symmetrization of the Cartan matrix $A$. This defines the scalar product between two generic roots, since any root can be written as a linear combination of simple roots.

The scalar product gives a natural notion of root length, i.e. $\alpha^2 := (\alpha \mid \alpha)$. In the finite dimensional case the roots are all spacelike, meaning $\alpha^2 > 0$; this reflects the fact that the scalar product has Euclidean signature. In the infinite-dimensional case this is no longer true: the spacelike roots are called *real roots*, the non-spacelike ones $\alpha^2 \leq 0$ are called *imaginary roots*. In particular, the lightlike roots $\alpha^2 = 0$ are called *null roots*.

The real roots are non-degenerate (i.e. the corresponding root spaces are one dimensional) as in the finite dimensional case, while the imaginary roots are not. The degeneracy of the imaginary roots is still an open problem, since a general closed formula for the multiplicity of an imaginary root for non-affine $\mathfrak{g}$ is yet to come.[2]

The discussion above was phrased in terms of complex Lie algebras. However, as is evident from (2), there is a natural basis in which the structure constants are real and therefore we can also consider $\mathfrak{g}$ over the real numbers in what is called the *split form* by simply taking the same generators and only real combinations of their multi-commutators. This split real Kac–Moody algebra will be denoted by $\mathfrak{g}$ in the following and from now on all discussions are for real Lie algebras. Furthermore we restrict attention to symmetrizable Cartan matrices.

## 2.2 Maximal Compact Subalgebras

We introduce the *Cartan–Chevalley involution* $\omega \in \mathrm{Aut}(\mathfrak{g})$, defined by

$$\omega(e_i) = -f_i, \quad \omega(f_i) = -e_i, \quad \omega(h_i) = -h_i; \tag{8}$$

its action on a root space is therefore $\omega(\mathfrak{g}_\alpha) = \mathfrak{g}_{-\alpha}$. Moreover it is linear and satisfies

$$\omega(\omega(x)) = x, \quad \omega([x, y]) = [\omega(x), \omega(y)]. \tag{9}$$

The latter relation allows to extend the involution from (8) to the whole algebra. The involutory, or *maximal compact* subalgebra, $K(\mathfrak{g})$ of $\mathfrak{g}$ is the set of fixed points under the Cartan-Chevalley involution

$$K(\mathfrak{g}) := \{x \in \mathfrak{g} \mid \omega(x) = x\}. \tag{10}$$

The name arises from the analogy with the finite dimensional split real case, where for example $K(\mathfrak{sl}(n, \mathbb{R})) = \mathfrak{so}(n, \mathbb{R})$: here $\mathfrak{so}(n, \mathbb{R})$ is the Lie algebra of the maximal compact subgroup $SO(n, \mathbb{R})$ of $SL(n, \mathbb{R})$ and the involution is simply $\omega(x) = -x^T$.

The combination $x_i := e_i - f_i$ is the only one which is invariant under $\omega$ in each triple (corresponding to the compact $\mathfrak{so}(2, \mathbb{R}) \subset \mathfrak{sl}(2, \mathbb{R})$), therefore it

---

[2]In fact, the multiplicities are not even known in closed form for the simplest such algebra with Cartan matrix

$$A = \begin{pmatrix} 2 & -3 \\ -3 & 2 \end{pmatrix}.$$

represents an element of $K(\mathfrak{g})$. An explicit presentation of the algebra $K(\mathfrak{g})$ was found by Berman [26], whose result we quote in the following theorem (in its version from [20]):

**Theorem** *Let $\mathfrak{g}$ be a symmetrizable Kac–Moody algebra with generalised Cartan matrix $A$ and let $x_i = e_i - f_i$. Then the maximal compact subalgebra $K(\mathfrak{g})$ is generated by the elements $\{x_1, \ldots, x_n\}$ that satisfy the following relations:*

$$P_{-A_{ij}}(\mathrm{ad}_{x_i})x_j = 0, \tag{11}$$

*where*

$$P_m(t) = \begin{cases} (t^2 + m^2)(t^2 + (m-2)^2) \cdots (t^2 + 1) & \text{if } m \text{ is odd}, \\ (t^2 + m^2)(t^2 + (m-2)^2) \cdots (t^2 + 4)t & \text{if } m \text{ is even}. \end{cases} \tag{12}$$

The theorem drastically simplifies for a simply-laced algebra:

**Corollary** *If $\mathfrak{g}$ is a simply-laced Kac–Moody algebra, then the generators $\{x_1, \ldots, x_n\}$ of $K(\mathfrak{g})$ satisfy*

$$[x_i, [x_i, x_j]] + x_j = 0 \quad \text{if the vertices } i, j \text{ are connected by a single edge}, \tag{13a}$$

$$[x_i, x_j] = 0 \quad \text{otherwise}. \tag{13b}$$

To illustrate these general statements we present some examples, which give a flavor of how the relations (11) are concretely realised. Consider the symmetric algebra $AE_3$, whose Dynkin diagram is in Fig. 1 and whose Cartan matrix is

$$A[AE_3] = \begin{pmatrix} 2 & -1 & 0 \\ -1 & 2 & -2 \\ 0 & -2 & 2 \end{pmatrix}; \tag{14}$$

then its subalgebra $K(AE_3)$ is generated, thanks to (11), by $\{x_1, x_2, x_3\}$ that satisfy

$$[x_1, [x_1, x_2]] + x_2 = 0, \quad [x_2, [x_2, x_1]] + x_1 = 0, \quad [x_1, x_3] = 0, \tag{15a}$$

$$[x_2, [x_2, [x_2, x_3]]] + 4[x_2, x_3] = 0, \quad [x_3, [x_3, [x_3, x_2]]] + 4[x_3, x_2] = 0. \tag{15b}$$

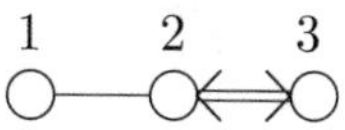

**Fig. 1** Dynkin diagram of $AE_3$

**Fig. 2** Dynkin diagram of
$G_2^{++}$

Let us work out explicitly the first relation in (15b): since $A_{23} = -2$, from (11) and the definition of the polynomial $P_m(t)$ we get

$$\left[(\mathrm{ad}_{x_2})^2 + 4\right](\mathrm{ad}_{x_2})x_3 = (\mathrm{ad}_{x_2})^3 x_3 + 4(\mathrm{ad}_{x_2})x_3 \overset{!}{=} 0,$$

which is precisely (15b). The same construction works for the other relations.

A second example is provided by $G_2^{++}$, whose Dynkin diagram is given in Fig. 2. The corresponding Cartan matrix is

$$A[G_2^{++}] = \begin{pmatrix} 2 & -1 & 0 & 0 \\ -1 & 2 & -1 & 0 \\ 0 & -1 & 2 & -1 \\ 0 & 0 & -3 & 2 \end{pmatrix}. \tag{16}$$

Therefore the maximal compact subalgebra $K(G_2^{++})$ is generated by $\{x_1, x_2, x_3, x_4\}$ that satisfy

$$[x_1, [x_1, x_2]] + x_2 = 0, \quad [x_2, [x_2, x_1]] + x_1 = 0, \tag{17a}$$

$$[x_2, [x_2, x_3]] + x_3 = 0, \quad [x_3, [x_3, x_2]] + x_2 = 0, \tag{17b}$$

$$[x_3, [x_3, x_4]] + x_4 = 0, \quad [x_4, [x_4, [x_4, [x_4, x_3]]]] + 10\,[x_4, [x_4, x_3]] + 9\,x_3 = 0, \tag{17c}$$

$$[x_1, x_3] = 0, \quad [x_1, x_4] = 0, \quad [x_2, x_4] = 0. \tag{17d}$$

These relations are easily derived from (11).

Finally consider the algebra $BE_{10}$, whose Dynkin diagram is in Fig. 3 (inverting the double arrow, we get the Dynkin diagram of the other rank-10 algebra $CE_{10}$, which however does not appear to be relevant for physics). The simply-laced part of

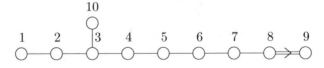

**Fig. 3** Dynkin diagram of $BE_{10}$

the algebra gives rise to the usual Berman relations (13) for $K(BE_{10})$, and the only one that is not of the type (13) is

$$[x_9, [x_9, [x_9, x_8]]] + 4[x_9, x_8] = 0. \tag{18}$$

Now we turn back to the general features of the maximal compact subalgebra $K(\mathfrak{g})$. The subalgebra $K(\mathfrak{g})$ of an infinite-dimensional Kac–Moody algebra is *not* of Kac–Moody type [27]. This means that there is no grading of $K(\mathfrak{g})$: being the $K(\mathfrak{g})$ generators a combination of elements from positive and negative root spaces

$$K(\mathfrak{g})_\alpha \equiv K(\mathfrak{g})_{-\alpha} \subset \mathfrak{g}_\alpha \oplus \mathfrak{g}_{-\alpha}, \tag{19}$$

$K(\mathfrak{g})$ possesses a *filtered structure*

$$[K(\mathfrak{g})_\alpha, K(\mathfrak{g})_\beta] \subset K(\mathfrak{g})_{\alpha+\beta} \oplus K(\mathfrak{g})_{\alpha-\beta}. \tag{20}$$

This means that we can associate a root to each $K(\mathfrak{g})$ generator, by keeping in mind that we are not really speaking of a 'root' of $K(\mathfrak{g})$, but of the structure given by (19). It also means that the established tools of representation theory (in particular, highest and lowest weight representations) are not available for $K(\mathfrak{g})$.

More specifically, below we will look for *spinorial*, i.e. double-valued representations of the compact subalgebra $K(\mathfrak{g})$. For this, what is known about Kac–Moody representation theory is of no help: although it is easy to generate representations of $K(\mathfrak{g})$ by decomposing representations of $\mathfrak{g}$ itself w.r.t. $K(\mathfrak{g}) \subset \mathfrak{g}$, these will not be spinorial, hence not suitable for the description of fermions. As it will turn out, so far only non-injective finite-dimensional, hence *unfaithful*, representations of this type are known, and it remains a major open problem to find faithful spinorial representations that are not obtained as tensor products of unfaithful spinorial representations of $K(\mathfrak{g})$ with representations of $\mathfrak{g}$.

A general property of $K(\mathfrak{g})$ defined through the Cartan–Chevalley involution (8) is that the restriction of the invariant bilinear form on $\mathfrak{g}$ to $K(\mathfrak{g})$ is of *definite* type even though the form on $\mathfrak{g}$ is indefinite for Kac–Moody algebras. To understand this, we note that the bilinear form on $\mathfrak{g}$ satisfies

$$(e_i, f_j) = \delta_{ij}, \tag{21}$$

which implies

$$(x_i, x_j) = -\delta_{ij}, \tag{22}$$

for all simple generators of $K(\mathfrak{g})$, so that the form is negative definite on them (for $K(\mathfrak{g})$ we obviously do not need the remaining relation $((h_i|h_j) = A_{ij})$. This property can be seen to extend to all of $K(\mathfrak{g})$. This is another reason why $K(\mathfrak{g})$ is referred to as the maximal compact subalgebra, as finite-dimensional compact (semi-simple) algebras are characterised by having a negative definite invariant form. Unlike for the finite dimensional case, the term 'compact' in the case of affine or indefinite Kac–Moody algebras or groups is used to highlight this *algebraic* property and does not imply a topological notion of compactness for the corresponding algebra or group (even though a topology can be defined w.r.t. the norm induced by (22)).

# 3   Spin Representations

We now address the question of constructing spin representations of $K(\mathfrak{g})$, i.e. representations of $K(\mathfrak{g})$ that are double-valued when lifted to a group representation. It turns out that a useful prior step is to introduce a convenient metric on the space of roots, called the DeWitt metric, which arises in canonical treatments of diagonal space-time metrics in general relativity [28, 29]. It is most easily written in the *wall basis* $e^{\mathrm{a}}$ on $\mathfrak{h}^*$, such that $\mathrm{a} = 1, \ldots, n$ lies in the same range as the index we have used for labelling the simple roots $\alpha_i$. We use the different font a to emphasise its property that the inner product takes the standard form

$$e^{\mathrm{a}} \cdot e^{\mathrm{b}} = G^{\mathrm{ab}}, \qquad G^{\mathrm{ab}} = \delta^{\mathrm{ab}} - \frac{1}{n-1}. \tag{23}$$

The DeWitt metric $G^{\mathrm{ab}}$ has Lorentzian signature $(1, n-1)$. A root $\alpha$ is expanded as $\alpha = \sum_{\mathrm{a}=1}^{n} \alpha_{\mathrm{a}} e^{\mathrm{a}}$ and in the examples below we will provide explicit lists of coefficients $\alpha_{\mathrm{a}}$ for the simple roots $\alpha_i$ such that their inner product gives back the symmetrised Cartan matrix (7) via $\alpha_i \cdot \alpha_j = B_{ij}$. Below we will also use letters $p_{\mathrm{a}} \equiv \alpha_{\mathrm{a}}$ to denote the root components, such that $\alpha \equiv (p_1, \cdots, p_d)$.

The spinor representations of $K(\mathfrak{g})$ will be constructed on a tensor product space $V \otimes W$ where $W$ carries the 'spinorial' part of the representation and $V$ the 'tensorial' part. The spaces $V$ we consider here are obtained within the $s$-fold tensor product of $\mathfrak{h}^*$. On $(\mathfrak{h}^*)^{\otimes s}$ we can introduce a multi-index $\mathcal{A} = (\mathrm{a}_1, \ldots, \mathrm{a}_s)$ and a bilinear form $G^{\mathcal{AB}} = G^{\mathrm{a}_1 \mathrm{b}_1} \cdots G^{\mathrm{a}_s \mathrm{b}_s}$. We shall typically consider only the case when $V$ is a symmetric tensor product, such that all indices in $G^{\mathrm{a}_1 \mathrm{b}_1} \cdots G^{\mathrm{a}_s \mathrm{b}_s}$ have to be symmetrised.

The space $W$ on the other hand will be taken to be an irreducible real spinor representation of the Euclidean $\mathfrak{so}(n)$ and we label its components by an index $\alpha$ whose range depends on the size of the real spinor. On $W$ there is also a bilinear form $\delta_{\alpha\beta}$.

Elements of $V \otimes W$ will then be denoted by $\phi_\alpha^{\mathcal{A}}$ and, for $\mathcal{A}$ referring to the $s$-fold symmetric product, the object $\phi_\alpha^{\mathcal{A}}$ is called a *spinor* of spin $\left(s + \frac{1}{2}\right)$; as already emphasised in [13] this terminology, which is inspired by $d = 4$ higher spin theories, should not be taken too literally in that it mixes the notion of space-time spin with that on DeWitt space. Consider now the relation

$$\left\{ \phi_\alpha^{\mathcal{A}}, \phi_\beta^{\mathcal{B}} \right\} = G^{\mathcal{AB}} \delta_{\alpha\beta}; \tag{24}$$

this equation is meant to generalise the standard Dirac bracket of the components of a gravitino field, see for instance [7]. The elements $\phi_\alpha^A$ are in this case thought of as 'second quantised' field operators with canonical anti-commutation relations. A particular way of realising this relation is as Clifford generators [24].

Because of the Clifford algebra defining relation (24), the $V \otimes W$ matrices have to be antisymmetric under simultaneous interchange of the two index pairs. Given the elements $X_{AB} = X_{BA}, Y_{CD} = -Y_{CD}$ in $V$, and $S^{\alpha\beta} = -S^{\beta\alpha}, T^{\gamma\delta} = T^{\delta\gamma}$ in $W$, these satisfy $X_{AB} S^{\alpha\beta} = -X_{BA} S^{\beta\alpha}$ and $Y_{CD} T^{\gamma\delta} = -Y_{DC} T^{\delta\gamma}$. The generic elements of the Clifford algebra over $V \otimes W$ can thus always be written as a sum $(\hat{A} + \hat{B})$, where

$$\hat{A} = X_{AB} S^{\alpha\beta} \, \phi_\alpha^A \phi_\beta^B \, , \quad \hat{B} = Y_{CD} T^{\gamma\delta} \, \phi_\gamma^C \phi_\delta^D \, . \tag{25}$$

The commutator of two such elements is conveniently evaluated in terms of the 'master relation'

$$\left[\hat{A}, \hat{B}\right] = \left([X, Y]_{AB} \{S, T\}^{\alpha\beta} + \{X, Y\}_{AB} [S, T]^{\alpha\beta}\right)\phi_\alpha^A \phi_\beta^B \, , \tag{26}$$

hence the form (25) is preserved.

We use (25) to write an ansatz for a bilinear operator $J(\alpha)$ associated to a root $\alpha$ of $K(\mathfrak{g})$, viz.

$$\hat{J}(\alpha) := X(\alpha)_{AB} \, \phi_\alpha^A \Gamma(\alpha)_{\alpha\beta} \phi_\beta^B \, . \tag{27}$$

If $\alpha$ is real, $\hat{J}(\alpha)$ refers to the essentially unique element in $K(\mathfrak{g})_\alpha$ while for imaginary roots $\alpha$ there are as many components to $\hat{J}(\alpha)$ as the root multiplicity of $\alpha$. The matrix $\Gamma(\alpha)$ is built out of the Euclidean Dirac gamma matrices corresponding to the $\ell = 0$ generators of $K(\mathfrak{g})$, potentially augmented by a *time-like* gamma matrix, in such a way that each element $\alpha$ of the root lattice is mapped to an element of the associated real Clifford algebra; its dimension depends on which space $W$ we are considering, and is related to the irreducible spinor representation of $\mathfrak{so}(n)$ of $\mathfrak{so}(1, n)$ underlying $W$. We shall be more explicit in the examples below. $X(\alpha)$ is called *polarisation tensor* and is required to elevate the spin-$\frac{1}{2}$ representation generated by $\Gamma(\alpha)$ to higher spin; again, we shall be more explicit below.

We can express the action of the simple generators $x_i \equiv J(\alpha_i)$ of $K(\mathfrak{g})$ on the spinor $\phi_\alpha^A$: this is given by

$$(J(\alpha)\phi)_\alpha^A \equiv \left[\hat{J}(\alpha), \phi\right]_\alpha^A = -2X(\alpha)^A{}_B \Gamma(\alpha)_\alpha^\beta \, \phi_\beta^B \, , \tag{28}$$

or

$$J(\alpha)_{AB}{}^{\alpha\beta} \equiv -2X(\alpha)_{AB} \Gamma(\alpha)^{\alpha\beta} \, . \tag{29}$$

When written without a hat, we think of $\hat{J}(\alpha)$ as the classical representation of the generator that acts simply by matrix multiplication on the components of $\phi_\alpha^A$. We will use this expression as our ansatz when for searching for spin representations. The strategy will be to look for suitable $X(\alpha)$ and $\Gamma(\alpha)$ such that $J(\alpha)$ satisfies the Berman relations (11).

## 3.1 The Simply-Laced Case

The ansatz (29) allows to translate the Berman relations (13) of a simply-laced algebra into relations for the matrices $X(\alpha)$ and $\Gamma(\alpha)$.

In the simply-laced case, we can associate a gamma matrix $\Gamma(\alpha)$ to each element $\alpha$ of the root lattice using a standard construction [30]. These matrices $\Gamma(\alpha)$ satisfy the relations

$$\Gamma(\alpha)\Gamma(\beta) = (-1)^{\alpha\cdot\beta}\Gamma(\beta)\Gamma(\alpha), \tag{30a}$$

$$\Gamma(\alpha)^T = (-1)^{\frac{1}{2}\alpha\cdot\alpha}\Gamma(\alpha), \tag{30b}$$

$$\Gamma(\alpha)^2 = (-1)^{\frac{1}{2}\alpha\cdot\alpha}\,\mathbb{1}\,. \tag{30c}$$

As we will see below, these relations can also be satisfied when the algebra is not simply-laced, provided one normalises the inner product such that the shortest roots have $\alpha^2 = 2$. These relations are absolutely crucial for the construction of higher spin representations. In particular, the relations (34) below would not make sense without (30).

Using the $\Gamma$-matrices we can construct a solution to the Berman relations (13) in the non-simply-laced case as follows. First, observe that the expressions (30) imply

$$[\Gamma(\alpha), [\Gamma(\alpha), \Gamma(\beta)]] = -4\Gamma(\beta) \quad \text{when } \alpha \cdot \beta = \mp 1, \tag{31a}$$

$$[\Gamma(\alpha), \Gamma(\beta)] = 0 \quad \text{when } \alpha \cdot \beta = 0. \tag{31b}$$

Thus, choosing $J(\alpha_i) = \frac{1}{2}\Gamma(\alpha_i)$ for all simple roots will solve the Berman relations (13) for the representation $x_i = J(\alpha_i)$. In terms of the ansatz (29), this means $X(\alpha) = -1/4$ for the polarisation and the space $V$ is one-dimensional. For simply-laced algebras $\mathfrak{g}$ one has also that all real roots are Weyl conjugate and we can choose

$$J(\alpha) = \frac{1}{2}\Gamma(\alpha), \tag{32}$$

for all real roots $\alpha$. This is the spin-$\frac{1}{2}$ representation constructed in [4, 5, 7, 20]. Its spinorial character follows from the relation

$$\exp(2\pi J(\alpha)) = -\mathbb{1}, \tag{33}$$

**Fig. 4** Dynkin diagram of
$AE_d$

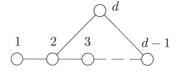

which will also hold for higher spin representations.

Now we move on to higher spin: in this case the space $W$ is more than one-dimensional and the polarisation tensor $X(\alpha)$ has a non-trivial structure. Making the ansatz (29) for $J(\alpha)$ we can then determine a sufficient condition for $X(\alpha)$ to solve the Berman relations (13). Upon using (30) and (26), we see that

$$\{X(\alpha), X(\beta)\} = \frac{1}{2}X(\alpha \pm \beta) \qquad \text{if } \alpha \cdot \beta = \mp 1, \tag{34a}$$

$$[X(\alpha), X(\beta)] = 0 \qquad \text{if } \alpha \cdot \beta = 0, \tag{34b}$$

is a sufficient condition when the roots range over the simple roots (or even all real roots). In this relation the polarisation tensors indices on the polarisation tensors are raised and lowered with $G^{AB}$ and its inverse in order to define the matrix products. The relations (34) are a sufficient condition for every simply-laced Kac–Moody algebra and they will be the main equations we solve. We now discuss higher spin representations for some simply-laced Kac–Moody algebras, namely $AE_d$, $E_{10}$ and $DE_{10}$.

### 3.1.1 $AE_d$ and $K(AE_d)$ $(d > 3)$

The indefinite Kac–Moody algebra $AE_d$, obtained by double extension of the Cartan type $A_{d-2}$, is the conjectured symmetry algebra for Einstein's gravity in $(d + 1)$ dimensions. For $d > 9$, these algebras are no longer hyperbolic, but of general Lorentzian type. This holds in particular for $AE_{10}$, which is a subalgebra of $E_{10}$ and is conjectured to govern the gravitational sector of 11-dimensional supergravity.

The $AE_d$ Dynkin diagram is depicted in Fig. 4. The corresponding Cartan matrix can be realised by simple roots in the wall basis. Because this algebra is associated with pure Einstein gravity in $(d + 1)$ space-time dimensions, the relevant DeWitt metric is

$$G^{ab} = \delta^{ab} - \frac{1}{d-1}, \tag{35}$$

and the simple roots in the wall basis are

$$\alpha_1 = (1, -1, 0, \ldots, 0), \tag{36a}$$

$$\alpha_2 = (0, 1, -1, 0, \ldots, 0), \tag{36b}$$

$$\vdots$$

$$\alpha_{d-1} = (0, \ldots, 0, 1, -1), \tag{36c}$$

$$\alpha_d = (0, 0, 1, \ldots, 1, 2). \tag{36d}$$

The spin-$\frac{1}{2}$ representation of $K(AE_d)$ can be constructed using the real matrices $\Gamma(\alpha)$ appearing in (30). For their explicit construction we need the *real* SO(1, $d$) Clifford algebra in $(d + 1)$-dimensional spacetime generated by $\{\Gamma_\mu, \Gamma_\nu\} = 2\eta_{\mu\nu}$.[3]

The SO(1, $d$) Clifford algebra contains the real element

$$\Gamma_* \equiv \Gamma_0 \Gamma_1 \cdots \Gamma_d = \frac{1}{(d+1)!} \epsilon^{\mu_0 \mu_1 \cdots \mu_d} \Gamma_{\mu_0} \Gamma_{\mu_1} \cdots \Gamma_{\mu_d}, \tag{37}$$

that satisfies

$$(\Gamma_*)^2 = (-1)^{\frac{d(d+1)}{2}+1}, \qquad \left\{ \begin{array}{l} d \text{ even} : [\Gamma_*, \Gamma_a] = 0 \\ d \text{ odd} : \{\Gamma_*, \Gamma_a\} = 0 \end{array} \right\}. \tag{38}$$

The *real* SO($d$) Clifford algebra in $d$ dimensions with $\{\Gamma_a, \Gamma_b\} = 2\delta_{ab}$ (see e.g. [32, Section 2]), which is associated with the $\mathfrak{so}(d) = K(A_{d-1})$ subalgebra of $K(AE_d)$, is obviously contained in the above Clifford algebra. The need for the *Lorentzian* Clifford algebra SO(1, $d$) is thus not immediately obvious from the above Dynkin diagram, but is necessitated by the extra node $\alpha_d$ associated with the dual graviton, as we will shortly explain. We also note that when $d \equiv 2 \mod 4$, the matrix $\Gamma_*$ is proportional to the identity.

After these preparations we define the matrices $\Gamma(\alpha)$ for the simple roots of $AE_d$:

$$\Gamma(\alpha_1) \equiv \Gamma_{12}, \ldots, \Gamma(\alpha_{d-1}) \equiv \Gamma_{d-1,d}, \tag{39}$$

for the first $(d - 1)$ roots, and

$$\Gamma(\alpha_d) \equiv \Gamma_{345 \cdots d-1} \Gamma_* = \Gamma_* \Gamma_{345 \cdots d-1} = (-1)^{\frac{d(d-1)}{2}} \Gamma_{012d}, \tag{40}$$

for the last simple root. Before motivating this definition from supergravity, we first check the Berman relations with the identification (32). The relations for the line spanned by the roots $\alpha_1, \ldots, \alpha_{d-1}$ correspond to the usual $\mathfrak{so}(d)$ algebra written in terms of gamma matrices and therefore are automatically satisfied. The non-trivial key relation to check is therefore the one involving nodes $d - 1$ and $d$ of diagram 4.

---

[3]We use the mostly plus signature for Minkowski space-time and $\epsilon^{01 \cdots d} = +1$ with Lorentz indices $\mu, \nu, \ldots \in \{0, 1, \ldots, d\}$. The $\Gamma$-matrices are real $n_d \times n_d$ matrices with $n_3 = 4$, $n_4 = n_5 = 8$, $n_6 = \cdots = n_9 = 16$ and $n_{10} = 32$ (see e.g. [31, 32]; for yet higher $d$ the numbers follow from Bott periodicity $n_{d+8} = 16n_d$ [33]).

This leads to

$$
\begin{aligned}
[\Gamma(\alpha_d), [\Gamma(\alpha_d), \Gamma(\alpha_{d-1})]] &= (\Gamma_*)^2 \left[ \Gamma_{34\ldots d-1}, \left[ \Gamma_{34\ldots d-1}, \Gamma_{d-1,d} \right] \right] \\
&= 2 (\Gamma_*)^2 \left[ \Gamma_{34\ldots d-1}, \Gamma_{34\ldots d-2,d} \right] \\
&= 4 (\Gamma_*)^2 (-1)^{d(d-3)/2} \Gamma_{d-1,d} \\
&= -4 \, \Gamma(\alpha_d),
\end{aligned}
$$

where the important point is that the property of $\Gamma_*$ is always such that this sign works out to be negative—showing explicitly the necessity of going Lorentzian, since otherwise the Berman relations would not hold. In summary, we have verified all Berman relations and constructed the spin-$\frac{1}{2}$ representation of $K(AE_d)$ for all $d > 3$ in terms of the associated Lorentzian Clifford algebra.

The above definition of $\Gamma(\alpha)$ can be extended to the whole root lattice by multiplication. For instance, for real roots $\alpha \equiv (p_1, \cdots, p_d)$ this leads to the simple formula

$$
\Gamma(\alpha) = (\Gamma_1)^{p_1} \cdots (\Gamma_d)^{p_d} (\Gamma_*)^{\frac{1}{d-1} \sum_{a=1}^d p_a}. \tag{41}
$$

These matrices satisfy (30), which is crucial for reducing the problem of finding higher spin representations to solve for the polarisation tensors.

Let us briefly motivate the definition (40) from supergravity. The group $AE_d$ is conjectured to be related to symmetries of gravity in $d + 1$ space-time dimensions and, where it exists, also of the corresponding supergravity theory. Decomposing the adjoint of $AE_d$ under its $A_{d-1} \cong \mathfrak{sl}(d)$ subalgebra reveals the gravitational fields: at level $\ell = 0$ one has the adjoint of $\mathfrak{gl}(d)$ corresponding to the graviton and at level $\ell = 1$ one finds the dual graviton that is a hook-tensor representation of type $(d - 2, 1)$ under $\mathfrak{gl}(d)$, meaning that it can be represented by a tensor $h_{a_1 \ldots a_{d-2}, a_0}$ with anti-symmetry in the first $d-2$ indices ($h_{[a_1 \ldots a_{d-2}], a_0} = h_{a_1 \ldots a_{d-2}, a_0}$) and Young irreducibility constraint $h_{[a_1 \ldots a_{d-2}, a_0]} = 0$. From the partial match of the bosonic gravitational dynamics to an $AE_d / K(AE_d)$ geodesic one knows this field to be related to (the traceless part of) the spin connection via a duality equation of the form [3, 27, 34][4]

$$
\omega_{a_0 bc} = \frac{1}{(d-2)!} \epsilon_{0bca_1 \ldots a_{d-2}} \partial_0 h_{a_1 \ldots a_{d-2}, a_0}. \tag{42}
$$

The indices $a, b$ here are spatial vector indices of $\mathfrak{so}(d)$. The root vector for $\alpha_d$ corresponds to the component $h_{34\ldots d, d}$.

---

[4]More precisely, one should consider the coefficients of anholonomy but for our discussion here this difference does not matter.

Assuming now a supersymmetry variation for a vector-spinor $\delta_\epsilon \Psi_\mu = \left(\partial_\mu + \frac{1}{4}\omega_{\mu\rho\sigma}\Gamma^{\rho\sigma}\right)\epsilon$ for the standard redefined 0-component of the vector-spinor leads to [7]

$$\delta_\epsilon\left(\Psi_0 - \Gamma_0\Gamma^a\Psi\right) = \frac{1}{4}\omega_{0ab}\Gamma^{ab}\epsilon - \frac{1}{4}\Gamma_0\Gamma^a\omega_{abc}\Gamma^{bc}\epsilon + \cdots, \tag{43}$$

where the ellipsis denotes terms that contain partial derivative and components of the spin connection that are irrelevant for the present discussion. Using the 'dictionary', according to which $\omega_{0ab}$ multiplies the level $\ell = 0$ generator of $K(AE_d)$ acting on the spin-$\frac{1}{2}$ field $\epsilon$, while $\partial_0 h_{a_1...a_{d-2},a_0}$ multiplies the level $\ell = 1$ generators of $K(AE_d)$ acting on $\epsilon$ we deduce that the first $d - 1$ roots of $K(AE_d)$ should be represented by matrices $\Gamma_{ab}$ as we claimed in (39) and that the generator for the root $\alpha_d$ is represented by

$$\epsilon^{0123...d}\,\Gamma_{012d} \propto \Gamma_{345...d-1}\Gamma_*, \tag{44}$$

as claimed in (40), where we also used the fact that we only consider the traceless part of the spin connection. We recall that when $d \equiv 2 \mod 4$, the matrix $\Gamma_*$ is proportional to the identity and can then be eliminated from this formula. This happens for example for $AE_{10} \subset K(E_{10})$ consistent with the known formulas for the spin-$\frac{1}{2}$ field of $K(E_{10})$, see [7] and the next section.

In order to find higher-spin representations of $K(AE_d)$ we can thus look for solutions of (34); the ansatz for $X(\alpha)$ depends on the number of indices $s$ that determine the spin $\left(s + \frac{1}{2}\right)$. For the spin-$\frac{3}{2}$ case we have a two-index tensor $X(\alpha)_{ab}$; the unique solution to (34) consistent with the symmetries is

$$X(\alpha)_a{}^b = -\frac{1}{2}\alpha_a\alpha^b + \frac{1}{4}\delta_a{}^b. \tag{45}$$

The spin-$\frac{5}{2}$ solution is

$$X(\alpha)_{ab}{}^{cd} = \frac{1}{2}\alpha_a\alpha_b\alpha^c\alpha^d - \alpha_{(a}\delta_{b)}{}^{(c}\alpha^{d)} + \frac{1}{4}\delta_a^{(c}\delta_b^{d)}. \tag{46}$$

The spin-$\frac{5}{2}$ representation is not irreducible: the space of elements of the form $G_{ab}\phi_\alpha^{ab}$ forms a subrepresentation isomorphic to the spin-$\frac{1}{2}$ representation and splits off as a direct summand. It is worth noting that these expressions do *not* depend on the rank $d$ of the algebra: the coefficients are universal.

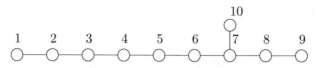

**Fig. 5** Dynkin diagram of $E_{10}$

In the spin-$\frac{7}{2}$ representation we find a different situation, since now there is an explicit dependence on the rank $d$ (and hence on the dimension of the associated physical theory):

$$X(\alpha)_{\text{abc}}^{\text{def}} = -\frac{1}{3}\alpha_a\alpha_b\alpha_c\alpha^d\alpha^e\alpha^f + \frac{3}{2}\alpha_{(a}\alpha_b\delta_{c)}^{(d}\alpha^e\alpha^{f)} - \frac{3}{2}\alpha_{(a}\delta_b^{(d}\delta_{c)}^e\alpha^{f)} + \frac{1}{4}\delta_{(a}^{(d}\delta_b^e\delta_{c)}^{f)}$$
$$+ \frac{14 + d \pm 2\sqrt{6(8+d)}}{(2+d)^2}\alpha_{(a}G_{bc)}G^{(de}\alpha^{f)}$$
$$- \frac{6 \pm \sqrt{6(8+d)}}{6(2+d)}\left(\alpha_{(a}\alpha_b\alpha_{c)}G^{(de}\alpha^{f)} + \alpha_{(a}G_{bc)}\alpha^{(d}\alpha^e\alpha^{f)}\right).$$

$$(47)$$

Let us also mention that the above formulas are valid for real $\alpha$, and not just the simple roots, again by Weyl conjugacy (since the Weyl group also acts appropriately on the polarisation tensors). One can now try to extend these results to still higher spins by making the most general ansatz for a spin-$\frac{9}{2}$ representation in the four-fold symmetric tensor product of $\mathfrak{h}^*$. However, we have not been able so far to find any non-trivial solution to (34), and the same holds for yet higher spins $s = \frac{11}{2}, \frac{13}{2}, \cdots$. This clearly indicates the need for more general ansätze.

### 3.1.2 $E_{10}$ and $K(E_{10})$

The maximally extended hyperbolic Kac–Moody algebra $E_{10}$ is conjectured to be a fundamental symmetry of M-theory, see [3].[5] Besides its intimate relation with maximal supergravity, it is also the universal such algebra which contains all other simply-laced hyperbolic Kac–Moody algebras [35].

The $E_{10}$ Dynkin diagram is shown in Fig. 5. The DeWitt metric is

$$G^{\text{ab}} = \delta^{\text{ab}} - \frac{1}{9},$$

$$(48)$$

---

[5]An alternative proposal based on the non-hyperbolic 'very extended' Kac–Moody algebra $E_{11}$ was put forward in [2].

and the simple roots in the wall basis are

$$\alpha_1 = (1, -1, 0, \ldots, 0), \tag{49a}$$

$$\alpha_2 = (0, 1, -1, 0, \ldots, 0), \tag{49b}$$

$$\vdots$$

$$\alpha_9 = (0, \ldots, 0, 0, 1, -1), \tag{49c}$$

$$\alpha_{10} = (0, \ldots, 0, 1, 1, 1). \tag{49d}$$

We use the real $32 \times 32$ gamma matrices that satisfy $\{\Gamma_a, \Gamma_b\} = 2\delta_{ab}$ $(a, b = 1, \ldots, 10)$ to define

$$\Gamma(\alpha) := (\Gamma_1)^{p_1} \cdots (\Gamma_{10})^{p_{10}}. \tag{50}$$

or equivalently,

$$\Gamma(\alpha_1) = \Gamma_{12}, \ldots, \Gamma(\alpha_9) = \Gamma_{9\,10}, \Gamma(\alpha_{10}) = \Gamma_{89\,10}. \tag{51}$$

Then it is easy to check that the generators $J(\alpha) = \frac{1}{2}\Gamma(\alpha)$ satisfy the Berman relations (13).

The polarisation tensors of the spin $\frac{3}{2}$, $\frac{5}{2}$ and $\frac{7}{2}$ representations for $K(E_{10})$ are exactly the same as in the $K(AE_d)$ case for $d = 10$; in particular, the spin-$\frac{7}{2}$ polarisation tensor is

$$X(\alpha)_{abc}^{\ \ \ def} = -\frac{1}{3}\alpha_a\alpha_b\alpha_c\alpha^d\alpha^e\alpha^f + \frac{3}{2}\alpha_{(a}\alpha_b\delta_{c)}^{(d}\alpha^e\alpha^{f)} - \frac{3}{2}\alpha_{(a}\delta_b^{(d}\delta_{c)}^e\alpha^{f)} + \frac{1}{4}\delta_{(a}^{(d}\delta_b^e\delta_{c)}^{f)}$$

$$+ \frac{1}{12}(2 \pm \sqrt{3})\,\alpha_{(a}G_{bc)}G^{(de}\alpha^{f)}$$

$$- \frac{1}{12}(1 \pm \sqrt{3})\left(\alpha_{(a}\alpha_b\alpha_c)G^{(de}\alpha^{f)} + \alpha_{(a}G_{bc)}\alpha^{(d}\alpha^e\alpha^{f)}\right). \tag{52}$$

These representations have been studied in [13] and [15].

### 3.1.3 $DE_{10}$ and $K(DE_{10})$

We next analyze the spinor representations of $K(DE_{10})$, the involutory subalgebra of $DE_{10}$ which is related to the pure type I supergravity in 10 dimensions [36].

The $DE_{10}$ Dynkin diagram is given in Fig. 6, and its DeWitt metric is

$$G^{ab} = \delta^{ab} - \frac{1}{8}, \quad G^{10,a} = G^{a,10} = 0, \quad G^{10,10} = 2, \tag{53}$$

**Fig. 6** Dynkin diagram of $DE_{10}$

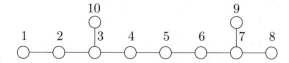

where a, b $= 1, \ldots, 9$; in this case the metric takes a non-standard form because of the presence of the dilaton in the supergravity theory associated to $G^{10,10}$.

The ten simple roots are

$$\alpha_1 = (1, -1, 0, \ldots |0), \tag{54a}$$

$$\alpha_2 = (0, 1, -1, 0, \ldots |0), \tag{54b}$$

$$\vdots$$

$$\alpha_8 = (0, \ldots, 0, 1, -1|0), \tag{54c}$$

$$\alpha_9 = (0, \ldots, 0, 1, 1|1/2), \tag{54d}$$

$$\alpha_{10} = (0, 0, 0, 1, \ldots, 1 | -1/2). \tag{54e}$$

To find the spin-$\frac{1}{2}$ representation we make use of the $16 \times 16$ real SO(9) gamma matrices $\Gamma_a$ $(a = 1, \ldots, 9)$ and define (for $\alpha \equiv (p_1, \ldots, p_9|p_{10})$)

$$\Gamma(\alpha) := (\Gamma_1)^{p_1} \cdots (\Gamma_9)^{p_9}, \tag{55}$$

so that, in particular, $\Gamma(\alpha_{10}) = \Gamma_{456789}$. Importantly, the dilaton coordinate (the tenth one) does not affect the representation. It is easily checked that $J(\alpha) = \frac{1}{2}\Gamma(\alpha)$ satisfies the Berman relations for a simply laced algebra (13). The polarisation tensors for the higher spin representations are the same of the $E_{10}$ case, $DE_{10}$ being a simply-laced algebra.

We note that the above definition implies $\Gamma(\alpha_8) = \Gamma(\alpha_9) = \Gamma_{89}$. This degeneracy follows from the fact that $DE_{10}$ is associated with type I (half maximal) supergravity, and there are only 16, rather than 32, real spinor components. It means that this 16-dimensional spinor representation of $K(DE_{10})$ does not distinguish the two spinor nodes of the Dynkin diagram. From the point of view of space-time rotations this is related to the fact that the spatial rotations SO(9) sit inside SO(9)$\times$ SO(9) $\subset K(DE_{10})$ as the *diagonal* subgroup, and thus treat both nodes in the same way. One could also introduce a bit artificially a 16-component spinor representation where $\Gamma(\alpha_9) = -\Gamma(\alpha_8)$ that is superficially of different chirality, but this change of sign can be undone by simply redefining the generators of $K(DE_{10})$ and thus is of no real significance. As usual assigning chirality is a convention that only becomes meaningful in relation to other chiral objects.

## 3.2   The Non-simply-laced Case

When the algebra $\mathfrak{g}$ is not simply-laced we have to solve different relations than (13). This is, we look for generators $J(\alpha)$ that satisfy the original Berman relations (11).

### 3.2.1   $AE_3$ and $K(AE_3)$

The Dynkin diagram of $AE_3$ is given in Fig. 1 and its Cartan matrix in (14). It was first studied in detail in [37]. We notice that this algebra is non-simply-laced but symmetric: this means that the Berman relations are symmetric, and they are given in (15).

The metric is

$$G^{\mathrm{ab}} = \delta^{\mathrm{ab}} - \frac{1}{2}, \tag{56}$$

and the simple roots in the wall basis are

$$\alpha_1 = (1, -1, 0), \tag{57a}$$

$$\alpha_2 = (0, 1, -1), \tag{57b}$$

$$\alpha_3 = (0, 0, 2). \tag{57c}$$

We construct the gamma matrices $\Gamma(\alpha)$ starting from the real symmetric $4 \times 4$ gamma matrices $\Gamma_a$ ($a = 1, \ldots, 3$) that satisfy $\{\Gamma_a, \Gamma_b\} = 2\delta_{ab}$. For any root $\alpha \equiv (p_1, p_2, p_3)$ we deduce the matrices $\Gamma(\alpha)$ as a special case of (40):

$$\Gamma(\alpha) = (\Gamma_1)^{p_1}(\Gamma_2)^{p_2}(\Gamma_3)^{p_3}(\Gamma_*)^{(p_1+p_2+p_3)/2}. \tag{58}$$

This definition is chosen in such a way that the relations (30) are satisfied for any root $\alpha$, as can be verified. Then, the matrices for the simple roots are $\Gamma(\alpha_1) = \Gamma_{12}$, $\Gamma(\alpha_2) = \Gamma_{23}$ and $\Gamma(\alpha_3) = \Gamma_*$. It is easy to see that these matrices generate the algebra $\mathfrak{u}(2)$.

Now it is easy to check that the generators $J(\alpha) = \frac{1}{2}\Gamma(\alpha)$ give rise to a spin-$\frac{1}{2}$ representation as in the simply-laced case, since they satisfy the Berman relations (15). Because $\Gamma_*$ commutes with all elements generated from the simple roots, there exists in this particular case a non-trivial *central element* in the second quantised description which is given by

$$\mathcal{Z} = \phi^{\mathrm{a}}_\alpha G_{\mathrm{ab}}(\Gamma_*)_{\alpha\beta}\phi^{\mathrm{b}}_\beta. \tag{59}$$

For higher $d$ such a central element exists whenever $\Gamma_*$ is anti-symmetric (because the fermion operators *anti*-commute), which for $AE_d$ is the case for $d = 3, 7, \cdots$. However, it does not exist for $K(E_{10})$. The existence of this central element plays an important role in the analysis of supersymmetric quantum cosmology in a Bianchi-type truncation [21, 22], where it helps in particular in diagonalising the Hamiltonian and in elucidating the structure of the quartic fermion terms (which are often ignored in discussions of supersymmetric quantum cosmology).

In order to find higher spin representations, we need to rewrite the Berman relations (15) in terms of the polarisation tensors $X(\alpha)$: from the ansatz (29) and the relations (30) among the $\Gamma(\alpha)$'s, the Eq. (15) turn into

$$4\{X(\alpha_1), \{X(\alpha_1), X(\alpha_2)\}\} - X(\alpha_2) = 0, \tag{60a}$$

$$4\{X(\alpha_2), \{X(\alpha_2), X(\alpha_1)\}\} - X(\alpha_1) = 0, \tag{60b}$$

$$\{X(\alpha_2), \{X(\alpha_2), \{X(\alpha_2), X(\alpha_3)\}\}\} - \{X(\alpha_2), X(\alpha_3)\} = 0, \tag{60c}$$

$$\{X(\alpha_3), \{X(\alpha_3), \{X(\alpha_3), X(\alpha_2)\}\}\} - \{X(\alpha_3), X(\alpha_2)\} = 0, \tag{60d}$$

$$[X(\alpha_1), X(\alpha_3)] = 0. \tag{60e}$$

There are several solutions to these relations. The spin-$\frac{3}{2}$ and spin-$\frac{5}{2}$ representations are exactly the same as in the simply-laced case. The novelty of $K(AE_3)$ stands in the spin-$\frac{7}{2}$, which gives *two* different representations:

$$
\begin{aligned}
X(\alpha)_{abc}{}^{def} &= -\frac{1}{3}\alpha_a\alpha_b\alpha_c\alpha^d\alpha^e\alpha^f + \frac{3}{2}\alpha_{(a}\alpha_b\delta_{c)}^{(d}\alpha^e\alpha^{f)} - \frac{3}{2}\alpha_{(a}\delta_b^{(d}\delta_{c)}^e\alpha^{f)} + \frac{1}{4}\delta_{(a}^{(d}\delta_b^e\delta_{c)}^{f)} \\
&\quad + \frac{1}{25}(17 \pm 2\sqrt{66})\,\alpha_{(a}G_{bc)}G^{(de}\alpha^{f)} \\
&\quad - \frac{1}{30}(6 \pm \sqrt{66})\left(\alpha_{(a}\alpha_b\alpha_{c)}G^{(de}\alpha^{f)} + \alpha_{(a}G_{bc)}\alpha^{(d}\alpha^e\alpha^{f)}\right),
\end{aligned}
\tag{61}
$$

and

$$
\begin{aligned}
X(\alpha)_{abc}{}^{def} &= -\frac{1}{3}\alpha_a\alpha_b\alpha_c\alpha^d\alpha^e\alpha^f + \frac{3}{2}\alpha_{(a}\alpha_b\delta_{c)}^{(d}\alpha^e\alpha^{f)} - \frac{3}{2}\alpha_{(a}\delta_b^{(d}\delta_{c)}^e\alpha^{f)} + \frac{1}{4}\delta_{(a}^{(d}\delta_b^e\delta_{c)}^{f)} \\
&\quad + \frac{1}{50}(19 \pm 4\sqrt{21})\,\alpha_{(a}G_{bc)}G^{(de}\alpha^{f)} \\
&\quad - \frac{1}{30}(6 \pm \sqrt{21})\left(\alpha_{(a}\alpha_b\alpha_{c)}G^{(de}\alpha^{f)} + \alpha_{(a}G_{bc)}\alpha^{(d}\alpha^e\alpha^{f)}\right).
\end{aligned}
\tag{62}
$$

Note that the first solution can be obtained from the $K(AE_d)$ case by setting $d = 3$ (compare with Eq. (47)), while the second expression is completely new. Its relation to gravity is unknown to date.

### 3.2.2 $G_2^{++}$ and $K(G_2^{++})$

$G_2^{++}$ plays the role of the symmetry algebra for 5-dimensional supergravity, similarly as $E_{10}$ does for 11-dimensional supergravity [38]. There are indeed many similarities between the two theories [39]; nevertheless, the underlying symmetry algebras $G_2^{++}$ and $E_{10}$ exhibit very different properties.

The $G_2^{++}$ Dynkin diagram is in Fig. 2, and its Cartan matrix in (16). This algebra is not only non-simply-laced, it is also non-symmetric. Nonetheless $G_2^{++}$ is symmetrizable, which means that it is possible to find a diagonal matrix $D$ such that $B = DA$ is symmetric. The choice $D = \text{diag}(3, 3, 3, 1)$ leads to

$$B = \begin{pmatrix} 6 & -3 & 0 & 0 \\ -3 & 6 & -3 & 0 \\ 0 & -3 & 6 & -3 \\ 0 & 0 & -3 & 2 \end{pmatrix}; \tag{63}$$

since the scalar product will be introduced with respect to the symmetrised Cartan matrix $B$, the fourth root $\alpha_4$ will be interpreted as the 'short' root with length 2. The Berman relations are given in (17).

Because the associated supergravity theory lives in five dimensions [40, 41], the DeWitt metric is

$$G^{ab} = \delta^{ab} - \frac{1}{3}. \tag{64}$$

However, in order to ensure that the root lengths are always even and not fractional, we rescale this metric by a factor of 3, viz.

$$G^{ab} \rightarrow \tilde{G}^{ab} \equiv 3G^{ab} = 3\delta^{ab} - 1. \tag{65}$$

A further advantage of this choice (which of course does not affect the Cartan matrix of $G_2^{++}$) is that relations (30) remain in force.

The simple roots in the wall basis are given by

$$\alpha_1 = (1, -1, 0, 0), \tag{66a}$$

$$\alpha_2 = (0, 1, -1, 0), \tag{66b}$$

$$\alpha_3 = (0, 0, 1, -1), \tag{66c}$$

$$\alpha_4 = (0, 0, 0, 1). \tag{66d}$$

For the construction of the associated $\Gamma(\alpha_i)$ we use the very same 8-by-8 matrices as for $AE_4$, the only difference being the assignment for the simple root $\alpha_4$:

$$\Gamma(\alpha_1) = \Gamma_{12}, \quad \Gamma(\alpha_2) = \Gamma_{23}, \quad \Gamma(\alpha_3) = \Gamma_{34}, \quad \Gamma(\alpha_4) = \Gamma_4\Gamma_*, \tag{67}$$

where $\Gamma_*$ was defined in (37) and commutes with $\Gamma_a$. As before its presence is necessary to satisfy the quadrilinear relation in (17). We recall that the R symmetry of simple supergravity in five dimensions is $USp(2)$ [41]. As we shall see, this R symmetry will be enlarged to $USp(4)$ by the extension to $K(G_2^{++})$. We also note that a proposal for including the $USp(2)$ R symmetry into the algebra by using a non-split real form of an extension of $G_2^{++}$ was made in [42].

In order to investigate the higher spin representations, we write the Berman relations for the polarisation tensors $X(\alpha)$: they are

$$4\{X(\alpha_1), \{X(\alpha_1), X(\alpha_2)\}\} - X(\alpha_2) = 0, \tag{68a}$$

$$4\{X(\alpha_2), \{X(\alpha_2), X(\alpha_1)\}\} - X(\alpha_1) = 0, \tag{68b}$$

$$4\{X(\alpha_2), \{X(\alpha_2), X(\alpha_3)\}\} - X(\alpha_3) = 0, \tag{68c}$$

$$4\{X(\alpha_3), \{X(\alpha_3), X(\alpha_2)\}\} - X(\alpha_2) = 0, \tag{68d}$$

$$4\{X(\alpha_3), \{X(\alpha_3), X(\alpha_4)\}\} - X(\alpha_4) = 0, \tag{68e}$$

$$16\{X(\alpha_4), \{X(\alpha_4), \{X(\alpha_4), \{X(\alpha_4), X(\alpha_3)\}\}\}\}$$
$$-40\{X(\alpha_4), \{X(\alpha_4), X(\alpha_3)\}\} + 9X(\alpha_3) = 0, \tag{68f}$$

$$[X(\alpha_1), X(\alpha_3)] = [X(\alpha_1), X(\alpha_4)] = [X(\alpha_2), X(\alpha_4)] = 0. \tag{68g}$$

Up to now, we have considered ansätze for $X(\alpha)$ in which the coefficients are the same for all the simple roots; in this case, because of the short root $\alpha_4$, it is natural to assume a different behaviour according to the length of the roots. This means that we expect to find different coefficients for $X(\alpha_4)$ if compared to the other polarisation tensors. The spin-$\frac{3}{2}$ representation is indeed found to be

$$X(\alpha)_a{}^b = \mp\frac{1}{2}\alpha_a\alpha^b \pm \frac{1}{4}\delta_a{}^b \quad \text{if } \alpha \neq \alpha_4, \tag{69a}$$

$$X(\alpha)_c{}^d = \pm\frac{3}{2}\alpha_c\alpha^d \mp \frac{3}{4}\delta_c{}^d \quad \text{if } \alpha = \alpha_4. \tag{69b}$$

It is important to stress that the first solution is consistent with the results from the simply-laced case: indeed $\alpha_1, \alpha_2, \alpha_3$ form a simply-laced part of $K(G_2^{++})$.

### 3.2.3 $BE_{10}$ and $K(BE_{10})$

As a last example we consider the algebra $K(BE_{10})$, which is related to the low energy limit of the heterotic and type I superstrings [43]. One can here proceed in the usual way, by looking for gamma matrices that satisfy the Berman relations and then look for the polarisation tensors. It is indeed possible to find at least one matrix representation that fullfills the Berman relations, which however suffers from the same degeneracy as $DE_{10}$: the contribution of the extended root $\alpha_9$ (refer to the Dynkin diagram, Fig. 3) has to be trivially realised if one wants to obtain a representation consistent with the physical theory. This unusual feature, however, does not lead to inconsistencies or to a trivialization of the full algebra: the Berman relation (18) makes possible to have $x_9 = 0$ without affecting the other generators. This is a peculiar property of this particular relation, since $x_9$ appears in both commutators in (18).

Once one has fixed the generator associated to $\alpha_9$ to be zero, then the spin representation is easily found: "removing" the ninth node we are left with the affine algebra $E_9$, and the $K(E_9)$ representations are known [17, 18]. They are given by the usual ansatz for $\Gamma(\alpha)$ (the same as for $DE_{10}$), built out of the $16 \times 16$ Dirac matrices. The polarisation tensors are then the standard ones for a simply-laced algebra.

Finally, let us comment on the consistency of the representation by focusing on a particular subalgebra of $BE_{10}$, namely $DE_{10}$. The embedding $DE_{10} \subset BE_{10}$ is suggested by the embedding of pure type I supergravity into type I supergravity with additional matter couplings. This embedding can be directly realised by comparison of the Dynkin diagrams, Figs. 3 and 6: let $E_i$, $H_i$, $F_i$ be the $DE_{10}$ generators and $e_i$, $h_i$, $f_i$ the $BE_{10}$ ones. By letting

$$E_i = e_i, \quad H_i = h_i, \quad F_i = f_i \qquad \text{for } i = 1, \ldots, 8, 10, \tag{70a}$$

$$E_9 = \frac{1}{2}[e_9, [e_9, e_8]], \quad H_9 = h_8 + h_9, \quad F_9 = \frac{1}{2}[f_9, [f_9, f_8]], \tag{70b}$$

one can check that the $DE_{10}$ Chevalley–Serre relations are satisfied by virtue of the $BE_{10}$ ones.

By using the explicit form of the $K(BE_{10})$ generators $x_i = e_i - f_i$ we find that

$$[x_9, [x_9, x_8]] = [e_9, [e_9, e_8]] - [f_9, [f_9, f_8]] - 2x_8,$$

which, by virtue of the embedding (70), is equivalent to

$$X_9 = \frac{1}{2}[x_9, [x_9, x_8]] + x_8, \tag{71}$$

where $X_i = E_i - F_i$ are the $K(DE_{10})$ generators. It is clear, from the last expression, that $x_9 = 0$ implies $X_9 = x_8$: this is consistent with the results from Sect. 3.1.3, where the gamma matrices relative to the roots $\alpha_8$ and $\alpha_9$ are equal, $\Gamma(\alpha_8) = \Gamma(\alpha_9) = \Gamma_{89}$, which means that the corresponding generators are equivalent. We here see the same degeneracy that we already encountered for $DE_{10}$, in accord with the fact that $BE_{10}$ is relevant for type I supergravity with (vector multiplet) matter couplings.

# 4    Quotients

The spin representations constructed above are unfaithful, since they are finite-dimensional representations of an infinite-dimensional Lie algebra. The unfaithfulness implies the existence of non-trivial ideals i for the algebra, where i is spanned by those combinations of $K(\mathfrak{g})$ elements which annihilate all vectors in the given representation. For each i this entails the existence of the quotient Lie algebra

$$\mathfrak{q} = K(\mathfrak{g})/\mathfrak{i}. \tag{72}$$

Growing with the spin of the representation the algebra gets more and more faithful, so that the quotient becomes bigger (and the ideal smaller). For instance, in the case of the spin-$\frac{1}{2}$ representation of $K(E_{10})$ the quotient is isomorphic to $\mathfrak{so}(32)$, as can be easily shown by using

$$\mathfrak{q} \cong \operatorname{Im}\rho, \tag{73}$$

where $\rho : K(\mathfrak{g}) \rightarrow \operatorname{End}(V \otimes W)$ is the representation map. It then follows immediately that $\mathfrak{q} \cong \mathfrak{so}(32)$ because the matrices $\Gamma(\alpha)$ for real roots $\alpha$ close into the antisymmetric 32-by-32 matrices. Things are not so easy for the higher spin representations, due to the dimension of the representations and to the presence of non-trivial polarisation tensors.

In general, questions about the quotient algebra can be asked in terms of the representation matrices because these give a faithful image of $\mathfrak{q}$. However, we will encounter two very unusual features. One is that the quotient algebra $\mathfrak{q}$ in general is *not* a subalgebra of $K(\mathfrak{g})$. The other is that, as a finite-dimensional Lie algebra, $\mathfrak{q}$ is in general *non-compact* even though it descends from the maximal compact subalgebra $K(\mathfrak{g}) \subset \mathfrak{g}$!

## 4.1    Spin-$\frac{1}{2}$ Quotients in $AE_d$

From the description of the simple generators for $K(AE_d)$ in the spin-$\frac{1}{2}$ representation in (39) and (40) it is also possible to determine the quotient algebras for the various $d$. The quotient depends on the properties of the gamma matrices of $\mathfrak{so}(1, d)$ that are known to exhibit Bott periodicity, i.e., the properties (Majorana, Weyl, etc.) are periodic in $d$ with period 8.

The relevant quotient algebra in terms of gamma matrices is generated by $\Gamma_{ab}$ ($\ell = 0$), $\Gamma_{0abc}$ ($\ell = 1$) and repeated commutation for the various cases leads to antisymmetric matrices according to

$$\ell = 0 \qquad\qquad : \qquad\qquad \Gamma_{[2]} \tag{74a}$$

$$\ell = 1 \qquad\qquad : \qquad\qquad \Gamma_0\Gamma_{[3]} \tag{74b}$$

$$\ell = 2 \subset [\ell = 1, \ell = 1] \qquad\qquad : \qquad\qquad \Gamma_{[6]} \tag{74c}$$

$$\ell = 3 \subset [\ell = 2, \ell = 1] \qquad : \qquad \Gamma_0 \Gamma_{[7]} \qquad \text{(74d)}$$

$$\ell = 4 \subset [\ell = 3, \ell = 1] \qquad : \qquad \Gamma_{[10]} \qquad \text{(74e)}$$

$$\vdots \qquad\qquad\qquad\qquad\qquad\qquad \vdots$$

We have only listed the new matrices at each step and for a given $d$ one has to truncate away those matrices that vanish by having too many antisymmetric indices. $\Gamma_{[k]}$ here denotes the $k$-fold antisymmetric product of gamma matrices.

Based on this sequence of generators one can determine the quotient Lie algebras as

$$\mathfrak{q}_{1/2}(K(AE_3)) \cong \mathfrak{u}(2) \qquad\qquad \text{acting on } \mathbb{R}^4 \qquad\qquad \text{(75a)}$$

$$\mathfrak{q}_{1/2}(K(AE_4)) \cong \mathfrak{usp}(4) \cong \mathfrak{so}(5) \qquad \text{acting on } \mathbb{R}^8 \qquad\qquad \text{(75b)}$$

$$\mathfrak{q}_{1/2}(K(AE_5)) \cong \mathfrak{usp}(4) \oplus \mathfrak{usp}(4) \qquad \text{acting on } \mathbb{R}^8 \oplus \mathbb{R}^8 \qquad \text{(75c)}$$

$$\mathfrak{q}_{1/2}(K(AE_6)) \cong \mathfrak{spin}(9) \qquad\qquad \text{acting on } \mathbb{R}^{16} \qquad\qquad \text{(75d)}$$

$$\mathfrak{q}_{1/2}(K(AE_7)) \cong \mathfrak{u}(8) \qquad\qquad\qquad \text{acting on } \mathbb{R}^{16} \qquad\qquad \text{(75e)}$$

$$\mathfrak{q}_{1/2}(K(AE_8)) \cong \mathfrak{so}(16) \qquad\qquad \text{acting on } \mathbb{R}^{16} \qquad\qquad \text{(75f)}$$

$$\mathfrak{q}_{1/2}(K(AE_9)) \cong \mathfrak{so}(16) \oplus \mathfrak{so}(16) \qquad \text{acting on } \mathbb{R}^{16} \oplus \mathbb{R}^{16} \qquad \text{(75g)}$$

$$\mathfrak{q}_{1/2}(K(AE_{10})) \cong \mathfrak{so}(32) \qquad\qquad \text{acting on } \mathbb{R}^{32} \qquad\qquad \text{(75h)}$$

All spinorial representations are double valued as it should be. As we have indicated in the above list, in the cases when the number of space dimensions is $d = 5 + 4k$ (for $k \in \mathbb{Z}_{\geq 0}$), one may naively end up with a reducible spinor space when constructing the Clifford algebra. For these values of $d$ one has that $(\Gamma_*)^2 = +\mathbb{1}$ and can use this to define chiral spinors by projecting to the $\Gamma_* = \pm \mathbb{1}$ eigenspaces. The irreducible spin-$\frac{1}{2}$ representation then just corresponds to a single summand of the space given in the list. This is the case in particular for $K(AE_9)$ that embeds into $K(DE_{10})$ where we constructed the 16-dimensional spin-$\frac{1}{2}$ representation in Sect. 3.1.3.

The occurrence of unitary algebras for $d = 3 + 4k$ is due to the fact that there the element $\Gamma_*$ belongs to the quotient and since $(\Gamma_*)^2 = -\mathbb{1}$ this can be used as a complex structure that is respected by all elements of the quotient, leading to a unitary algebra.

We also note some coincidences between quotient algebras that have a physical interpretation in terms of truncations of certain mulitplets in supergravity:

$$\mathfrak{q}_{1/2}(K(G_2^{++})) \cong \mathfrak{q}_{1/2}(K(AE_4)) \cong \mathfrak{usp}(4), \qquad\qquad \text{(76a)}$$

$$\mathfrak{q}_{1/2}(K(DE_{10})) \cong \mathfrak{q}_{1/2}(K(AE_9)) \cong \mathfrak{q}_{1/2}(K(BE_{10})) \cong \mathfrak{so}(16), \qquad \text{(76b)}$$

$$\mathfrak{q}_{1/2}(K(AE_{10})) \cong \mathfrak{q}_{1/2}(K(E_{10})) \cong \mathfrak{so}(32). \qquad\qquad \text{(76c)}$$

Finally, we note that for all these spin-$\frac{1}{2}$ representations *only the real root spaces are represented faithfully*, but not the root spaces associated with imaginary roots.

## 4.2   Quotient Algebras for $K(E_{10})$

For $K(E_{10})$, we present a more detailed analysis of the quotient Lie algebras for the higher spin representations. The various representations of $K(E_{10})$ that we have presented above admit an invariant bilinear form $(\phi, \psi)$, whose specific form depends on the representation. The form is invariant w.r.t. the action of the algebra elements, or equivalently

$$(x \cdot \phi, \psi) + (\phi, x \cdot \psi) = 0, \tag{77}$$

where $x\cdot$ represents the action of a $K(E_{10})$ element on the representation space. The bilinear form is symmetric in its arguments.

The abstract expression (77) can be written explicitly in terms of the generators of the representation, and the general result is

$$\eta J + J^T \eta = 0, \tag{78}$$

where $\eta \equiv G^{AB}\delta_{\alpha\beta}$ is the bilinear form and $J \equiv J(\alpha)$ is the root generator. Plugging the ansatz (29) into (78), one finds for an antisymmetric $\Gamma$ (e.g. for a real root)

$$GX - X^T G = 0, \tag{79a}$$

which means *symmetric X*; on the converse, for a symmetric $\Gamma$ (e.g. for a null root)

$$GX + X^T G = 0, \tag{79b}$$

i.e. *antisymmetric X*.

We note that not every involutory subalgebra of an indefinite Kac–Moody algebra will admit such an invariant form. For example, for $K(E_9)$ and $K(E_{11})$ there are no such invariant bilinear forms [7]. We also note that for $E_{11}$ one normally considers an involution different from (8), called the 'temporal' involution [2, 44, 45] such that the fixed-point algebra has an indefinite bilinear form and in particular contains a Lorentz subalgebra $\mathfrak{so}(1, 10)$ rather than the Euclidean $\mathfrak{so}(11)$. As these satisfy (78), we immediately deduce that $\mathfrak{q} \subset \mathfrak{so}(\eta)$, where $\eta$ is the bilinear form on the representation space that is left invariant by the action of $K(\mathfrak{g})$. In the following, we shall analyse this in some detail for the case of $E_{10}$.

In the spin-$\frac{1}{2}$ case, the bilinear form is $\eta = \delta_{\alpha\beta}$ and its signature is $(32, 0)$: the quotient algebra must therefore be a subalgebra of $\mathfrak{so}(32)$. By inspection of the explicit representation matrices in terms of anti-symmetric gamma matrices one easily checks that any $\mathfrak{so}(32)$ matrix (in the fundamental representation) is in the image of the spin-$\frac{1}{2}$ representation and therefore the quotient algebra is isomorphic to $\mathfrak{so}(32)$. At the group level, the corresponding isomorphism is $K(E_{10}) = SO(32)$.[6]

For the higher spin representations, it is less easy to compute explicitly a basis for the image of the representation (which is isomorphic to the quotient by (73)). The dimension of the expected quotient is usually big, and it is a very hard computational problem to find all the independent matrices which form a basis for the quotient; furthermore the dimension of the representation itself is quite large, which means a huge size for the matrices.

In order to overcome these problems, we make use of the following procedure: given the $\Gamma(\alpha)$, the constraint (78) can be solved for the $X(\alpha)$. If the solutions $X$ are realised in the algebra $K(E_{10})$, then they provide a basis (together with $\Gamma(\alpha)$) for the image of the representation. Hence, by (73), they provide a basis for the quotient. From a practical point of view, we have to choose a basis of $\Gamma$ matrices (which is made of 1024 elements), and then solve the constraint (78) with respect to $X$: therefore we have to check that all the solutions are realised in the representation space.

### 4.2.1 Spin-$\frac{3}{2}$

The bilinear form is given by

$$(\phi, \psi) = \phi^a G_{ab} \psi^b, \tag{80}$$

whose signature is (288, 32). The dimension of $\mathfrak{so}(288, 32)$ is 51 040, that we can decompose into $J = X \otimes \Gamma$ as follows. The $\Gamma(\alpha)$ range over the full Clifford algebra in ten Euclidean dimensions; this gives a basis of $2^{10} = 1024$ matrices. Of these 1024 matrices, 496 are antisymmetric (corresponding to $\mathfrak{so}(32)$ realised by real roots $\alpha$ and the only relevant ones for spin 1/2) and 528 are symmetric. In order for $J = X \otimes \Gamma$ to be antisymmetric with respect to (77), the antisymmetric $\Gamma(\alpha)$ have to be paired with symmetric polarisation tensors $X(\alpha)$ and vice versa, see (79). Symmetric matrices $\Gamma(\alpha)$ are obtained for $\alpha^2 \in 4\mathbb{Z}$, so unlike spin-$\frac{1}{2}$ they require null roots for their realization. Altogether this leads to $51\,040 = 496 \times 55 + 528 \times 45$ elements of $\mathfrak{so}(288, 32)$, where now also elements of the null root spaces are represented faithfully.

For the proof we have to verify that the basis dimensions are realised in the representation space, and this can be done by keeping a $\Gamma(\alpha)$ fixed and analysing all $X(\beta)$ such that $\Gamma(\beta) = \Gamma(\alpha)$ (there are infinitely many such $\beta$'s for $E_{10}$). Actually, it suffices to verify this property just for a smaller set of representative

---

[6] For $K(E_{11})$, the corresponding quotient Lie algebra is $\mathfrak{sl}(32)$ [16].

choices of $\Gamma(\alpha)$ as roots that are related by Weyl transformations do not have to be considered separately, as Weyl transformations rotate both the roots and the associated polarisation tensors. For real roots of $E_{10}$ it is in fact enough to do this for one simple root $\alpha_i$ because all (infinitely many) real roots are Weyl conjugate.

This analysis can be done on a computer and we find that all the solutions from the constraint (78) are realised in the representation space: this means that the quotient algebra of the spin-$\frac{3}{2}$ representation is

$$\mathfrak{q}_{3/2}(K(E_{10})) \cong \mathfrak{so}(288, 32) \,. \tag{81}$$

This result is an interesting outcome: the maximal *compact* subalgebra $K(E_{10})$ gives rise to the *non-compact* quotient $\mathfrak{so}(288, 32)$. This surprising feature is shared by all known spin $> \frac{1}{2}$ representations, and has no analog in the case of finite-dimensional Lie algebras, as will be further discussed below.

### 4.2.2 Spin-$\frac{5}{2}$ and Spin-$\frac{7}{2}$

The very same procedure can be applied to the spin-$\frac{5}{2}$ representation: here, the invariant bilinear form is

$$(\phi, \psi) = \phi^{\mathrm{ab}} G_{\mathrm{ac}} G_{\mathrm{bd}} \psi^{\mathrm{cd}}, \tag{82}$$

which has signature $(1472, 288)$. Hence, the expected quotient algebra is $\mathfrak{so}(1472, 288)$: again, a non-compact quotient. Following our general procedure we solve the constraint (78) for $X$ (with the aid of the $\Gamma$'s basis), and the solution we find is a set of 1540 symmetric $X$ matrices associated to the antisymmetric $\Gamma$'s and 1485 antisymmetric $X$'s associated to the symmetric $\Gamma$'s, which combine to a basis of $\mathfrak{so}(1472, 288)$.

Again, we now would have to check that the solutions from (78) are all realised in the algebra. To be sure, we have so far not been able to generate a complete basis for the polarisation tensors in the algebra by iterating commutators, so we do not yet have a definite proof whether all solutions of the constraint are actually realised in $K(E_{10})$ or not. However, the number of independent elements in $K(E_{10})$ generated with our procedure is already so large that the resulting quotient algebra cannot be anything else but

$$\mathfrak{q}_{5/2}(K(E_{10})) \cong \mathfrak{so}(1472, 288) \,. \tag{83}$$

where now also elements of root spaces associated to timelike imaginary roots with $\alpha^2 = -2$ are represented faithfully. Finally, the conjectured quotient for the spin-$\frac{7}{2}$ representation is $\mathfrak{so}(5568, 1472)$ but to verify this claim would require even more extensive checks than for spin-$\frac{5}{2}$.

## 4.3   General Remarks on the Quotients

We found that the quotients for the spin-$\frac{3}{2}$ and spin-$\frac{5}{2}$ representations are non-compact, in marked contrast with the fact that they originate from a compact algebra, namely $K(E_{10})$. This is a special feature that can occur only for infinite-dimensional algebras.[7]

If $\mathfrak{g}$ is a finite-dimensional simple and simply-laced algebra of rank at least 2, its maximal compact subalgebra $K(\mathfrak{g})$ is semi-simple, admitting no non-trivial solvable ideals. Representations of $K(\mathfrak{g})$ are then completely reducible by the standard theory of semi-simple compact Lie algebras, and this applies in particular to the adjoint representation $K(\mathfrak{g})$ itself. Therefore, given the existence of an ideal $\mathfrak{i} \subset K(\mathfrak{g})$, this is a representation of $K(\mathfrak{g})$ and, by complete reducibility, must split off as a direct summand of $K(\mathfrak{g})$, i.e. $K(\mathfrak{g}) = \mathfrak{i} \oplus \mathfrak{q}$ as a sum of $K(\mathfrak{g})$ representations. In this decomposition we have already indicated that $K(\mathfrak{g})/\mathfrak{i} \cong \mathfrak{q}$ is the remaining summand. An explicit construction of $\mathfrak{q}$ as the orthogonal complement of $\mathfrak{i}$ in the (negative) definite invariant form on $K(\mathfrak{g})$ is mentioned at the end of Sect. 2.2. The bilinear form on $\mathfrak{q}$ as a subalgebra of $K(\mathfrak{g})$ is obtained by restriction and therefore preserves the definiteness property: any subalgebra of $K(\mathfrak{g})$ must be compact. An instance where this happens is the Cartan type $D_n$. The split real form is $\mathfrak{g} = \mathfrak{so}(n, n)$ with $K(\mathfrak{g}) = \mathfrak{so}(n) \oplus \mathfrak{so}(n)$, showing explicitly that $K(\mathfrak{g})$ is semi-simple, has ideals and that all subalgebras (= quotients) are compact.

For infinite-dimensional $K(\mathfrak{g})$, it is not known whether complete reducibility still holds and one can therefore not repeat the same argument as above. Our investigations of the explicit unfaithful representations show that it can happen that an ideal $\mathfrak{i}$ does *not* split off as a direct summand and therefore the quotient Lie algebra $\mathfrak{q} = K(\mathfrak{g})/\mathfrak{i}$ need not be a subalgebra of $K(\mathfrak{g})$. Whether or not $\mathfrak{q}$ admits an invariant bilinear form and whether this form is definite is a question that has then to be answered separately as this form does not follow from that on $K(\mathfrak{g})$ by restriction. If $\mathfrak{i}$ should split off as a direct summand one could construct the quotient again as the orthogonal complement of $\mathfrak{i}$ as in the finite-dimensional case. Below we shall use this construction to argue that for $K(E_9)$ the quotient cannot be a subalgebra (given by finite linear combinations of $K(\mathfrak{g})$ elements) but rather is distributional in nature. This shows that $\mathfrak{i}$ does not split off in that case.

This result also applies to $K(E_{10})$: for instance, the quotient $\mathfrak{so}(288, 32)$ is not contained in $K(E_{10})$ as a subalgebra. Even though the *image* of the $K(E_{10})$ generators in the *unfaithful* representation behaves as $\mathfrak{so}(288, 32)$ matrices, the generators themselves in $K(E_{10})$ do not obey the $\mathfrak{so}(288, 32)$ algebra. If one wanted to write combinations of $K(E_{10})$ that behave like $\mathfrak{so}(288, 32)$ in $K(E_{10})$, one would require non-convergent infinite sums of generators, whose commutators could not be meaningfully evaluated.

---

[7]We are indebted to R. Köhl for discussions on this point.

It is also worth noting that the $K(E_{10})$ ideals for increasing spin are not necessarily embedded into one another for increasing spin even though this might appear to be the case from the quotient algebras that we have determined above and the increasing faithfulness of the representation. Let us assume that $i_{3/2} \subset i_{1/2}$; then $\mathfrak{q}_{3/2}$ has to act on the spin-$\frac{1}{2}$ representation as well.[8] But there is no way $\mathfrak{so}(288, 32)$ can act on the 32 components of the Dirac representation spinor: this implies that the ideals are not contained in each other. (This is in contrast with what happens for $K(E_9)$, where the ideals are in fact contained in each other, see [18] and below.)

A more manageable illustration of the fact that the quotient by the ideal $i$ is generally *not* a subalgebra of the infinite-dimensional involutory algebra, unlike in the finite-dimensional case, is provided by the example of the affine Kac–Moody algebra $E_9$ and its involutory subalgebra $K(E_9)$, and the action of the latter on the fermions of maximal $D = 2$ supergravity. Namely, as shown in [17, 18, 46], to which we refer for further details, the involutory subalgebra $K(E_9)$ admits an explicit realization via the $E_9$ loop algebra; more precisely, it is realised by elements

$$x(h) = \frac{1}{2}h^{IJ}(t)X^{IJ} + h^A(t)Y^A, \tag{84}$$

where $(X^{IJ}, Y^A)$ are the $120 + 128$ generating matrices of $E_8$ and $t \in \mathbb{C}$ is the spectral parameter. The transformation (84) belongs to $K(E_9)$ Lie algebra if the (meromorphic) functions $\left(h^{IJ}(t), h^A(t)\right)$ obey the constraints

$$h^{IJ}\left(\frac{1}{t}\right) = h^{IJ}(t), \quad h^A\left(\frac{1}{t}\right) = -h^A(t), \tag{85}$$

as well as the reality constraint exhibited in [17]. For the Dirac ($=$ spin-$\frac{1}{2}$) representation the action of $x(h)$ on the $(16 + 16)$ chiral spinor components $\varepsilon^I_\pm$ is obtained by evaluation of the functions $\left(h^{IJ}(t), h^A(t)\right)$ at the distinguished points $t = \pm 1$ in the spectral parameter plane [46][9]

$$x(h) \cdot \varepsilon^I_+ = h^{IJ}(1)\,\varepsilon^J_+, \quad x(h) \cdot \varepsilon^I_- = h^{IJ}(-1)\,\varepsilon^J_-. \tag{86}$$

Hence the quotient algebra is

$$\mathfrak{q}_{1/2}(K(E_9)) \cong \mathfrak{so}(16)_+ \oplus \mathfrak{so}(16)_-, \tag{87}$$

which is manifestly not contained in either $K(E_9)$ or $E_9$. The spin-$\frac{3}{2}$ representation consists of the 128 spinor components $\chi^{\dot{A}}_\pm$ and the gravitino components $\psi^I_{2\pm}$ (we here adopt the superconformal gauge, otherwise we would have an extra $(16 + 16)$

---

[8]We thank G. Bossard for pointing this out to us.

[9]We note that each choice $t = \pm 1$ yields an irreducible 16-component chiral spinor of $K(E_9)$ but we treat the two choices together as they result from the decomposition of the 32 components of the $K(E_{10})$ Dirac spinor.

components $\psi_{\pm}^I$, see [17, 18][10]). The action of $x(h)$ is now realised by evaluation of $h$ and its first derivatives at $t = \pm 1$ [17]; more precisely,

$$x(h) \cdot \psi_{2\pm}^I = h^{IJ}(\pm 1) \, \psi_{2\pm}^J,$$

$$x(h) \cdot \chi_{\pm}^{\dot{A}} = \frac{1}{4} h^{IJ}(\pm 1) \, \Gamma_{\dot{A}\dot{B}}^{IJ} \, \chi_{\pm}^{\dot{B}} \mp \frac{1}{2} (h')^A(\pm 1) \, \Gamma_{A\dot{A}}^I \psi_{2\pm}^I. \tag{88}$$

The quotient algebra in the spin-$\frac{3}{2}$ representation is thus

$$\mathfrak{q}_{3/2}(K(E_9)) \cong \left[ \mathfrak{so}(16)_+ \oplus \mathfrak{so}(16)_- \right] \oplus \left[ \mathbb{R}_+^{128} \oplus \mathbb{R}_-^{128} \right], \tag{89}$$

which is a semi-direct sum and again not a subalgebra of $K(E_9)$. This is consistent with the appearance of non-compact quotients in $K(E_{10})$ (which contains $K(E_9)$), as there is no way to embed a translation group into a compact group (and indeed, the above transformations are contained in $\mathfrak{so}(288, 32)$). Furthermore, as shown in [18], there is an inclusion of ideals $\mathfrak{i}_{3/2} \subset \mathfrak{i}_{1/2}$ for $K(E_9)$ and this is consistent with the fact that (89) can act on the $(16 + 16)$ spin-$\frac{1}{2}$ components by having a trivial action of the translation part $\mathbb{R}_+^{128} \oplus \mathbb{R}_-^{128}$ on the spinor. We also note that the spin-$\frac{3}{2}$ representation is indecomposable, in that the representation matrices of (88) take on a triangular form

$$x(h) \cdot \begin{pmatrix} \chi^{\dot{A}} \\ \psi_2^I \end{pmatrix} \sim \begin{pmatrix} \mathfrak{so}(16)_{\text{spin.}} & \mathbb{R}_{128 \times 16}^{128} \\ 0 & \mathfrak{so}(16)_{\text{fund.}} \end{pmatrix} \begin{pmatrix} \chi^{\dot{A}} \\ \psi_2^I \end{pmatrix}. \tag{90}$$

The indecomposability means that although $\psi_2^I$ is a subrepresentation of the spin-$\frac{3}{2}$ representation it does not split off as a direct summand, showing again the failure of complete reducibility for $K(\mathfrak{g})$ in the Kac–Moody case.

The difficulty of quotienting out the ideal can be traced back to the distributional nature of the complement of the ideal in $K(E_9)$ for both examples. The ideals $\mathfrak{i}_{1/2}$ and $\mathfrak{i}_{3/2}$, respectively, are spanned by those elements $x(h)$ in (84) for which either $h^{IJ}(\pm 1) = 0$, or $h^{IJ}(\pm 1) = 0$ and $(h')^A(\pm 1) = 0$. Consequently, the complement of either ideal would have to have $\delta$-function-like support at $t = \pm 1$ in the spectral parameter plane, and thus cannot be represented in the form (84) with smooth functions $h$ [18]. We anticipate that the difficulties for $K(E_{10})$ are of a similar, but more severe nature, probably requiring a substantial generalisation of known concepts of distribution theory, as it is not clear whether the formally divergent series can be handled by established tools of distribution theory, as was the case for $K(E_9)$.

---

[10]With these extra components we would indeed recover the counting of the spin-$\frac{3}{2}$ representation of $K(E_{10})$: $2 \times (16 + 128 + 16) = 320$.

## 5  $K(E_{10})$ and Standard Model Fermions

We now come to a tantalizing, but much more speculative feature of the finite-dimensional spin-$\frac{3}{2}$ spinor representation of $K(E_{10})$, and a feature that may eventually allow to link up the abstract mathematical theory developed here to Standard Model physics. Namely, for $E_{10}$ it turns out that the spin-$\frac{3}{2}$ representation, when viewed from the point of view of four-dimensional physics, can be interpreted as the combination of eight massive gravitinos and 48 spin-$\frac{1}{2}$ matter fermions, a remarkable coincidence with the fact that the observed quarks and leptons in the Standard Model (including right-handed neutrinos) come in three families of 16 spin-$\frac{1}{2}$ fermions: $48 = 3 \times 16$!

This coincidence was already foreshadowed by the fact that maximal $N = 8$ supergravity, in four dimensions, after complete breaking of supersymmetry, is characterised by 48 spin-$\frac{1}{2}$ fermions. As first stressed by Gell-Mann in [47], this number matches that of the 48 fermions of the Standard Model if one includes three right-handed neutrinos; moreover, there is a way of putting the known quarks and leptons into representations of the residual symmetry group $SU(3) \times U(1)$ of $N = 8$ supergravity remaining after the gauge group $SO(8)$ is broken, *provided* the supergravity $SU(3)$ is identified with the diagonal subgroup of the color group $SU(3)_c$ and a putative family symmetry group $SU(3)_f$. However, there remained a mismatch in the electric charge assignments by $\pm\frac{1}{6}$ that was only recently fixed: namely, as shown in [10], the correct charges are obtained by deforming the subgroup $U(1)$ in the way explained below that is not compatible with the supergravity theory itself. The required deformation is, however, contained in $K(E_{10})$ [11] acting on the spin-$\frac{3}{2}$ representation of $K(E_{10})$! The deformation of the $U(1)$ subgroup of $SU(3) \times U(1)$ required to rectify the mismatch of electric charges is obtained by acting on the tri-spinors $\chi^{ijk}$ of $N = 8$ supergravity which transform in the $\mathbf{56} \equiv \mathbf{8} \wedge \mathbf{8} \wedge \mathbf{8}$ of $SU(8)$, with the 56-by-56 matrix $\exp\left(\frac{1}{6}\omega\mathcal{I}\right)$, where

$$\mathcal{I} := \frac{1}{2}\Big(T \wedge \mathbb{1} \wedge \mathbb{1} + \mathbb{1} \wedge T \wedge \mathbb{1} + \mathbb{1} \wedge \mathbb{1} \wedge T + T \wedge T \wedge T\Big); \tag{91}$$

here the matrix $T$ is defined by

$$T := \begin{pmatrix} 0 & 1 & 0 & 0 & 0 & 0 & 0 & 0 \\ -1 & 0 & 0 & 0 & 0 & 0 & 0 & 0 \\ 0 & 0 & 0 & 1 & 0 & 0 & 0 & 0 \\ 0 & 0 & -1 & 0 & 0 & 0 & 0 & 0 \\ 0 & 0 & 0 & 0 & 0 & 1 & 0 & 0 \\ 0 & 0 & 0 & 0 & -1 & 0 & 0 & 0 \\ 0 & 0 & 0 & 0 & 0 & 0 & 0 & 1 \\ 0 & 0 & 0 & 0 & 0 & 0 & -1 & 0 \end{pmatrix}. \tag{92}$$

This matrix belongs to $\mathfrak{so}(8)$ and represents an imaginary unit, with $T^2 = -\mathbb{1}$, which in turn implies $\mathcal{I}^2 = -\mathbb{1}$. The presence of the triple product $T \wedge T \wedge T$ in $\mathcal{I}$ is the reason why $\mathcal{I}$ is *not* an $\mathfrak{su}(8)$ element, hence does not belong to the R symmetry of $N = 8$ supergravity (confirming the long known fact that this theory by itself cannot account for the charge assignments of the Standard Model fermions). Enlarging the SU(8) symmetry to $K(E_{10})$, however, one can show [11] that in fact $\mathcal{I}$ does belong to the infinite-dimensional algebra $K(E_{10})$ or, to be completely precise, to the associated quotient q. Importantly, it is not possible to construct the element (91) without the over-extended root of $E_{10}$, the root system of any regular subalgebra such as $E_9$ being too restricted. This could make $E_{10}$ a crucial ingredient in connecting supergravity to the Standard Model particles: no finite-dimensional or affine subalgebra can possibly achieve this.

Since $\mathcal{I}$ does not belong to the R symmetry SU(8), one can now ask which bigger group is generated by commuting $\mathcal{I}$ with SU(8) in all possible ways in the **56** representation of SU(8). Indeed, by repeated commutation of $\mathcal{I}$ with elements of $\mathfrak{su}(8)$, one ends up with the bigger algebra $\mathfrak{su}(56)$, which should thus be viewed as a subalgebra of $\mathfrak{so}(288, 32)$. To show this we take the standard $\mathfrak{su}(8)$ basis in the fundamental representation, made of by 28 antisymmetric matrices (nothing but the $\mathfrak{so}(8)$ basis) and 35 symmetric matrices multiplied by the imaginary unit. These matrices belong to the **8** representation; we pass to the **56** representation of $\mathfrak{su}(8)$ by means of the antisymmetrised product used in (91). Then we commute $\mathcal{I}$ with the 63 elements of the basis; we collect the new independent elements from the commutators, and redo the procedure again until we stabilise the set of matrices. Proceeding in this way, after four iterations we reach a stable set, i.e. no new elements are produced by the commutators. This new set has dimension 3135 and corresponds to $\mathfrak{su}(56)$, since it is made of by $56 \times 56$ anti-hermitian and traceless matrices.

Because $\mathcal{I} \in K(E_{10})$, the group SU(56) is therefore contained in the closure of SU(8) and $\mathcal{I}$, and it is thus a subgroup of the vector spinor quotient Spin(288, 32)$/\mathbb{Z}_2$ (at the level of algebra, the quotient is $\mathfrak{so}(288, 32)$). Let us also mention that the very same calculation can be done with $\mathfrak{so}(8)$: starting from the standard basis (28 antisymmetric matrices) in the **8** representation, we build the **56** representation and start to commute with $\mathcal{I}$. Proceeding in this way, we reach 1540 independent antisymmetric $56 \times 56$ matrices to obtain the desired $\mathfrak{so}(56)$ basis.

Perhaps even more importantly, it has been shown in recent work [12] that actually the full (chiral) Standard Model group SU(3)$_c \times$ SU(2)$_w \times$ U(1)$_Y$ can be embedded into this unfaithful representation of $K(E_{10})$, thus in particular undoing the descent to the diagonal subgroup SU(3) of the color and family groups, with a hypothetical family symmetry SU(3)$_f$ that does not commute with the electroweak symmetries (because the upper and lower components of the would-be electroweak doublets must be assigned to opposite representations of SU(3)$_f$ [47]), That such an embedding is possible is not entirely unexpected in view of the fact that the above SU(56) acts *chirally* in four dimensions and via its subgroup SU(48) is large enough to contain the Standard Model groups as subgroups (as well as a novel family symmetry). So the more intricate open question and the true challenge is to understand whether and why this symmetry breaking might be preferred in $K(E_{10})$. We also reiterate that the eight massive gravitinos are an integral part of the spin-$\frac{3}{2}$ representation, and thus must likewise be incorporated into the physics. Indeed, a possible interpretation as (Planck mass) Dark Matter candidates was suggested for them in [48], together with possible experimental tests of this proposal. However, the main challenge that remains is to see how $K(E_{10})$ can 'unfold' in terms of bigger and increasingly less unfaithful $K(E_{10})$ representations to give rise to actual space-time fermions, explaining how the above groups can be elevated to *bona fide* gauge symmetries in space and time (we note that explaining the emergence of bosonic space-time fields and symmetries from $E_{10}$ likewise remains an open problem). There remains a long way to go!

**Acknowledgements** H. Nicolai would like to thank V. Gritsenko and V. Spiridonov for their hospitality in Dubna. His work has received funding from the European Research Council (ERC) under the European Union's Horizon 2020 research and innovation programme (grant agreement No 740209). We are grateful to G. Bossard, R. Köhl and H.A. Samtleben for comments and discussions.

# References

1. B. Julia, Group disintegrations. Conf. Proc. **C8006162**, 331–350 (1980)
2. P.C. West, $E_{11}$ and M theory. Class. Quant. Grav. **18**, 4443–4460 (2001). arXiv:hep-th/0104081 [hep-th]. http://dx.doi.org/10.1088/0264-9381/18/21/305
3. T. Damour, M. Henneaux, H. Nicolai, $E_{10}$ and a 'small tension expansion' of M theory. Phys. Rev. Lett. **89**, 221601 (2002). arXiv:hep-th/0207267 [hep-th]. http://dx.doi.org/10.1103/PhysRevLett.89.221601
4. S. de Buyl, M. Henneaux, L. Paulot, Hidden symmetries and Dirac fermions. Class. Quant. Grav. **22**, 3595–3622 (2005). arXiv:hep-th/0506009 [hep-th]. http://dx.doi.org/10.1088/0264-9381/22/17/018
5. T. Damour, A. Kleinschmidt, H. Nicolai, Hidden symmetries and the fermionic sector of eleven-dimensional supergravity. Phys. Lett. **B634**, 319–324 (2006). arXiv:hep-th/0512163 [hep-th]. http://dx.doi.org/10.1016/j.physletb.2006.01.015
6. S. de Buyl, M. Henneaux, L. Paulot, Extended $E_8$ invariance of 11-dimensional supergravity. J. High Energy Phys. **02**, 056 (2006). arXiv:hep-th/0512292 [hep-th]. http://dx.doi.org/10.1088/1126-6708/2006/02/056
7. T. Damour, A. Kleinschmidt, H. Nicolai, $K(E_{10})$, supergravity and fermions. J. High Energy Phys. **08**, 046 (2006). arXiv:hep-th/0606105 [hep-th]. http://dx.doi.org/10.1088/1126-6708/2006/08/046
8. A. Kleinschmidt, H. Nicolai, IIA and IIB spinors from $K(E_{10})$. Phys. Lett. **B637**, 107–112 (2006). arXiv:hep-th/0603205 [hep-th]. http://dx.doi.org/10.1016/j.physletb.2006.04.007
9. A. Kleinschmidt, Unifying R-symmetry in M-theory, in *15th International Congress on Mathematical Physics (ICMP06) Rio de Janeiro, Brazil, August 6–11, 2006* (2007). arXiv:hep-th/0703262 [hep-th]. https://doi.org/10.1007/978-90-481-2810-5

10. K.A. Meissner, H. Nicolai, Standard model fermions and $N = 8$ supergravity. Phys. Rev. **D91**, 065029 (2015). `arXiv:1412.1715` [hep-th]. http://dx.doi.org/10.1103/PhysRevD.91. 065029

11. A. Kleinschmidt, H. Nicolai, Standard model fermions and $K(E_{10})$. Phys. Lett. **B747**, 251–254 (2015). `arXiv:1504.01586` [hep-th]. http://dx.doi.org/10.1016/j.physletb.2015. 06.005

12. K.A. Meissner, H. Nicolai, Standard model fermions and infinite-dimensional R-symmetries. Phys. Rev. Lett. **121**(9), 091601 (2018). `arXiv:1804.09606` [hep-th]. http://dx.doi. org/10.1103/PhysRevLett.121.091601

13. A. Kleinschmidt, H. Nicolai, On higher spin realizations of $K(E_{10})$. J. High Energy Phys. **08**, 041 (2013). `arXiv:1307.0413` [hep-th]. http://dx.doi.org/10.1007/JHEP08(2013)041

14. A. Kleinschmidt, H. Nicolai, N.K. Chidambaram, Canonical structure of the $E_{10}$ model and supersymmetry. Phys. Rev. **D91**(8), 085039 (2015). `arXiv:1411.5893` [hep-th]. http://dx.doi.org/10.1103/PhysRevD.91.085039

15. A. Kleinschmidt, H. Nicolai, Higher spin representations of $K(E_{10})$, in *Proceedings, International Workshop on Higher Spin Gauge Theories: Singapore, Singapore, November 4–6, 2015* (2017), pp. 25–38. `arXiv:1602.04116` [hep-th]. http://dx.doi.org/10.1142/ 9789813144101_0003

16. P.C. West, $E_{11}$, SL(32) and central charges. Phys. Lett. **B575**, 333–342 (2003). `arXiv:hep-th/0307098` [hep-th]. http://dx.doi.org/10.1016/j.physletb.2003.09.059

17. H. Nicolai, H. Samtleben, On $K(E_9)$. Q. J. Pure Appl. Math. **1**, 180–204 (2005). `arXiv:hep-th/0407055` [hep-th]. http://dx.doi.org/10.4310/PAMQ.2005.v1.n1.a8

18. A. Kleinschmidt, H. Nicolai, J. Palmkvist, $K(E_9)$ from $K(E_{10})$. J. High Energy Phys. **06**, 051 (2007). `arXiv:hep-th/0611314` [hep-th]. http://dx.doi.org/10.1088/1126-6708/ 2007/06/051

19. T. Damour, C. Hillmann, Fermionic Kac-Moody billiards and supergravity. J. High Energy Phys. **08**, 100 (2009), `arXiv:0906.3116` [hep-th]. http://dx.doi.org/10.1088/1126- 6708/2009/08/100

20. G. Hainke, R. Köhl, P. Levy, Generalized spin representations. Münster J. Math. **8**(1), 181–210 (2015). With an appendix by Max Horn and Ralf Köhl. http://dx.doi.org/10.17879/ 65219674985

21. T. Damour, P. Spindel, Quantum supersymmetric cosmology and its hidden Kac–Moody structure. Class. Quant. Grav. **30**, 162001 (2013), `arXiv:1304.6381` [gr-qc]. http:// dx.doi.org/10.1088/0264-9381/30/16/162001

22. T. Damour, P. Spindel, Quantum supersymmetric Bianchi IX cosmology. Phys. Rev. **D90**(10), 103509 (2014). `arXiv:1406.1309` [gr-qc]. http://dx.doi.org/10.1103/PhysRevD.90. 103509

23. D. Ghatei, M. Horn, R. Köhl, S. Weiß, Spin covers of maximal compact subgroups of Kac– Moody groups and spin-extended Weyl groups. J. Group Theory **20**(3), 401–504 (2017). http:// dx.doi.org/10.1515/jgth-2016-0034

24. R. Lautenbacher, R. Köhl, Extending generalized spin representations. J. Lie Theory **28**, 915–940 (2018). e-prints. `arXiv:1705.00118` [math.RT]

25. V.G. Kac, *Infinite Dimensional Lie Algebras*, 2nd edn. (Cambridge University Press, Cambridge, 1990)

26. S. Berman, On generators and relations for certain involutory subalgebras of Kac-Moody Lie algebras. Comm. Algebra **17**(12), 3165–3185 (1989). http://dx.doi.org/10.1080/ 00927878908823899

27. A. Kleinschmidt, H. Nicolai, Gradient representations and affine structures in $AE_n$. Class. Quant. Grav. **22**, 4457–4488 (2005). `arXiv:hep-th/0506238` [hep-th]. http://dx.doi. org/10.1088/0264-9381/22/21/004

28. B.S. DeWitt, Quantum theory of gravity. 1. The canonical theory. Phys. Rev. **160**, 1113–1148 (1967). [**3**, 93 (1987)]. http://dx.doi.org/10.1103/PhysRev.160.1113

29. T. Damour, S. de Buyl, M. Henneaux, C. Schomblond, Einstein billiards and overextensions of finite dimensional simple Lie algebras. J. High Energy Phys. **08**,

030 (2002). arXiv:hep-th/0206125 [hep-th]. http://dx.doi.org/10.1088/1126-6708/2002/08/030

30. P. Goddard, D.I. Olive, Kac-Moody and Virasoro algebras in relation to quantum physics. Int. J. Mod. Phys. **A1**, 303 (1986). [86 (1986)]. http://dx.doi.org/10.1142/S0217751X86000149

31. J. Polchinski, *String Theory. Vol. 2: Superstring Theory and Beyond*. Cambridge Monographs on Mathematical Physics (Cambridge University Press, Cambridge, 2007). http://dx.doi.org/10.1017/CBO9780511618123

32. B. de Wit, A.K. Tollsten, H. Nicolai, Locally supersymmetric D = 3 nonlinear sigma models. Nucl. Phys. **B392**, 3–38 (1993). arXiv:hep-th/9208074 [hep-th]. http://dx.doi.org/10.1016/0550-3213(93)90195-U

33. M.F. Atiyah, R. Bott, A. Shapiro, Clifford modules. Topology **3**, S3–S38 (1964). http://dx.doi.org/10.1016/0040-9383(64)90003-5

34. T. Damour, M. Henneaux, H. Nicolai, Cosmological billiards. Class. Quant. Grav. **20**, R145–R200 (2003). arXiv:hep-th/0212256 [hep-th]. http://dx.doi.org/10.1088/0264-9381/20/9/201

35. S. Viswanath, Embeddings of hyperbolic Kac-Moody algebras into $E_{10}$. Lett. Math. Phys. **83**(2), 139–148 (2008). http://dx.doi.org/10.1007/s11005-007-0214-7

36. C. Hillmann, A. Kleinschmidt, Pure type I supergravity and $DE_{10}$. Gen. Rel. Grav. **38**, 1861–1885 (2006). arXiv:hep-th/0608092 [hep-th]. http://dx.doi.org/10.1007/s10714-006-0352-8

37. A.J. Feingold, I.B. Frenkel, A hyperbolic Kac-Moody algebra and the theory of Siegel modular forms of genus 2. Math. Ann. **263**(1), 87–144 (1983). http://dx.doi.org/10.1007/BF01457086

38. S. Mizoguchi, K. Mohri, Y. Yamada, Five-dimensional supergravity and hyperbolic Kac-Moody algebra $G_2^H$. Class. Quant. Grav. **23**, 3181–3194 (2006), arXiv:hep-th/0512092 [hep-th]. http://dx.doi.org/10.1088/0264-9381/23/9/026

39. S. Mizoguchi, N. Ohta, More on the similarity between D = 5 simple supergravity and M theory. Phys. Lett. **B441**, 123–132 (1998). arXiv:hep-th/9807111 [hep-th]. http://dx.doi.org/10.1016/S0370-2693(98)01122-8

40. A.H. Chamseddine, H. Nicolai, Coupling the SO(2) supergravity through dimensional reduction. Phys. Lett. **96B**, 89–93 (1980). [Erratum: Phys. Lett. B **785**, 631 (2018). arXiv:1808.08955]. http://dx.doi.org/10.1016/0370-2693(80)90218-X,10.1016/j.physletb.2018.05.029

41. E. Cremmer, Supergravities in 5 dimensions, in *Superspace and Supergravity, Proceedings of Nuffield Workshop in Cambridge (UK), June 16–July 12, 1980*, ed. by S. Hawking, M. Rocek (Cambridge University Press, Cambridge, 1980)

42. A. Kleinschmidt, D. Roest, Extended symmetries in supergravity: the semi-simple case. J. High Energy Phys. **07**, 035 (2008). arXiv:0805.2573 [hep-th]. http://dx.doi.org/10.1088/1126-6708/2008/07/035

43. T. Damour, M. Henneaux, $E_{10}$, $BE_{10}$ and arithmetical chaos in superstring cosmology. Phys. Rev. Lett. **86**, 4749–4752 (2001). arXiv:hep-th/0012172 [hep-th]. http://dx.doi.org/10.1103/PhysRevLett.86.4749

44. F. Englert, L. Houart, $G^{+++}$ invariant formulation of gravity and M theories: exact BPS solutions. J. High Energy Phys. **01**, 002 (2004). arXiv:hep-th/0311255 [hep-th]. http://dx.doi.org/10.1088/1126-6708/2004/01/002

45. A. Keurentjes, $E_{11}$: sign of the times. Nucl. Phys. **B697**, 302–318 (2004). arXiv:hep-th/0402090 [hep-th]. http://dx.doi.org/10.1016/j.nuclphysb.2004.06.058

46. H. Nicolai, Two-dimensional gravities and supergravities as integrable system. Lect. Notes Phys. **396**, 231–273 (1991). http://dx.doi.org/10.1007/3-540-54978-1_12

47. M. Gell-Mann, From renormalizability to calculability?, in *Shelter Island 1983, Proceedings, Quantum Field Theory and The Fundamental Problems Of Physics*, ed. by R. Jackiw, N.N. Khuri, S. Weinberg, E. Witten (Dover Publications, New York, 1986), pp. 3–23

48. K.A. Meissner, H. Nicolai, Planck mass charged gravitino dark matter (2019) arXiv:1809.01441 [hep-ph]

# BPS Spectra and Invariants for Three- and Four-Manifolds

**Du Pei**

## Contents

1   Introduction........................................................................ 218
2   Three-Manifolds and the Categorification of WRT Invariants ............................ 219
    2.1   Fivebranes on Three-Manifolds and Categorification of WRT Invariant ............. 221
    2.2   Examples ................................................................... 231
    2.3   $M_3 = L(p, 1)$ ............................................................. 235
    2.4   More Examples ............................................................. 238
3   Four-Manifolds and Topological Modular Forms .......................................... 238
    3.1   From Six to Two Dimensions, via Four-Manifolds ................................. 242
    3.2   Examples ................................................................... 259
    3.3   What's Not Included ......................................................... 266
References ............................................................................ 266

**Abstract** In these notes, I discuss how to obtain new invariants of three- and four-manifolds from BPS spectra of quantum field theories. The main ingredients are supersymmetric theories in six-dimensions and their twisted compactifications. We will also see, in concrete examples, how several well-known invariants of three- and four-manifolds can be realized as partition functions of supersymmetric gauge theories and computed using localization technique. These notes summarizes author's lecture given at the Winter School on "Partition Functions and Automorphic Forms" at JINR, Russia in February 2018 and are based on joint works with Sergei Gukov, Pavel Putrov and Cumrun Vafa.

D. Pei (✉)
Center of Mathematical Sciences and Applications, Harvard University, Cambridge, MA, USA
e-mail: dpei@cmsa.fas.harvard.edu

© Springer Nature Switzerland AG 2020                        217
V. A. Gritsenko, V. P. Spiridonov (eds.), *Partition Functions
and Automorphic Forms*, Moscow Lectures 5,
https://doi.org/10.1007/978-3-030-42400-8_5

# 1 Introduction

Can we hear the shape of a drum? Much like harmonics of a musical instrument, spectra of quantum systems contain wealth of useful information. Of particular interest are supersymmetric or the so-called BPS states which, depending on the problem at hand, can manifest themselves either as minimal surfaces, or solutions to partial differential equations, or other "extremal" objects. Thus, a spectrum of BPS states in Calabi–Yau compactifications can be used to reconstruct the geometry of the Calabi–Yau space itself and, as we explain in this paper, spectra of BPS states play a similar role in low-dimensional topology.

One fruitful approach to constructing numerical invariants of three- and four-manifolds, as well as the corresponding homological invariants of three-manifolds, is based on gauge theory. The famous examples are Donaldson–Witten and Seiberg–Witten (SW) invariants of four-manifolds, and corresponding instanton and monopole [1] Floer homologies of three-manifolds. All of them were extensively studied in the mathematical literature and have appropriate rigorous definitions that go back to the previous century. The numerical invariants are realized in terms of counting solutions to certain partial differential equations, while the homological invariants build on the ideas of Andreas Floer [2]. In particular, Seiberg–Witten invariants of four-manifolds had great success distinguishing some homeomorphic but non-diffeomorphic four-manifolds. And, in the world of three-manifolds, the so-called Heegaard Floer homology constructed by Ozsvath and Szabo [3] gives a much simpler and non-gauge theoretic definition of a homological invariant, which is believed to be equivalent to the monopole Floer homology.

A seemingly different class of three-manifold invariants, the so-called Witten–Reshetikhin–Turaev (WRT) invariants [4, 5], comes from a different type of TQFT, which sometimes is called "of Schwarz type" to distinguish it from the TQFTs "of cohomological type" mentioned in the previous paragraph [6]. At the turn of the century, however, the distinction between the two types started to blur and the recent works such as [7–9] suggest it may even go away completely in the future. In fact, the first hints for this go back to the early work [10–13] that relates Seiberg–Witten theory in three dimensions to Chern–Simons theory with $U(1|1)$ super gauge group. The latter provides a much simpler invariant compared to the usual Chern–Simons theory, say, with $SU(2)$ gauge group, due to cancellations between bosonic and fermionic contributions. Therefore, if one can find a 4d TQFT that categorifies $SU(2)$ Chern–Simons theory in 3d, similar to how 4d SW theory categorifies 3d SW theory, it would help a great deal with the classification problem of smooth four-manifolds. The first step in constructing such categorification is, of course, to find a homological invariant of three-manifolds whose (equivariant) Euler characteristic gives the WRT invariant.

In Sect. 2, we will tackle this problem by providing a physical definition of new homological invariants $\mathcal{H}_a(M_3)$ of three-manifolds (possibly, with knots) labeled by abelian flat connections. The physical system in question involves a 6d fivebrane theory on $M_3$ times a 2-disk, $D^2$, whose Hilbert space of BPS states plays the role of a basic building block in categorification of various partition functions of 3d $\mathcal{N} = 2$ theory $T[M_3]$: $D^2 \times S^1$ half-index, $S^2 \times S^1$ superconformal index, and $S^2 \times S^1$ topologically twisted index. The first partition function is labeled by a choice of boundary condition and provides a refinement of Chern-Simons (WRT) invariant. A linear combination of them in the unrefined limit gives the analytically continued WRT invariant of $M_3$. The last two can be factorized into the product of half-indices. We show how this works explicitly for many examples, including lens spaces, circle fibrations over Riemann surfaces, and plumbed three-manifolds.

In Sect. 3, we will study a new class of invariants for four-manifolds. We build a connection between topology of smooth four-manifolds and the theory of topological modular forms by considering topologically twisted compactification of 6d $(1, 0)$ theories on four-manifolds. The effective 2d theory has $(0, 1)$ supersymmetry, and the "topological Witten genus" of this 2d theory produces a new invariant of the four-manifold, valued in the ring of weakly holomorphic topological modular forms. We describe basic properties of this map and present a few simple examples.

## 2 Three-Manifolds and the Categorification of WRT Invariants

The existence of a categorification of the WRT invariant was envisioned by Crane and Frenkel [14] more than 20 years ago, and the first evidence came with the advent of knot homology [15–17] which categorifies WRT invariants of knots and links (realized by Wilson lines in CS theory) in $S^3$, also known as the colored Jones polynomial. The physical understanding of a homological approach to HOMFLY polynomials was independently initiated in the physics literature in [18], which later led to the physical interpretation of Khovanov–Rozansky homology as certain BPS Hilbert spaces [19–22] (see e.g. [23, 24] for an overview and an extensive list of references).

The physical construction suggests that there should be a homological invariant that categorifies (in a certain sense) the WRT invariant of general three-manifolds with knots inside. Namely, such homological invariant can be understood as the BPS sector of the Hilbert space of the 6d $\mathcal{N} = (2, 0)$ theory (that is a theory describing dynamics of coincident M5-branes in M-theory) on $M_3 \times D^2 \times \mathbb{R}$ with a certain supersymmetry preserving background along $M_3 \times D^2$. Equivalently, if one first reduces the 6d theory on $M_3$, it can be understood as the BPS Hilbert space of the effective 3d $\mathcal{N} = 2$ theory $T[M_3]$ on $D^2 \times \mathbb{R}$. On the other hand, if one first compactifies on $D^2$, one does not get an ordinary 4d gauge theory on $M_3 \times \mathbb{R}$ like 4d SW gauge theory[1]:

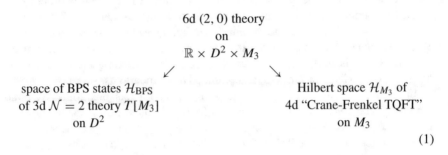

$$\text{(1)}$$

There is another natural SUSY-preserving background on which one can quantize $T[M_3]$. Since the IR physics of $T[M_3]$ is governed by a non-trivial 3d $\mathcal{N} = 2$ SCFT, one can consider its radial quantization and study its Hilbert space on $S^2 \times \mathbb{R}$. This should provide us with another non-trivial homological invariant of $M_3$ which should have roughly the same level of complexity as the BPS Hilbert space on $D^2 \times \mathbb{R}$ which categorifies the WRT invariant, but with several advantages due to the presence of operator-state correspondence and no need to specify a boundary condition at $\partial D^2 = S^1$.

The set of boundary conditions that one can put at $\partial D^2 \cong S^1$ in the path integral can be understood as follows. Let us represent $D^2$ as an elongated cigar, which asymptoically looks like $S^1 \times \mathbb{R}$. After compactification of the stack of fivebranes on $S^1$ we obtain 5d maximally supersymmetric gauge theory. The supersymmetric vacua of such theory on $M_3 \times \mathbb{R}$ (where $\mathbb{R}$ is the original time direction) are given by flat connections[2] on $M_3$. The number of such supersymetric vacua is the same as the number of boundary conditions (cf. [21, 26–29]). As was found in [9], the subset of such boundary conditions that corresponds to abelian flat connections plays an important role; here we summarize various clues that all point to a special role of abelian flat connections:

- charges of BPS states in the target space theory [18];
- relation between homological invariants of different rank [8];
- resurgent analysis of complex Chern-Simons theory [25];
- similar analysis of 3d $\mathcal{N} = 2$ theories [9, Sec. 2.2];
- Langlands duality for flat connections on three-manifolds [9, Sec. 2.10].

Moreover, in [9], we gave many examples of $q$-series invariants for various three-manifolds which exhibit integrality, and in many cases we also write the

---

[1]Roughly speaking, the effective 4d theory is an infinite KK-like tower on 4d gauge theories. However one needs to appropriately sum it up. The decategorified counterpart of such summation was studied in [25].

[2]The choice of such flat connections should not be confused with the choice of flat connections in CS theory on $M_3$. As will be explained later in detail they are related by S-transform.

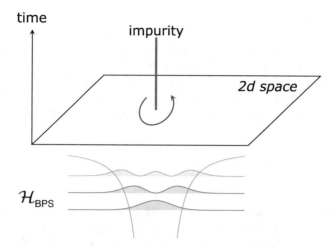

**Fig. 1** The space of BPS states in 3d $\mathcal{N} = 2$ theory on $\mathbb{R} \times D^2$ with an impurity, relevant to the physical realization of Heegaard Floer homology $HF(M_3)$, monopole Floer homology $HM(M_3)$, as well as categorification of WRT invariants of three-manifolds with knots

corresponding homological invariants. All of these invariants are labeled by (connected components of) abelian flat connections; we do not have any such example associated with a non-abelian flat connection.

In [9] we studied the relation between such homological invariants and their decategorified counterparts—superconformal indices. The structure described above, of course, will also manifest itself at the level of partition functions, i.e. indices, once we compactify the time on $S^1$. For some of the examples in [9] for there are no cancellations among states in computing the refined index. In those cases, the homological invariants are faithfully captured by the refined index computation.

In this section, we will be working with closed three-manifold, although most of the results have natural extensions to three-manifolds with knots inside by studying partition function of 3d $\mathcal{N} = 2$ theory $T[M_3]$ in the presence of line operators, cf. Fig. 1. For more details, please see [9, Sec. 4].

## 2.1 Fivebranes on Three-Manifolds and Categorification of WRT Invariant

The goal of this subsection is to introduce the key players and their interrelation. To keep the discussion simple and concrete, we choose the gauge group to be $G = SU(2)$.

### 2.1.1 Preliminaries

Before we present a mathematically-friendly summary of our proposal and the physics behind it, we need to introduce some notations, especially those relevant to abelian flat connections that will be central in our discussion.

Consider a closed and connected three-manifold $M_3$, with $\partial M_3 = \emptyset$. In order to present the results in full generality, it will be useful to consider the linking pairing on the torsion part of $H_1(M_3, \mathbb{Z})$:

$$
\begin{aligned}
\ell k : \mathrm{Tor}\, H_1(M_3) \otimes \mathrm{Tor}\, H_1(M_3) & \longrightarrow & \mathbb{Q}/\mathbb{Z} \\
[a] \otimes [b] & \longmapsto & \#(a \cap b')/n
\end{aligned}
\tag{2}
$$

where $b'$ is a 2-chain such that $\partial b' = nb$ for an integer $n$. Such $b'$ and $n$ exist because $[b]$ is torsion. As usual, $\#(a \cap b')$ denotes the number of intersection points counted with signs determined by the orientation. Note that the linking form provides an isomorphism between $\mathrm{Tor}\, H_1(M_3)$ and its Pontryagin dual $(\mathrm{Tor}\, H_1(M_3))^* \equiv \mathrm{Hom}(\mathrm{Tor}\, H_1(M_3), U(1))$ via the pairing $e^{2\pi i \ell k(\cdot, \cdot)}$.

The $\mathbb{Z}_2$ Weyl group acts on the elements $a \in \mathrm{Tor}\, H_1(M_3, \mathbb{Z})$ via $a \mapsto -a$. The set of orbits is the set of connected components of abelian flat $SU(2)$ connections on $M_3$ (i.e., connections in the image of $\rho : \mathcal{M}_{\mathrm{flat}}(M_3, U(1)) \to \mathcal{M}_{\mathrm{flat}}(M_3, SU(2))$ from the embedding $U(1) \subset SU(2)$),[3]

$$
\mathrm{Tor}\, H_1(M_3, \mathbb{Z})/\mathbb{Z}_2 \;\cong\; \pi_0 \mathcal{M}_{\mathrm{flat}}^{\mathrm{ab}}(M_3, SU(2)) .
\tag{3}
$$

It is also useful to introduce a shorthand notation for the stabilizer subgroup:

$$
\mathcal{W}_a \;\equiv\; \mathrm{Stab}_{\mathbb{Z}_2}(a) \;=\; \begin{cases} \mathbb{Z}_2, \ a = -a \,, \\ 1, \ \ \text{otherwise} \,. \end{cases}
\tag{4}
$$

### 2.1.2 $D^2 \times S^1$ Partition Function of $T[M_3]$ and WRT Invariant

Now we are ready to present a slightly generalized and improved version of the results from [8, sec. 6].

**Categorification of WRT Invariant**
Let $Z_{SU(2)_k}[M_3]$ be the partition function of $SU(2)$ Chern-Simons theory with "bare" level $(k - 2)$ on $M_3$, also known as the WRT invariant. We use the standard

---

[3]It is in fact $(\mathrm{Tor}\, H_1(M_3, \mathbb{Z}))^* /\mathbb{Z}_2$ that is canonically identified with components of abelian flat connections. However, as the distinction between $\mathrm{Tor}\, H_1(M_3, \mathbb{Z})$ and its dual is only important in Sect. 2.1.2, we will use the same set of labels $\{a, b, \ldots\}$ for elements in both groups.

"physics" normalization where

$$Z_{SU(2)_k}\left[S^2 \times S^1\right] = 1, \tag{5}$$

and

$$Z_{SU(2)_k}\left[S^3\right] = \sqrt{\frac{2}{k}} \sin\left(\frac{\pi}{k}\right). \tag{6}$$

The following conjecture was proposed in [8].[4]

*Conjecture 1* The WRT invariant can be decomposed into the following form:

$$Z_{SU(2)_k}[M_3] = (i\sqrt{2k})^{b_1(M_3)-1} \sum_{\substack{a,b \in \\ \text{Tor } H_1(M_3,\mathbb{Z})/\mathbb{Z}_2}} e^{2\pi i k \ell k(a,a)} S_{ab} \widehat{Z}_b(q)\Big|_{q \to e^{\frac{2\pi i}{k}}} \tag{7}$$

with[5,6]

$$\widehat{Z}_b(q) \in 2^{-c} q^{\Delta_b} \mathbb{Z}[[q]] \qquad \Delta_b \in \mathbb{Q}, \qquad c \in \mathbb{Z}_+ \tag{8}$$

convergent in $|q| < 1$ and

$$S_{ab} = \frac{e^{4\pi i \ell k(a,b)} + e^{-4\pi i \ell k(a,b)}}{|\mathcal{W}_a|\sqrt{|\text{Tor } H_1(M_3,\mathbb{Z})|}}. \tag{9}$$

In other words, we claim the existence of new three-manifold invariants $\widehat{Z}_a$, which admit $q$-series expansion with integer powers and integer coefficients (hence, more suitable for categorification) and from which the WRT invariant can be reconstructed via (7). While the formal mathematical definition of the invariants $\widehat{Z}_a$ is waiting to be discovered, they admit a physics definition that will be reviewed below and can be independently computed via techniques of resurgent analysis. In particular, each term

$$\sum_{b \in \text{Tor } H_1(M_3,\mathbb{Z})} e^{2\pi i k \ell k(a,a)} S_{ab} \widehat{Z}_b(q)\Big|_{q \to e^{\frac{2\pi i}{k}}} \equiv e^{2\pi i k \ell k(a,a)} Z_a(q) \tag{10}$$

---

[4]A related conjecture was made in [30]. However it did not include the $S$-transform, which is crucial for restoring integrality and categorification.

[5]The constant positive integer $c$ depends only on $M_3$ and in a certain sense measures its "complexity". In many simple examples $c = 0$, and the reader is welcome to ignore $2^{-c}$ factor which arises from some technical subtleties. Its physical origin was explained in [9].

[6]Later in the text we will sometimes use slightly redefined quantities, $\widehat{Z}_a(q) \to q^{\Delta} \widehat{Z}_a(q)$, where $\Delta$ is a common, $a$ independent rational number.

in the sum (7) is a certain resummation of the perturbative (in $\frac{2\pi i}{k}$ or, equivalently, in $(1 - q)$) expansions of the WRT invariant around the corresponding abelian flat connection $a$ [25].

In order to avoid unnecessary technical complications, in the rest of this survey we assume that Tor $H_1(M_3, \mathbb{Z})$ has no $\mathbb{Z}_2$ factors.[7] Under this assumption, the $S$-matrix satisfies

$$\sum_b S_{ab} S_{bc} = \delta_{ac}. \tag{11}$$

Moreover, as will be discussed in detail below, physics predicts the existence of a $\mathbb{Z} \times \mathbb{Z} \times$ Tor $H_1(M_3, \mathbb{Z})/\mathbb{Z}_2$ graded homological invariant of $M_3$:

$$\mathcal{H}_{D^2}[M_3] = \bigoplus_{a \in \text{Tor } H_1(M_3, \mathbb{Z})/\mathbb{Z}_2} \mathcal{H}_a[M_3], \qquad \mathcal{H}_a[M_3] = \bigoplus_{\substack{i \in \mathbb{Z} + \Delta_a, \\ j \in \mathbb{Z}}} \mathcal{H}_a^{i,j} \tag{12}$$

which categorifies the $q$-series $\widehat{Z}_a(q)$. Namely,[8]

$$\widehat{Z}_a(q) = \sum_{\substack{i \in \mathbb{Z} + \Delta_a, \\ j \in \mathbb{Z}}} q^i (-1)^j \dim \mathcal{H}_a^{i,j}. \tag{13}$$

Because of their close relation to homological invariants, we usually refer to $\widehat{Z}_a(q)$ as *homological blocks*. The vector space $\mathcal{H}_{D^2}[M_3]$ can be interpreted as the closed three-manifold analog of Khovanov-Rozansky knot homology. From this point of view, $e^{4\pi i \ell k(a, \cdot)}$ can be understood as the variable associated with Tor $H_1(M_3, \mathbb{Z})/\mathbb{Z}_2$ grading and enters into the decomposition (7) much like $q$, the variable for one of the $\mathbb{Z}$-gradings (the "$q$-grading"). Note that the label $a$ in $\mathcal{H}_a^{i,j}$ is reminiscent of the Spin$^c$ structure in Heegaard/monopole Floer homologies of three-manifolds. This fact, of course, is not an accident and plays an important role in the relation between Heegaard/monopole Floer homologies of three-manifolds and the categorification of WRT invariants [8].

Next, we describe the physics behind the Conjecture 1. (A mathematically inclined reader may skip directly to Conjecture 2.)

---

[7]Recall that Tor $H_1(M_3, \mathbb{Z})$, as a finitely generated abelian group, can be decomposed into Tor $H_1(M_3, \mathbb{Z}) = \prod_i \mathbb{Z}_{p_i}$. We ask for a fairly weak condition that $\mathbb{Z}_2$ doesn't appear in this decomposition. In other words, $M_3$ is a $\mathbb{Z}_2$-homology sphere. Equivalently, there is a unique Spin structure on $M_3$, so that there is no ambiguity in specifying Nahm-pole boundary condition for $\mathcal{N} = 4$ $SU(2)$ SYM on $M_3 \times \mathbb{R}_+$ [21].

[8] The presence of $1/2$ factors that produce $2^{-c}$ overall factor in (8) can be interpreted as presence of factors $\cong \mathbb{C}[x]$ with $\deg_q x = 0$ in $\mathcal{H}_a[M_3]$. The $q$-graded Euler charecteristic of $\mathbb{C}[x]$ is naively divergent: $1 + 1 + 1 + \ldots$, but its zeta-regularization gives $1/2$.

## Physics Behind the Proposal

From physics point of view, the homological invariants $\mathcal{H}_a[M_3]$ can be realized by the following M-theory geometry,

$$
\begin{array}{lccccc}
N \text{ fivebranes}: & \mathbb{R} \times & M_3 & \times & D^2 & \\
& & \cap & & \cap & \\
\text{space-time}: & \mathbb{R} \times & T^*M_3 & \times & TN & \quad (14) \\
& & \circlearrowleft & & \circlearrowleft & \\
\text{symmetries}: & & \text{``}U(1)_N\text{''} & & U(1)_q \times U(1)_R. &
\end{array}
$$

or, equivalently, any of its dual descriptions (some of which will be discussed below). Here, the first two lines summarize the geometry of the fivebranes and their ambient space, whereas the last line describes their symmetries. The reason "$U(1)_N$" appears in quotes is that it is a symmetry of our physical system only when $M_3$ is a Seifert three-manifold, unlike the "universal" symmetry group $U(1)_q \times U(1)_R$.

In order to preserve supersymmetry for a general metric on $M_3$, it has to be embedded in the geometry of ambient space-time as a supersymmetric (special Lagrangian) cycle $M_3 \subset CY_3$ which, according to McLean's theorem, always looks like $M_3 \subset T^*M_3$ near the zero section. Equivalently, the geometry $T^*M_3$ represents a partial topological twist on the fivebrane world-volume, upon which three of the scalar fields on the world-volume become sections of the cotangent bundle of $M_3$. As a result, one can first reduce the 6d $(2, 0)$ theory—the world-volume theory of M5-branes—on $M_3$ to obtain a 3d $\mathcal{N} = 2$ SCFT usually denoted as $T[M_3; G]$, where $G = U(N)$ or $SU(N)$, $N$ being the number of M5-branes. All SUSY-protected objects like partition functions, index and BPS spectra of the resulting theory $T[M_3; G]$ do not depend[9] on the metric of $M_3$, and give rise to numerical as well as homological invariants of $M_3$.

Similarly, in order to preserve supersymmetry of the brane system (14) along the other world-volume directions of the fivebranes, one needs to introduce a SUSY-preserving background along $D^2$. Moreover, it needs to be done in a way that preserves the rotation symmetry $U(1)_q \times U(1)_R$ and allows to keep track of the corresponding quantum numbers (spins) of BPS states, as required for categorification. The suitable background can be described in a number of equivalent ways: as the Omega-background along $TN \cong \mathbb{R}^4_{q,t}$, in which $D^2 \cong \mathbb{R}^2$ is embedded as a linear subspace, or as a $U(1)_q \times U(1)_R$ invariant Lagrangian submanifold (the "cigar") $D^2$ in the Taub–NUT space $TN$ where one keeps track of the spin with respect to the rotation symmetry, cf. Fig. 2. To emphasize that one keeps track of both spins under $U(1)_q \times U(1)_R$ symmetry, the adjective *refined* is often added to the invariant, BPS state, or other object under consideration.

---

[9]A "folk theorem" states that any continuous deformation of the metric on $M_3$ results in a $Q$-exact term of the supergravity background. However, there may be dependence on discrete data such as the Atiyah 2-framing [31].

**Fig. 2** Another
representation of the
background in Fig. 1

(time)   x

When $M_3$ is a Seifert manifold, the brane system (14) enjoys an extra symmetry $U(1)_N$ that appears in a degeneration limit of the metric on $M_3$ and can be used to redefine the R-symmetry of the SCFT $T[M_3]$. When $M_3$ is $\Sigma \times S^1$, the symmetry $U(1)_N \times U(1)_R$ is further enhanced to the $SU(2)_N \times SU(2)_R$ R-symmetry of the 3d $\mathcal{N} = 4$ theory $T[\Sigma \times S^1]$; when $M_3$ is a generic Seifert manifold, one combination gives the R-symmetry of the 3d $\mathcal{N} = 2$ theory $T[M_3]$ which we denote as $U(1)_R$, while another is a flavor symmetry $U(1)_\beta$, see [8, sec. 3.4] for details.

After reduction on $M_3$, the system (14) gives a theory $T[M_3; G]$ in space-time $D^2 \times \mathbb{R}$, illustrated in Fig. 1, and we can consider its Hilbert space with a certain boundary condition at $\partial D^2 = S^1$. For $N = 2$ and $G = SU(2)$, these boundary conditions turn out to be labeled by $a \in \mathrm{Tor}\, H_1(M_3, \mathbb{Z})/\mathbb{Z}_2$. Then, we arrive at a set of doubly-graded homological invariants of $M_3$ labeled by $a$,

$$\mathcal{H}_a[M_3] = \mathcal{H}_{T[M_3]}(D^2; a) = \bigoplus_{\substack{i \in \mathbb{Z}+\Delta_a, \\ j \in \mathbb{Z}}} \mathcal{H}_a^{i,j}, \tag{15}$$

given by the BPS sector of the Hilbert space of $T[M_3; SU(2)]$. This is the subspace annihilated by two of the four supercharges of the 3d $\mathcal{N} = 2$ supersymmetry (different choices are related by automorphisms of the superconformal algebra, resulting in isomorphic BPS spaces). The grading $i$ counts the charge under the $U(1)_q$ rotation of $D^2$ and the "homological" grading $j$ corresponds to R-charge of the $U(1)_R$ R-symmetry. When $M_3$ is Seifert, the $U(1)_\beta$ symmetry will give rise to the third grading on $\mathcal{H}_a[M_3]$.

One can understand $\mathcal{H}_a[M_3] = \mathcal{H}_{T[M_3]}(D^2; a)$ as the massless multi-particle BPS spectrum of $T[M_3]$ with a label $a \in \mathrm{Tor}\, H_1(M_3, \mathbb{Z})/\mathbb{Z}_2$ being a discrete charge. From the M-theory point of view, the BPS particles of $T[M_3]$ arise from M2-branes ending on the pair of M5-branes that realize the $A_1$ 6d $\mathcal{N} = (2, 0)$ theory. The boundaries of M2-branes wrap 1-cycles $(\widetilde{a}, -\widetilde{a})$ so that $[\widetilde{a}] = a \in \mathrm{Tor}\, H_1(M_3, \mathbb{Z})/\mathbb{Z}_2$. This is similar to the counting of BPS states in [18]. Note, however, that BPS particles that arise from M2-branes ending on a non-torsion 1-cycles of $M_3$ have mass and do not enter into the IR BPS spectrum. Therefore, it is elements in $\mathrm{Tor}\, H_1(M_3, \mathbb{Z})/\mathbb{Z}_2$ that give rise to physical boundary conditions which specifies a superselection sector labeled by this brane charge, resulting in a physical BPS Hilbert space $\mathcal{H}_a$ with integrality property. In contrast, a flat connection, given by an element of $(\mathrm{Tor}\, H_1(M_3, \mathbb{Z}))^*/\mathbb{Z}_2$ doesn't correspond to any physical boundary conditions. Instead, it is a linear combination of physical boundary conditions leading to a mixture of different charge sectors.

From this point of view, the S-transform in Conjecture 1 carries out the change of basis between *charges* (valued in $\mathrm{Tor}\, H_1(M_3, \mathbb{Z})/\mathbb{Z}_2$) and *holonomies* (valued in $(\mathrm{Tor}\, H_1(M_3, \mathbb{Z}))^*/\mathbb{Z}_2$) using the natural pairing between them via the "Aharonov–Bohm phase." More precisely, the M2-branes ending on M5-branes produce particles in the effective 3d theory carrying electric-magnetic charge $b \in \mathrm{Tor}\, H_1(M_3, \mathbb{Z})/\mathbb{Z}_2$.[10] On the other hand, $a \in (\mathrm{Tor}\, H_1(M_3, \mathbb{Z}))^*/\mathbb{Z}_2$ specified the holomony, and, by viewing $(\mathrm{Tor}\, H_1(M_3, \mathbb{Z}))^*$ as the group of characters of $\mathrm{Tor}\, H_1(M_3, \mathbb{Z})$, we have

$$S_{ab} \propto \sum_{\mathbb{Z}_2 \text{ orbit of } a} \chi_a(b) = \sum_{\mathbb{Z}_2 \text{ orbit of } a} e^{4i\pi \ell k(a,b)}. \tag{16}$$

Geometrically, $S_{ab}$ is the trace of the holonomy of the flat connection labeled by $a$ along the 1-cycle representing homology class $b$.

Alternatively, one can understand the boundary conditions in the type IIB duality frame of the brane system (14), where S-transform can be interpreted as the S-duality of type IIB string theory. Indeed, a quotient of the 11-dimensional space-time by a circle action $U(1)_q$ lands us in type IIA string theory, which can be further T-dualized along the "time" circle $S^1$. (Equivalently, these two dualities can be combined into one step, which is the standard M-theory/type IIB duality.) The resulting system involves D3-branes ending on a D5-brane, and S-duality of type IIB string theory maps it into a stack of D3-branes ending on an NS5-brane [21]. Note, that the natural choice of boundary conditions at infinity for a system of D3-branes ending on an NS5-brane is an arbitrary (not necessarily abelian) $SL(2, \mathbb{C})$ flat connection on $M_3$. Such choice of a flat connection corresponds to considering analytically continued $SU(2)$ Chern-Simons theory on the Lefschetz thimble associated to that flat connection. However, in general, the corresponding partition function is not continuous in a disk $|q| < 1$ due to Stokes phenomena. Instead, in (10) we consider quantities $Z_a(q)$ which are labeled by *abelian* flat connections only, and analytic inside $|q| < 1$. As explained in [25], for a given value of $\arg k$ one can express $Z_a(q)$ as a linear combination of the Feynman path integral on Lefschetz thimbles. If one were to write the S-transform in the basis corresponding to Lefschetz thimbles, instead of $Z_a(q)$, the S-matrix would be $k$-dependent.

By definition, each homological block $\widehat{Z}_a(q)$ is the graded Euler characteristics (25) of $\mathcal{H}_a$ that can be computed as an supersymmetric partition function of $T[M_3]$ on $D^2 \times_q S^1$ with an $\mathcal{N} = (0, 2)$ supersymmetric boundary condition $a$ and metric corresponding to rotation of the disk $D^2$ by $\arg q$ when we go around the $S^1$,

$$\widehat{Z}_a(q) = Z_{T[M_3]}(D^2 \times_q S^1; a). \tag{17}$$

---

[10]Note that in the case of $N$ M5-branes wrapping $M_3$, the M2-branes produce states charged under the magnetic $U(1)^N/S_N$ (not necessarily $U(N)$) symmetry, as explained in [18].

If one knows the Lagrangian description of $T[M_3]$, this partition function can be computed using localization, see e.g. [32].

Note, if the Lagrangian description of $T[M_3]$ contains chiral multiplets charged under the gauge symmetry, carrying zero R-charge (at the unitarity bound) and with Neumann boundary conditions, then the integral computing the $D^2 \times_q S^1$ partition function (i.e. the half-index) will be singular in general. For example, for $g$ adjoint chiral multiplets the partition function has the following form:

$$\frac{1}{2} \int_{|z|=1} \frac{dz}{2\pi i z} \frac{f(z,q)}{(1-z^2)^{g-1}(1-z^{-2})^{g-1}}, \quad f(z,q) = f(z^{-1},q) \in \mathbb{Z}[z,z^{-1}][[q]].$$
(18)

A natural way to regularize it is to take the principle value prescription:

$$\frac{1}{2} \text{ v.p.} \int_{|z|=1} \frac{dz}{2\pi i z} \frac{f(z,q)}{(1-z^2)^{g-1}(1-z^{-2})^{g-1}} \equiv$$
$$\frac{1}{4} \left( \int_{|z|=1+\epsilon} + \int_{|z|=1-\epsilon} \right) \frac{dz}{2\pi i z} \frac{f(z,q)}{(1-z^2)^{g-1}(1-z^{-2})^{g-1}}.$$
(19)

As we will see in many examples, this regularization prescription is in agreement with the relation between $\widehat{Z}_a(q) = Z_{T[M_3]}(D^2 \times_q S^1; a)$ and the WRT invariant. This is also the source of the $2^{-c}$ factor in (8).[11]

### 2.1.3 Superconformal Index of $T[M_3]$ and Its Factorization

Although the new invariants $\widehat{Z}_a(q)$ and their categorification $\mathcal{H}_a[M_3]$ have a direct connection to the WRT invariant via (7), from the viewpoint of 3d $\mathcal{N} = 2$ theory $T[M_3; SU(2)]$, it is not the most natural or simplest object to consider. The main reason is that, in principle, there are infinitely many possible boundary conditions that could be considered.[12] Identifying the correct subset that appears in (17) may be subtle, yet possible, as we shall see in concrete examples.

---

[11] Such factor can appear even without singularity, when there is only $g = 1$ adjoint chiral with R-charge 0. It cancels the contribution of the vector multiplet, leaving

$$\frac{1}{2} \int \frac{dz}{2\pi i z} f(z,q) = \frac{1}{2}[z^0 \text{ coefficient of } f(z,q)],$$
(20)

where one can take $f(z,q) = 1$, as an example.

[12] If we understand the boundary condition as a coupling of the 3d theory to a 2d theory living on the boundary, then the infinite number of possibilities can be seen, for example, from the fact that we can always introduce a 2d theory decoupled from the bulk.

The more natural object is the superconformal index of $T[M_3]$ or, equivalently, the partition function on $S^2 \times S^1$ with a certain supersymmetry preserving background:

$$\mathcal{I}(q) \equiv \mathrm{Tr}_{\mathcal{H}_{S^2}}(-1)^F q^{R/2+J_3} = Z_{T[M_3]}(S^2 \times_q S^1) \tag{21}$$

where $\mathcal{H}_{S^2}$ is the space of BPS states of 3d $\mathcal{N} = 2$ SCFT $T[M_3]$ or, equivalently, $Q$-cohomology of all physical local operators, $F$ is the fermion number, $R$ is the generator of the $U(1)_R$ R-symmetry and $J_3$ is the Cartan generator of the $SO(3)$ isometry of $S^2$. By construction, this index has the desired integrality property[13] and can be categorified,

$$\mathcal{H}_{S^2} = \bigoplus_{i,j} \mathcal{H}_{S^2}^{i,j}, \qquad \mathcal{I}(q) = \sum_{i,j \in \mathbb{Z}} q^i (-1)^j \dim \mathcal{H}_{S^2}^{i,j} \in \mathbb{Z}[[q]]. \tag{22}$$

Another advantage of $\mathcal{H}_{S^2}$ compared to $\mathcal{H}_{D^2}$ is that it has a natural ring structure which is given by multiplication of BPS operators. One of the statements of 3d/3d correspondence is that the partition function $Z_{T[M_3]}(S^2 \times_q S^1)$ computes the partition function of complex Chern-Simons on $M_3$ with real part of the "level" being 0 and analytically continued imaginary part $\propto \tau \equiv \log q/(2\pi i)$. Motivated by the topological/anti-topological fusion [26] and its recent 3d incarnation [27–29, 33], we would like to make the following conjecture:

*Conjecture 2* For $\widehat{Z}_a(q)$ as in Conjecture 1, the following holds

$$\mathcal{I}(q) = \sum_{a \in \mathrm{Tor}\, H_1(M_3, \mathbb{Z})/\mathbb{Z}_2} |\mathcal{W}_a| \widehat{Z}_a(q) \widehat{Z}_a(q^{-1}) \qquad \in \mathbb{Z}[[q]] \tag{23}$$

where $\widehat{Z}_a(1/q)$ is an appropriate extension[14] of $\widehat{Z}_a(q)$ to the region $|q| > 1$.

---

[13]For reasons similar to the ones described at the end of the previous section, in general a negative power of 2 can appear as an overall factor. We omit it in some generic formulas to avoid clutter and instead focus attention on the conceptual structure. As mentioned earlier and as we shall see in examples, the effect of such fractions on categorification can typically be traced to the existence of bosonic modes with zero $q$-grading.

[14]For a generic three-manifold $M_3$, the analytic continuation of the series $\widehat{Z}_a(q) \in \mathbb{Z}[[q]]$ convergent in $|q| < 1$ may not exist outside $|q| = 1$, at least in the standard way. However, one possible way to define it for general $M_3$ is

$$\widehat{Z}_a(q^{-1}) \equiv \widehat{Z}_a(q)\big|_{M_3 \to \overline{M}_3}, \tag{24}$$

where $\overline{M}_3$ denotes $M_3$ with the reversed orientation. Note that therefore $\mathcal{I}(q)$, unlike $\widehat{Z}_a(q)$, is insensitive to the orientation.

### 2.1.4  Further Refinement

As we discussed above, for Seifert three-manifolds the theory $T[M_3]$ has an extra
$U(1)_\beta$ flavor symmetry. In particular, this is the case for $M_3 = L(p, 1)$, which
will serve as an important example to us later. This flavor symmetry in $T[M_3]$
results in the presence of an extra $\mathbb{Z}$-grading in homological invariants $\mathcal{H}_{D^2}$ and $\mathcal{H}_{S^2}$
considered in the previous sections and a possibility to consider the corresponding
refined indices (equivariant Euler characteristics):

$$\widehat{Z}_a(q; t) = \sum_{\substack{i \in \mathbb{Z} + \Delta_a, \\ j, \ell \in \mathbb{Z}}} q^i t^\ell (-1)^j \dim \mathcal{H}_a^{i, j; \ell}, \tag{25}$$

and

$$\mathcal{I}(q; t) = \sum_{i, j, \ell \in \mathbb{Z}} q^i t^\ell (-1)^j \dim \mathcal{H}_{S^2}^{i, j; \ell}. \tag{26}$$

where, from the physics point of view, $\ell$ is the $U(1)_\beta$ charge. The refined indices
obviously provide more information about the underlying vector spaces and, as we
will see in examples, can be used sometimes to compute (conjecturally) the full
homological invariants via the "homological-flavor locking" phenomenon.

The refined version of the factorization formula (23) reads

$$\mathcal{I}(q; t) = \sum_{a \in \text{Tor } H_1(M_3, \mathbb{Z})/\mathbb{Z}_2} |W_a| \widehat{Z}_a(q; t) \widehat{Z}_a(q^{-1}; t^{-1}). \tag{27}$$

### 2.1.5  Topologically Twisted Index of $T[M_3]$

Another interesting invariant of $M_3$ which can be realized as an observable of $T[M_3]$
and has a categorification by construction is the topologically twisted index on $S^2 \times S^1$ [34]. Namely, one can consider the 3d $\mathcal{N} = 2$ theory $T[M_3]$ with a background
value of the $U(1)_R$-symmetry connection equal to the spin connection on $S^2$. In
terms of the effective 2d $\mathcal{N} = (2, 2)$ theory obtained by compactifying $T[M_3]$ on
$S^1$, this is the familiar A-twist along the $S^2$.

The topologically twisted index and the underlying homological invariant of $M_3$
have the same fugacities/gradings as the superconformal index:

$$\mathcal{I}_{\text{top}}(q; t) = \sum_{i, j, \ell \in \mathbb{Z}} q^i t^\ell (-1)^j \dim \mathcal{H}_{S^2_{\text{top}}}^{i, j; \ell}. \tag{28}$$

As in the superconformal index $\mathcal{I}(q; t)$, here the parameter $q$ plays the role of
the Omega-background parameter corresponding to rotating $S^2$ along one of the
axes. However, in general the topologically twisted index $\mathcal{I}_{\text{top}}(q; t)$ has a much

simpler structure (as will be explained later) compared to the superconformal index $\mathcal{I}(q)$. Namely, $\mathcal{I}_{\text{top}}(q;t)$ is a rational function of $q,t$ (i.e. it has a form of the index of quantum mechanics with two supercharges), whereas $\mathcal{I}(q,t)$ can be as transcendental as, say, a quantum dilogarithm or (Jacobi) mock modular form. Nevertheless, the topologically twisted index is expected to have a similar factorization into homological blocks:

$$\mathcal{I}_{\text{top}}(q,t) = \sum_{a \in \text{Tor } H_1(M_3,\mathbb{Z})/\mathbb{Z}_2} |W_a| \widehat{Z}_a(q;t) \widehat{Z}_a(q^{-1};t). \qquad (29)$$

The difference from (27) is due to the fact that the supersymmetric backround chosen for superconformal $S^2 \times S^1$ index can be interpreted as doing topological A-twist along one of the $D^2 \times S^1$ halves and anti-A-twist along the other $D^2 \times S^1$ half,

$$\mathcal{I}(q;t) = \left( \overline{A} - \text{twist} \quad \right)\!\!\left( \quad A - \text{twist} \right) \qquad (30)$$

whereas the background for the topologically twisted $S^2 \times S^1$ index is such that the same A-twist is performed along both halves:

$$\mathcal{I}_{\text{top}}(q;t) = \left( A - \text{twist} \quad \right)\!\!\left( \quad A - \text{twist} \right). \qquad (31)$$

## 2.2 Examples

In this subsection, we present many examples, illustrating the general proposal outlined previously. In particular, our goal is two-fold: first, we wish to use the proposed physics definition of the new homological invariants $\mathcal{H}_a[M_3]$ to compute them in concrete examples, to the extent that one can start exploring the structure of the results and explicitly test the Conjectures 1 and 2, which is our second goal. We start with the simplest three-manifold and gradually move to the study of more complex ones.

### 2.2.1  $M_3 = S^3$

For $M_3 = S^3$ and $G = U(N)$, the 3d $\mathcal{N} = 2$ theory $T[S^3]$ is a 3d $\mathcal{N} = 2$ Chern-Simons theory with gauge group $U(N)$ at level 1 and an adjoint chiral multiplet $\phi$, whose R-charge is equal to 2 [35]. This theory is dual (the duality is usually referred to as the "duality appetizer" [7, 36, 37]) to a system of $N$ free chirals, making it simple to analyze. The R-charges of the free chirals are given by $R = 2, 4, \ldots, 2N$, and they have charges $1, 2, \ldots, N$ under the $U(1)_\beta$ flavor symmetry that rotates the original adjoint chiral $\phi$ by a phase.

As a result, the superconformal index of the theory $T[S^3]$ can be expressed a simple product,

$$\mathcal{I}_{U(N)}(q, t) = \prod_{i=1}^{N} \frac{(t^{-i}q^{1-i}; q)_\infty}{(t^i q^i; q)_\infty}, \tag{32}$$

where, as usual, the $q$-Pochhammer symbol is defined as

$$(z; q)_n = \prod_{j=0}^{n-1}(1 - zq^j). \tag{33}$$

Since $H_1(S^3) = 0$, there is only one homological block $\widehat{Z}_0(q, t)$, realized as the $S^1 \times D^2$ partition function of the $N$ free chirals with Neumann boundary conditions,

$$\widehat{Z}_0(q, t) = \prod_{i=1}^{N} \frac{1}{(t^i q^i; q)_\infty}. \tag{34}$$

In the terminology of [8] this is an *unreduced* homological block.[15] In order to relate it to the WRT invariant of $S^3$ as in Conjecture 1, before taking the unrefined limit $t \to 1$, one has to divide by the contribution of the Cartan components of the adjoint chiral $\phi$. The result looks like

$$\widehat{Z}_0(q) = (tq; q)_\infty^N \, \widehat{Z}_0(q, t)\Big|_{t \to 1} = \prod_{j=1}^{N-1}(1 - q^j)^{N-j}. \tag{35}$$

In the case of $G = SU(2)$, the dual theory consists of just one free chiral multiplet with R-charge 4 and $U(1)_\beta$ charge 2, whose index is

$$\mathcal{I}_{SU(2)}(q, t) = \frac{(t^{-2}q^{-1}; q)_\infty}{(t^2 q^2; q)_\infty}. \tag{36}$$

and we have

$$\widehat{Z}_0(q, t) = \frac{1}{(t^2 q^2; q)_\infty}, \qquad \widehat{Z}_0(q) = (1 - q^2). \tag{37}$$

---

[15] Although the terminology "reduced" and "unreduced" here is very similar to the one used in knot homology, there is no direct connection.

Using the standard prescription for extending the quantum dilogarithm outside $|q| < 1$,[16]

$$(x; q^{-1})_\infty \propto \frac{1}{(xq; q)_\infty}, \tag{38}$$

we have, for $G = U(N)$,[17]

$$\widehat{Z}_0(1/q, 1/t) = \prod_{i=1}^{N} \frac{1}{(t^{-i}q^{-i}; 1/q)_\infty} = \frac{1}{N!} \prod_{i=1}^{N} (t^{-i}q^{1-i}; q)_\infty. \tag{40}$$

We see that the superconformal index (32) indeed admits a factorization $\grave{a}$ $la$ (27),

$$\mathcal{I}(q, t) = |\mathcal{W}_0| \, \widehat{Z}_0(q, t) \widehat{Z}_0(q^{-1}, t^{-1}), \tag{41}$$

with only one homological block in this case.

### Categorifying the Index

As $T[S^3, U(N)]$ is dual to a system of free chirals, its space of BPS states on $S^2$—which gives the homological invariant associated with $M_3 = S^3$—factorizes as product

$$\mathcal{H}(S^3) = \mathcal{H}_2 \otimes \mathcal{H}_4 \otimes \ldots \otimes \mathcal{H}_{2N}, \tag{42}$$

where $\mathcal{H}_{2i}$ is the BPS Hilbert space of a single 3d $\mathcal{N} = 2$ chiral multiplet with R-charge $2i$ and $U(1)_\beta$ flavor charge $i$. Thus, we now turn to the problem of categorifying the index of a free chiral multiplet.

Recall that for a 3d $\mathcal{N} = 2$ chiral multiplet $\Phi$ of R-charge $r$ and flavor charge $f$, the superconformal index—equal to the equivariant Euler characteristic of the BPS Hilbert space $\mathcal{H}_{r,f}$—is given by

$$\mathcal{I}_r(q, t) = \chi_{q,t}(\mathcal{H}_{r,f}) = \sum_{i,j,\ell \in \mathbb{Z}} q^i t^\ell (-1)^j \dim \mathcal{H}_{r,f}^{i,j;\ell} = \prod_{j=0}^{\infty} \frac{1 - t^{-f} q^{1-r/2+j}}{1 - t^f q^{r/2+j}}. \tag{43}$$

---

[16]Note that this is not an ordinary analytic continuation. The latter actually does not exists because $|q| = 1$ is a natural boundary.

[17]Note that there is an ambiguous overall constant in (38). We fix it in (40) by requiring that the unrefined quantities are related by the ordinary analytic continuation. Namely,

$$\widehat{Z}_0(1/q, 1/t) \cdot (t^{-1}q^{-1}; q^{-1})_\infty^N \Big|_{t \to 1} = \widehat{Z}_0(1/q, 1/t) \cdot (1/t; q)_\infty^{-N} \Big|_{t \to 1} = \widehat{Z}_0(1/q) = \prod_{j=1}^{N-1} (1 - q^{-j})^{N-j}. \tag{39}$$

Each term in this product corresponds to a generator of the BPS Hilbert space $\mathcal{H}_{r,f}$. One can identify the denominator of (43) as the contribution of the bosonic modes $\partial^j \phi$, and the numerator as the contribution of the fermionic modes $\partial^j \overline{\psi}_+$. Then $\mathcal{H}_{r,f}$ is freely generated by these generators as a supercommutative algebra,

$$\mathcal{H}_{r,f} = \mathbb{C}[x_i, \xi_i] \cong \Omega^\bullet(\mathrm{Sym}^\infty(\mathbb{C})) \tag{44}$$

with an infinite set of even generators $x_0, x_1, x_2, \ldots$ and odd generators $\xi_0, \xi_1, \xi_2, \ldots$ coming from

$$x_i \quad \longleftrightarrow \quad \partial^j \phi, \tag{45}$$
$$\xi_i \quad \longleftrightarrow \quad \partial^j \overline{\psi}_+.$$

In fact, this result has already appeared in math and physics literature (see e.g. [8, 38] and references therein), and it is isomorphic to the colored HOMFLY-PT homology of the unknot.

The charges of the generators are summarized below

|  | $R$ | $F$ | $j_3$ | $\frac{R}{2} + j_3$ |
|---|---|---|---|---|
| $\phi$ | $r$ | $1$ | $0$ | $\frac{r}{2}$ |
| $\overline{\psi}_+$ | $1-r$ | $-1$ | $\frac{1}{2}$ | $1-\frac{r}{2}$ |
| $\partial_+$ | $0$ | $0$ | $1$ | $1$ |
| $\partial^j \phi$ | $r$ | $1$ | $j$ | $\frac{r}{2}+j$ |
| $\partial^j \overline{\psi}_+$ | $1-r$ | $-1$ | $\frac{1}{2}+j$ | $1-\frac{r}{2}+j$ |

Here $R$ and $F$ are generators of the R-symmetry $U(1)_R$ and the flavor symmetry $U(1)_\beta$, and $j_3$ generates a $U(1)$ subgroup of the $SU(2)$ isometry of the $S^2$. Using this data, one can also obtain the Poincaré polynomial of $\mathcal{H}_{r,f}^{i,j;\ell}$

$$\mathcal{P}_r(q, z, t) = \sum_{i,j,\ell \in \mathbb{Z}} q^i z^j t^\ell \dim \mathcal{H}_{r,f}^{i,j;\ell} = \prod_{j=0}^{\infty} \frac{1 + z^{1-r} t^{-f} q^{1-r/2+j}}{1 - z^r t^f q^{r/2+j}}. \tag{46}$$

One interesting observation of [8] is that, for some simple three-manifolds $M_3$, it is often the case that one can trade the $\ell$-grading for the homological $j$-grading, making the Poincaré polynomial effectively computable. This phenomenon can be seen at the level of $\mathcal{H}_{r,f}$ for a single chiral. Indeed, the $(i, j, \ell)$-degrees of the generators are not independent, and under the substitution of variables $z^r \to t^f$ and $z \to -1$, the Poincaré polynomial for the first two gradings becomes (43),

$$\mathcal{P}_r(q, z, 1) = \prod_{j=0}^{\infty} \frac{1 + z^{1-r} q^{1-r/2+j}}{1 - z^r q^{r/2+j}} \quad \leadsto \quad \mathcal{P}_r(q, -1, t) = \mathcal{I}_r(q, t) = \prod_{j=0}^{\infty} \frac{1 - t^{-f} q^{1-r/2+j}}{1 - t^f q^{r/2+j}}.$$

The homology (44) obtained from the index has the same form as the homology on $D^2 \times_q S^1$ found in [8]. This could be justified by the following argument. The geometry $S^1 \times S^2$ is conformally flat, and the SUSY variation is obtain from that on flat space. The latter, in turn, is equivalent to partially A-twisted 3d $\mathcal{N} = 2$ theory, which is precisely the setting of [8].

## 2.3 $M_3 = L(p, 1)$

### 2.3.1 Refined Superconformal Index

We now move to the case of $M_3 = L(p, 1) \cong S^3/\mathbb{Z}_p$. The theory $T[L(p, 1), U(N)]$ is an $\mathcal{N} = 2$ Chern-Simons theory at level $p$, with a chiral multiplet $\Phi$ in the adjoint representation with R-charge $R(\Phi) = 2$ [35, 39, 40]. Its superconformal index is given by (see e.g. [41])

$$
\mathcal{I}_{U(N)}(q, t) = \sum_{m_1 \geqslant \cdots \geqslant m_N \in \mathbb{Z}} \frac{1}{|W_m|} \int_{|z_i|=1} \prod_j \frac{dz_j}{2\pi i z_j} \prod_i (z_i)^{pm_i}
$$

$$
\prod_{i \neq j}^N t^{-|m_i - m_j|/2} q^{-R|m_i - m_j|/4} \times
$$

$$
\left(1 - q^{|m_i - m_j|/2} \frac{z_i}{z_j}\right) \prod_{i \neq j}^N \frac{\left(\frac{z_j}{z_i} t^{-1} q^{|m_i - m_j|/2 + 1 - R/2}; q\right)_\infty}{\left(\frac{z_i}{z_j} t q^{|m_i - m_j|/2 + R/2}; q\right)_\infty} \times \left[\frac{(t^{-1} q^{1-R/2}; q)_\infty}{(t q^{R/2}; q)_\infty}\right]^N .
$$

$$(47)$$

Here $R$ stands for the R-charge of the adjoint chiral multiplet $\Phi$ and the fugacity $t$ is associated to the $U(1)_\beta$ flavor symmetry which acts on the adjoint chiral multiplet via $\Phi \mapsto e^{i\theta}\Phi$. Using some computer algebra (e.g. Mathematica) one can calculate explicitly $\mathcal{I}(q, t)$ as a series in $q$ up to a relatively high order. The coefficients are polynomials in $t$, that is

$$
\mathcal{I}(q, t) \in \mathbb{Z}[t][[q]].
$$

$$(48)$$

For gauge group $G = SU(2)$, the expression for the index simplifies to

$$
\mathcal{I}_{SU(2)}(q, t) = \frac{1}{2} \sum_{m \in \mathbb{Z}} \int \frac{dz}{2\pi i z} z^{2pm} t^{-2|m|} q^{-2|m|} (1 - z^{\pm 2} q^{|m|})
$$

$$
\times \frac{(1/t; q)_\infty (z^{\pm 2} q^{|m|}/t; q)_\infty}{(qt; q)_\infty (z^{\pm 2} q^{|m|+1} t; q)_\infty}
$$

$$(49)$$

where we use the standard notation

$$f(z^{\pm 2}) \equiv f(z^2)f(z^{-2}).\tag{50}$$

The *unreduced* homological blocks can be calculated using the following formula [8]:

$$\widehat{Z}_a(q,t) = Z_{T[M_3]}(D^2 \times_q S^1; a) =$$

$$\frac{1}{|\mathcal{W}_a|} \frac{1}{(tq;q)_\infty^N} \int_{|z_i|=1} \prod_{i=1}^N \frac{dz_i}{2\pi i z_i} \prod_{i\neq j} \frac{(z_i/z_j;q)_\infty}{(z_i/z_j tq;q)_\infty} \Theta_a^{\mathbb{Z}^N,p}(z;q),\tag{51}$$

where $\Theta_a^{\mathbb{Z}^N;p}(z;q)$ is the theta function of the rank-$N$ lattice $\mathbb{Z}^N$ with quadratic form $p \cdot \mathrm{Id}$:

$$\Theta_a^{\mathbb{Z}^N;p}(z,q) = \sum_{n\in p\mathbb{Z}^N+a} q^{\sum_{i=1}^N n_i^2/2p} \prod_{i=1}^N z_i^{n_i}.\tag{52}$$

We would like to check that the following relation holds:

$$\mathcal{I}_{U(N)}(q,t) = \sum_{a\in\mathbb{Z}_p^N/S_N} |\mathcal{W}_a|\widehat{Z}_a(q,t)\widehat{Z}_a(1/q,1/t).\tag{53}$$

Compared to the case $p = 1$ ($M_3 = S^3$) considered in Sect. 2.2.1, there are now multiple homological blocks. Another technical complication is that the formula (51) only defines $\widehat{Z}_a(q,t)$ for $|q| < 1$, since the theta function is only given in terms of series convergent in $|q| < 1$, with no canonical analytic continuation outside of the unit disk.

This problem was resolved in [9, Sec. 3.2.4]. For now, let us note that in the unrefined case ($t = 1$) such problem does not appear because

$$\widehat{Z}_a(q) = \left.\frac{\widehat{Z}_a(q,t)}{(tq;q)_\infty^N}\right|_{t\to 1} = \frac{1}{|\mathcal{W}_a|} \int_{|z_i|=1} \prod_{i=1}^N \frac{dz_i}{2\pi i z_i} \prod_{i\neq j}(1-z_i/z_j) \Theta_a^{\mathbb{Z}^N,p}(z;q) \quad \in \mathbb{Z}[q]$$

$$\tag{54}$$

is just a polynomial and is obviously well-defined for any $|q| < \infty$. Note, that the factorization of the superconformal index in the unrefined case was essentially checked in [7].

### 2.3.2 Topologically Twisted Index of $T[L(p, 1)]$

For group $G = SU(2)$, the topologically twisted index (refined by angular momentum) of $T[L(p, 1)]$ reads [34]:

$$\mathcal{I}_{\text{top}}(q, t) = \frac{1}{2} \sum_{m \in \mathbb{Z}_{\text{JK}}} \int \frac{dz}{2\pi i z} \frac{z^{2(p+2)m} q^{-m} (1 - z^2 q^m)(1 - z^{-2} q^m)}{(z^2 t q^{1-m}; q)_{2m-1} (z^{-2} t q^{1+m}; q)_{-2m-1} (tq; q)_{-1}},$$

(55)

where, as usual, the Pochhammer symbol with negative integer in the subscript is defined via the following identity:

$$(x; q)_n = \frac{1}{(xq^n; q)_{-n}}.$$

(56)

The contour in (55) is chosen according to the Jeffrey–Kirwan residue prescription. Namely, we either choose poles at $z = 0, z = \pm\sqrt{t} \, q^{\cdots}$ or at $z = \infty, z = \pm 1/\sqrt{t} \, q^{\cdots}$. The result is strikingly simple:

$$\mathcal{I}_{\text{top}}(q, t) = \begin{cases} \frac{1}{(t^2 q^2; q)_{-3}} \equiv (t^2/q; q)_3, & p = 1, \\ 1 - t^4, & p = 2, \\ 1 - t^3, & p \geq 3. \end{cases}$$

(57)

As expected, the result for $p = 1$ is in agreement with the dual description by a free chiral multiplet with R-charge 4. Note that for $p > 1$ the result turns out to be $q$-independent. In fact, for large $p$, the twisted index of $T[L(p, 1); G]$ with a general Lie group $G$ becomes

$$\mathcal{I}_{\text{top}} \xrightarrow{p \gg 1} P_{-t}(G),$$

(58)

where $P_{-t}(G)$ denotes the Poincaré polynomial of $G$ in variable $-t$. For $G = SU(2)$,

$$P_{-t}(G) = 1 - t^3$$

(59)

and, for $SU(3)$, one has [42]:

$$\mathcal{I}_{\text{top}} \xrightarrow{p \gg 1} 1 - t^3 - t^5 + t^8 = P_{-t}(SU(3)).$$

(60)

Additionally, the BPS spectrum can be identified with the cohomology of $G$,

$$\mathcal{H}^{\bullet}_{\text{tw-BPS}} = H^{\bullet}(G).$$

(61)

An argument for this relation mentioned above was essentially given in Section 5 of [43] after Proposition 3 using the language of algebraic stacks, which one can also translate into an argument using low energy effective supersymmetric quantum mechanics.

Although the topological index looks much simpler compared to the superconformal index, it should also admit a factorization into the homological blocks via (29). For $G = SU(2)$, we would like to check

$$\mathcal{I}_{\text{top}}(q, t) = \sum_{a \in \mathbb{Z}_p / \mathbb{Z}_2} |W_a| \widehat{Z}_a(q, t) \widehat{Z}_a(1/q, t). \tag{62}$$

It is easy to see how this works for $p = 1$ ($M_3 = S^3$):

$$\widehat{Z}_0(q, t) \cdot \widehat{Z}_0(1/q, t) = \frac{1}{(t^2 q^2; q)_\infty} \cdot \frac{1}{(t^2 q^{-2}; q^{-1})_\infty} = \frac{(t^2 q^{-1}; q)_\infty}{(t^2 q^2; q)_\infty} = (t^2/q; q)_3. \tag{63}$$

In the case $p > 1$, again, the problem of defining $\widehat{Z}_a(1/q, t)$ arises. It was resolved for both superconformal and topologically twisted index in Section 3.2.4 of [9].

## 2.4  More Examples

In this note, due to limitation of space, we only discussed a few examples of three-manifolds that are the simplest. In [9], we studied several rather big classes of three-manifolds including the total space of a degree-$p$ circle bundle over a Riemann surface, and three-manifolds labeled by plumbing graphs satisfying certain properties.

## 3  Four-Manifolds and Topological Modular Forms

The existence of non-trivial superconformal theories in six dimensions has been one of the major discoveries of the past few decades in string theory. These come in two classes depending on the number of supersymmetries: $(2, 0)$ or $(1, 0)$. The case of $(2, 0)$ has been the most studied one and it comes in ADE-types [44]. The A-type is realized by parallel M5 branes in M-theory [45]. The $(1, 0)$ is far more extensive in variety and a recent classification of them has been proposed in [46–48]. They are related to singularities in elliptic Calabi-Yau threefolds.[18] These theories are interesting in their own right in six dimensions as novel quantum systems which are

---

[18]When the elliptic fibration is trivial this gives the ADE type singularities leading to the special case with enhanced $(2, 0)$ supersymmetry.

decoupled from gravity. Moreover there has been tremendous activity by studying their compactifications down to four and three dimensions to get novel theories even in lower dimensions.

On the other hand $(2,0)$ theories have been used to obtain invariants for manifolds in four and three dimensions [8, 9, 21, 39, 49–53]. Namely one considers topologically twisted theories by embedding an $SU(2)_+$ part of spin connection $SU(2)_+ \times SU(2)_-$ for four-manifolds or the $SU(2)$ holonomy for the three manifold, with an $SU(2)$ subgroup of R-symmetry group which is $SO(5)$, leading to supersymmetric theories in lower dimensions. In particular for the generic four-manifolds this leads to is $(0, 2)$ in two dimensions and for the case of three-manifolds leading to $\mathcal{N} = 2$ theories in 3d. However there are far more six dimensional $(1, 0)$ theories and it is natural to ask what kinds of invariants do they lead to when compactifying them to lower dimensions. Unlike the $(2, 0)$ case, the R-symmetry for these theories is exactly $SU(2)$ so it is the most economical one to allow defining topologically twisted theories for four- and three-manifolds! Twisted compactifications of them on four-manifolds lead to $(0, 1)$ supersymmetric theories in 2d and $\mathcal{N} = 1$ supersymmetric theories in 3d.[19] Topological aspects of the resulting $2d$ and $3d$ theories can be viewed as invariants associated to 4- and three-manifolds. In other words we associate to each three-manifold a 3d $\mathcal{N} = 1$ supersymmetric quantum field theory

$$M_3 \rightsquigarrow T[M_3], \tag{64}$$

where $T$ labels a particular 6d (1,0) Theory. Similarly to each three-manifold it associates a 2d $(0, 1)$ theory,

$$M_4 \rightsquigarrow T[M_4]. \tag{65}$$

Evaluating an elliptic genus [54] of $T[M_4]$ thus should produce an invariant of $M_4$ valued in the ring of modular forms MF. The ordinary elliptic genus is believed to be extendable to the so-called topological Witten genus $\sigma$, valued in the ring $\pi_*$TMF of (stable) homotopy groups of the spectrum of topological modular forms TMF. Thus, naively, we have the following diagram:

$$
\left\{ \begin{array}{c} \text{Spin} \\ \text{4-manifolds} \end{array} \right\} \Big/ \sim \text{diffeomorphism} \xrightarrow{\;T\;} \left\{ (0,1)\ \text{theories} \right\} \Big/ \sim \text{SUSY deformations} \xrightarrow{\;\sigma\;} \pi_*\text{TMF}
$$

$$
\text{elliptic genus} \searrow \qquad \downarrow
$$

$$
\text{MF} \subset \mathbb{Z}[[q]] \tag{66}
$$

---

[19]If it is a Kähler manifold, the supersymmetry of the 2d theory will be enhanced to $(0, 2)$ while for hyper-Kähler manifolds it leads to $(0, 4)$ theories.

However, the 6d $(1, 0)$ theories are typically richer than their $(2, 0)$ counterparts:
They typically come equipped with additional global symmetries. This allows one
to turn on background gauge fields in the flavor group $G$. In the 4d case this amounts
to choosing an instanton background and in 3d case to a flat bundle in $G$.[20] Due
to various technicalities, discussed in detail in the main part of [55], one needs to
refine the statement as follows. Let the flavor symmetry of the 6d theory be $G$, and
the maximal order (modulo—in the multiplicative sense—perfect squares) of the
elements of the defect group [56] be $N_0$. This is related to the analog notion of "spin
structure" for the 6d $(1, 0)$ theory and is needed to define the partition function of
$(1, 0)$ theories on six-manifolds. Fix a subgroup $G' \subset G$ for which we have turned
on the background field on the four-manifold and let

$$G_{2d} := \text{Centralizer}_G(G') \tag{67}$$

be its centralizer subgroup. This corresponds to the group which can be viewed as
the surviving portion of the flavor symmetry group in 2d. Then there is a following
diagram of maps:

$$
\left\{
\begin{array}{c}
\text{Spin} \\
\text{4-manifolds} \\
\text{w/ maps to } BG'
\end{array}
\right\}_{\sim \text{ diffeomorphism, homotopy}}
\xrightarrow{T}
\left\{
\begin{array}{c}
\text{relative} \\
\text{(0,1) theories} \\
\text{w/ symmetry } G_{2d}
\end{array}
\right\}_{\sim \text{ SUSY deformations}}
\xrightarrow{\sigma}
\pi_* \text{TMF}_{G_{2d}}(\Gamma_0(N_0))
$$

$$
\searrow_{\text{equivariant elliptic genus}} \qquad \qquad \downarrow
$$

$$
\text{MF}_{G_{2d}}(\Gamma_0(N_0)) \subset R(G_{2d})[[q]] \tag{68}
$$

where $\text{MF}_{G_{2d}}(\Gamma_0(N_0))$ is the ring of weakly holomorphic $G_{2d}$-equivariant modular
forms with respect to congruence subgroup $\Gamma_0(N_0)$ and $R(G_{2d})$ is the representation
ring of $G_{2d}$. In general, we do not expect the map $\sigma$, the equivariant topological
Witten genus, to be defined for all 2d theories obtained in this way, but only for those
theories whose space of bosonic zero-modes is $G_{2d}$-compact (i.e. fixed loci under
the $G_{2d}$-action are all compact). However, as we will see later, the composition $\sigma \circ T$
is still expected to be defined for (almost) all four-manifolds if a generic map to $BG'$
is chosen. This is because twisting the 2d partition function with flavor symmetry
$G_{2d}$ will get rid of zero modes and renders the path-integral finite.

To get invariants for the three manifolds in this way, the easiest thing to consider
is $M_3 \times S^1$ and turn on instanton flavor background leading again to $(0, 1)$ theory
in 2d for each choice of instanton background.[21]

Finally the main question is how do we actually compute the corresponding
modular forms for arbitrary $M_4$ with some $G$-bundle on it. It turns out that this is

---

[20] We can use abelian subfactors in $G$ and turn on constant fluxes in addition to the above choices.
[21] If $G$ is trivial, there are other ways, discussed in [55], where we can obtain invariants for $M_3$.

not easy, unfortunately. Nevertheless for some special cases of the 6d theory and for some special $M_4$ such as product of a pair of Riemann surfaces, we have managed to compute it. The idea is to use the knowledge of the 4d $\mathcal{N} = 1$ theory obtained for these theories when we compactify from 6d to 4d on a Riemann surface [57–59], and use this to compute the partition function on $T^2$ times another Riemann surface. Part of the difficulty in computing the partition functions is that the 6d $(1, 0)$ theories do not typically have a tangible field theoretic description. However, compactifying them on a circle, and going to 5d, they do seem to have convenient gauge theoretic descriptions. Viewing this circle as one of the two circles of elliptic genus torus $T^2$, we formulate the necessary computation to obtain Witten genus from this 5d perspective. Even though we have not used this picture to do explicit computations, we believe this may hold the key to more general approach to such computations in the future.

From a mathematical perspective, since 6d $(1, 0)$ theories are classified by singularities of elliptic Calabi–Yau threefolds, we can summarize the maps by saying that each singular elliptic Calabi–Yau $ECY$ gives rise to a map from four-manifolds (which of course includes $M_3 \times S^1$) to topological modular forms:

$$ECY : \quad \{\text{4-manifolds}\} \to \pi_*(\text{TMF}).$$

If resulting 6d $(1, 0)$ theory is relative or has flavor symmetries, the target should of course be $\pi_*\text{TMF}_{G_{2d}}(\Gamma_0(N_0))$. As we will explain later, this map can be upgraded into a functor between categories. Although this does not necessarily provide stronger invariants, it incorporates three-manifolds into the story in an interesting way.

In a sense the new invariants we associate to four-manifolds are extensions of Donaldson's invariants: If we reverse the order of compactification and first compactify on $T^2$ and then on the four-manifold, the theory has $\mathcal{N} = 2$ supersymmetry in four dimensions. It differs from the usual twist studied by Witten [60], in that it includes extra degrees of freedom coming from six dimensions. These extra fields would lead to a modular partition function instead of what one has in the case of Donaldson theory. In that sense they have a feature more similar to $\mathcal{N} = 4$ Yang–Mills theory which leads to modular partition functions [49]. On the other hand they categorify the four manifold invariants, in the sense that if we compactify the 6d $(1, 0)$ theory on a four manifold the Hilbert space of supersymmetric states of the $(0, 1)$ theory in 2d together with all the residual flavor symmetries which act on them. Thus we end up with a rich class of invariants for four manifolds which from the physics perspective is rather interesting and one expects them to lead to new mathematical insights in understanding invariants for four-manifolds.

## 3.1   From Six to Two Dimensions, via Four-Manifolds

Given an arbitrary 6d (1,0) superconformal theory, one can consider it on a six-manifold of the form

$$M_4 \times T_\tau^2 \tag{69}$$

where $M_4$ is an oriented Spin four-manifold and $T_\tau^2$ is a flat 2-torus with complex structure parameter $\tau$. We work in the Euclidean signature, so that the Riemannian holonomy group is contained in an $SO(4)_E \times U(1)_E$ subgroup of $SO(6)_E$. Along the four-manifold $M_4$, we perform a partial topological twist to preserve at least one supercharge. The supersymmetry algebra of the 6d theory contains an $SU(2)_R$ R-symmetry. The topological twist is then realized by identifying the $SU(2)_R$ principle bundle with the $SU(2)_+$ factor of $\mathrm{Spin}(4)_E = SU(2)_+ \times SU(2)_-$ bundle, the lift of the $SO(4)_E$ orthonormal tangent frame bundle of $M_4$. After the twist, the supercharges transform as

$$(\mathbf{4}, \mathbf{2}) \qquad \rightarrow \quad ((\mathbf{2}, \mathbf{1})_{+\frac{1}{2}} \oplus (\mathbf{1}, \mathbf{2})_{-\frac{1}{2}}, \mathbf{2}) \quad \rightarrow (\mathbf{1}, \mathbf{1})_{+\frac{1}{2}} \oplus (\mathbf{3}, \mathbf{1})_{+\frac{1}{2}} \oplus (\mathbf{2}, \mathbf{2})_{-\frac{1}{2}}$$

$$\mathrm{Spin}(6)_E \times SU(2)_R \supset \mathrm{Spin}(4)_E \times U(1)_E \times SU(2)_R \supset SU(2)_{\mathrm{diag}} \times SU(2)_- \times U(1)_E. \tag{70}$$

The supercharge in the $(\mathbf{1}, \mathbf{1})_{+\frac{1}{2}}$ representation becomes a scalar on the $M_4$, and therefore defines a globally constant supersymmetry transformation along $M_4$ of general holonomy. By taking the size of $M_4$ to be small compared to the size of $T_\tau^2$, one obtains an effective 2d theory which, by analogy with [39], we shall denote as $T[M_4]$ without making explicit the dependence on the choice of the parent 6d theory. From the viewpoint of this 2d theory, the supercharge in the $(\mathbf{1}, \mathbf{1})_{+\frac{1}{2}}$ representation will be the supercharge of the two-dimensional (0, 1) supersymmetry algebra.

While in general the physical theory $T[M_4]$ may depend on the conformal class of the metric on $M_4$, in this survey we will focus on supersymmetry-protected quantities that are invariant under diffeomorphism of $M_4$ and, therefore, are independent of the choice of the metric. One such invariant that will play a central role in this section is the partition function of the 6d theory on $M_4 \times T_\tau^2$ with an odd Spin structure on $T_\tau^2$. As it is protected by supersymmetry, one indeed expects it to depend only on the diffeomorphism class of $M_4$, which is completely determined by the topology and smooth structure.

In particular, this partition function should not depend on the relative size of $M_4$ and $T_\tau^2$. This allows different interpretations of the 6d partition function:

$$Z_{6d}[M_4 \times T_\tau^2] = Z_{T[M_4]}[T_\tau^2] = Z_{5d}[M_4 \times S^1](q) = Z_{4d}[M_4](\tau). \tag{71}$$

The second quantity in (71) is the elliptic genus of $T[M_4]$,

$$Z_{T[M_4]}[T_\tau^2] := \mathrm{Tr}_R (-1)^F q^{L_0}, \tag{72}$$

given by a trace over the Hilbert space of $T[M_4]$ on a circle in the Ramond sector [54]. Here, as usual, $q = \exp(2\pi i \tau)$. The third quantity in (71) is the partition function of the effective 5d theory obtained by compactification of the 6d theory on a circle with fixed value $q$ of the holonomy of $U(1)$ Kaluza-Klein (KK) graviphoton symmetry along $S^1$. Finally, $Z_{4d}[M_4](\tau)$ is the partition function of the effective 4d theory obtained by further compactifying the 5d theory on a circle, *with all KK modes included*. The 4d theory has $\mathcal{N} = 2$ supersymmetry and is topologically twisted on $M_4$. The 4d R-symmetry is $SU(2)_R$ where $SU(2)_R$ is inherited from the parent 6d theory. Therefore, from the 4d point of view the topological twist is the usual Donaldson–Witten twist performed by identifying the $SU(2)_R$ with the $SU(2)_+$ factor of $\mathrm{Spin}(4)_E = SU(2)_+ \times SU(2)_-$ space-time symmetry. The same conclusion, of course, follows from the fact that 4d $\mathcal{N} = 2$ theories (unlike $\mathcal{N} = 4$) have an essentially unique topological twist on manifolds of generic holonomy.[22]

Generically the 4d theory is not superconformal and does not have extra $U(1)_r$ R-symmetry. Due to the KK modes it has hypers with mass of order $\tau$ which explicitly brake it. The parameter $\tau$ can also appear in 4d theory as a holomorphic gauge coupling constant. In certain special cases one can expect an effective 4d description in terms of a superconformal 4d theory where $\tau$ in (71) plays the role of a holomorphic exactly marginal coupling.

Note that, apart from the usual elliptic genus (72), it is believed that there exist other quantities invariant under SUSY-preserving deformations of 2d $(0, 1)$ theories valued in finite cyclic subgroups of the commutative ring $\pi_* \mathrm{TMF}$ [61, 62]. Here, TMF is the spectrum of a generalized cohomology theory known as "topological modular forms" and $\pi_*$ denotes stable homotopy groups (see e.g. [63]). Roughly speaking, the commutative ring $\pi_* \mathrm{TMF}$ can be understood as an extension of the subring of weakly holomorphic modular forms, where the ordinary elliptic genus takes values, by the ideal of $\pi_* \mathrm{TMF}$ generated by all torsion elements.

Specifically, for each 2d $(0, 1)$ theory with the gravitational anomaly

$$c_R - c_L = \frac{d}{2} \tag{73}$$

there should be an invariant, which we call the *topological Witten genus*,[23] valued in the abelian group $\pi_d \mathrm{TMF}$. The value of the invariant in the free part of $\pi_d \mathrm{TMF}$ coincides with the usual elliptic genus and can be non-zero only when $d \equiv 0 \bmod 4$, but there are in addition torsion-valued invariants. The simplest example is the mod-

---

[22] Note, twisting the $SU(2)_-$ subgroup of $\mathrm{Spin}(4)_E$ is related to this twist by an orientation reversal.

[23] In the mathematical literature, "topological Witten genus" usually refers to the map

$$\Omega_*^{\mathrm{String}} \cong \pi_*(\mathrm{MString}) \to \pi_*(\mathrm{tmf})$$

induced by the String-orientation of tmf: MString → tmf. Our version of the topological Witten genus map reduces to this definition when we take the $(0, 1)$ theory to be the sigma-model with a string manifold as the target.

2 index first mentioned in [54], while the more general torsion-valued invariants are currently understood only in terms of 2d $(0, 1)$ sigma-models with compact target space. We expect that such invariants, much like the usual elliptic genus discussed earlier, all have appropriate counterparts on the four-manifold side. Since they are still associated to an elliptic curve, we expect that on the 5d/4d side one would need to consider certain observables of the same 5d/4d theory or its mild modification (such as e.g. orbifolding). Such observables then should produce torsion-valued invariants of smooth four-manifolds.

### 3.1.1   Flux Compactifications of 6d $(1, 0)$ Theories

For a generic 6d theory and a four-manifold $M_4$, the most basic partition function (71) is often divergent due to presence of non-compact bosonic zero-modes. As we illustrate explicitly in Sect. 3.1.5, these non-compact zero modes originate from 6d bosonic zero modes and then persist in 2d and 5d/4d descriptions as well. Thus, from the 5d/4d perspective, they make the partition function $Z_{5d}[M_4 \times S^1](q) = Z_{4d}[M_4](\tau)$ ill-defined because the moduli spaces of the solutions of the corresponding BPS equations are non-compact. This problem, of course, also appears at the level of the 2d partition function, the elliptic genus (72), because such non-compact zero-modes are present in the 2d theory $T[M_4]$ as well. The presence of non-compact bosonic zero-modes also makes the torsion-valued invariants mentioned earlier ill-defined. And, even when there are no such bosonic zero-modes, there might be fermionic zero-modes which can force that partition function to vanish. However, all such issues can be fixed by turning on non-trivial background fields for global symmetries.

An important feature that distinguishes 6d $(1, 0)$ theories from 6d $(2, 0)$ theories is that the former generically have non-trivial flavor symmetries. Using these symmetries, one can define more general compactifications on $M_4 \times T_\tau^2$ by turning on background vector fields for the global symmetry $G$, while still preserving at least one real supercharge. Similar general backgrounds were considered for compactifications of 6d $(1, 0)$ theories on Riemann surfaces in [58, 59, 64]. On a Riemann surface one can turn on fluxes for an abelian subgroup of the global symmetry $G$ as well as the holonomies consistent with the chosen fluxes. On four-manifolds there is a new feature: namely, apart from turning on fluxes and holonomies, one can also consider configurations with non-trivial instanton number for a non-abelian subgroup of $G$. In order to preserve supersymmetry, the background field should be self-dual. This is the same condition as imposed on dynamical gauge fields of 4d $\mathcal{N} = 2$ theories by supersymmetric localization, except in the present context they are fixed background fields and are not integrated over.

### 3.1.2 Anomaly Polynomial Reduction

Some basic information about the effective 2d theory can be obtained from its anomaly polynomial $I_4$. It can be obtained by integrating the degree-8 anomaly polynomial $I_8$ of the 6d theory over $M_4$. The anomaly polynomials of general 6d theories were studied in [65, 66]. Various explicit examples can be found in [67].

The anomaly polynomial of a generic 6d (1,0) theory with global symmetry $G$ has the following form,

$$I_8 = \alpha c_2(R)^2 + \beta c_2(R) p_1(T) + \gamma p_1(T)^2 + \delta p_2(T) + I_8^{(\text{flavor})}, \tag{74}$$

where $c_2(R)$ is the second Chern class for the R-symmetry bundle, $p_1(T)$ and $p_2(T)$ are Pontryagin classes of the tangent bundle, and $\alpha, \beta, \gamma, \delta$ are real coefficients. In the above formula, we explicitly separated the contribution coming from the 't Hooft anomalies for the global symmetry $G$, possibly mixed with other symmetries,

$$I_8^{(\text{flavor})} = \omega_4^{(1)}(G) + \omega_2^{(1)}(G) c_2(R) + \omega_2^{(2)}(G) p_1(T). \tag{75}$$

Here, $\omega_n^{(i)}(G)$ can be understood as elements of $H^{2n}(BG, \mathbb{Q})$,[24] so that the whole anomaly polynomial is $I_8 \in H^8(BSO \times BG \times BSU(2)_R, \mathbb{Q})$ and defines a Chern–Simons-like invertible TQFT in seven dimensions.

The anomaly polynomial of the effective 2d theory $T[M_4]$ can be obtained by integrating the above mentioned 8-form over a four-manifold. When no background for the flavor symmetry $G$ is turned on, and when the four-manifold $M_4$ is of generic holonomy, the anomaly polynomial of $T[M_4]$ reads

$$I_4 = \frac{c_R - c_L}{24} p_1(T) + \omega_2^{(2d)}(G) \tag{76}$$

with

$$c_R - c_L = 18 \cdot (\beta - 8\gamma - 4\delta)\sigma + 12\beta\chi \tag{77}$$

and

$$\omega_2^{(2d)}(G) = -\frac{2\chi + 3\sigma}{4} \omega_2^{(1)}(G) + 3\sigma\, \omega_2^{(2)}(G) \in H^4(BG, \mathbb{Q}), \tag{78}$$

where $\chi$ and $\sigma$ denote Euler characteristic and signature of $M_4$.

---

[24]When $G = SU(N)$ or $U(N)$ they coincide with Chern classes up to factors, $\omega_n^{(i)} \propto c_n^{(i)}$.

### 3.1.3   Turning on Flavor Symmetry Background

In this section we consider the effect of turning on a non-trivial bundle for a subgroup of the flavor symmetry $G' \subset G$. In particular, as we explain below, in general it modifies the anomaly polynomial of the effective 2d theory. For calculation of anomalies, only the isomorphism class of a $G'$-gauge bundle over $M_4$ matters. Equivalently, the relevant information is given by the homotopy class of the map to the classifying space of $BG'$:

$$\mu : M_4 \longrightarrow BG'. \tag{79}$$

Note that for a $U(1)_i$ subgroup of $G'$ the choice of the homotopy class of the map to $BU(1) = B^2\mathbb{Z}$ is equivalent to the choice of flux $c_1 (U(1)_i) \in H^2(M_4, \mathbb{Z})$. For any simple Lie group $G_j$ inside $G'$ the choice of the map to $BG_j$ in particular involves the choice of the instanton number in $H^4(M_4, \mathbb{Z})$. It is given by the pullback of the free generator of $H^4(BG_j, \mathbb{Z}) \cong \mathbb{Z}$. Note, that supersymmetry requires a positivity condition on the instanton numbers (defined with the proper sign), and unless it is satisfied, the supersymmetry protected quantities, such as the partition function, will vanish.

As in the introduction, let

$$G_{2d} := \text{Centralizer}_{G'}(G) \tag{80}$$

be the centralizer subgroup of $G'$ in $G$. It has a meaning of the flavor symmetry that remains unbroken in the effective 2d theory after turning on a generic background for the $G'$-bundle on $M_4$. Sometimes, we will denote the effective 2d theory as $T[(M_4, \mu)]$ in order to emphasize the dependence on the topological type of a background flavor symmetry bundle. Thus, the compactification of the 6d theory on four-manifolds defines the following map,

$$\begin{aligned} T : \{\text{4-manifolds with } G' \text{ bundles}\} &\longrightarrow \{\text{2d } (0, 1) \text{ theories with symmetry } G_{2d}\} \\ (M_4, \mu) &\longmapsto T[(M_4, \mu)] \end{aligned}. \tag{81}$$

In what follows we determine explicitly the anomaly polynomial of $T[(M_4, \mu)]$. Since $G'$ and $G_{2d} := \text{Centralizer}_G(G')$ are two commuting subgroups of $G$, the multiplication map

$$G_{2d} \times G' \hookrightarrow G \times G \longrightarrow G \tag{82}$$

is a homomorphism and therefore induces a continuous map

$$\phi : BG_{2d} \times BG' \to BG. \tag{83}$$

The maps $\mu$, and $\phi$ can be used to construct the following map reducing cohomological grading by 4:

$$\Phi = \left( \text{id}_{H^*(BG_{2d}, \mathbb{Q})} \otimes \left( \int_{M_4} \circ \mu^* \right) \right) \circ \phi^* : H^*(BG, \mathbb{Q}) \to H^{*-4}(BG_{2d}, \mathbb{Q}) \tag{84}$$

where

$$\mu^* : H^*(BG', \mathbb{Q}) \longrightarrow H^*(M_4, \mathbb{Q}) \tag{85}$$

and

$$\int_{M_4} : H^*(M_4, \mathbb{Q}) \to \mathbb{Q} \tag{86}$$

is the pairing with the fundamental class of $M_4$, supported in degree 4. We will also need the following map preserving cohomological grading:

$$\Psi = \left( \text{id}_{H^*(BG_{2d}, \mathbb{Q})} \otimes \epsilon^* \right) \circ \phi^* : H^*(BG, \mathbb{Q}) \to H^*(BG_{2d}, \mathbb{Q}) \tag{87}$$

where

$$\epsilon : \text{pt} \longrightarrow BG' \tag{88}$$

with the pullback being the projection on the unit in the cohomology ring,

$$\epsilon^* : H^*(BG', \mathbb{Q}) \longrightarrow H^*(\text{pt}, \mathbb{Q}) \cong \mathbb{Q}. \tag{89}$$

The anomaly polynomial of the effective 2d theory on $M_4$ of generic holonomy is then modified to

$$I_4 = \frac{c_R - c_L}{24} p_1(T) + \omega_2^{(2d)}(G_{2d}) \tag{90}$$

with

$$c_R - c_L = 18 \cdot (\beta - 8\gamma - 4\delta)\sigma + 12 \beta \chi + 24 \Phi(\omega_2^{(2)}(G)) \tag{91}$$

and

$$\omega_2^{(2d)}(G_{2d}) = -\frac{2\chi + 3\sigma}{4} \Psi(\omega_2^{(1)}(G)) + 3\sigma \Psi(\omega_2^{(2)}(G)) + \Phi(\omega_4^{(1)}(G)) \in H^4(BG_{2d}, \mathbb{Q}). \tag{92}$$

In a more formal way, the relation between $I_8$ and $I_4$ can be described as follows. Assuming the coefficients of the anomaly polynomials are rational numbers, the anomaly polynomial of the 6d theory can be understood as a $\mathbb{Q}$-valued bordism invariant of Spin manifolds with $SU(2)_R$ and $G$-bundles,

$$I_8 \in \mathrm{Hom}\left(\Omega_8^{\mathrm{Spin}}(BSU(2)_R \times BG), \mathbb{Q}\right), \tag{93}$$

where, as usual, $\Omega_d^{\xi}(X)$ denotes $d$-dimensional bordism group of manifolds equipped with a $\xi$-structure and a map to $X$. The elements of the group are represented by pairs $(M_d, \alpha)$ where $M_d$ is a $d$-manifold equipped with a $\xi$-structure and $\alpha : M_d \to X$. Similarly,

$$I_4 \in \mathrm{Hom}\left(\Omega_4^{\mathrm{Spin}}(BG_{2d}), \mathbb{Q}\right). \tag{94}$$

Then, consider the map

$$\begin{aligned}
\Theta_{(M_4,\mu)} : \Omega_*^{\mathrm{Spin}}(BG_{2d}) &\to \Omega_{*+4}^{\mathrm{Spin}}(BSU(2)_R \times BG) \\
(M_d, \alpha) &\mapsto (M_d \times M_4, \lambda \times \phi(\alpha \times \mu))
\end{aligned} \tag{95}$$

where $\lambda : M_4 \to SU(2)_R$ is the map determined by the topological twisting procedure, that is the projection onto the first component of the Spin structure map $M_4 \to B\mathrm{Spin}(4) = BSU(2)_+ \times BSU(2)_-$. Let $\Theta_{(M_4,\mu)}^*$ denote the induced map

$$\Theta_{(M_4,\mu)}^* : \mathrm{Hom}\left(\Omega_{*+4}^{\mathrm{Spin}}(BSU(2)_R \times BG), \mathbb{Q}\right) \to \mathrm{Hom}\left(\Omega_*^{\mathrm{Spin}}(BG_{2d}), \mathbb{Q}\right). \tag{96}$$

Then, the relation between the two anomaly polynomials can be concisely written as

$$I_4 = \Theta_{(M_4,\mu)}^*(I_8). \tag{97}$$

The anomaly polynomials describe only perturbative 't Hooft anomalies of the respective theories. The explicit relation between non-perturbative anomalies in 6d and 2d can be obtained in a similar way.

One of the nice bonus features of the flavor symmetry background, already mentioned earlier, is that it helps to regularize the partition function (71) which otherwise might be ill-defined due to bosonic zero-modes. When this happens, the extension of the elliptic genus to the value of the topological Witten genus in $\pi_*\mathrm{TMF}$ may also be ill-defined. There is, however, a simple toy example of the map from four-manifolds to $\pi_*\mathrm{TMF}$ that avoids this problem and is completely well-defined, even without flavor symmetry backgrounds.

### 3.1.4  A Toy Model

There is a simple map

$$\mathfrak{T} : \{\text{4-manifolds}\} \longrightarrow \pi_* \text{TMF} \tag{98}$$

where the right-hand side is actually the ordinary, familiar version of the TMF (i.e. non-equivariant and level-1). This map is quite simple and depends only on the topology of $M_4$. It vanishes on many four-manifolds, but at the same time the image contains some non-trivial torsion elements and has simple behavior under the connected sum.

Specifically, the map is given by post-composing the map

$$M_4 \longmapsto \text{2d } (0, 1) \text{ lattice CFT with } \Gamma := H^2(M_4, Z) / \text{Tor } H^2(M_4, Z) \tag{99}$$

with the topological Witten genus map. The $(0, 1)$ lattice SCFT above contains $b_2^-$ left-moving real compact bosons, $b_2^+$ right-moving real compact bosons and their super-partners, which are $b_2^+$ right-moving real fermions. The compact bosons are valued in the $H^2(M_4, \mathbb{R})/\Gamma$ torus with chirality determined by the $\pm 1$ eigenvalue of the Hodge star acting on $H^2(M_4, \mathbb{R})$, viewed as the space of harmonic 2-forms. This is the direct $(0, 1)$ analogue of the $(0, 2)$ lattice CFT considered in [51]. Note that, while the 2d theory depends on the conformal class of the metric on $M_4$, theories associated to homeomorphic four-manifolds can be continuously connected while preserving $(0, 1)$ supersymmetry and, therefore, have the same TMF class. The composition with the topological Witten genus gives a topological modular form of degree determined by $b_2^\pm$ of the manifold,

$$M_4 \longmapsto \mathfrak{T}[M_4] \in \pi_d \text{TMF}, \qquad d = 3b_2^+ - 2b_2^-. \tag{100}$$

Although the map (99) does not arise from any physical 6d theory, it exhibits many qualitative features of the full-fledged map in (68). This is because the 2d lattice SCFT can be understood as a subsector of the 2d theory produced by compactification of a single 6d $(1, 0)$ tensor multiplet, which will be discussed in detail in Sect. 3.1.5. More precisely, this is the sector that arises from the reduction of the self-dual 2-form field, arguably the most non-trivial ingredient of 6d theories! At the same time, unlike 2d theories $T[M_4]$ that arise from compactification of a full 6d SCFTs, the lattice SCFT above has no non-compact bosonic zero-modes and is always an absolute Spin-theory. The absoluteness follows from the fact that the lattice $H^2(M_4, \mathbb{Z})$ is self-dual for closed four-manifolds. So, there is a unique partition function for a given Spin structure on $T_\tau^2$. Another simplification compared to the map $M_4 \mapsto T[M_4]$ is that the definition of the map (99) requires neither Spin structure on $M_4$, nor smooth structure.

In certain cases the lattice SCFT can be given a nice compact supersymmetric sigma-model description. For example, when $M_4 = S^2 \times S^2$, it can be described as a $(0, 1)$ sigma-model with target $S^1$ with an odd Spin structure, cf. [51]. The size of

$S^1$ is given by the ratio of the sizes of $S^2$'s in $M_4$. The topological Witten genus in this case is given by the value of a mod-2 index of the Dirac operator on the target space. The value in this case is non-trivial and is given by the generator

$$\eta \in \pi_1 \text{tmf} \simeq \mathbb{Z}_2 \subset \pi_1 \text{TMF} \simeq \pi_1 \text{tmf}\left[\Delta^{-24}\right]. \tag{101}$$

Another example worth mentioning is $M_4 = \mathbb{CP}^2$. The $(0, 1)$ lattice SCFT in this case can be described in terms of the following free fields: a compact real chiral (right-moving or, equivalently, holomorphic) boson $\phi$ and a real right-moving fermion $\psi_3$. In terms of these fields, the supercharge can be written as $Q = \bar{\partial}\phi\psi_3$. According to the well-known bosonisation, a theory of a free compact real chiral boson is equivalent to a theory of one free complex right-moving fermion $\psi \equiv \psi_1 + i\psi_2$. Moreover, under the bosonisation map, the fields are related in such a way that $\bar{\partial}\phi = \psi_1\psi_2$. Therefore, we arrive at a theory of three right-moving fermions with the supercharge $Q = \psi_1\psi_2\psi_3$. The value of the topological Witten genus for the latter is believed to be,

$$\nu \in \pi_3 \text{tmf} \cong \mathbb{Z}_{24}, \tag{102}$$

where $\nu$ is the generator of the $\mathbb{Z}_{24}$ group.[25] This fact can be understood as follows. The theory of three free real fermions with the supercharge $Q = \psi_1\psi_2\psi_3$ can be interpreted as the $(0, 1)$ supersymmetric $SU(2)$ WZW [68, 69] with zero bosonic level (the total level of the affine $SU(2)$ symmetry differs by $+2$ from the bosonic one). In the UV such theory can be described as an $S^3$ sigma-model with a String-structure given by the generator $[S^3] \in H^3(S^3, \mathbb{Z}) \cong \mathbb{Z}$ or, in the string terminology, one unit of NS-NS flux (see e.g. [70]). Here we use the fact that on oriented $S^3$ the space of String-structures can be canonically identified with $H^3(S^3, \mathbb{Z})$.

Note that the map $\mathfrak{T}$, unlike the full-fledged map $T$ given by twisted compactification of a 6d SCFT, is multiplicative in the $\pi_* \text{TMF}$ ring under the connected sum operation. This is because connected sum operation corresponds to stacking 2d lattice SCFTs up to continuous deformation, i.e.,

$$\mathfrak{T}(M_4 \# M_4') = \mathfrak{T}(M_4) \cdot \mathfrak{T}(M_4'). \tag{103}$$

Note that we have the relations $\eta^3 = 12\nu$, $\eta^4 = 0$, so in particular taking connected sum with $S^2 \times S^2$ is a nilpotent operation of order-4. The same is true for the $\mathbb{CP}^2 \#$ operation, since $\nu^4 = 0$.

---

[25] We thank E. Witten for pointing this out to us.

### 3.1.5 $T[M_4]$ for 6d $(1, 0)$ Hyper, Vector and Tensor Multiplets

In this section we consider the KK reduction of three basic 6d $(1, 0)$ multiplets—tensor, hyper, and vector—on a four-manifold. For the sake of technical simplicity we assume that the homology of $M_4$ has no torsion. Then, compactification of free 6d $(1, 0)$ multiplets on $M_4$ with topological twist described earlier gives the following 2d $(0,1)$ content:

#### 6d $(1, 0)$ Tensor Multiplet

- $(0, 1)$ $\Gamma = H^2(M_4, \mathbb{Z})$ lattice CFT described in detail in Sect. 3.1.4. Note that in case when $H^2(M_4, \mathbb{Z})$ has torsion, the lattice SCFT will be stacked with a 2d TQFT, which is a Tor $H^2(M_4, Z)$-finite-group gauge theory.
- $b_1$ $(0, 1)$ vector multiplets (equivalent to Fermi multiplets on-shell).
- $b_0$ $(0, 1)$ chiral multiplet ($b_0 = 1$ for connected $M_4$).

As one can see, the presence of $b_0$ non-compact chiral multiplets makes the elliptic genus, and, more generally, topological Witten genus ill-defined. On the other hand, if one reduces the tensor multiplet on $T_\tau^2$, it will produce a 4d $\mathcal{N} = 2$ vector multiplet.

For completeness, let us also write the formula for the resulting gravitational anomaly in 2d, which can be obtained by combining the contributions from the fields above,

$$\Delta(c_R - c_L)_{(tensor)} = \frac{3}{2}b_2^+ - b_2^- + \frac{1}{2}b_0 - \frac{1}{2}b_1 = \frac{\chi + 5\sigma}{4}. \tag{104}$$

Equivalently it can be obtained by integration of the 6d anomaly polynomial

$$I_8^{(tensor)} = \frac{c_2(R)^2}{24} + \frac{c_2(R)p_1(T)}{48} + \frac{23p_1(T)^2 - 116p_2(T)}{5760} \tag{105}$$

over $M_4$, as described in Sect. 3.1.2.

#### 6d $(1, 0)$ Hyper Multiplet

- $\sigma > 0 \Rightarrow \sigma/4$ $(0, 1)$ Fermi multiplets,
- $\sigma < 0 \Rightarrow |\sigma|/4$ $(0, 1)$ chiral multiplets.

The field content depends on the sign of the signature of $M_4$, and when $\sigma = 0$, the 2d theory is trivial.

The results above are given by counting harmonic spinors on a four-manifold. Note that, naively, we get $h^\pm$ copies 2d $(0,1)$ Fermi/chiral multiplet, where $h^\pm$ denote the number of chiral/anti-chiral harmonic spinors. However, since each pair of Fermi and chiral multiplets can be given a mass, only the difference $h^+ - h^-$ matters and its value is determined by the index theorem. Note, in this simple example one can see explicitly that, in order to define a map from four-manifolds to 2d $(0, 1)$ theories one needs four-manifolds to be Spin; otherwise, the signature is not divisible by 4 in general. For smooth Spin manifolds, $\sigma \in 16\,\mathbb{Z}$ by Rokhlin theorem, and the formulas above make sense.

Again, for $\sigma < 0$ and without flavor symmetry backgrounds, the presence of $|\sigma|/4$ non-compact chiral multiplets will make the topological Witten genus ill-defined.

Its gravitational anomaly in 2d is

$$\Delta(c_R - c_L)_{(\text{hyper})} = -\frac{\sigma}{8} \tag{106}$$

and can be equivalently derived by integrating the 6d anomaly polynomial

$$I_8^{(\text{hyper})} = \frac{7p_1(T)^2 - 4p_2(T)}{5760} \tag{107}$$

over $M_4$.

## 6d $(1, 0)$ Vector Multiplet

- $b_2^-$ $(0, 1)$ Fermi multiplets.
- $b_1$ $(0, 1)$ compact chiral multiplets. The compactness follows from the fact that the scalar fields are given by holonomies of the vector field on the four-manifold. After taking into account large gauge transformations, they are effectively valued in $H^1(M_4, \mathbb{R})/H^1(M_4, \mathbb{Z}) \cong T^{b_1}$.
- $b_0$ $(0, 1)$ vector multiplets (equivalent to Fermi multiplets on shell).

Unlike the cases of tensor and hyper-multiplets, here one finds no non-compact bosonic zero-modes for any $M_4$. If one reverses the order of compactification, the 6d vector multiplet on $T_\tau^2$ gives a 4d $\mathcal{N} = 2$ vector multiplet, same as in the case of 6d tensor multiplet.

The corresponding gravitational anomaly is

$$\Delta(c_R - c_L)_{(\text{vector})} = -\frac{1}{2}b_2^+ + \frac{1}{2}b_1 - \frac{1}{2}b_0 = -\frac{\chi + \sigma}{4}. \tag{108}$$

Equivalently, it can be obtained by integrating the 6d anomaly polynomial

$$I_8^{(\text{vector})} = -\frac{c_2(R)^2}{24} - \frac{c_2(R)p_1(T)}{48} - \frac{7p_1(T)^2 - 4p_2(T)}{5760} \tag{109}$$

over $M_4$.

### 3.1.6 A Cure for Non-compactness: Equivariant Partition Functions

Consider first non-compact zero modes that originate from 6d hyper multiplets. A nice feature of 6d theories is that they generically have flavor symmetries that act non-trivially on the Higgs branch. Suppose no background flavor symmetry fields are turned on along $M_4$. By computing the partition function on $T_\tau^2 \times M_4$ equivariantly with respect to the unbroken flavor symmetry $G_{2d}$, one can hope that is will become finite. This will indeed be the case if there are no fixed points under the $G_{2d}$-action at the infinite boundary of the Higgs branch. By the equivariant partition function on $T_\tau^2 \times M_4$ we mean the partition function with a non-trivial holonomy of the background $G_{2d}$-gauge fields along the time circle of $T_\tau^2$ turned on. In particular, for each $U(1)_i$ subgroup of $G_{2d}$ this will give a $U(1)$-valued parameter $x_i$ (which can be naturally analytically continued to a $\mathbb{C}^*$-valued parameter). This will modify the partition function of the 2d theory $T[M_4]$ on the torus from (72) to

$$Z_{T[(M_4,\mu)]}[T_\tau^2](\{x_i\}; q) := \text{Tr}_R(-1)^F q^{L_0} \prod_i x_i^{h_i} \tag{110}$$

where $h_i$ denote the weights of the $U(1)_i$-action on the Hilbert space of 2d theory on a circle. Here and in the rest of this survey, the subscript "R" stands for the Ramond sector of the Hilbert space on a circle. When it is well-defined, the right-hand side of (110) gives an element of $\mathbb{Z}[x][[q]]$. Naively, one would expect the result to be a multi-variable weak Jacobi modular form. However, in general this will be spoiled by the fact that the theory $T[M_4]$ should be regarded as a relative theory. We will address this in more detail in Sect. 3.1.7.

The Hilbert space at each $q$-degree carries a representation of $G_{2d}$, and turning on a holonomy $g \in G_{2d}$ gives a $q$-series with coefficients in the ring of characters of $G_{2d}$. Therefore, the equivariant elliptic genus, when well-defined, can be understood as a map

$$\text{EG}_{G_{2d}} : \left\{ \begin{array}{c} \text{2d } (0, 1) \text{ theories} \\ \text{with } G_{2d} \text{ symmetry} \end{array} \right\} \longrightarrow R(G_{2d})[[q]]$$

$$\text{a theory} \longmapsto \text{Tr}_R(-1)^F g\, q^{L_0} = \sum_{R,m} c_{R,m} \chi_R(g) q^m \tag{111}$$

where $R(G)$ is the representation ring of $G_{2d}$, $\chi_R$ are characters of irreducible representations of $G_{2d}$, and each coefficient $c_{R,m} \in \mathbb{Z}_{\geq 0}$ counts the multiplicity of $R$ in the BPS Hilbert space with $q$-degree $m$. Composing it with the map (81) we get:

$$
\left\{ \begin{array}{l} \text{4-manifolds } M_4 \\ \text{with } G'\text{-bundle} \end{array} \right\} \xrightarrow{T} \left\{ \begin{array}{l} \text{2d } (0,1) \text{ theories } T[M_4] \\ \text{with } G_{2d} \text{ symmetry} \end{array} \right\} \xrightarrow{\text{EG}_{G_{2d}}} R(G_{2d})[[q]].
\tag{112}
$$

We propose that the second map can be refined by replacing the equivariant elliptic genus with an appropriately defined *equivariant topological Witten genus* valued in the ring of $G_{2d}$-equivariant topological modular forms. However, in order to make a more precise statement one needs to address the issue of possible relativeness of the 2d theories in the image of the map $T$.

The non-compact zero modes originating from tensor multiplets are more subtle, since the flavor symmetry does not act on them. However, we would like to point out that free tensor multiplets actually are not present in 6d SCFTs, so *a priori* it is not obvious if they would contribute or not. One can hope that similarly to the compactification on 6d $(2,0)$ theories they do not actually contribute to supersymmetric configurations, as was argued in [49].

### 3.1.7  Defect Group, Relativeness and Modularity Level

In this section, we will for a moment ignore the technicalities associated with zero modes and background flavor symmetry bundles , and instead address a different and, in a sense, completely independent technical complication. This complication comes from the fact that many 6d $(1,0)$ SCFTs should be understood as *relative* theories, rather than *absolute* ones (see [71] for a general framework and discussion of such relative theories). A relative theory can be understood as a theory leaving on the boundary of a non-invertible TQFT. The partition function of a $d$-dimensional relative theory on a manifold $M_d$ is not a number, but rather a vector in the Hilbert space of the $d+1$-dimensional TQFT on $M_d$. In the case of a 6d $(1,0)$ theory, the corresponding 7d TQFT is an abelian 3-form Chern–Simons theory with action

$$
\int \sum_{ij} \Omega_{ij} C_i d C_j
\tag{113}
$$

where $\Omega$ is the symmetric Dirac pairing matrix on the charge lattice of self-dual strings $\Lambda_{\text{string}}$. In a way, the relation between 6d theory of self-dual 2-forms to 7d 3-form Chern–Simons theory is analogous to the relation between chiral WZW in 2d and 3d Chern–Simons theory. The essential information about this 7d TQFT is captured by the defect group

$$
\mathcal{C} := \Lambda_{\text{string}}^* / \Lambda_{\text{string}} \cong \text{Coker}\,\Omega.
\tag{114}
$$

It was proposed in [56] that, for a 6d SCFT with an F-theory realization, the defect group can be identified with the first cohomology of the three-dimensional link of the singularity in the base in the conformal limit. Note that the defect group comes equipped with a perfect bilinear pairing

$$\ell k : \mathcal{C} \otimes \mathcal{C} \to \mathbb{Q}/\mathbb{Z} \tag{115}$$

which is inherited from the symmetric Dirac pairing (i.e. intersection form) on $\Lambda_{\text{string}}$. It can also be identified with the linking pairing on the first cohomology of the link of the singularity.

On a general six-manifold $M_6$, $\ell k$ together with the intersection pairing on cohomology defines a non-degenerate antisymmetric form on $H^3(M_6, \mathcal{C})$, and the partition function of the 6d SCFT will be labeled by elements of a Lagrangian subgroup of $H^3(M_6, \mathcal{C})$, which, by definition, is a subgroup maximally isotropic with respect to the pairing. We are interested in the case $M_6 = M_4 \times T_\tau^2$. Suppose, for simplicity, that $M_4$ is simply-connected and fix a basis in $H^1(T_\tau^2, \mathbb{Z})$. Then, there is natural choice of a Lagrangian subgroup isomorphic to $H^2(M_4, \mathcal{C})$. And, the partition function of the 6d theory can be defined as a vector labeled by a discrete flux [56, 72, 73],

$$Z_a^{\text{4d}}[M_4](\tau) := Z_a^{\text{6d}}[M_4 \times T_\tau^2], \qquad a \in H^2(M_4, \mathcal{C}). \tag{116}$$

A simple example of this phenomenon is when the 6d SCFT is a $(2, 0)$ theory of type $A_1$. Then $\mathcal{C} = \mathbb{Z}_2$ and the discrete flux $a \in H^2(M_4, \mathbb{Z}_2)$ can be identified with the 't Hooft flux of $SU(2)$ $\mathcal{N} = 4$ 4d SYM on $M_4$ (the second Stiefel–Whitney class $w_2$ of the corresponding $SU(2)/\mathbb{Z}_2 = SO(3)$ principle bundle).

Under the change of basis on $H^1(T_\tau^2, \mathbb{Z})$ the vector of partition functions transforms as follows (up to an overall extra phase corresponding to gravitational anomaly determined by anomaly polynomials):

$$S : Z_a^{\text{4d}}[M_4](-1/\tau) = \sum_{b \in H^2(M_4, \mathcal{C})} e^{2\pi i \langle a, b \rangle} Z_a^{\text{4d}}[M_4](\tau),$$

$$\tag{117}$$

$$T : Z_a^{\text{4d}}[M_4](\tau + 1) = e^{\pi i \langle a, a \rangle} Z_a^{\text{4d}}[M_4](\tau),$$

where

$$\langle \cdot, \cdot \rangle : H^2(M_4, \mathcal{C}) \otimes H^2(M_4, \mathcal{C}) \to \mathbb{Q}/\mathbb{Z} \tag{118}$$

is defined by composition of intersection form on the second cohomology of $M_4$ with perfect pairing on $\mathcal{C}$ (115). Note that the diagonal pairing $\langle a, a \rangle$ is well-defined modulo $2\mathbb{Z}$ when $M_4$ is Spin.

Since the partition function of the 6d theory on $M_4 \times T_\tau^2$ can be also interpreted as the elliptic genus of the effective 2d theory $T[M_4]$, it follows that the 2d theory is also relative,

$$Z_a^{\text{4d}}[M_4](\tau) = Z_{T[M_4],a}[T_\tau^2] := \text{Tr}_{\mathcal{H}_a^R}(-1)^F q^{L_0}. \tag{119}$$

The corresponding 3d TQFT is the compactification of 7d 3-form Chern–Simons theory on $M_4$ which is the 3d Abelian Chern–Simons theory with (118) being Dirac pairing on the anyons. The relativeness of the 2d $(0, 1)$ theory corresponds to the fact the elliptic genus is not a modular form but rather a vector valued modular form transforming under $SL(2, \mathbb{Z})$ according to (117).

Instead of dealing with vector-valued modular forms one can just consider $Z_0^{4d}[M_4](\tau)$, the partition function with vanishing discrete flux. For example, when the 6d theory is the $(2, 0)$ theory of type $A_1$, this is the partition function of 4d $\mathcal{N} = 4$ SYM on $M_4$ with gauge group $SU(2)$, which is known to be a modular form for the $\Gamma_0(2)$ congruence subgroup of $SL(2, \mathbb{Z})$. In general, $Z_0^{4d}[M_4](\tau)$ is a modular form for $\Gamma_0(N_0)$, where $N_0$ is the smallest positive integer that annihilates $\mathcal{C}$, i.e. $N_0 \cdot a = 0$ for all $a \in \mathcal{C}$. $N_0$ is also the maximal order of elements in $\mathcal{C}$ and can be understood more concretely as follows. The defect group is (non-canonically) isomorphic to a product of finite cyclic groups,

$$\mathcal{C} \cong \prod_i \mathbb{Z}_{p_i}. \tag{120}$$

The perfect pairing (115) on $\mathcal{C}$ is then zero on a pair of elements from two different cyclic factors, while for elements from the same factor it is given by

$$\begin{aligned} \mathbb{Z}_{p_i} \otimes \mathbb{Z}_{p_i} &\longrightarrow \mathbb{Q}/\mathbb{Z} \\ a \otimes b &\longmapsto q_i ab/p_i \mod 1 \end{aligned} \tag{121}$$

where $q_i$ is coprime with $p_i$. Then,

$$N_0 = \mathrm{LCM}(\{p_i\}), \tag{122}$$

is the least common multiple of all $p_i$'s. It follows that $T^{N_0}$ acts trivially on all components of the vector-valued partition function,

$$T^{N_0} : Z_a^{4d}[M_4](\tau) \mapsto Z_a^{4d}[M_4](\tau + N_0) = e^{\pi i N_0 \langle a, a \rangle} Z_a^{4d}[M_4](\tau) = Z_a^{4d}[M_4](\tau). \tag{123}$$

where we used the fact that the intersection pairing on a spin four-manifold in even. The zero-flux partition function $Z_0^{(4d)}[M_4](\tau)$ is then invariant under $T$ and $ST^{N_0}S$, the elements generating $\Gamma_0(N_0) \subset SL(2, \mathbb{Z})$.

To summarize, we get an invariant of $T[M_4]$ under supersymmetry-preserving deformations valued in $\mathrm{MF}(N_0) := \mathrm{MF}(\Gamma_0(N_0))$, the ring of modular forms of level $N_0$. These are modular forms invariant (up to a factor determined by weight) under the $\Gamma_0(N_0)$ congruence subgroup. Much as for absolute 2d $(0, 1)$ theories, where the usual elliptic genus can be refined by the topological elliptic genus valued in $\pi_*\mathrm{TMF}$, for relative $T[M_4]$ we expect to have a topological Witten genus valued in $\pi_*\mathrm{TMF}(N_0)$, where $\mathrm{TMF}(N_0) := \mathrm{TMF}(\Gamma_0(N_0))$ is the spectrum of topological weakly holomorphic modular forms of level $N_0$ (see e.g. [74, 75]).

Composing it with the map $T$, the compactification of a given 6d SCFT on $M_4$, and ignoring the issues with non-compactness we get ("naive version"):

$$
\left\{ \begin{array}{c} \text{Spin} \\ \text{4-manifolds} \end{array} \right\} \xrightarrow{\ T\ } \left\{ \begin{array}{c} \text{relative} \\ \text{2d } (0,1) \text{ theories} \end{array} \right\} \xrightarrow{\ \sigma\ } \pi_* \mathrm{TMF}(N_0)
$$

with $EG$ going to

$$
\mathrm{MF}(N_0) \subset \mathbb{Z}[[q]] \, .
$$

Turning on a non-trivial flavor symmetry background on $M_4$ and replacing the maps by their equivariant version, we arrive at the refined version of the map (112):

$$
\left\{ \begin{array}{c} \text{Spin 4-manifolds} \\ \text{with } G'\text{-bundles} \end{array} \right\} \xrightarrow{\ T\ } \left\{ \begin{array}{c} \text{relative } (0,1) \text{ theories} \\ \text{with } G_{2d} \text{ symmetry} \end{array} \right\} \xrightarrow{\ \sigma\ } \pi_* \mathrm{TMF}_{G_{2d}}(N_0)
$$

with $EG_{G_{2d}}$ going to

$$
\mathrm{MF}_{G_{2d}}(N_0) \subset R(G_{2d})[[q]] \, .
$$

This is the diagram that we have seen in the beginning of this section.

Finally, let us note that theories with the defect group, such that in the decomposition (120)

$$
p_i = k_i^2, \ k_i \in \mathbb{Z}_+ \quad \text{for all } i \tag{124}
$$

can effectively be made absolute by considering a linear combination

$$
\widetilde{Z}^{4d}[M_4](\tau) := \sum_{a \in H^2(M_4, \mathcal{C}')} Z_a^{4d}[M_4](\tau), \tag{125}
$$

where

$$
\mathcal{C}' = \bigoplus_i \mathbb{Z}_{k_i} \tag{126}
$$

is a subgroup of $\mathcal{C}$ on which the induced pairing is trivial. This ensures that $\widetilde{Z}$ above is invariant under the full $SL(2, \mathbb{Z})$.

More generally, let $p_i = p_i' k_i^2$ where $k_i \in \mathbb{Z}$ and $p_i'$ is an integer with no perfect square factors. Then one can redefine

$$
N_0 = \mathrm{LCM}(\{p_i'\}_i) \tag{127}
$$

and construct a modular form w.r.t. $\Gamma_0(N_0)$.

### 3.1.8 Three-Manifolds

The most straightforward way to produce invariants of three-manifolds in this framework is to consider four-manifolds of the form $M_3 \times S^1$. Without any flavor symmetry background, this invariant is not very interesting because the 2d supersymmetry is enhanced to $\mathcal{N} = (1, 1)$. However, for a general flavor symmetry background, that is a map

$$M_3 \times S^1 \to BG' \tag{128}$$

the supersymmetry will still be $(0, 1)$ and the (topological) Witten genus will be non-trivial. In particular, this will be the case when the map above is not homotopic to a product of maps $M_3 \to BG_1$ and $S^1 \to BG_2$, where $G_{1,2}$ are two commuting subgroups of $G'$. Therefore, for a fixed 6d $(1, 0)$ theory with flavor symmetry $G$ one produces invariants of three-manifolds equipped with the map (128) valued in $G_{2d} := \text{Centralizer}_G(G')$-equivariant TMF.

Another way to produce non-trivial invariants is to consider four-manifolds associated to three-manifolds via

$$M_4 = (M_3 \times S^1)\#Z_4 \tag{129}$$

where $Z_4$ is some fixed "canonical" four-manifold, for example $Z_4 = \mathbb{CP}^2$, or $\overline{\mathbb{CP}}^2$, or $S^2 \times S^2$. Then, even with a trivial flavor symmetry background the effective 2d theory will have generically $(0, 1)$ supersymmetry and the corresponding (topological) Witten genus is expected to be non-trivial.

In fact, under mild assumptions, the connected sum defines a commutative algebra inside $\text{TMF}_*$ and each $M_3$ gives rise to a module, which should be a rather powerful invariant for $M_3$. The multiplication in this algebra is almost always different from the ring multiplication, as the latter is realized by taking the disjoint union of four-manifolds, not the connected sum.[26]

Instead of considering invariants of standalone three-manifolds, one can also consider invariants of four-manifolds with three-manifold boundaries. Compactification of a 6d theory on a four-manifold with boundary then produces a 3d effective $\mathcal{N} = 1$ theory with a boundary condition which breaks supersymmetry to $(1, 0)$ in 2d. Different four-manifolds with the same boundary correspond to different boundary conditions in the same 3d theory $T[M_3]$. This is analogous to the setup considered in [39] where compactification of a 6d $(2, 0)$ theory on a four-manifold with boundary gives an effective 3d $\mathcal{N} = 2$ theory with $(0, 2)$ boundary condition. As in [76] and *op. cit.* one can use this point of view to interpret gluing of four-manifolds along a common boundary in terms of "sandwiching" three-dimensional $\mathcal{N} = 1$ theories.

---

[26]In the toy model discussed in Sect. 3.1.4, the two actually coincide, but the toy model doesn't come from any 6d theory. Instead, it is obtained by taking a *subsector* of the 6d free tensor multiplet on $M_4$.

In particular, it means there must exist many non-trivial 2d $(0, 1)$ dualities which correspond to 4d Kirby moves.

Another interesting question is whether the cutting-and-gluing mentioned above is *funtorial*. In other words, given a 6d $(1, 0)$ theory, can one upgrade the map

$$T : \quad \{\text{4-manifolds}\} \rightarrow \pi_*(\text{TMF}) \tag{130}$$

into a functor

$$\mathcal{T} : \quad \text{Cob}_4 \rightarrow \pi_*(\text{TMF}) - \text{mod}, \tag{131}$$

where the left-hand side is the category of smooth four-dimensional cobordisms and the right-hand side is the category of modules over the ring $\pi_*(\text{TMF})$? This is a "4d TQFT over the ring $\pi_*(\text{TMF})$" that associates each three-manifold a $\pi_*(\text{TMF})$-module, each cobordism a map between modules, and each closed four-manifolds an elements in $\pi_*(\text{TMF})$.

## 3.2 Examples

In this section, we give some examples illustrating various aspects of the general program outline in previous sections.

### 3.2.1 $N$ M5-Branes Probing $\mathbb{C}^2/\mathbb{Z}_k$ Singularity

While more general 6d $(1, 0)$ SCFTs can be constructed via F-theory, we will first focus on a two-parameter family that can be realized as the world-volume theory of $N$ M5-branes probing a $\mathbb{R} \times \mathbb{C}^2/\mathbb{Z}_k$ singularity (here $\mathbb{Z}_k$ acts as rotations by opposite phases on $\mathbb{C}^2$). The topologically twisted compactification of the 6d theory on a spin four-manifold $M_4$ then can be realized geometrically in M-theory as follows

$$
\begin{array}{ccc}
N \text{ fivebranes:} & \mathbb{R}^2 \times M_4 \\
& \cap \quad\;\; \cap \\
\text{space-time :} & \mathbb{R}^3 \times X_8 \; .
\end{array}
\tag{132}
$$

$X_8$ here is a local $\text{Spin}(7)$-holonomy space given by the total space of a fibration

$$
\begin{array}{c}
\mathbb{C}^2/\mathbb{Z}_k \rightarrow X_8 \\
\downarrow \\
M_4
\end{array}
\tag{133}
$$

obtained by identifying the $SU(2)_+$ factor in $\text{Spin}(4)_{M_4} = SU(2)_+ \times SU(2)_-$ holonomy group of $M_4$ with the $SU(2)$ isometry group of $\mathbb{C}^2/\mathbb{Z}_k$. The four-manifold $M_4$ then is Cayley cycle in $X_8$.

The geometry involving a spin three-manifold $M_3$ is similar—$M_3$ will now be embedded in a seven-dimensional $G_2$-manifold as an associative cycle, with $\mathbb{Z}_k$ singularity fibered along it by identifying $Spin(3)_{M_3} = SU(2)$ holonomy of $M_3$ with $SU(2)$ isometry of $\mathbb{C}^2/\mathbb{Z}_k$:

$$
\begin{array}{c}
\mathbb{C}^2/\mathbb{Z}_k \rightarrow X_7 \\
\downarrow \\
M_3
\end{array}
\tag{134}
$$

For $k = 1$ such setup which provides a correspondence between three-manifolds and 3d $\mathcal{N} = 2$ theories was considered in [77].

To set up the convention, we now describe more explicitly the twisting procedure. We use $\phi_1, \ldots, \phi_5$ to parametrize the $\mathbb{C}^2 \times \mathbb{R}_5$ transverse space and to denote the corresponding five scalars coming on the M5-brane world-volume. The $\mathbb{R}_5$ direction is not used for topological twist, while the isometry group of $\mathbb{C}^2 = \mathbb{R}^4$ is $SO(4)$ with the double cover being $\text{Spin}(4) = SU(2)_+ \times SU(2)_-$. The two copies of $SU(2)$ act on

$$
\begin{pmatrix} \phi_1 + i\phi_2 & i\phi_3 + \phi_4 \\ i\phi_3 - \phi_4 & \phi_1 - i\phi_2 \end{pmatrix}
\tag{135}
$$

by multiplication of the left and right. And $\mathbb{Z}_k$ is generated by the left-multiplication of

$$
\begin{pmatrix} e^{2\pi i/k} & 0 \\ 0 & e^{-2\pi i/k} \end{pmatrix}.
\tag{136}
$$

For generic values of $k$, the commutant of $\mathbb{Z}_k$ in $\text{Spin}(4)$ is $SU(2)_+$,[27] which is identified with the R-symmetry group of the 6d $(1, 0)$ theory. After the topological twist, $\phi_{1,\ldots,4}$ will transform under a complex two dimensional (real four dimensional) representation of $SU(2)_+$ factor in $\text{Spin}(4)_{M_4}$ (or $SU(2) = \text{Spin}(3)_{M_3}$), while $\phi_5$ will remain a singlet.

Notice that even for $k = 1$ this M-theory setup is different (and generically preserves half as much supersymmetry) from the one usually used to describe 6d $(2,0)$ theories topologically twisted on four- and three-manifolds, where 5-branes wrap coassociative cycle in a $G_2$-manifold and a Lagrangian cycle in a Calabi-Yau three-fold respectively.

---

[27]Notice that for $k > 1$, the commutant of $\mathbb{Z}_k$ in $\text{Spin}(5)$ lives inside the $\text{Spin}(4)$ subgroup, this ensures that the $\mathbb{R}_5$ direction is not needed for the topological twist.

## $(N, k) = (2, 1)$ Theory on $M_4 = \Sigma_1 \times \Sigma_2$

When $k = 1$ the supersymmetry is actually $(2, 0)$. However, effectively one can consider the $(2, 0)$ theory as $(1, 0)$ with flavor symmetry $G = SU(2)_f$. This follows from decomposition of the $(2, 0)$ R-symmetry group as $\mathrm{Spin}(5)_R \supset SU(2)_R \times SU(2)_f$ where $SU(2)_R$ is $(1, 0)$ R-symmetry.

In other words, we consider a 6d $(2, 0)$ theory on $M_4 \times T_\tau^2$, with Donaldson–Witten twist that uses only $SU(2)_R$ subgroup, i.e. the one that can be generalized to any 6d $(1, 0)$ theory. Also, when this twist is applied to $(2, 0)$ the resulting effective 2d theory $T[M_4]$ has the same amount of SUSY as for compactifications of $(1, 0)$, so, in a sense, for this twist there is no qualitative difference between 6d $(2, 0)$ and $(1, 0)$ theories.

Using the general formula for the anomaly polynomial of $(N, k)$ theories, one finds the central charges with trivial $SU(2)_f$ flavor symmetry background on $M_4 = \Sigma_1 \times \Sigma_2$ to be

$$
\begin{aligned}
c_L &= 41(g_1 - 1)(g_2 - 1), \\
c_R &= 42(g_1 - 1)(g_2 - 1).
\end{aligned}
\tag{137}
$$

The case of $M_4 = \Sigma_1 \times \Sigma_2$ can be studied by first compactifying it on $\Sigma_1$ and then computing $\Sigma_2 \times T^2$ index of the effective theory. The effective 4d $\mathcal{N} = 1$ is a particular member of the family of 4d $\mathcal{N} = 1$ theories obtained by wrapping M5-branes on the zero section of $L_p \times L_q$ bundle over $\Sigma_1$ such that $c_1(L_n) = n$ and $p + q = 2 - 2g_1$. This has been considered in the literature and in certain cases there is a Lagrangian description. The case when $p = q = 1 - g_1$ corresponds to the case when the $SU(2)_f$ background on $\Sigma_1$ is trivial. If it also trivial along $\Sigma_2$ it means that $G' = 1$, $G_{2d} = G = SU(2)_f$ in term of the general setup. When $p \neq q$, this corresponds to turning on non-trivial flux of $G' = U(1)_f \subset G = SU(2)_f$ along $\Sigma_1$. If the flavor symmetry background along $\Sigma_2$ remains inside the same subgroup, then the unbroken 2d flavor symmetry group is $G_{2d} = G' = U(1)_f$.

The case of two M5-branes gives the $A_1$ $(2, 0)$ theory after decoupling the center of mass motion. As was mentioned before, one can first compactify the 6d $(2, 0)$ theory on $\Sigma_1$ to get an effective 4d $\mathcal{N} = 1$ theory described, for example, in [57]. Compared to the usual class $\mathcal{S}$ theories, here one has trinions colored by $\pm$. The trinions of the same type are glued by $\mathcal{N} = 2$ vector multiplet and trinions of different types are glued by $\mathcal{N} = 1$ vector multiplets. In general the resulting theory has $U(1)_r$ and $U(1)_f$ non-anomalous flavor symmetry and there is a way to identify the correct IR R-symmetry (i.e. how it mixes with $U(1)_f$). For the twist corresponding to the trivial flux along $\Sigma_1$ there are equal number of $\pm$ trinions in the generalized quiver. In this case $U(1)_f$ enhances to $SU(2)_f$. The case of having different numbers of $\pm$ trinions corresponds to having a non-trivial $U(1)_f$ flux along $\Sigma_1$ given by the difference of these numbers.

Denote the flux of $G' = U(1)_f \subset G = SU(2)_f$ along $\Sigma_i$ as $n_i \in \mathbb{Z}$. Let us first pick

$$n_1 = (g_1 - 1), \qquad n_2 = -(g_2 - 1) \tag{138}$$

For the manifold of this particular type ($M_4 = \Sigma_1 \times \Sigma_2$) the resulting twist (for R- and flavor symmetries together) can be also interpreted as Kapustin-Witten (aka "Langlands") twist of 4d $N = 4$ SYM, i.e. the one where we twist the full $SO(4)$ on $M_4$.

The result for the partition function is the following (Note that the symmetry $g_1 \leftrightarrow g_2$ is a non-trivial self-consistency test since the calculation treats $\Sigma_1$ and $\Sigma_2$ in a very asymmetric fashion):

$$Z^{6d}[\Sigma_1 \times \Sigma_2 \times T_\tau^2] = A(q, v)^{3(g_1-1)(g_2-1)} \in R(G_{2d})[[q]] \tag{139}$$

where

$$A(q, v) := (v^2 - 1/v^2) \cdot (1 - 8q + (26 - 1/v^4 - v^4)q^2 + 8(-6 + 1/v^4 + v^4)q^3 +$$
$$(78 - 27/v^4 - 27v^4)q^4 + 8(-20 + 7/v^4 + 7v^4)q^5 + \ldots) \tag{140}$$

and $v$ is the $U(1)_f = G_{2d}$ fugacity. Note that with the choice of fluxes (138) 2d theory actually has (2,2) symmetry, but turning on $v$ breaks it to (0,2)).

$$\left. \frac{A(q, v)}{(v^2 - 1/v^2)} \right|_{v \to 1} = \eta(q)^8 / \eta(q^2)^4 \tag{141}$$

Consider now a different background:

$$n_i = +(g_i - 1), \quad \text{for } i = 1, 2. \tag{142}$$

The unbroken 2d flavor symmetry is again $G_{2d} = G' = U(1)_f \subset SU(2)_f = G$. The result for the partition function reads

$$Z^{6d}[\Sigma_1 \times \Sigma_2 \times T_\tau^2] = B(q, v)^{(g_1-1)(g_2-1)} \in R(G_{2d})[[q]] \tag{143}$$

where

$$B(q, v) = \frac{(v - 1/v)^3}{(v + 1/v)} \cdot (1 - 12(v^2 + 2 + 1/v^2)q +$$
$$(414 + 75/v^4 + 284/v^2 + 284v^2 + 75v^4)q^2 + O(q^3)). \tag{144}$$

Note that in this case the overall factor

$$\frac{(v - 1/v)^3}{v + 1/v} \tag{145}$$

has infinite series in $v$, unlike for $n_1 = (g_1 - 1)$, $n_2 = -(g_2 - 1)$ case.

Finally consider the case $g_1 = 2$ with zero flux. One possible way to realize the corresponding 4d $\mathcal{N} = 1$ theory is by the following content:

- $\mathcal{N} = 1$ $SU(2)_{i=1,2,3}$ vector multiplets
- Chirals in $(2, 2, 2, 2, +1/2)$ representation of $SU(2)_+ \times SU(2)_- \times SU(2)_3 \times SU(2)_f \times U(1)_r$

Here the normalization of R-charge is such that supercharges have charge 1. The fact that it is half-integer for chirals gives some technical difficulties, so that later we have to consider condition $(g_2 - 1) \in 2\mathbb{Z}$. Note that naively $SU(2)_f$ is part of $U(2)_f = (U(1)_t \times SU(2)_f)/\mathbb{Z}_2$ global symmetry, but its diagonal $U(1)_t$ is anomalous (there are $c_1(U(1)_t)c_2(SU(2)_i)$ terms in the anomaly polynomial). If one tries to calculate $\Sigma_2 \times T_\tau^2$ partition function of this theory by the same method, the following problem arises: the corresponding Bethe equations are degenerate, i.e. have infinite number of solutions.

To circumvent this problem one can choose some deformation. Consider the following way to lift the degeneracy: formally turn on the fugacity corresponding to the anomalous $U(1)_t$. Even though $U(1)_t$ is anomalous symmetry in 2d, it is a valid symmetry of effective quantum mechanics obtained by compactification on $S^1$, even if we keep all KK modes. In particular, it is a valid $U(1)_t$ global symmetry of 3d $\mathcal{N} = 2$ theory obtained by putting the above 4d $\mathcal{N} = 1$ on $S^1$. Mathematically, this means that there is a corresponding grading on the vector space, but not on the chiral algebra (i.e. no corresponding 2d $U(1)_t$ currents). For $g_1 = 2$, $g_2 = 3$ the partition function with this deformation has the following $q$-expansion:

$$Z^{6d}[\Sigma_1 \times \Sigma_2 \times T_\tau^2]/2^8 =$$

$$[729 \cdot \mathbf{1} + t^2(3898 \cdot \mathbf{1} + 3990 \cdot \mathbf{3}) + t^4(11978 \cdot \mathbf{1} + 24495 \cdot \mathbf{3} + 8713 \cdot \mathbf{5}) + \ldots]$$

$$+ q[-(13832 \cdot \mathbf{1} + 11889 \cdot \mathbf{3}) + t^2(4287704 \cdot \mathbf{1} + 6395508 \cdot \mathbf{3} + 2108956 \cdot \mathbf{5}) + \ldots]$$

$$+ q^2[t^{-2}(25250 \cdot \mathbf{1} + 25230 \cdot \mathbf{3}) + \ldots]$$

$$+ \ldots \in R(G_{2d} \times U(1)_t)[[q]] \tag{146}$$

where $\mathbf{d}$ is the character of $G = G_{2d} = SU(2)_f$ representation of dimension $d$.

## $(N, k) = (2, 2)$ Theory on $M_4 = \Sigma_1 \times \Sigma_2$

As in the case of $(N, k) = (2, 1)$ theory, to calculate the partition function of 6d $(N, k) = (2, 2)$ theory on $\Sigma_1 \times \Sigma_2 \times T_\tau^2$ one can first reduce the theory on $\Sigma_1$ and then calculate $\Sigma_2 \times T_\tau^2$ topologically twisted index [78] of the effective 4d $\mathcal{N} = 1$ theory. The effective 4d theory is not Lagrangian per se, but can be constructed from pieces that have Lagrangian description [58]. In particular, such construction involves taking certain couplings to infinity and gauging on a global symmetries which are not present in UV but appear in the IR. Still this description is sufficient for calculation of the index using localization. In other words, one needs to generalize the calculation of $S^3 \times S^1$ superconformal index done in [58] to the case of $\Sigma_2 \times T_\tau^2$ index. The latter case is technically more involved. In particular it requires solving a system of rather complicated algebraic (at finite order in $q$) Bethe ansatz equations. Again, one has to require $(g_2 - 1) \in 2\mathbb{Z}$ in order to apply localization.

For closed $\Sigma_{1,2}$ with no flavor symmetry fluxes the reduction of the anomaly polynomial gives the following formula for the central charges:

$$\begin{aligned} c_L &= 134(g_1 - 1)(g_2 - 1), \\ c_R &= 132(g_1 - 1)(g_2 - 1). \end{aligned} \tag{147}$$

In general we would like to turn on some fluxes along $\Sigma_1 \times \Sigma_2$. The 6d theory has flavor symmetry $SO(7) \supset SU(2) \times SU(2) \times U(1)$ with maximal torus $U(1)_\beta \times U(1)_\gamma \times U(1)_t$ (in the notations of [58]). The effective 4d theory obtained by compactification of the 6d theory on $\Sigma_1$ with fluxes w.r.t. $U(1)_\beta \times U(1)_\gamma \times U(1)_t$ can be again described by gluing together certain trinion theories $T_A^\pm$ and $T_B^\pm$. Each trinion corresponds to a sphere with three ("maximal") punctures. Each puncture breaks $SO(7)$ flavor symmetry down to a certain $SU(2) \times U(1)^2$ subgroup. There are punctures of two different "colors" corresponding to different embeddings. The trinions $T_A^\pm$ have all three punctures of the same color while $T_B^\pm$ have punctures of different color. The theories $T_A^+$ and $T_B^+$ correspond to spheres supporting fluxes (1/4,1/4,1) and (-1/4,1/4,1) respectively. The theories $T_A^-$ and $T_B^-$ correspond to spheres with opposite fluxes and differ by charge conjugation. Each puncture also introduces $SU(2)^2$ global symmetry, so that gluing punctures together corresponds to gauging the diagonal of $SU(2)^2 \times SU(2)^2$ symmetry (after introducing extra matter depending on the type of punctures). The trinion theories $T_A^\pm$ and $T_B^\pm$ can be build from building blocks that have Lagrangian description. The description involves three copies of $SU(2)$ gauge groups. We direct the reader to [58] for the details.

Consider for example genera $g_1 = 2$ and $g_2 = 3$. If we have zero fluxes $(0, 0, 0)$ (i.e. $G' = 1$) along $\Sigma_1$, then one encounters again the problem that the Bethe equations, that arise in calculation of the partition function of the effective 4d $\mathcal{N} = 1$ on $\Sigma_2 \times T_\tau^2$, are degenerate.

If we instead take fluxes $(1/2, 1/2, 2)$ along $\Sigma_1$ (this preserves $SU(2)_{\text{diag}} \times U(1)^2$ flavor symmetry in 4d and can be realized, for example, by gluing two copies of $T_A^+$ theories), the Bethe equations are non-degenerate, and one can get a finite answer when the fugacities $(\beta, \gamma, t)$ of unbroken part of $SO(7)$ are turned on.

Suppose the fluxes along $\Sigma_2$ are $(B, \Gamma, 0)$. For generic $B, \Gamma$ the unbroken symmetry is $G_{2d} = U(1)^3$ but when $B = \Gamma$, it is $G_{2d} = SU(2)_{\text{diag}} \times U(1)^2$, with $SU(2)_{\text{diag}}$ fugacity being $\sqrt{\beta/\gamma}$). Then (by summing over $2^{12}$ solutions of Bethe equations) we get the answer of the following form:

$$Z^{6d}[\Sigma_1 \times \Sigma_2 \times T_\tau^2] =$$

$$t^{16} \frac{(t^4 \gamma^8)^\Gamma (t^4 \beta^8)^B}{(1 - \beta^2 \gamma^2)^8} \times ((\beta\gamma)^4 + O(t))$$

$$+q(4(\beta^8 \gamma^4 - 3\beta^6 \gamma^6 - 2\beta^6 \gamma^2 + \beta^4 \gamma^8 - 8\beta^4 \gamma^4 + \beta^4 - 2\beta^2 \gamma^6 - 3\beta^2 \gamma^2 + \gamma^4) + O(t)) + O(q^2))$$

$$\in R(G_{2d})[[q]] \qquad (148)$$

One can also consider $\Sigma_1$ to be just a basic building block, i.e. $\Sigma_1 = S^2 \setminus 3\text{pt}$, a pair of pants. Suppose that all three punctures are of the same color so that the effective 4d theory is $T_A^+$. The result for the partition function of the 6d theory, depends on $q$ the nome of 2-torus, $t, \beta, \gamma$, the fugacities of $SU(2) \times U(1)^2 \subset SO(7)$ global symmetry, and on $SU(2)^2$ fugacities $u_{1,2}, v_{1,2}, z_{1,2}$ associated to each puncture. In particular we have:

$$Z^{6d}[(S^2 \setminus 3\text{pt}) \times \Sigma_2 \times T_\tau^2](q; u_{1,2}, v_{1,2}, z_{1,2}; t, \beta, \gamma) = \frac{t^8 \beta^2 \gamma^2}{(1 - \beta^2 \gamma^2)} + O(t^{10}) + O(q)$$

$$(149)$$

for $g_2 = 3$ and zero flavor fluxes along $\Sigma_2$.

### 3.2.2 E-String Theory

The calculation of the partition function on $M_4 = \Sigma_1 \times \Sigma_2$ with possible fluxes can be done similarly to the case of $(N, k) = (2, 2)$ theory, by using the results of [59]. Consider for example the case $g_1 = 1$, $g_2 = 2$ with one unit of flux for $G' = U(1) \subset G = E_8$ along $\Sigma_1$. The flux breaks $E_8$ flavor symmetry down to $E_7 \times U(1)$. The compactification on $\Sigma_1$ first produces the 4d $\mathcal{N} = 1$ theory that has Lagrangian description with $SU(2)^2$ gauge symmetry, $SU(8) \times U(1)$ flavor symmetry and the chiral multiplets in the following representations:

| | $SU(2)$ | $SU(2)$ | $SU(8)$ | $U(1)$ |
|---|---|---|---|---|
| $\Phi_1$ | $2$ | $1$ | $8$ | $-1/2$ |
| $\Phi_2$ | $1$ | $2$ | $\overline{8}$ | $-1/2$ |
| $B_{1,2}$ | $2$ | $2$ | $1$ | $+1$ |
| $F_{1,2}$ | $1$ | $1$ | $1$ | $-2$ |

$$(150)$$

There are also the following terms turned on in the superpotential: $\Phi_1\Phi_2 B_1$, $\Phi_1\Phi_2 B_2$, $B_1^2 F_1$, $B_2^2 F_2$ with obvious projections on the trivial representation. The $SU(8)$ flavor symmetry is enhanced to $E_7$ in the IR.

Calculation of the $T_\tau^2 \times \Sigma_2$ topologically twisted index of this theory yields:

$$
Z^{6d}[\Sigma_1 \times \Sigma_2 \times T_\tau^2] = (\mathbf{1}\,t^4 + \mathbf{56}\,t^6 + \ldots)
$$
$$
- (2 \cdot \mathbf{1} + 2 \cdot \mathbf{56}\,t^2 + \ldots)q + (\mathbf{1}\,t^{-4} + \mathbf{56}\,t^2 + \ldots)q^2 + \ldots \in R(G_{2d})[[q]]
$$
$$
(151)
$$

where $t$ is the $U(1)$ flavor fugacity and $\mathbf{d}$ denotes representation of $E_7$ of dimension $d$.

### 3.3 What's Not Included

Due to limitation of space, we have to omit much material, such as the connection to known invariants of four-manifolds, the relation between physics of 2d $(0, 1)$ theories and the theory of topological modular forms, quantization of coefficients of anomaly polynomials, and possible mathematical ways to define the new invariants. We refer the interested readers to the paper [55] for more details.

## References

1. P. Kronheimer, T. Mrowka, *Monopoles and Three-Manifolds*, vol. 10 (Cambridge University Press, Cambridge, 2007)
2. A. Floer, An instanton-invariant for 3-manifolds. Commun. Math. Phys. **118**(2), 215–240 (1988)
3. P. Ozsváth, Z. Szabó, Holomorphic disks and topological invariants for closed three-manifolds. Ann. Math. **159**, 1027–1158 (2004)
4. E. Witten, Quantum field theory and the Jones polynomial. Commun. Math. Phys. **121**, 351–399 (1989)
5. N. Reshetikhin, V.G. Turaev, Invariants of 3-manifolds via link polynomials and quantum groups. Invent. Math. **103**(3), 547–597 (1991)
6. D. Birmingham, M. Blau, M. Rakowski, G. Thompson, Topological field theory. Phys. Rep. **209**, 129–340 (1991)
7. D. Pei, K. Ye, A 3d-3d appetizer. J. High Energy Phys. **2016**, 008 (2016). arXiv:1503.0480
8. S. Gukov, P. Putrov, C. Vafa, Fivebranes and 3-manifold homology (2016). arXiv:1602.0530
9. S. Gukov, D. Pei, P. Putrov, C. Vafa, BPS spectra and 3-manifold invariants (2017). arXiv:1701.0656
10. L. Rozansky, H. Saleur, S and T matrices for the super U(1,1) WZW model: application to surgery and three manifolds invariants based on the Alexander-Conway polynomial. Nucl. Phys. **B389**, 365–423 (1993) hep-th/9203069
11. D. Chang, I. Phillips, L. Rozansky, R matrix approach to quantum superalgebras $su_q(m/n)$. J. Math. Phys. **33**, 3710–3715 (1992). hep-th/9207075

12. L. Rozansky, H. Saleur, Reidemeister torsion, the Alexander polynomial and U(1,1) Chern-Simons Theory. J. Geom. Phys. **13**, 105–123 (1994). hep-th/9209073
13. G. Meng, C.H. Taubes, SW = Milnor torsion. Math. Res. Lett. **3**, 661–674 (1996)
14. L. Crane, I.B. Frenkel, Four-dimensional topological quantum field theory, Hopf categories, and the canonical bases. J. Math. Phys. **35**(10), 5136–5154 (1994). Topology and physics
15. M. Khovanov, A categorification of the Jones polynomial. Duke Math. J. **101**, 359 (1999)
16. M. Khovanov, L. Rozansky, Matrix factorizations and link homology. Fundam. Math. **199**(1), 1–91 (2008)
17. I. Frenkel, C. Stroppel, J. Sussan, Categorifying fractional Euler characteristics, Jones-Wenzl projectors and $3j$-symbols. Quantum Topol. **3**(2), 181–253 (2012)
18. H. Ooguri, C. Vafa, Knot invariants and topological strings. Nucl. Phys. **B577**, 419–438 (2000). hep-th/9912123
19. S. Gukov, A.S. Schwarz, C. Vafa, Khovanov-Rozansky homology and topological strings. Lett. Math. Phys. **74**, 53–74 (2005). hep-th/0412243
20. S. Gukov, Gauge theory and knot homologies. Fortsch. Phys. **55**, 473–490 (2007). arXiv:0706.2369
21. E. Witten, Fivebranes and Knots (2011). arXiv:1101.3216
22. M. Aganagic, S. Shakirov, Knot homology and refined Chern-Simons index. Commun. Math. Phys. **333**(1), 187–228 (2015). arXiv:1105.5117
23. S. Chun, S. Gukov, D. Roggenkamp, Junctions of surface operators and categorification of quantum groups (2015). arXiv:1507.0631
24. S. Nawata, A. Oblomkov, Lectures on knot homology (2015). arXiv:1510.0179
25. S. Gukov, M. Marino, P. Putrov, Resurgence in complex Chern-Simons theory (2016). arXiv:1605.0761
26. S. Cecotti, C. Vafa, Topological antitopological fusion. Nucl. Phys. **B367**, 359–461 (1991)
27. S. Pasquetti, Factorisation of N = 2 theories on the squashed 3-sphere. J. High Energy Phys. **04**, 120 (2012). arXiv:1111.6905
28. C. Beem, T. Dimofte, S. Pasquetti, Holomorphic blocks in three dimensions. J. High Energy Phys. **12**, 177 (2014). arXiv:1211.1986
29. S. Cecotti, D. Gaiotto, C. Vafa, $tt^*$ geometry in 3 and 4 dimensions. J. High Energy Phys. **05**, 055 (2014). arXiv:1312.1008
30. K. Hikami, Decomposition of Witten–Reshetikhin–Turaev invariant: linking pairing and modular forms. Chern-Simons Gauge Theory **20**, 131–151 (2011)
31. M. Atiyah, On framings of 3-manifolds. Topology **29**(1), 1–7 (1990)
32. Y. Yoshida, K. Sugiyama, Localization of 3d $\mathcal{N} = 2$ supersymmetric theories on $S^1 \times D^2$ (2014). arXiv:1409.6713
33. M. Blau, G. Thompson, Chern-Simons theory with complex gauge group on Seifert fibred 3-manifolds (2016). arXiv:1603.0114
34. F. Benini, A. Zaffaroni, A topologically twisted index for three-dimensional supersymmetric theories. J. High Energy Phys. **07**, 127 (2015). arXiv:1504.0369
35. S. Gukov, D. Pei, Equivariant Verlinde formula from fivebranes and vortices. arXiv:1501.0131
36. D. Jafferis, X. Yin, A Duality Appetizer (2011). arXiv:1103.5700
37. A. Kapustin, H. Kim, J. Park, Dualities for 3d theories with tensor matter. J. High Energy Phys. **12**, 087 (2011). arXiv:1110.2547
38. E. Gorsky, S. Gukov, M. Stosic, Quadruply-graded colored homology of knots (2013). arXiv:1304.3481
39. A. Gadde, S. Gukov, P. Putrov, Fivebranes and 4-manifolds (2013). arXiv:1306.4320
40. H.-J. Chung, T. Dimofte, S. Gukov, P. Sułkowski, 3d-3d correspondence revisited. J. High Energy Phys. **04**, 140 (2016). arXiv:1405.3663
41. Y. Imamura, S. Yokoyama, Index for three dimensional superconformal field theories with general R-charge assignments. J. High Energy Phys. **04**, 007 (2011). arXiv:1101.0557
42. S. Gukov, D. Pei, W. Yan, K. Ye, Equivariant Verlinde algebra from superconformal index and Argyres-Seiberg duality (2016). arXiv:1605.0652

43. J.E. Andersen, S. Gukov, D. Pei, The Verlinde formula for Higgs bundles (2016). arXiv:1608.0176

44. E. Witten, Some comments on string dynamics, in *Future Perspectives in String Theory. Proceedings, Conference, Strings'95, Los Angeles, March 13–18* (1995), pp. 501–523. hep--th/9507121

45. A. Strominger, Open p-branes. Phys. Lett. **B383**, 44–47 (1996). hep-th/9512059, 116(1995)

46. J.J. Heckman, D.R. Morrison, C. Vafa, On the classification of 6D SCFTs and generalized ADE orbifolds. J. High Energy Phys. **05**, 028 (2014). arXiv:1312.5746. Erratum: JHEP06,017(2015)

47. L. Bhardwaj, Classification of 6d $\mathcal{N} = (1, 0)$ gauge theories. J. High Energy Phys. **11**, 002 (2015). arXiv:1502.0659

48. J.J. Heckman, D.R. Morrison, T. Rudelius, C. Vafa, Atomic classification of 6D SCFTs. Fortsch. Phys. **63**, 468–530 (2015). arXiv:1502.0540

49. C. Vafa, E. Witten, A strong coupling test of S-duality. Nucl. Phys. **B431**, 3–77 (1994). hep-th/9408074

50. A. Kapustin, E. Witten, Electric-magnetic duality and the geometric langlands program. Commun. Num. Theor. Phys. **1**, 1–236 (2007). hep-th/0604151

51. M. Dedushenko, S. Gukov, P. Putrov, Vertex algebras and 4-manifold invariants (2017). arXiv:1705.0164

52. T. Dimofte, D. Gaiotto, S. Gukov, Gauge theories labelled by three-manifolds. Commun. Math. Phys. **325**, 367–419 (2014). arXiv:1108.4389

53. T. Dimofte, D. Gaiotto, S. Gukov, 3-Manifolds and 3d indices. Adv. Theor. Math. Phys. **17**(5), 975–1076 (2013). arXiv:1112.5179

54. E. Witten, Elliptic genera and quantum field theory. Commun. Math. Phys. **109**, 525 (1987)

55. S. Gukov, D. Pei, P. Putrov, C. Vafa, 4-manifolds and topological modular forms (2018). arXiv:1811.0788

56. M. Del Zotto, J.J. Heckman, D.S. Park, T. Rudelius, On the defect group of a 6D SCFT. Lett. Math. Phys. **106**(6), 765–786 (2016). arXiv:1503.0480

57. I. Bah, C. Beem, N. Bobev, B. Wecht, Four-dimensional SCFTs from M5-Branes. J. High Energy Phys. **06**, 005 (2012). arXiv:1203.0303

58. S.S. Razamat, C. Vafa, G. Zafrir, 4d $\mathcal{N} = 1$ from 6d (1, 0). J. High Energy Phys. **04**, 064 (2017). arXiv:1610.0917

59. H.-C. Kim, S.S. Razamat, C. Vafa, G. Zafrir, E-string theory on Riemann surfaces. Fortsch. Phys. **66**(1), 1700074 (2018). arXiv:1709.0249

60. E. Witten, Topological quantum field theory. Commun. Math. Phys. **117**, 353 (1988)

61. S. Stolz, P. Teichner, *What Is an Elliptic Object?*. London Mathematical Society Lecture Note Series (Cambridge University Press, Cambridge, 2004), pp. 247–343

62. S. Stolz, P. Teichner, Supersymmetric field theories and generalized cohomology (2011). e-prints. arXiv:1108.0189

63. M.J. Hopkins, Topological modular forms, the Witten genus, and the theorem of the cube, in *Proceedings of ICM, 1994, Birkhauser, Zurich* (1995), pp. 554–565

64. H.-C. Kim, S.S. Razamat, C. Vafa, G. Zafrir, Compactifications of ADE conformal matter on a torus (2018). arXiv:1806.0762

65. K. Ohmori, H. Shimizu, Y. Tachikawa, K. Yonekura, Anomaly polynomial of general 6d SCFTs. Prog. Theor. Exp. Phys. **2014**(10), 103B07 (2014). arXiv:1408.5572

66. C. Cordova, T.T. Dumitrescu, K. Intriligator, Anomalies, renormalization group flows, and the a-theorem in six-dimensional (1, 0) theories. J. High Energy Phys. **10**, 080 (2016). arXiv:1506.0380

67. K. Ohmori, H. Shimizu, Y. Tachikawa, K. Yonekura, 6d $\mathcal{N} = (1, 0)$ theories on $T^2$ and class S theories: Part I. J. High Energy Phys. **07**, 014 (2015). arXiv:1503.0621

68. P.D. Vecchia, V. Knizhnik, J. Petersen, P. Rossi, A supersymmetric Wess-Zumino lagrangian in two dimensions. Nucl. Phys. B **253**, 701–726 (1985)

69. Y. Kazama, H. Suzuki, New N=2 superconformal field theories and superstring compactification. Nucl. Phys. **B321**, 232–268 (1989)

70. V. Braun, S. Schafer-Nameki, Supersymmetric WZW models and twisted K-theory of SO(3). Adv. Theor. Math. Phys. **12**(2), 217–242 (2008). hep-th/0403287
71. D.S. Freed, C. Teleman, Relative quantum field theory. Commun. Math. Phys. **326**, 459–476 (2014). arXiv:1212.1692
72. Y. Tachikawa, On the 6d origin of discrete additional data of 4d gauge theories. J. High Energy Phys. **05**, 020 (2014). arXiv:1309.0697
73. E. Witten, Geometric Langlands From Six Dimensions (2009). arXiv:0905.2720
74. M. Mahowald, C. Rezk, Topological modular forms of level 3. Pure Appl. Math. Q. **5**(2), 853–872 (2009)
75. M. Hill, T. Lawson, Topological modular forms with level structure. Invent. Math. **203**(2), 359–416 (2016)
76. B. Feigin, S. Gukov, VOA[$M_4$] (2018). arXiv:1806.0247
77. J. Eckhard, S. Schafer-Nameki, J.-M. Wong, An $\mathcal{N} = 1$ 3d-3d correspondence. *J. High Energy Phys.* **07**, 052 (2018). arXiv:1804.0236
78. F. Benini, A. Zaffaroni, Supersymmetric partition functions on Riemann surfaces. Proc. Symp. Pure Math. **96**, 13–46 (2017). arXiv:1605.0612

# Introduction to the Theory of Elliptic Hypergeometric Integrals

**Vyacheslav P. Spiridonov**

## Contents

1  Introduction.................................................................... 272
2  Elliptic Hypergeometric Integrals ............................................ 273
3  Properties of the Elliptic Gamma Function ................................... 277
4  The Elliptic Beta Integral .................................................... 282
5  An Elliptic Extension of the Euler–Gauss Hypergeometric Function ....................... 286
6  Multiple Elliptic Hypergeometric Integrals................................... 293
7  Rarefied Elliptic Hypergeometric Integrals .................................. 296
8  An Integral Bailey Lemma.................................................... 300
9  Connection with Four Dimensional Superconformal Indices ............................ 307
References ...................................................................... 315

**Abstract** We give a brief account of the key properties of elliptic hypergeometric integrals—a relatively recently discovered top class of transcendental special functions of hypergeometric type. In particular, we describe an elliptic generalization of Euler's and Selberg's beta integrals, elliptic analogue of the Euler–Gauss hypergeometric function and some multivariable elliptic hypergeometric functions on root systems. The elliptic Fourier transformation and corresponding integral Bailey lemma technique is outlined together with a connection to the star-triangle relation and Coxeter relations for a permutation group. We review also the interpretation of elliptic hypergeometric integrals as superconformal indices of four dimensional supersymmetric quantum field theories and corresponding applications to Seiberg type dualities.

V. P. Spiridonov (✉)
Laboratory of Theoretical Physics, Joint Institute for Nuclear Research, Dubna, Moscow Region, Russia

National Research University Higher School of Economics, Moscow, Russia

© Springer Nature Switzerland AG 2020
V. A. Gritsenko, V. P. Spiridonov (eds.), *Partition Functions and Automorphic Forms*, Moscow Lectures 5,
https://doi.org/10.1007/978-3-030-42400-8_6

# 1  Introduction

The Euler–Gauss hypergeometric function [1] is one of the most useful classical special functions. Its most popular definition is given by the $_2F_1$-series:

$$F(a, b; c; x) := {}_2F_1(a, b; c; x) = \sum_{n=0}^{\infty} \frac{(a)_n (b)_n}{n!(c)_n} x^n, \qquad |x| < 1, \tag{1.1}$$

where $(a)_n = a(a+1)\ldots(a+n-1)$ is the Pochhammer symbol. Alternatively, it can be defined by the Euler integral representation

$$F(a, b; c; x) = \frac{\Gamma(c)}{\Gamma(c-b)\Gamma(b)} \int_0^1 t^{b-1}(1-t)^{c-b-1}(1-xt)^{-a} dt, \tag{1.2}$$

where $\mathrm{Re}(c) > \mathrm{Re}(b) > 0$ and $x \notin [1, \infty[$, or the Barnes integral representation

$$F(a, b; c; x) = \frac{\Gamma(c)}{\Gamma(a)\Gamma(b)} \int_{-i\infty}^{i\infty} \frac{\Gamma(a+u)\Gamma(b+u)\Gamma(-u)}{\Gamma(c+u)}(-x)^u du, \tag{1.3}$$

where $\Gamma(x)$ is the Euler gamma function

$$\Gamma(x) = \int_0^\infty t^{x-1} e^{-t} dt, \qquad \mathrm{Re}(x) > 0.$$

In (1.3) the poles of the integrand $u = -a - k, -b - k$ and $u = k, k \in \mathbb{Z}_{\geq 0}$, are separated by the integration contour.

The function $F(a, b; c; x)$ satisfies a special differential equation called the hypergeometric equation:

$$x(1-x)y''(x) + (c - (a+b+1)x)y'(x) - ab y(x) = 0, \tag{1.4}$$

determining the solution analytic around the regular singular point $x = 0$. This is the second order differential equation with three regular singularities fixed at $x = 0, 1, \infty$ by linear fractional transformation.

For $x = 1$ the value of function $F(a, b; c; 1)$ (1.2) can be computed due to the explicit evaluation of Euler's beta integral:

$$\int_0^1 t^{x-1}(1-t)^{y-1} dt = \frac{\Gamma(x)\Gamma(y)}{\Gamma(x+y)}, \qquad \mathrm{Re}(x), \mathrm{Re}(y) > 0. \tag{1.5}$$

Restrictions on the parameters indicated above lead to well defined functions, they may be relaxed by analytic continuation.

All these exact formulas and related ones were generalized in many different ways. We mention the most essential developments:

- extension to higher order hypergeometric functions $_{n+1}F_n$,
- $q$-deformation of plain hypergeometric functions,
- extension of univariate to multivariable special functions,
- elliptic deformation of all above functions.

It is the last step which will be our main subject in these notes. It represents a relatively recent development in the theory of special functions with the basic results obtained around 2000. For describing the most general elliptic hypergeometric functions one has to use integral representations [49], since the infinite series of the corresponding type are not well defined. Note however, that the first examples of elliptic hypergeometric functions emerged in the terminating series form as particular elliptic function solutions of the Yang–Baxter equation [20] which were constructed in a case-by-case manner in [9].

The most interesting elliptic hypergeometric integrals are associated with two independent root systems related in a remarkable way to supersymmetric quantum field theories, where these integrals emerge as superconformal indices [38]. The first root system determines their structure as matrix integrals over the Haar measure of a particular compact Lie group (the gauge group in field theory), and the second one is related to a Lie group of symmetry transformations of functions in parameters (the flavor group in field theory). There are many exact relations between such integrals, a large number of which are still in a conjectural form.

We shall not try to cover all aspects of the theory, but consider some introductory material at the elementary level and give a brief review of more recent developments. There are other surveys on this subject [41, 42, 55], where some of the skipped topics are discussed. A deep algebraic geometry point of view on the functions of interest is given in [35].

## 2  Elliptic Hypergeometric Integrals

The very first basic example of elliptic hypergeometric integrals was discovered in [49]. Let us start from the conceptual definition of such integrals introduced in [51]. For simplicity we limit its consideration only to the univariate case.

The key property of the univariate elliptic hypergeometric integrals is that they are defined as contour integrals

$$I := \int_C \Delta(u)du,$$

whose kernel $\Delta(u)$ satisfies a first order finite difference equation

$$\Delta(u + \omega_1) = f(u; \omega_2, \omega_3)\Delta(u),  \tag{2.1}$$

where the coefficient $f(u; \omega_2, \omega_3)$ is an elliptic function with periods $\omega_2$ and $\omega_3$, and $\omega_{1,2,3}$ are some incommensurate complex numbers. Incommensurability means that $\sum_{k=1}^{3} n_k \omega_k \neq 0$ for $n_k \in \mathbb{Z}$.

Elliptic functions form a particular beautiful family of special functions [2]. Let us remind that they are defined as the meromorphic doubly-periodic functions:

$$f(u + \omega_2) = f(u + \omega_3) = f(u), \quad \text{Im}(\omega_2/\omega_3) \neq 0.$$

Consider their general structure before discussing solutions of the defining equation (2.1). For that we need an infinite product

$$(z; p)_\infty := \prod_{j=0}^{\infty}(1 - zp^j), \quad |p| < 1, \, z \in \mathbb{C}.$$

With its help we define a Jacobi theta function as

$$\theta(z; p) := (z; p)_\infty (pz^{-1}; p)_\infty, \quad z \in \mathbb{C}^\times.$$

It has important symmetry properties:

$$\theta(pz; p) = \theta(z^{-1}; p) = -z^{-1}\theta(z; p).  \tag{2.2}$$

Using the Jacobi triple product identity one can write the Laurent series expansion

$$\theta(z; p) = \frac{1}{(p; p)_\infty} \sum_{k \in \mathbb{Z}}(-1)^k p^{k(k-1)/2} z^k.  \tag{2.3}$$

For convenience we provide the standard odd Jacobi theta function definition:

$$\theta_1(u|\tau) = -\theta_{11}(u) = -\sum_{k \in \mathbb{Z}} e^{\pi i \tau(k+1/2)^2} e^{2\pi i(k+1/2)(u+1/2)}  \tag{2.4}$$

$$= ip^{1/8}e^{-\pi i u}(p; p)_\infty \theta(e^{2\pi i u}; p), \quad p = e^{2\pi i \tau}.$$

For interested readers we suggest small calculational tasks like the following one. *Exercise:* find zeros of $\theta(z; p)$, verify (2.2), deduce the general quasiperiodicity relation for $\theta(p^k z; p)$, $k \in \mathbb{Z}$, and prove identity (2.3). The latter problem can be solved by computing the $z$-series expansion for the symmetric finite product $\prod_{k=1}^{n}(1 - zp^{k-1/2})(1 - z^{-1}p^{k-1/2})$ (using the $z \to pz$ functional equation for that) and taking the limit $n \to \infty$.

There is nice factorized representation of the elliptic functions in terms of the Jacobi theta functions. Denote

$$p := e^{2\pi i \omega_3/\omega_2}, \quad z := e^{2\pi i u/\omega_2}.$$

Then, according to the theorem established by Abel and Jacobi, one can write up to a multiplicative constant,

$$h(z; p) := f(u; \omega_2, \omega_3) = \prod_{k=1}^{m} \frac{\theta(t_k z; p)}{\theta(w_k z; p)}, \qquad \prod_{k=1}^{m} t_k = \prod_{k=1}^{m} w_k. \qquad (2.5)$$

Indeed, the periodicity $f(u + \omega_2) = f(u)$ of this meromorphic function is evident. Since the shift $u \to u + \omega_3$ is equivalent to $z \to pz$, one has $f(u + \omega_3)/f(u) = h(pz)/h(z) = \prod_{k=1}^{m}(w_k/t_k) = 1$. So, we have an elliptic function with $m$ poles (or zeros) in the fundamental parallelogram of periods $(\omega_2, \omega_3)$. Vice versa, given an elliptic function with $m$ poles and zeros at the points fixed by parameters $t_k$ and $w_k$, we can divide it by (2.5) and see that the resulting function is bounded on $\mathbb{C}$ (it is doubly periodic and has no poles), and by the Liouville theorem it is constant. The parameter $m$ is called the order of elliptic functions and we call the linear constraint $\prod_{k=1}^{m} t_k = \prod_{k=1}^{m} w_k$ the balancing condition (it explains the origin of the old notion of balancing in the theory of hypergeometric functions [21, 50]).

It is convenient to use compact notation

$$\theta(a_1, \ldots, a_k; p) := \theta(a_1; p) \cdots \theta(a_k; p), \quad \theta(at^{\pm 1}; p) := \theta(at; p)\theta(at^{-1}; p).$$

Then the "addition" formula for theta functions takes the form

$$\theta(xw^{\pm 1}, yz^{\pm 1}; p) - \theta(xz^{\pm 1}, yw^{\pm 1}; p) = yw^{-1}\theta(xy^{\pm 1}, wz^{\pm 1}; p). \qquad (2.6)$$

The proof of this relation is rather easy. The ratio of the left- and right-hand sides satisfies equation $h(px) = h(x)$ (i.e., it is $p$-elliptic) and represents a bounded function of the variable $x \in \mathbb{C}^{\times}$. Therefore it does not depend on $x$ according to the Liouville theorem, but for $x = w$ the equality is evident.

Let us turn now to the elliptic hypergeometric integrals. In terms of the multiplicative coordinate $z = e^{2\pi i u/\omega_2}$ elliptic functions are determined by the equation $h(pz) = h(z)$. Now we demand that the integrand $\Delta(u) =: \rho(z)$ is a meromorphic function of $z \in \mathbb{C}^{\times}$, which is an additional strong restriction. Then it is convenient to introduce a second base variable $q := e^{2\pi i \omega_1/\omega_2}$, so that the shift $u \to u + \omega_1$ becomes equivalent to the multiplication $z \to qz$. Changing the integration variable, we come to the following definition of elliptic hypergeometric integrals:

$$I_{EHI} = \int \rho(z)\frac{dz}{z}, \qquad \rho(qz) = h(z; p)\rho(z), \quad h(pz) = h(z), \qquad (2.7)$$

where the explicit form of $h(z; p)$ is given in (2.5).

Because of the factorized form of $h(z; p)$, for solving the equation $\rho(qz) = h(z; p)\rho(z)$ it is sufficient to solve the linear first order $q$-difference equation with a simple theta function coefficient

$$\gamma(qz) = \theta(z; p)\gamma(z). \qquad (2.8)$$

One can check that a particular solution of (2.8) is given by the function

$$\gamma(z) = \Gamma(z; p, q) := \prod_{j,k=0}^{\infty} \frac{1 - z^{-1}p^{j+1}q^{k+1}}{1 - zp^j q^k}, \qquad |p|, |q| < 1, \qquad (2.9)$$

which is called the (standard) elliptic gamma function. Note that the original Eq. (2.8) does not impose the constraint $|q| < 1$, whereas in (2.9) we have such an additional restriction.

The problem of generalizing the Euler gamma function was considered by Barnes, who defined the multiple gamma functions of arbitrary order [3], and Jackson [23], who introduced the basic versions of gamma functions. Although the function (2.9) is related to their considerations, its usefulness was established only in modern time after the work of Ruijsenaars [43], where the term "elliptic gamma function" was introduced. A further systematic investigation of this function was performed by Felder and Varchenko [18] who discovered its $SL(3, \mathbb{Z})$ symmetry transformations (they also pointed out that this function appeared implicitly already in Baxter's work on the eight vertex model [4]). In [51] the author constructed the modified elliptic gamma function, which gives a solution of Eq. (2.8) in the regime $|q| = 1$ (it is meromorphic in $u \propto \log z$, not $z$). It will be described in the next section.

*Exercise*: derive the solution (2.9) from scratch by iterations using the factorized form of $\theta(z; p)$.

Changing the variable $z \to t_k z, w_k z$ in (2.8), we find solutions of the equation defining $\rho(z)$ for each theta function factor in $h(z; p)$. The final result is evident now: the general univariate elliptic hypergeometric integral has the form

$$I_{EHI}(\underline{t}, \underline{w}; p, q) = \int \prod_{k=1}^{m} \frac{\Gamma(t_k z; p, q)}{\Gamma(w_k z; p, q)} \frac{dz}{z}, \qquad \prod_{k=1}^{m} t_k = \prod_{k=1}^{m} w_k, \qquad (2.10)$$

where one has to specify the contour of integration. The typical choice is a closed contour encircling the essential singularity point $z = 0$, e.g. the unit circle $\mathbb{T}$. Surprisingly, these functions generalize all previously known univariate ordinary and $q$-hypergeometric functions. They depend on $2m$ complex variables $t_k$ and $w_k$ subject to one constraint. We shall not describe explicitly the limits to lower level hypergeometric objects, but only indicate how it can be done. Note that for $0 < |p|, |q| < 1$ it is not possible to simplify functions (2.10) by taking parameters to zero or infinity. Therefore all well defined degenerations require limits to the boundary values of bases.

*Exercise:* investigate the $p \to 0$ limit of (2.10) for fixed parameters and when some of the parameters behave as powers of $p$.

Consider the uniqueness of the derived expression for $I_{EHI}$. Evidently, solutions of Eq. (2.8) are defined up to the multiplication by an arbitrary elliptic function of some order $l$, whose general form was fixed in (2.5). However, one can write

$$\prod_{j=1}^{l} \frac{\theta(a_j z; p)}{\theta(b_j z; p)} = \prod_{j=1}^{l} \frac{\Gamma(qa_j z; p, q)\Gamma(b_j z; p, q)}{\Gamma(a_j z; p, q)\Gamma(qb_j z; p, q)}, \qquad \prod_{j=1}^{l} a_j = \prod_{j=1}^{l} b_j,$$

and see that the right-hand side expression can be absorbed to the kernel in (2.10) by extension of the set of parameters $\{t_k\} \to \{t_k, qa_j, b_j\}$ and $\{w_k\} \to \{w_k, a_j, qb_j\}$ without violating the balancing condition. Therefore (2.10) can be considered as a general solution. As to the initial equation (2.1), its solutions can be multiplied by arbitrary function of period $\omega_1$ which cannot be fixed without imposing additional constraints.

# 3 Properties of the Elliptic Gamma Function

For describing properties of the elliptic gamma function we take the same $\omega_{1,2,3}$ as in the previous section and introduce three bases

$$p = e^{2\pi i \omega_3/\omega_2}, \quad q = e^{2\pi i \omega_1/\omega_2}, \quad r = e^{2\pi i \omega_3/\omega_1}$$

and their particular modular partners

$$\tilde{p} = e^{-2\pi i \omega_2/\omega_3}, \quad \tilde{q} = e^{-2\pi i \omega_2/\omega_1}, \quad \tilde{r} = e^{-2\pi i \omega_1/\omega_3}.$$

The first relation we draw attention to is an evident symmetry in bases

$$\Gamma(z; p, q) = \Gamma(z; q, p),$$

which looks quite unexpected taking into account how asymmetrically the bases $p$ and $q$ enter Eq. (2.8). Due to this symmetry one actually has two finite-difference equations

$$\Gamma(qz; p, q) = \theta(z; p)\Gamma(z; p, q), \quad \Gamma(pz; p, q) = \theta(z; q)\Gamma(z; p, q).$$

Poles and zeros of the elliptic gamma function form a two-dimensional array of geometric progressions

$$z_{\text{poles}} = p^{-j}q^{-k}, \quad z_{\text{zeros}} = p^{j+1}q^{k+1}, \quad j, k \in \mathbb{Z}_{\geq 0}.$$

The inversion relation for $\Gamma(z; p, q)$ has the form

$$\Gamma(z; p, q) = \frac{1}{\Gamma(\frac{pq}{z}; p, q)}, \tag{3.1}$$

and there is a useful normalization condition $\Gamma(\sqrt{pq}; p, q) = 1$.

The quadratic transformation

$$\Gamma(z^2; p, q) = \Gamma(\pm z, \pm q^{1/2}z, \pm p^{1/2}z, \pm(pq)^{1/2}z; p, q)$$

can be established by a direct analysis of the infinite products. Here and below we use the conventions

$$\Gamma(t_1, \ldots, t_k; p, q) := \Gamma(t_1; p, q) \cdots \Gamma(t_k; p, q),$$
$$\Gamma(\pm z; p, q) := \Gamma(z; p, q)\Gamma(-z; p, q),$$
$$\Gamma(tz^{\pm k}; p, q) := \Gamma(tz^k; p, q)\Gamma(tz^{-k}; p, q).$$

The limiting relation

$$\lim_{z \to 1}(1 - z)\Gamma(z; p, q) = \frac{1}{(p; p)_\infty(q; q)_\infty} \tag{3.2}$$

is required for residue calculus and reduction of integrals to terminating elliptic hypergeometric series (non-terminating such series do not converge).

Taking the logarithm of the infinite product (2.9), expanding the logarithms of individual factors $\log(1 - x) = -\sum_{n=1}^{\infty} x^n/n$, and changing the summation order yields the following representation

$$\Gamma(z; p, q) = \exp\left(\sum_{n=1}^{\infty} \frac{1}{n} \frac{z^n - (pq/z)^n}{(1 - p^n)(1 - q^n)}\right), \tag{3.3}$$

which converges for $|pq| < |z| < 1$ and is very useful for quantum field theory purposes.

Denote $p = e^{-\delta}$ and consider the limit $\delta \to 0$. The leading asymptotics takes the form

$$\Gamma(z; p, q) = \exp\left(\frac{1}{\delta}E_2(z; q) - \frac{1}{2}\log\theta(z; q)\right)(1 + O(\delta)), \tag{3.4}$$

where

$$E_2(z; q) = \sum_{n=0}^{\infty} \text{Li}_2(q^n z) - \sum_{n=1}^{\infty} \text{Li}_2(q^n/z), \quad \text{Li}_2(z) = \sum_{n=1}^{\infty} \frac{z^n}{n^2}.$$

$\mathrm{Li}_2(z)$ is known as Euler's dilogarithm function and $E_2(z; q)$ is directly related to the elliptic dilogarithm function, which recently emerged in the computation of a sunset Feynman diagram [6] as the difference $\hat{E}(z; q) = E(z; q) - E(-z; q)$. The latter function emerges in the asymptotics of the ratio $\Gamma(z; p, q)/\Gamma(-z; p, q)$. A different relation between the elliptic gamma function with $p = q$ and the elliptic dilogarithm was described in [30].

We shall need also the second order generalization of the elliptic gamma function

$$\Gamma(z; p, q, t) = \prod_{j,k,l=0}^{\infty} (1 - zp^j q^k t^l)(1 - z^{-1} p^{j+1} q^{k+1} t^{l+1}), \quad |t|, |p|, |q| < 1, \ z \in \mathbb{C}^{\times}.$$

It is related to function (2.9) via the difference equation

$$\Gamma(qz; p, q, t) = \Gamma(z; p, t)\Gamma(z; p, q, t), \tag{3.5}$$

and its inversion relation has the form $\Gamma(pqtz; p, q, t) = \Gamma(z^{-1}; p, q, t)$.

A solution of the key equation (2.8) in the domain $|q| > 1$ is easily found to be

$$\gamma(z) = \frac{1}{\Gamma(q^{-1}z; p, q^{-1})} = \Gamma(pz^{-1}; p, q^{-1}).$$

As to the regime $|q| = 1$, one has to abandon meromorphicity of solutions of (2.8) in $z$ and look for an analytical function of $u$ solving the finite-difference equation

$$f(u + \omega_1) = \theta(e^{2\pi i u/\omega_2}; p) f(u) \tag{3.6}$$

valid for $\omega_1/\omega_2 \in \mathbb{R}$. The function $f(u) = \Gamma(e^{2\pi i u/\omega_2}; p, q)$ solving this equation for $|q| < 1$ satisfies two more equations

$$f(u + \omega_2) = f(u), \qquad f(u + \omega_3) = \theta(e^{2\pi i u/\omega_2}; q) f(u).$$

For incommensurate $\omega_i$ these three equations define function $f(u)$ uniquely up to multiplication by a constant. This follows from the Jacobi theorem stating that nontrivial functions cannot have three incommensurate periods.

For $|q| < 1$ the general solution of (3.6) has the form $f(u)\varphi(u)$ with arbitrary periodic function $\varphi(u + \omega_1) = \varphi(u)$. It appears that for a special choice of $\varphi(u)$ this product defines an analytic function of $u$ even for $\omega_1/\omega_2 > 0$. Such a choice has been found in [51], where the following modified elliptic gamma function was introduced:

$$G(u; \boldsymbol{\omega}) := \Gamma(e^{2\pi i u/\omega_2}; p, q)\Gamma(re^{-2\pi i u/\omega_1}; \tilde{q}, r). \tag{3.7}$$

This function satisfies (3.6) and two other equations

$$G(u + \omega_2) = \theta(e^{2\pi i u/\omega_1}; r)G(u),\tag{3.8}$$

$$G(u + \omega_3) = \frac{\theta(e^{2\pi i u/\omega_2}; q)}{\theta(e^{-2\pi i u/\omega_1}; \tilde{q})}G(u) = e^{-\pi i B_{2,2}(u;\omega_1,\omega_2)}G(u),\tag{3.9}$$

where $B_{2,2}$ is a second order Bernoulli polynomial

$$B_{2,2}(u; \omega_1, \omega_2) = \frac{u^2}{\omega_1 \omega_2} - \frac{u}{\omega_1} - \frac{u}{\omega_2} + \frac{\omega_1}{6\omega_2} + \frac{\omega_2}{6\omega_1} + \frac{1}{2}.$$

Here the exponential multiplier in (3.9) emerges from the modular transformation law for the theta function

$$\theta\left(e^{-2\pi i \frac{u}{\omega_1}}; e^{-2\pi i \frac{\omega_2}{\omega_1}}\right) = e^{\pi i B_{2,2}(u;\omega)}\theta\left(e^{2\pi i \frac{u}{\omega_2}}; e^{2\pi i \frac{\omega_1}{\omega_2}}\right).\tag{3.10}$$

*Exercise:* derive this relation from the modular transformation laws for the Jacobi $\theta_1$-function

$$\theta_1(u/\tau | - 1/\tau) = -i\sqrt{-i\tau}\, e^{\pi i u^2/\tau}\theta_1(u|\tau)\tag{3.11}$$

and the Dedekind $\eta$-function

$$\eta(-1/\tau) = (-i\tau)^{1/2}\eta(\tau), \quad \eta(\tau) = e^{\frac{\pi i \tau}{12}}\left(e^{2\pi i \tau}; e^{2\pi i \tau}\right)_\infty.\tag{3.12}$$

Now one can check that the same three Eqs. (3.6), (3.8), and (3.9) and the normalization condition $G(\sum_{k=1}^3 \omega_k/2; \omega) = 1$ are satisfied by the following function

$$G(u; \omega) = e^{-\frac{\pi i}{3} B_{3,3}(u;\omega)}\Gamma(e^{-2\pi i \frac{u}{\omega_3}}; \tilde{r}, \tilde{p}),\tag{3.13}$$

where $|\tilde{p}|, |\tilde{r}| < 1$, and $B_{3,3}(u; \omega)$ is the third order Bernoulli polynomial

$$B_{3,3}(u; \omega) = \frac{(u - \sum_{m=1}^3 \frac{\omega_m}{2})((u - \sum_{m=1}^3 \frac{\omega_m}{2})^2 - \frac{1}{4}\sum_{m=1}^3 \omega_m^2)}{\omega_1 \omega_2 \omega_3}.$$

Since the solution of this set of equations is unique (from the nonexistence of triply periodic functions and given normalization), we conclude that the functions (3.7) and (3.13) coincide.

However, from expression (3.13), the function $G(u; \omega)$ is seen to remain a well-defined meromorphic function of $u$ even for $\omega_1/\omega_2 > 0$. Indeed, if the latter ratio is real, one can take both $\omega_1$ and $\omega_2$ real (since the parameters enter only in ratios).

Then one will have simultaneously $|\tilde{r}| < 1$ and $|\tilde{p}| < 1$ guaranteeing convergence of infinite products in (3.13) only if $\omega_1/\omega_2 > 0$, which gives $|q| = 1$. The equality of (3.7) and (3.13) is directly related to a special modular transformation for the elliptic gamma function from the SL(3, $\mathbb{Z}$)-group [18].

The function $G(u; \omega)$ satisfies the reflection relation $G(a; \omega)G(b; \omega) = 1$, $a + b = \sum_{k=1}^{3} \omega_k$. From (3.13) it is not difficult to see the symmetry $G(u; \omega_1, \omega_2, \omega_3) = G(u; \omega_2, \omega_1, \omega_3)$.

The multiple Bernoulli polynomials described above are generated by the following expansion:

$$\frac{x^m e^{xu}}{\prod_{k=1}^{m}(e^{\omega_k x} - 1)} = \sum_{n=0}^{\infty} B_{m,n}(u; \omega_1, \ldots, \omega_m) \frac{x^n}{n!}$$

emerging in the theory of Barnes multiple gamma function [3].

Let us take the limit $\text{Im}(\omega_3) \to +\infty$ and assume that $\text{Re}(\omega_1), \text{Re}(\omega_2) > 0$. Then $\text{Im}(\omega_3/\omega_1), \text{Im}(\omega_3/\omega_2) \to +\infty$ and $p, r \to 0$. As a result, the expression (3.7) reduces to

$$G(u; \omega) \underset{p,r \to 0}{=} \frac{(e^{2\pi i u/\omega_1} \tilde{q}; \tilde{q})_\infty}{(e^{2\pi i u/\omega_2}; q)_\infty}.$$

From the representation (3.13) one obtains a singular relation

$$G(u; \omega) \underset{p,r \to 0}{=} e^{\frac{\pi i}{2} B_{2,2}(u, \omega_1, \omega_2)} \lim_{\text{Im}(\frac{\omega_3}{\omega_1}), \text{Im}(\frac{\omega_3}{\omega_2}) \to +\infty} e^{-\pi i \omega_3 \frac{2u - \omega_1 - \omega_2}{12\omega_1\omega_2}}$$

$$\times \Gamma(e^{-2\pi i \frac{u}{\omega_3}}; e^{-2\pi i \frac{\omega_1}{\omega_3}}, e^{-2\pi i \frac{\omega_2}{\omega_3}}).$$

For $\text{Re}(\omega_1), \text{Re}(\omega_2) > 0$ and $\omega_3 \to +i\infty$ this result can be rewritten as an asymptotic relation

$$\Gamma(e^{-2\pi vu}; e^{-2\pi v\omega_1}, e^{-2\pi v\omega_2}) \underset{v \to 0^+}{=} e^{-\pi \frac{2u - \omega_1 - \omega_2}{12 v \omega_1 \omega_2}} \gamma^{(2)}(u; \omega_1, \omega_2), \tag{3.14}$$

where

$$\gamma^{(2)}(u; \omega_1, \omega_2) := e^{-\frac{\pi i}{2} B_{2,2}(u; \omega_1, \omega_2)} \frac{(e^{2\pi i u/\omega_1} \tilde{q}; \tilde{q})_\infty}{(e^{2\pi i u/\omega_2}; q)_\infty}, \tag{3.15}$$

is the standard hyperbolic gamma function.

*Exercise:* derive the infinite product representation (3.15) from the integral representation

$$\gamma^{(2)}(u; \omega_1, \omega_2) = \exp\left(-\text{p.v.} \int_{\mathbb{R}} \frac{e^{ux}}{(e^{\omega_1 x} - 1)(e^{\omega_2 x} - 1)} \frac{dx}{x}\right) \tag{3.16}$$

with appropriate restrictions on the parameters needed for convergence. Here "p.v." means "principal value", i.e. an average of integrals with the contours passing infinitesimally above and below the singular point $x = 0$.

In particular, note that for $\mathrm{Re}(\omega_1), \mathrm{Re}(\omega_2) > 0$ and $0 < \mathrm{Re}(u) < \mathrm{Re}(\omega_1) + \mathrm{Re}(\omega_2)$ integral in (3.16) converges and defines $\gamma^{(2)}(u; \omega_1, \omega_2)$ as an analytic function of $u$ even for $\mathrm{Im}(\omega_1/\omega_2) = 0$, when $|q| = 1$. The limiting relation (3.14) was rigorously established first in a different way by Ruijsenaars [43]. Its uniformity was proven by Rains in [33]. The hyperbolic gamma function plays a crucial role in the construction of $q$-hypergeometric functions in the regime $|q| = 1$ [24]. It was introduced in quantum field theory by Faddeev under the name modular (or noncompact) quantum dilogarithm [15, 17]. In a similar sense, the elliptic gamma function has a meaning of a "quantum" deformation of the elliptic dilogarithm function.

## 4 The Elliptic Beta Integral

One of the differences from ordinary hypergeometric functions and their $q$-deformations consists in the fact that it is not straightforward to construct an equation which is satisfied by the general elliptic hypergeometric function (2.10). In order to find elliptic analogues of the relations described in the introduction one has to impose additional structural constraints on the corresponding parameters. A basic germ, a kind of the cornerstone for building constructive identities for such integrals is provided by the evaluation of univariate elliptic beta integral [49].

Let complex parameters $p, q, t_j, j = 1, \ldots, 6$, satisfy the constraints $|p|, |q|, |t_j| < 1$ and the balancing condition

$$\prod_{j=1}^{6} t_j = pq.$$

Then the following integral identity holds true

$$\kappa \int_{\mathbb{T}} \frac{\prod_{j=1}^{6} \Gamma(t_j x^{\pm 1}; p, q)}{\Gamma(x^{\pm 2}; p, q)} \frac{dx}{x} = \prod_{1 \le j < k \le 6} \Gamma(t_j t_k; p, q), \qquad (4.1)$$

where $\mathbb{T}$ is the unit circle of positive orientation and

$$\kappa = \frac{(p; p)_\infty (q; q)_\infty}{4\pi i}.$$

We sketch the proof of this statement suggested in [53]. Note first that the integrand has poles at the points $z = t_j q^a p^b$, $j = 1, \ldots, 6, a, b \in \mathbb{Z}_{\ge 0}$, converging to zero, and their reciprocals $z = t_j^{-1} q^{-a} p^{-b}$, diverging to infinity. The integration contour $\mathbb{T}$ separates these sets of poles.

Now we apply the gamma function inversion

$$\Gamma(t_6 x; p, q) = \frac{1}{\Gamma(pq/(t_6 x); p, q)} = \frac{1}{\Gamma(Ax^{-1}; p, q)}, \qquad A := \prod_{m=1}^{5} t_m,$$

and rewrite the integral evaluation as

$$I(t_1, \ldots, t_5; p, q) = \kappa \int_{\mathbb{T}} \rho(x; t_1, \ldots, t_5; p, q) \frac{dx}{x} = 1,$$

where

$$\rho(x; t_1, \ldots, t_5; p, q) = \frac{\prod_{j=1}^{5} \Gamma(t_j x^{\pm 1}, t_j^{-1} A; p, q)}{\Gamma(x^{\pm 2}, Ax^{\pm 1}; p, q) \prod_{1 \le i < j \le 5} \Gamma(t_i t_j; p, q)}.$$

This kernel function satisfies the $q$-difference equation

$$\rho(x; qt_1) - \rho(x; t_1) = g(q^{-1} x) \rho(q^{-1} x; t_1) - g(x) \rho(x; t_1) \qquad (4.2)$$

with

$$g(x) = \frac{\prod_{m=1}^{5} \theta(t_m x; p)}{\prod_{m=2}^{5} \theta(t_1 t_m; p)} \frac{\theta(t_1 A; p)}{\theta(x^2, xA; p)} \frac{t_1}{x}.$$

Dividing (4.2) by the $\rho$-function, one comes to the following elliptic functions identity

$$\frac{\theta(t_1 x, t_1 x^{-1}; p)}{\theta(Ax, Ax^{-1}; p)} \prod_{m=2}^{5} \frac{\theta(At_m^{-1}; p)}{\theta(t_1 t_m; p)} - 1 = \frac{t_1 \theta(t_1 A; p)}{x \theta(x^2; p) \prod_{m=2}^{5} \theta(t_1 t_m; p)}$$

$$\times \left( \frac{x^4 \prod_{m=1}^{5} \theta(t_m x^{-1}; p)}{\theta(Ax^{-1}; p)} - \frac{\prod_{m=1}^{5} \theta(t_m x; p)}{\theta(Ax; p)} \right).$$

which we suggest to prove as an exercise (compare the poles in $x$ and their residues in the parallelogram of periods of the left- and right-hand side expressions and verify the identity for a particular value of $x$).

Integrating Eq. (4.2) over $x \in \mathbb{T}$ one obtains the relation

$$I(qt_1) - I(t_1) = \left( \int_{q^{-1}\mathbb{T}} - \int_{\mathbb{T}} \right) g(x) \rho(x; t_1) \frac{dx}{x}.$$

Consideration of the poles of the function $g(x)\rho(x; t_1)$ shows that for $|t_1| < |q|$ it does not have singularities inside the annulus bounded by $\mathbb{T}$ and the circle of radius $|q|^{-1}$ denoted as $q^{-1}\mathbb{T}$. As a result, the right-hand side expression in the above relation vanishes and the equality $I(qt_1) = I(t_1)$ emerges in a natural way. After permuting $p$ and $q$ in the above considerations and imposing the additional constraint $|t_1| < |p|$, it becomes possible to write $I(pt_1) = I(t_1)$. Now, the Jacobi theorem on the absence of periodic functions with three incommensurate periods and $t_k$-permutational symmetry show that $I$ does not depend on parameters, $I = I(p, q)$.

In order to compute this constant, one can consider the limit $t_1t_2 \to 1$ when two pairs of residues pinch the contour of integration. After crossing a pair of poles and picking up the residues, one can see that the integral part vanishes in this limit, and the contribution of residues sums exactly to the needed value $I = 1$.

The derived elliptic beta integral evaluation represents a unique relation due to the following facts. First of all, it represents an elliptic extension of Newton's binomial theorem $_1F_0(a; x) = (1 - x)^{-a}$ and its $q$-analogue

$$_1\varphi_0(t; q; x) = \sum_{n=0}^{\infty} \frac{(t; q)_n}{(q; q)_n} x^n = \frac{(tx; q)_\infty}{(x; q)_\infty}, \quad (t; q)_n = \prod_{k=0}^{n-1}(1 - tq^k).$$

After setting $t = q^a$, for $q \to 1^-$ one has $_1\varphi_0(q^a; q; x) \to {}_1F_0(a; x)$. This yields the useful relation

$$\lim_{q\to 1^-} \frac{(q^a x; q)_\infty}{(x; q)_\infty} = (1 - x)^{-a}. \tag{4.3}$$

As to the terminating series version of the binomial theorem, its elliptic analogue is given by the Frenkel–Turaev sum [20], which can be obtained by a reduction of (4.1). To derive this sum let us take the limit $t_4t_5 \to q^{-N}$ for some positive integer $N$. More precisely, let us take parameter $t_5$ from inside $\mathbb{T}$ to outside such that $|pt_5|, |q^{N+1}t_5| < 1 < |t_5|$ and keep other parameters inside $\mathbb{T}$ in generic positions. Formula (4.1) will remain intact if we replace the contour $\mathbb{T}$ by $C$ which separates sequences of poles converging to zero from their reciprocals. Consideration of the poles related to the parameters $t_4$ and $t_5$ shows that if $t_4t_5 \to q^{-N}$ then $4(N + 1)$ poles start to pinch pairwise two parts of the contour $C$ lying outside and inside $\mathbb{T}$. As a result both, the left- and right-hand side expressions in (4.1) start to diverge.

To compute the limiting formula, resolve the balancing condition $t_6 = pq/\prod_{k=1}^{5} t_k$ and denote $\rho_E(z, \underline{t}) = \prod_{m=1}^{5} \Gamma(t_m z^{\pm 1}; p, q)/\Gamma(z^{\pm 2}, \prod_{k=1}^{5} t_k z^{\pm 1}; p, q)$. Let us force the contour $C$ to cross $2(N + 1)$ poles $z = (t_5 q^k)^{\pm 1}$, $k = 0, \dots, N$. Then the Cauchy theorem states that:

$$\kappa \int_C \rho_E(z, \underline{t}) \frac{dz}{z} = \kappa \int_{\mathbb{T}} \rho_E(z, \underline{t}) \frac{dz}{z}$$

$$+ \frac{\prod_{m=1}^{4} \Gamma(t_m t_5^{\pm 1}; p, q)}{\Gamma(t_5^{-2}, \prod_{k=1}^{5} t_k t_5^{\pm 1}; p, q)} \sum_{n=0}^{N} \frac{\theta(t_5^2 q^{2n}; p)}{\theta(t_5^2; p)} \prod_{m=0}^{5} \frac{\theta(t_m t_5)_n}{\theta(q t_m^{-1} t_5)_n} q^n. \tag{4.4}$$

Here we denoted $t_0 = q/\prod_{m=1}^{5} t_m = t_6/p$ and used the elliptic Pochhammer symbol

$$\theta(t)_n = \prod_{j=0}^{n-1} \theta(tq^j; p) = \frac{\Gamma(tq^n; p, q)}{\Gamma(t; p, q)}, \qquad \theta(t_1, \ldots, t_k)_n := \prod_{j=1}^{k} \theta(t_j)_n.$$

The residues are computed using the limiting relation (3.2).

Now we take the desired limit $t_5 t_4 \to q^{-N}$. The integral over the unit circle $\mathbb{T}$ stays finite, since the integrand is nonsingular on $\mathbb{T}$, whereas the sum of residues and the value of the original integral diverge. Dividing expression (4.4) and its evaluation (4.1) by $\Gamma(t_4 t_5; p, q)$, for $t_5 t_4 = q^{-N}$ one obtains the Frenkel–Turaev sum

$$_{10}V_9(t_5^2; t_0 t_5, t_1 t_5, t_2 t_5, t_3 t_5, q^{-N}) = \frac{\theta(q t_5^2, \frac{q}{t_1 t_2}, \frac{q}{t_1 t_3}, \frac{q}{t_2 t_3})_N}{\theta(\frac{q}{t_1 t_2 t_3 t_5}, \frac{q t_5}{t_1}, \frac{q t_5}{t_2}, \frac{q t_5}{t_3})_N}. \tag{4.5}$$

Here we use general notation for the very-well poised elliptic hypergeometric series introduced in [50]

$$_{m+1}V_m(t_0; t_1, \ldots, t_{m-4}; q, p) = \sum_{n=0}^{\infty} \frac{\theta(t_0 q^{2n}; p)}{\theta(t_0; p)} \prod_{k=0}^{m-4} \frac{\theta(t_k)_n}{\theta(q t_0 t_k^{-1})_n} q^n \tag{4.6}$$

with the balancing condition $\prod_{k=1}^{m-4} t_k = t_0^{(m-5)/2} q^{(m-7)/2}$ and the assumption that the series terminates because one of the parameters has the form $t_j = q^{-N}$. For $p \to 0$ the series (4.6) with fixed parameters reduces to the very-well poised balanced $_{m-1}\varphi_{m-2}$ series [21]. The original derivation of (4.5) in [20] is completely different from the given one which was suggested in [67]. Multivariable extensions of the elliptic hypergeometric series were considered for the first time in [71].

*Exercise*: verify the above derivation of (4.5) by completing all the details.

The next important property of the integral (4.1) is that it represents the top known generalization of the Euler beta integral (1.5). In particular, in the limit $p \to 0$, taken for fixed $t_1, \ldots, t_5$ and $t_6 \propto p$, one obtains the Rahman $q$-beta integral [31]. Subsequent turning one of the parameters to zero yields the Askey–Wilson $q$-beta integral whose reduction to (1.5) was explicitly described in [21]. More complicated degenerations of the elliptic beta integral are considered in [46].

Integral (4.1) serves as the measure for a biorthogonality relation of specific two-index functions, defined as products of two $_{12}V_{11}$ series, which generalize the Askey–Wilson, Jacobi and other classical orthogonal polynomials [51]. These functions comprise also Rahman's continuous biorthogonal rational functions [31]. The discrete measure analogues of these functions based on the Frenkel–Turaev sum were defined in [64].

Integral (4.1) is a germ for constructing infinitely many elliptic hypergeometric integrals admitting exact evaluation. It generates an elliptic Fourier transform

[52, 63], associated with an integral generalization of the Bailey chains techniques [70], integral operator realization of Coxeter relations [10], the star-triangle relation [5], and the Yang–Baxter equation. Identity (4.1) emerged in four dimensional supersymmetric quantum field theory as an equality of superconformal indices of two specific models [14]. Some of these unique features of the elliptic beta integral are described in more detail in the following.

The very first proof of formula (4.1) was obtained using the contiguous relation for integrals (5.5) and expansion in small $p$ [49], when the limiting $t_j = 0$ points enter the domain of analyticity of the expansion coefficients. A further refinement of such expansion arguments was suggested in [36], when the equality of formal series in $p$ in the left- and right-hand sides of identities is reached by establishing their rationality and coincidence on an infinite discrete set of parameter values. This gives another proof of the above formula based on the theory of Askey–Wilson polynomials and Frenkel–Turaev sum. The proof of [53] given above is self-contained—it does not require knowledge of any system of orthogonal functions and uses only a simple elliptic function identity.

# 5   An Elliptic Extension of the Euler–Gauss Hypergeometric Function

There are many generalizations of the $F(a, b; c; x)$ hypergeometric function. Let us describe the one related to the elliptic beta integral in a way as the beta function (1.5) is connected to (1.1). It is necessary to take two base variables $p, q$, $|p|, |q| < 1$, and eight parameters $t_1, \ldots, t_8 \in \mathbb{C}^\times$ satisfying the balancing condition $\prod_{j=1}^8 t_j = p^2 q^2$. Then, under additional constraints $|t_j| < 1$, an elliptic analogue of the Euler–Gauss hypergeometric function is defined by the integral [51]

$$V(\underline{t}) \equiv V(t_1, \ldots, t_8; p, q) := \kappa \int_{\mathbb{T}} \frac{\prod_{j=1}^8 \Gamma(t_j z^{\pm 1}; p, q)}{\Gamma(z^{\pm 2}; p, q)} \frac{dz}{z}. \tag{5.1}$$

By deforming the integration contour it is possible to partially relax the constraints on the parameters. Analytic continuation of (5.1) is achieved by increasing the absolute values of parameters and computing the residues of the integrand poles, so that the analytically continued function becomes a sum of the integral over some fixed contour and residues of the poles crossed by the contour. From this procedure one can see that the $V$-function is meromorphic for all values of parameters $t_j \in \mathbb{C}^\times$, when the contour of integration is not pinched which may happen for $t_j t_k = q^{-a} p^{-b}$, $a, b \in \mathbb{Z}_{\geq 0}$. It appears that the potential singularities from $t_j^2 = q^{-a} p^{-b}$ do not contribute and the product $\prod_{1 \leq j < k \leq 8} (t_j t_k; p, q)_\infty V(\underline{t})$ becomes a holomorphic function of the parameters [32]. For particular values of the parameters $t_j$ the $V$-function has delta-function type singularities [56, 62]. We remark that the expression (5.1) can be reduced to both Euler and Barnes type integral representations for $F(a, b; c; x)$.

The hypergeometric function (1.2) is reduced to Euler's beta integral for $a = 0$. In a similar way, its elliptic counterpart (5.1) reduces to the elliptic beta integral if a pair of parameters is constrained as $t_j t_k = pq$, $j \neq k$, as follows from the inversion relation (3.1).

Consider now symmetry transformations for the $V$-function. An evident symmetry is the possibility to permute bases $p$ and $q$. For describing symmetries in the parameters we remind some simplest facts from the theory of root systems and corresponding Weyl groups. Consider $\mathbb{R}^n$ with an orthonormal basis $e_i \in \mathbb{R}^n$, $\langle e_i, e_j \rangle = \delta_{ij}$. For any $x \in \mathbb{R}^n$ define its reflection with respect to the hyperplane orthogonal to some $v \in \mathbb{R}^n$:

$$ x \to S_v(x) = x - \frac{2\langle v, x \rangle}{\langle v, v \rangle} v, \qquad S_v^2 = 1. $$

If $x = const \cdot v$, then $S_v(x) = -x$. For $\langle v, x \rangle = 0$ one has $S_v(x) = x$.

Define $R$ as some set of vectors $\alpha_1, \ldots, \alpha_m \in \mathbb{R}^n$, forming a basis. If for any $\alpha, \beta \in R$, $S_\alpha(\beta) \in R$, then $R$ is called a root system. The reflections $W = \{S_\alpha\}$ form a finite subgroup of the rotation group $O(n)$. The vectors $\alpha_j$ are called the roots and the dimensionality of the space where they are defined is the rank of the root system.

If for all $\alpha, \beta \in R$ one has the integrality $2\langle \alpha, \beta \rangle / \langle \alpha, \alpha \rangle \in \mathbb{Z}$, then $R$ is called the crystallographic root system and $W$—the Weyl group.

If the only multiples of a root $\alpha$ in $R$ are $\pm \alpha$ then $R$ is called reduced and it is known to be related to a semi-simple Lie algebra. For such cases there exist four irreducible (i.e. indecomposable to direct sums) infinite classical series of root systems: $A_n$, $B_n$, $C_n$, $D_n$, and five exceptional cases: $G_2$, $F_4$, $E_6$, $E_7$, $E_8$.

Let us describe a few examples of root systems used in the following.

1. $A_n$ system: take $\mathbb{E} \in \mathbb{R}^{n+1}$ orthogonal to $\sum_{j=1}^{n+1} e_j$, i.e. for $u = \sum_{j=1}^{n+1} u_j e_j \in \mathbb{E}$ one has $\sum_{j=1}^{n+1} u_j = 0$. Then $R_{A_n} = \{e_i - e_j, \ i \neq j\}|_{i,j=1,\ldots,n+1}$ and the Weyl group is the permutation group $W(A_n) = S_{n+1}$.

2. $C_n$ system: take in $\mathbb{R}^n$ the roots $R_{C_n} = \{\pm 2e_i; \pm e_i \pm e_j, \ i < j\}|_{i,j=1,\ldots,n}$ and $W = S_n \times \mathbb{Z}_2^n$. The only non-reduced root system is $R_{BC_n}$, which contains the roots of the $C_n$ system and additionally $\{\pm e_1, \ldots, \pm e_n\}$.

3. $E_7$ system: take $\mathbb{E} \in \mathbb{R}^8$ orthogonal to $\sum_{j=1}^{8} e_j$, as for the $A_7$ root system. Then $R_{E_7} = \{e_j - e_k, \ j \neq k; \ \frac{1}{2} \sum_{l=1}^{8} \mu_j e_j, \ \mu_j = \pm 1$ with four values $\mu_j = 1\}|_{j,k=1,\ldots,8}$, and $W(E_7)$ is a particular finite group of order $72 \cdot 8!$.

As we will see, elliptic hypergeometric integrals are naturally related to the root systems in two qualitatively different ways.

The $V$-function is evidently invariant under the $S_8$-group of permutations of parameters $t_j$. It is the Weyl group of the $A_7$ root system. Consider now the double integral

$$\kappa \int_{\mathbb{T}^2} \frac{\prod_{j=1}^4 \Gamma(a_j z^{\pm 1}, b_j w^{\pm 1}; p, q)\, \Gamma(c z^{\pm 1} w^{\pm 1}; p, q)}{\Gamma(z^{\pm 2}, w^{\pm 2}; p, q)} \frac{dz}{z} \frac{dw}{w},$$

where complex parameters $a_j, b_j, c \in \mathbb{C}^\times$ are constrained as $|a_j|, |b_j|, |c| < 1$ and satisfy the balancing conditions

$$c^2 \prod_{j=1}^4 a_j = c^2 \prod_{j=1}^4 b_j = pq.$$

Since we integrate over compact domains, the order of integrations does not matter. The integrals over $z$ or $w$ are separately computable due to the key formula (4.1). Taking these integrals in the different order we come to the following transformation formula:

$$V(ca_1, \ldots, ca_4, b_1, \ldots, b_4) = \prod_{1 \le j < k \le 4} \frac{\Gamma(b_j b_k; p, q)}{\Gamma(a_j a_k; p, q)} V(a_1, \ldots, a_4, cb_1, \ldots, cb_4),$$

which can be rewritten in a more symmetric form

$$V(\underline{t}) = \prod_{1 \le j < k \le 4} \Gamma(t_j t_k, t_{j+4} t_{k+4}; p, q)\, V(\underline{s}), \tag{5.2}$$

where the parameters $t_j$ and $s_j$ are related to each other as

$$\begin{cases} s_j = \sqrt{\frac{pq}{t_1 t_2 t_3 t_4}}\, t_j, \ j = 1, 2, 3, 4 \\ s_j = \sqrt{\frac{pq}{t_5 t_6 t_7 t_8}}\, t_j, \ j = 5, 6, 7, 8 \end{cases}$$

and satisfy the constraints $|t_j|, |s_j| < 1$ matching the integration contour $\mathbb{T}$ on both sides of (5.2).

The function $V(\underline{t})$ appeared for the first time during the derivation of this fundamental relation in [51]. Let us write $t_j = e^{2\pi i x_j}(pq)^{1/4}$ and $s_j = e^{2\pi i y_j}(pq)^{1/4}$, $j = 1, \ldots, 8$. From the balancing condition we find $\sum_{j=1}^8 x_j = \sum_{j=1}^8 y_j = 0$. Now it is not difficult to see that the transformation of parameters in (5.2) is equivalent to the relation $y_j = x_j - \frac{\mu}{4} \sum_{k=1}^4 (x_k - x_{k+4})$ with $\mu = 1$ for $j = 1, \ldots, 4$ and $\mu = -1$ for $j = 5, \ldots, 8$, which precisely corresponds to the reflection $y = S_v(x)$ for the vector $v = (\sum_{k=1}^4 e_i - \sum_{k=5}^8 e_i)/2$ of the canonical length $\langle v, v \rangle = 2$. Permuting in $\binom{8}{4} = 70$ nontrivial ways the basis vectors in this $v$ one comes to the roots of the exceptional root system $E_7$ extending the $A_7$ root system, as described above.

Now one can consider all admissible $W(E_7)$ Weyl group reflections acting on the $V$-function. For instance, it is possible to repeat reflection (5.2) for the second time using the root $v = (e_3 + e_4 + e_5 + e_6 - e_1 - e_2 - e_7 - e_8)/2$ and permute in the resulting relation parameters in all possible ways. This yields the following symmetry transformation

$$V(\underline{t}) = \prod_{j,k=1}^{4} \Gamma(t_j t_{k+4}; p, q) \, V(T^{\frac{1}{2}}/t_1, \ldots, T^{\frac{1}{2}}/t_4, S^{\frac{1}{2}}/t_5, \ldots, S^{\frac{1}{2}}/t_8), \qquad (5.3)$$

where $T = t_1 t_2 t_3 t_4$, $S = t_5 t_6 t_7 t_8$ and one has the constraints $|T|^{1/2} < |t_j| < 1$, $|S|^{1/2} < |t_{j+4}| < 1$, $j = 1, 2, 3, 4$, in order to have $\mathbb{T}$ as the integration contour on both sides.

Finally, let us equate expressions on the right-hand sides of relations (5.2) and (5.3). After rewriting the resulting equality in terms of the parameters $s_j$, it takes the form

$$V(\underline{s}) = \prod_{1 \le j < k \le 8} \Gamma(s_j s_k; p, q) \, V(\sqrt{pq}/s_1, \ldots, \sqrt{pq}/s_8), \qquad (5.4)$$

where $|pq|^{1/2} < |s_j| < 1$ for all $j$. The key generating relation (5.2) was discovered in [51]. Transformations (5.3) and (5.4) were proved in a different way in [32], where the identification of these transformations with the group $W(E_7)$ was made.

Although these three identities for the $V$-function have different form, they are tied by the symmetry group. As it will be shown later on, the multiple elliptic hypergeometric integrals have transformations which can be considered as their separate generalizations, i.e. different elements of $W(E_7)$ may have individual multivariable extensions.

Let us identify parameters in (2.6) as $y = t_1, x = t_2, w = t_3$ and multiply this addition formula by $\rho(z; \underline{t}) = \prod_{j=1}^{8} \Gamma(t_j z^{\pm 1}; p, q)/\Gamma(z^{\pm 2}; p, q)$ with the balancing condition $\prod_{j=1}^{8} t_j = p^2 q$. Then we can write

$$t_3 \theta(t_2 t_3^{\pm 1}; p) \rho(z; q t_1, t_2, \ldots) + t_1 \theta(t_3 t_1^{\pm 1}; p) \rho(z; t_1, q t_2, \ldots)$$
$$+ t_2 \theta(t_1 t_2^{\pm 1}; p) \rho(z; t_1, t_2, q t_3, \ldots) = 0.$$

Integrating this relation over $z \in \mathbb{T}$ we obtain the following contiguous relation

$$\frac{t_1 V(q t_1)}{\theta(t_1 t_2^{\pm 1}, t_1 t_3^{\pm 1}; p)} + \frac{t_2 V(q t_2)}{\theta(t_2 t_1^{\pm 1}, t_2 t_3^{\pm 1}; p)} + \frac{t_3 V(q t_3)}{\theta(t_3 t_1^{\pm 1}, t_3 t_2^{\pm 1}; p)} = 0, \qquad (5.5)$$

where $V(qt_j)$ denotes the $V(\underline{t}; p, q)$-function with the parameter $t_j$ replaced by $qt_j$, so that the balancing condition takes the form indicated above.

Applying symmetry relations discussed in the previous section to the $V$-functions in (5.5) one obtains many differently looking identities. In particular, substitution of the third transformation (5.4) yields the contiguous relation

$$\frac{\prod_{j=4}^{8} \theta\left(t_1 t_j/q; p\right) V(t_1/q)}{t_1 \theta(t_2/t_1, t_3/t_1; p)} + \frac{\prod_{j=4}^{8} \theta\left(t_2 t_j/q; p\right) V(t_2/q)}{t_2 \theta(t_1/t_2, t_3/t_2; p)}$$

$$+ \frac{\prod_{j=4}^{8} \theta\left(t_3 t_j/q; p\right) V(t_3/q)}{t_3 \theta(t_1/t_3, t_2/t_3; p)} = 0, \quad (5.6)$$

where $\prod_{j=1}^{8} t_j = p^2 q^3$.

Consider now three equations: (1) the equation obtained from (5.5) after the replacement $t_1 \to q^{-1} t_1$, (2) the one obtained from (5.5) after the replacement $t_2 \to q^{-1} t_2$, and (3) the $t_3 \to q t_3$ transformed version of (5.6). Eliminating from them the functions $V(q^{-1} t_1, q t_3)$ and $V(q^{-1} t_2, q t_3)$ we come to the elliptic hypergeometric equation [54]:

$$\mathcal{A}(t_1, t_2, \ldots, t_8, q; p)\left(U(q t_1, q^{-1} t_2; p, q) - U(\underline{t}; p, q)\right) \quad (5.7)$$

$$+ \mathcal{A}(t_2, t_1, \ldots, t_8, q; p)\left(U(q^{-1} t_1, q t_2, ; p, q) - U(\underline{t}; p, q)\right) + U(\underline{t}; p, q) = 0,$$

where

$$\mathcal{A}(t_1, \ldots, t_8, q; p) = \frac{\theta(t_1/q t_3, t_3 t_1, t_3/t_1; p)}{\theta(t_1/t_2, t_2/q t_1, t_1 t_2/q; p)} \prod_{k=4}^{8} \frac{\theta(t_2 t_k/q; p)}{\theta(t_3 t_k; p)} \quad (5.8)$$

and

$$U(\underline{t}; p, q) := \frac{V(\underline{t}; p, q)}{\prod_{k=1}^{2} \Gamma(t_k t_3^{\pm 1}; p, q)}.$$

*Exercise:* verify that the coefficient $\mathcal{A}(t_1, \ldots, t_8, q; p)$ is invariant under the transformations $t_j \to p^{n_j} t_j$, $q \to p^n q$ for any $n_j, n \in \mathbb{Z}$ preserving the balancing condition $\prod_{j=1}^{8} t_j = p^2 q^2$.

Expressing $t_1$ in terms of $t_2$ (or vice versa) via the balancing condition, one sees that (5.7) is actually a second order $q$-difference equation in $t_2$ (or $t_1$). It shows that the elliptic hypergeometric integrals may emerge as solutions of particular finite-difference equations with elliptic function coefficients.

Since $\mathcal{A}(p^{-1} t_1, p t_2, \ldots) = \mathcal{A}(t_1, t_2, \ldots)$, the function $U(p^{-1} t_1, p t_2)$ defines the second independent solution of (5.7). Let us multiply (5.7) by $U(p^{-1} t_1, p t_2)$ and the equation for $U(p^{-1} t_1, p t_2)$ by $U(t_1, t_2)$ and subtract one from another. This yields

$$\mathcal{A}(t_1, t_2, \ldots t_8, q; p) D(p^{-1} t_1, q^{-1} t_2) = \mathcal{A}(t_2, t_1, t_3, \ldots, q; p) D(p^{-1} q^{-1} t_1, t_2),$$

$$(5.9)$$

where

$$D(t_1, t_2) = U(qpt_1, t_2)U(t_1, pqt_2) - U(qt_1, pt_2)U(pt_1, qt_2)$$

is the $t_1 \to pt_1$ and $t_2 \to qt_2$ renormalized version of the Casoratian (discrete Wronskian) with the balancing condition for $U$-function parameters $\prod_{j=1}^{8} t_j = pq$.

Let $t_2$ be an independent variable. Then $t_1 \propto 1/t_2$ due to the balancing condition. Therefore, after denoting $f(t_2) := D(t_1, t_2)$, relation (5.9) is nothing else than the following first order $q$-difference equation in $t_2$:

$$
\begin{aligned}
f(qt_2) &= \frac{A(pt_1, qt_2, t_3, \ldots, q; p)}{A(qt_2, pt_1, t_3, \ldots, q; p)} f(t_2) \\
&= -\frac{t_1}{qt_2} \frac{\theta(t_1/q^2 t_2, t_1/qt_3, t_1^{-1} t_3^{\pm 1}; p)}{\theta(t_2/t_1, t_2/t_3, q^{-1} t_2^{-1} t_3^{\pm 1}; p)} \prod_{k=4}^{8} \frac{\theta(t_2 t_k; p)}{\theta(t_1 t_k/q; p)} f(t_2).
\end{aligned}
$$

Its general solution has the form

$$D(t_1, t_2) = \varphi(t_2) \frac{\prod_{k=3}^{8} \Gamma(t_1 t_k, t_2 t_k)}{\Gamma(t_1/t_2, t_2/t_1)} \prod_{k=1}^{2} \frac{\Gamma(t_k^{-1} t_3^{\pm 1}; p, q)}{\Gamma(t_k t_3^{\pm 1}; p, q)}, \tag{5.10}$$

where $\varphi(qt_2) = \varphi(t_2)$. Since $D(t_1, t_2)$ is symmetric in $p$ and $q$, we can repeat the above consideration with permuted $p$ and $q$, which yields $\varphi(pt_2) = \varphi(t_2)$. By the Jacobi theorem, for incommensurate $p$ and $q$ this proves that $\varphi$ does not depend on $t_2$.

*Exercise:* compute the constant $\varphi$ by taking the limit $t_2 \to 1/t_3$ and using the residue calculus. Show that

$$\varphi = \frac{\prod_{3 \leq j < k \leq 8} \Gamma(t_j t_k; p, q)}{\Gamma(t_1^{-1} t_2^{-1}; p, q)},$$

which yields the following quadratic relation for the elliptic hypergeometric function [37]

$$V(pqt_1, t_2)V(t_1, pqt_2) - t_1^{-2} t_2^{-2} V(qt_1, pt_2)V(pt_1, qt_2) = \frac{\prod_{1 \leq j < k \leq 8} \Gamma(t_j t_k; p, q)}{\Gamma(t_1^{\pm 1} t_2^{\pm 1}; p, q)}. \tag{5.11}$$

Solutions of the elliptic hypergeometric equation (5.7) which we discussed so far are defined for $|q| < 1$. However, the equation itself does not assume such a constraint. In order to build its solutions in other domains of values of $q$ one can use symmetries of the Eq. (5.7) which are not symmetries of the described solutions. In particular, the following relation holds true

$$A\left(\frac{p^{1/2}}{t_1}, \ldots, \frac{p^{1/2}}{t_8}, q; p\right) = A\left(t_1, \ldots, t_8, q^{-1}; p\right).$$

This means that the scalings $t_j \rightarrow p^{1/2}/t_j$, $j = 1, \ldots, 8$, transform (5.7) to the same equation with the replacement of the base $q \rightarrow q^{-1}$. The inversion $q \rightarrow 1/q$ takes place also if one replaces $t_j \rightarrow p^{n_j}/t_j$ with integer $n_j$, $\sum_{j=1}^{8} n_j = 4$. So, in the regime $|q| > 1$ one obtains the following particular solution of (5.7) [37]

$$U_{|q|>1}(\underline{t}; q, p) = \frac{V(p^{1/2}/t_1, \ldots, p^{1/2}/t_8; p, q^{-1})}{\prod_{k=1}^{2} \Gamma(p/t_k t_3, t_3/t_k; p, q^{-1})}. \tag{5.12}$$

In order to obtain solutions of the elliptic hypergeometric equation on the unit circle $|q| = 1$, it is necessary to use the modified elliptic gamma function $G(u; \omega)$. Indeed, we can replace in the definitions of the elliptic beta integral and the $V$-function the function $\Gamma(z; p, q)$ by $G(u; \omega)$ and repeat all the considerations anew. Because the functional equations for these elliptic gamma functions are similar, one will obtain formulas analogous to those presented above. But from the representation (3.13) it follows that the difference between them lies only in the exponential factors containing the Bernoulli polynomials. As shown in [69], these factors can be removed reducing everything to a modular transformed version of the described above relations. By construction, such relations remain well defined even if $|q| = 1$. At the level of Eq. (5.7) one has the modular invariance

$$A(e^{2\pi i \frac{g_1}{\omega_2}}, e^{2\pi i \frac{g_2}{\omega_2}}, \ldots e^{2\pi i \frac{g_8}{\omega_2}}, e^{2\pi i \frac{\omega_1}{\omega_2}}; e^{2\pi i \frac{\omega_3}{\omega_2}}) = A(\ldots)\big|_{(\omega_2, \omega_3) \rightarrow (-\omega_3, \omega_2)}.$$

Therefore, a solution of (5.7) valid for $|q| = 1$ is obtained by using the described parametrization of variables and by making a particular modular transformation

$$U_{|q|=1}(\underline{t}; p, q) = U(\underline{t}; p, q)\big|_{(\omega_2, \omega_3) \rightarrow (-\omega_3, \omega_2)}.$$

Let us give another form of the elliptic hypergeometric equation. We single out the variable $x$ by setting $t_1 = cx$, $t_2 = c/x$ and denote

$$\varepsilon_1 = \frac{c}{t_3}, \quad \varepsilon_2 = \frac{\varepsilon_1}{q}, \quad \varepsilon_3 = ct_3 p^4, \quad \varepsilon_k = \frac{q}{ct_k}, \quad k = 4, \ldots, 8.$$

Since $c = \sqrt{t_1 t_2}$, one has the same balancing condition $\prod_{k=1}^{8} \varepsilon_k = p^2 q^2$. Evidently, scalings of parameters of the $U$-function in (5.7) are equivalent to the shifts $x \rightarrow q^{\pm 1} x$. After the replacement of $U(\underline{t})$ by some unknown function $f(x)$, (5.7) becomes a $q$-difference equation of the second order of the following symmetric form

$$A(x)\left(f(qx) - f(x)\right) + A(x^{-1})\left(f(q^{-1}x) - f(x)\right) + vf(x) = 0, \quad (5.13)$$

$$A(x) = \frac{\prod_{k=1}^{8} \theta(\varepsilon_k x; p)}{\theta(x^2, qx^2; p)}, \qquad v = \prod_{k=3}^{8} \theta\left(\frac{\varepsilon_k \varepsilon_1}{q}; p\right). \qquad (5.14)$$

Note that here $\varepsilon_k$-variables are constrained not only by the balancing condition, but also by the additional relation $\varepsilon_2 = \varepsilon_1/q$.

Clearly Eq. (5.13) has only $S_6$-symmetry in parameters $\varepsilon_k$, $k = 3, \ldots, 8$. However, as noticed by Zagier, the potential $\mathcal{A}$ from (5.8) itself can be written in a completely $S_8$-symmetric form. Indeed, denote

$$u_1 = \frac{t_1}{t_3}, \ u_2 = \frac{t_1}{qt_3}, \ u_3 = \frac{1}{t_1 t_3}, \ u_k = \frac{t_k t_2}{q}, \ k = 4, \ldots, 8, \ \lambda = \frac{t_2}{qt_3}.$$

Then one can write

$$\mathcal{A}(t_1, \ldots, t_8, q; p) = \frac{\lambda^2}{p^2} \prod_{k=1}^{8} \frac{\theta(u_k; p)}{\theta(v_k; p)}, \quad u_k v_k = \lambda, \quad \prod_{k=1}^{8} u_k = p^2 \lambda^4.$$

All $u_k$ variables are independent and $\lambda$ is determined by their product, i.e. the presence of the $S_8$ symmetry becomes evident.

Because of the distinguished role of the elliptic hypergeometric equation it is interesting to know all its roots of origin. It appears [54] that Eq. (5.13) is related to the eigenvalue problem $H\psi = E\psi$ for the restricted one particle Hamiltonian of the van Diejen model [66]. Namely, one has to take special eigenvalue $E = -v$ and impose two additional constraints on the parameters of the general model—the balancing condition and $\varepsilon_2 = \varepsilon_1/q$. Another place where this equation emerges in a natural way is the theory of elliptic Painlevé equation [44]. Namely, for a special restriction on the geometry of this equation it linearizes exactly to the elliptic hypergeometric equation [25]. In a related subject it emerges as the simplest rigid equation in the elliptic isomonodromy problem [34, 35]. A list of degenerations of the $V$-function to the lower level hypergeometric functions is considered in detail in [46, 65].

## 6   Multiple Elliptic Hypergeometric Integrals

There are many multiple integral generalizations of the elliptic beta integral evaluation and of the $V$-function. For all of them the integrands satisfy a set of linear $q$-difference equations of the first order in the integration variables with the elliptic function coefficients, similar to the univariate case.

We present the most useful examples of integrals associated with the root systems $C_n$ and $A_n$. In [68] it was suggested to distinguish two types of the multiple elliptic beta integrals: those for which the number of parameters depends on the rank of the root system were tagged as type I, and for type II this number is fixed. There is also a difference in the methods of proving their evaluation formulas.

So, the type I integral on the $C_n$ root system has the following form. Take $2n + 4$ complex parameters $t_1, \ldots, t_{2n+4}$ and bases $p, q$ with the absolute values $|p|, |q|, |t_j| < 1$, and impose the balancing condition $\prod_{j=1}^{2n+4} t_j = pq$. Then one has

$$
\kappa_n \int_{\mathbb{T}^n} \prod_{1 \le j < k \le n} \frac{1}{\Gamma(z_j^{\pm 1} z_k^{\pm 1}; p, q)} \prod_{j=1}^n \frac{\prod_{i=1}^{2n+4} \Gamma(t_i z_j^{\pm 1}; p, q)}{\Gamma(z_j^{\pm 2}; p, q)} \prod_{j=1}^n \frac{dz_j}{z_j}
$$

$$
= \prod_{1 \le i < j \le 2n+4} \Gamma(t_i t_j; p, q), \qquad \kappa_n = \frac{(p; p)_\infty^n (q; q)_\infty^n}{(4\pi i)^n n!}. \qquad (6.1)
$$

The simplest proof of this relation uses a direct generalization of the method described above for the univariate case. The ratio of the integrand and the right-hand side expression satisfies a linear difference equation in parameters and integration variables similar to (4.2). Other univariate arguments generalize as well [53], which yields (6.1). The original work [68], where this formula was suggested, contained only its partial justification. The first complete proof was given by Rains [32] using a different method and in a substantially more general setting. Namely, the following transformation formula was established in [32]

$$
I_n^{(m)}(t_1, \ldots, t_{2n+2m+4}) = \prod_{1 \le i < j \le 2n+2m+4} \Gamma(t_i t_j; p, q) \, I_m^{(n)} \left( \frac{\sqrt{pq}}{t_1}, \ldots, \frac{\sqrt{pq}}{t_{2n+2m+4}} \right)
$$

$$(6.2)$$

for the integrals

$$
I_n^{(m)}(t) = \kappa_n \int_{\mathbb{T}^n} \prod_{1 \le i < j \le n} \frac{1}{\Gamma(z_i^{\pm 1} z_j^{\pm 1}; p, q)} \prod_{j=1}^n \frac{\prod_{i=1}^{2n+2m+4} \Gamma(t_i z_j^{\pm 1}; p, q)}{\Gamma(z_j^{\pm 2}; p, q)} \frac{dz_j}{z_j},
$$

where $|t_j| < 1$ and $\prod_{j=1}^{2n+2m+4} t_j = (pq)^{m+1}$. As shown in [32], analytically the product $\prod_{1 \le k < l \le 2n+2m+4}(t_k t_l; p, q)_\infty I_n^{(m)}(t)$ is a holomorphic function of its parameters. Relation (6.2) can be considered as an elliptic analogue of the symmetry transformation for ordinary hypergeometric integrals established by Dixon [13]. Clearly it represents a multivariable extension of the third $V$-function symmetry transformation (5.4).

In [37] these integrals were written as determinants of univariate integrals

$$
I_n^{(m)}(t_1, \ldots, t_{2n+2m+4}) = \prod_{1 \le i < j \le n} \frac{1}{a_j \theta(a_i a_j^{\pm 1}; p) b_j \theta(b_i b_j^{\pm 1}; q)}
$$

$$\times \det_{1\leq i,j\leq n}\left(\kappa \int_{\mathbb{T}} \frac{\prod_{r=1}^{2n+2m+4}\Gamma(t_r z^{\pm 1}; p,q)}{\Gamma(z^{\pm 2}; p,q)}\prod_{k\neq i}\theta(a_k z^{\pm 1}; p)\prod_{k\neq j}\theta(b_k z^{\pm 1}; q)\frac{dz}{z}\right),$$

where $a_i, b_i$ are arbitrary auxiliary variables. Curiously, the Casoratian (5.11) emerges here as the required determinant for the choice $n = 2, m = 0$ and $a_i = b_i = t_i$, which yields the evaluation formula (6.1) for $n = 2$.

For the description of type II $C_n$ elliptic beta integral introduced in [67] one needs seven complex parameters $t, t_a$, $a = 1, \ldots, 6$, and bases $p$ and $q$ lying inside the unit disk $|p|, |q|, |t|, |t_a| < 1$, and satisfying the balancing condition $t^{2n-2}\prod_{i=1}^{6} t_i = pq$. Then the following integral evaluation holds true

$$\kappa_n \int_{\mathbb{T}^n}\prod_{1\leq j<k\leq n}\frac{\Gamma(t z_j^{\pm 1}z_k^{\pm 1}; p,q)}{\Gamma(z_j^{\pm 1}z_k^{\pm 1}; p,q)}\prod_{j=1}^{n}\frac{\prod_{i=1}^{6}\Gamma(t_i z_j^{\pm 1}; p,q)}{\Gamma(z_j^{\pm 2}; p,q)}\prod_{j=1}^{n}\frac{dz_j}{z_j}$$

$$= \prod_{j=1}^{n}\left(\frac{\Gamma(t^j; p,q)}{\Gamma(t; p,q)}\prod_{1\leq i<k\leq 6}\Gamma(t^{j-1}t_i t_k; p,q)\right). \tag{6.3}$$

As mentioned, the type II integral can be proved by a different method than the type I case [68]. Assuming that $t_6$ is a dependent variable, we denote the integral on the left-hand side of (6.3) as $I_n(t, t_1, \ldots, t_5)$ and consider the $(2n-1)$-fold integral

$$\int_{\mathbb{T}^{2n-1}}\prod_{1\leq j<k\leq n}\frac{1}{\Gamma(z_j^{\pm 1}z_k^{\pm 1}; p,q)}\prod_{l=1}^{n}\frac{\prod_{r=0}^{5}\Gamma(t_r z_l^{\pm 1}; p,q)}{\Gamma(z_l^{\pm 2}; p,q)}\frac{dz_l}{z_l}$$

$$\times \prod_{\substack{1\leq j\leq n\\1\leq k\leq n-1}}\Gamma(t^{1/2}z_j^{\pm 1}w_k^{\pm 1}; p,q)\prod_{1\leq j<k\leq n-1}\frac{1}{\Gamma(w_j^{\pm 1}w_k^{\pm 1}; p,q)}$$

$$\times \prod_{j=1}^{n-1}\frac{\Gamma(w_j^{\pm 1}t^{n-3/2}\prod_{s=1}^{5}t_s; p,q)}{\Gamma(w_j^{\pm 2}, w_j^{\pm 1}t^{2n-3/2}\prod_{s=1}^{5}t_s; p,q)}\frac{dw_j}{w_j},$$

where we introduced an auxiliary variable $t_0$ via the relation $t^{n-1}\prod_{r=0}^{5}t_r = pq$.

Integrals over $w_j$ or $z_j$ can be computed explicitly using the type I $C_n$-integral (6.3). Doing these integrations in different orders, one obtains the recurrence relation:

$$I_n(t, t_1, \ldots, t_5) = \frac{\Gamma(t^n; p,q)}{\Gamma(t; p,q)}\prod_{0\leq r<s\leq 5}\Gamma(t_r t_s; p,q)\, I_{n-1}(t, t^{1/2}t_1, \ldots, t^{1/2}t_5)$$

with known $n = 1$ initial condition. Resolving this recurrence one comes to the desired formula.

Expressing one of the parameters $t_i$ in terms of others using the balancing condition and taking the limit $p \to 0$ for fixed values of independent parameters, one reduces the above integrals to Gustafson's $C_n$ $q$-beta integrals from [22]. Relation (6.3) has a meaning of an elliptic extension of the Selberg integral evaluation formula [1, 19], which emerges as a result of its sequential degenerations.

Let us present also an elliptic beta integral of type I for the $A_n$ root system suggested in [51] and proven in [32] and [53]. Take $2n + 4$ parameters $t_m, s_m$, $m = 1, \ldots, n + 2$, and bases $p, q$ satisfying the constraints $|p|, |q|, |t_m|, |s_m| < 1$ and the balancing condition $ST = pq$, where $S = \prod_{m=1}^{n+2} s_m$ and $T = \prod_{m=1}^{n+2} t_m$. Then the following integral can be computed explicitly

$$\mu_n \int_{\mathbb{T}^n} \prod_{1 \le j < k \le n+1} \frac{1}{\Gamma(z_j z_k^{-1}, z_j^{-1} z_k; p, q)} \prod_{j=1}^{n+1} \prod_{m=1}^{n+2} \Gamma(s_m z_j, t_m z_j^{-1}; p, q) \prod_{j=1}^{n} \frac{dz_j}{z_j}$$

$$= \prod_{m=1}^{n+2} \Gamma(Ss_m^{-1}, Tt_m^{-1}; p, q) \prod_{k,m=1}^{n+2} \Gamma(s_k t_m; p, q), \qquad (6.4)$$

where $z_1 z_2 \cdots z_{n+1} = 1$ and

$$\mu_n = \frac{(p; p)_\infty^n (q; q)_\infty^n}{(2\pi i)^n (n + 1)!}.$$

Relations to the root systems emerge from the following observation. Combinations of the integration variables of the form $z_j^{\pm 1} z_k^{\pm 1}$, $j < k$, $z_j^{\pm 2}$ in (6.1), (6.3) and $z_j z_k^{-1}$, $j \ne k$, in (6.4) can be identified with formal exponentials of the roots $\pm e_j \pm e_k$, $j < k$, $\pm 2 e_j$ and $e_j - e_k$, $j \ne k$, of the $C_n$ and $A_n$ root systems, respectively.

## 7 Rarefied Elliptic Hypergeometric Integrals

Recently a further modification of the elliptic hypergeometric integrals has been introduced in [26, 27, 39, 58]. It emerged from considerations of supersymmetric quantum field theories on particular four dimensional space-time background $S^1 \times L(r, -1)_\tau$ involving a special lens space. The general squashed lens space $L(r, k)_\tau$ is obtained from the squashed three-dimensional sphere in the complex representation $|\tau z_1|^2 + |\tau^{-1} z_2|^2 = 1$ by identification of the points $(e^{2\pi i/r} z_1, e^{2\pi i k/r} z_2) \sim (z_1, z_2)$ for $k, r$ positive coprime integers $0 < k < r$. Let us describe briefly corresponding generalizations of the elliptic hypergeometric identities.

A proper extension of the elliptic gamma function, associated with a special lens space, is determined by two standard elliptic gamma functions with different bases

$$\gamma^{(r)}(z, m; p, q) := \Gamma(z p^m; p^r, pq) \Gamma(z q^{r-m}; q^r, pq),$$

where one has two integer parameters $r \in \mathbb{Z}_{>0}$ and $m \in \mathbb{Z}$. Using the double elliptic gamma function $\Gamma(z; p, q, t)$ with a special choice $t = pq$, one can write

$$
\gamma^{(r)}(z, m; p, q) = \frac{\Gamma(q^r z p^m; p^r, q^r, pq)}{\Gamma(z p^m; p^r, q^r, pq)} \frac{\Gamma(p^r z q^{r-m}; p^r, q^r, pq)}{\Gamma(z q^{r-m}; p^r, q^r, pq)}
$$

$$
= \frac{\Gamma((pq)^m q^{r-m} z; p^r, q^r, pq)}{\Gamma(q^{r-m} z; p^r, q^r, pq)} \frac{\Gamma((pq)^{r-m} p^m z; p^r, q^r, pq)}{\Gamma(p^m z; p^r, q^r, pq)}, \quad (7.1)
$$

which yields the "rarefied" product representation for $\gamma^{(r)}(z, m; p, q)$ of the form

$$
\gamma^{(r)}(z, m; p, q) = \prod_{k=0}^{m-1} \Gamma(q^{r-m} z (pq)^k; p^r, q^r) \prod_{k=0}^{r-m-1} \Gamma(p^m z (pq)^k; p^r, q^r),
$$

$$
(7.2)
$$

valid for $0 \le m \le r$ (similar expression exists for other values of $m$). This function is quasiperiodic in the discrete variable

$$
\gamma^{(r)}(z, m + r; p, q) = (-z)^{-m} q^{m(m+1)/2} p^{-m(m-1)/2} \gamma^{(r)}(z, m; p, q). \quad (7.3)
$$

The normalized function

$$
\Gamma^{(r)}(z, m; p, q) := \left(-\frac{z}{\sqrt{pq}}\right)^{\frac{m(m-1)}{2}} \left(\frac{p}{q}\right)^{\frac{m(m-1)(2m-1)}{12}} \gamma^{(r)}(z, m; p, q). \quad (7.4)
$$

was called in [58] the rarefied elliptic gamma function. For $r = 1$ independently of $m$ one has $\Gamma^{(1)}(z, m; p, q) = \Gamma(z; p, q)$, which provides a very convenient verification of identities involving $\Gamma^{(r)}(z, m; p, q)$.

Let us describe some properties of this function. The $p, q$ permutational symmetry changes to

$$
\Gamma^{(r)}(z, m; p, q) = \Gamma^{(r)}(z, -m; q, p). \quad (7.5)
$$

Instead of the plain difference equations one has simple recurrence relations

$$
\Gamma^{(r)}(qz, m + 1; p, q) = (-z)^m p^{\frac{m(m-1)}{2}} \theta(z p^m; p^r) \Gamma^{(r)}(z, m; p, q),
$$

$$
\Gamma^{(r)}(pz, m - 1; p, q) = (-z)^{-m} q^{\frac{m(m+1)}{2}} \theta(z q^{-m}; q^r) \Gamma^{(r)}(z, m; p, q). \quad (7.6)
$$

The inversion relation takes the form

$$
\Gamma^{(r)}(z, m; p, q) \Gamma^{(r)}(\tfrac{pq}{z}, -m; p, q) = 1, \quad (7.7)
$$

and the limiting relation needed for the residue calculus reads

$$\lim_{z\to 1}(1-z)\Gamma^{(r)}(z,0;p,q) = \lim_{z\to 1}(1-z)\gamma^{(r)}(z,0;p,q) = \frac{1}{(p^r;p^r)_\infty(q^r;q^r)_\infty}.$$

(7.8)

*Exercise:* verify all these relations.

The rarefied version of the elliptic beta integral has the following form. We take continuous parameters $t_1,\ldots,t_6,p,q$ and discrete ones $n_1,\ldots,n_6 \in \mathbb{Z}+\nu$, where $\nu = 0, \frac{1}{2}$, satisfying the constraints $|t_a|,|p|,|q| < 1$ and the balancing condition

$$\prod_{a=1}^{6} t_a = pq, \qquad \sum_{a=1}^{6} n_a = 0.$$

Then

$$\kappa^{(r)} \sum_{m\in\mathbb{Z}_r+\nu} \int_{\mathbb{T}} \rho^{(r)}(z,m;\underline{t},\underline{n})\frac{dz}{z} = \prod_{1\le a<b\le 6} \Gamma^{(r)}(t_a t_b, n_a+n_b; p,q),$$

(7.9)

where $\mathbb{T}$ is the positively oriented unit circle,

$$\kappa^{(r)} = \frac{(p^r;p^r)_\infty(q^r;q^r)_\infty}{4\pi i},$$

and the integrand has the form

$$\rho^{(r)}(z,m;\underline{t},\underline{n}) := \frac{\prod_{a=1}^{6}\Gamma^{(r)}(t_a z^{\pm 1}, n_a\pm m; p,q)}{\Gamma^{(r)}(z^{\pm 2},\pm 2m); p,q)}.$$

(7.10)

Here we use the compact notation

$$\Gamma^{(r)}(tz^{\pm 1},n\pm m;p,q) := \Gamma^{(r)}(tz,n+m;p,q)\Gamma^{(r)}(tz^{-1},n-m;p,q).$$

(7.11)

For $r = 1$ one gets relation (4.1) and the $r > 1$, $\nu = 0$ case of the evaluation (7.9) was established by Kels in [26], for $r > 1$, $\nu = \frac{1}{2}$ it was proven in [58] in the presented form and in [27] in the equivalent form of $A_1 \leftrightarrow A_0$ symmetry transformation.

A good calculational exercise is the proof of periodicity

$$\rho^{(r)}(z,m+r;\underline{t},\underline{n}) = \rho^{(r)}(z,m;\underline{t},\underline{n}),$$

(7.12)

because of which the sum over $m-\nu = 0,1,\ldots,r-1$ is equal to sums over any $r$ consecutive values of $m$. There is a particular symmetry between the terms in this sum following from the obvious relation

$$\rho^{(r)}(z,-m;\underline{t},\underline{n}) = \rho^{(r)}(z^{-1},m;\underline{t},\underline{n}).$$

Due to the $r$-periodicity in $m$ one has

$$c_{r-m} := \int_{\mathbb{T}} \rho^{(r)}(z, r-m; \underline{t}, \underline{n}) \frac{dz}{z} = \int_{\mathbb{T}} \rho^{(r)}(z^{-1}, m; \underline{t}, \underline{n}) \frac{dz}{z}$$

$$= \int_{\mathbb{T}} \rho^{(r)}(z, m; \underline{t}, \underline{n}) \frac{dz}{z} = c_m.$$

As a result the sum over $m$ in (7.9) can be written for $\nu = 0$ as

$$\sum_{m=0}^{r-1} c_m = \begin{cases} c_0 + c_{r/2} + 2 \sum_{m=1}^{r/2-1} c_m & \text{for even } r, \\ c_0 + 2 \sum_{m=1}^{(r-1)/2} c_m & \text{for odd } r, \end{cases} \tag{7.13}$$

and for $\nu = \frac{1}{2}$ as

$$\sum_{m=1/2}^{r-1/2} c_m = \begin{cases} 2 \sum_{m=1/2}^{r/2-1/2} c_m & \text{for even } r, \\ c_{r/2} + 2 \sum_{m=1/2}^{(r-2)/2} c_m & \text{for odd } r. \end{cases} \tag{7.14}$$

The type I multiple rarefied elliptic beta integral for the root system $C_n$ has the form

$$\kappa_n^{(r)} \sum_{m_1,\ldots,m_n \in \mathbb{Z}_k + \nu} \int_{\mathbb{T}^n} \rho_I(z_j, m_j; \underline{t}, \underline{n}) \prod_{j=1}^n \frac{dz_j}{z_j} = \prod_{1 \le a < b \le 2n+4} \Gamma(t_a t_b, n_a + n_b; p, q),$$
$$\tag{7.15}$$

where $\mathbb{T}$ is the unit circle of positive orientation, $\kappa_n^{(r)}$ is obtained from $\kappa_n$ after replacing $p, q \to p^r, q^r$, and the kernel is

$$\rho_I(z_j, m_j; \underline{t}, \underline{n}) := \prod_{1 \le j < k \le n} \frac{1}{\Gamma(z_j^{\pm 1} z_k^{\pm 1}, \pm m_j \pm m_k)} \prod_{j=1}^n \frac{\prod_{a=1}^{2n+4} \Gamma(t_a^{\pm 1} z_j, n_a \pm m_j)}{\Gamma(z_j^{\pm 2}, \pm 2m_j)},$$

where parameters $t_a, z_j \in \mathbb{C}^\times$, $n_a, m_j \in \mathbb{Z} + \nu$, satisfy the constraints $|t_a| < 1$ and the balancing condition

$$\prod_{a=1}^{2n+4} t_a = pq, \qquad \sum_{a=1}^{2n+4} n_a = 0. \tag{7.16}$$

The proof of the univariate case $n = 1$ can be adapted to the present situation by adjoining the peculiarities characteristic to the proof of type I integral (6.1) as well as the $r$-periodicity of the kernel in the discrete summation variables.

Similarly one can construct a computable rarefied type II $C_d$-integral, where for convenience we denoted the rank of the root system as $d$. For that it is necessary to take continuous parameters $t, t_a \in \mathbb{C}^\times, a = 1, \ldots, 6$, and bases $p, q$ such that

$|p|, |q|, |t|, |t_a| < 1$. Additionally, one needs eight discrete variables $n \in \mathbb{Z}, n_a \in \mathbb{Z} + \nu$, all together satisfying the balancing condition

$$t^{2d-2} \prod_{a=1}^{6} t_a = pq, \qquad 2n(d-1) + \sum_{a=1}^{6} n_a = 0. \tag{7.17}$$

Then

$$\kappa_d^{(r)} \sum_{m_1,\ldots,m_d \in \mathbb{Z}_r + \nu} \int_{\mathbb{T}^d} \prod_{1 \le j < k \le d} \frac{\Gamma(tz_j^{\pm 1} z_k^{\pm 1}, n \pm m_j \pm m_k)}{\Gamma(z_j^{\pm 1} z_k^{\pm 1}, \pm m_j \pm m_k)} \prod_{j=1}^{d} \frac{\prod_{a=1}^{6} \Gamma(t_a z_j^{\pm 1}, n_a \pm m_j)}{\Gamma(z_j^{\pm 2}, \pm 2m_j)} \frac{dz_j}{z_j}$$

$$= \prod_{j=1}^{d} \left( \frac{\Gamma(t^j, nj)}{\Gamma(t, n)} \prod_{1 \le a < b \le 6} \Gamma(t^{j-1} t_a t_b, n(j-1) + n_a + n_b) \right). \tag{7.18}$$

This formula is proved in a way similar to the $r = 1$ case (6.1), i.e. by considering a $(2d - 1)$-fold combination of summations and integrations of a specific function admitting usage of the rarefied type I $C_d$-formula (7.15) in two different sets of discrete summation and continuous integration variables which establishes a recurrence relation in the rank of the root system. For more details on these results, as well as generalizations of the $V$-function and elliptic hypergeometric equation, see [58]. Symmetry transformations for some multidimensional elliptic hypergeometric integrals were extended to the rarefied case in [27].

The rarefied hyperbolic hypergeometric integrals for general lens space were discussed in [12, 45]. In particular, in [45] a general univariate computable rarefied hyperbolic beta integral evaluation formula has been established.

## 8  An Integral Bailey Lemma

Using properties of the elliptic beta integral, the following integral transformation was introduced in [52]

$$\beta(w, t) = M(t)_{wz} \alpha(z, t) := \frac{(p; p)_{\infty}(q; q)_{\infty}}{4\pi i} \int_{\mathbb{T}} \frac{\Gamma(tw^{\pm 1} z^{\pm 1}; p, q)}{\Gamma(t^2, z^{\pm 2}; p, q)} \alpha(z, t) \frac{dz}{z} \tag{8.1}$$

with the assumption that $|tw^{\pm 1}| < 1$. The latter constraints can be relaxed by analytic continuation, e.g. by deforming the contour of integration, provided no singularities of the integrand are crossed during such a deformation. Pairs of functions connected by (8.1) were called integral elliptic Bailey pairs with respect to the parameter $t$. Using the evaluation formula (4.1) one can find a particular explicit Bailey pair $\alpha(z, t)$ and $\beta(z, t)$. Such a terminology emerged from the theory of

Bailey chains providing a systematic tool for constructing nontrivial identities for $q$-hypergeometric series [70]. In particular, it was targeted at the proof of Rogers-Ramanujan type identities. The definition (8.1) yielded the very first generalization of the Bailey chains technique from series to integrals.

As shown in [63], on the space of $A_1$-symmetric functions $f(z) = f(z^{-1})$ under particular constraints on the parameters and appropriate choice of the integration contours for analytically continued operators, the operators $M(t^{-1})_{wz}$ and $M(t)_{wz}$ become inversions one of the other. Passing to the real line integration one can use the generalized functions and symbolically write $M(t^{-1})M(t) = 1$, where 1 means an integral operator with the Dirac delta-function kernel [56, 57]. It is due to this $t \to t^{-1}$ inversion relation, which looks similar to the Fourier transform, that the transformation (8.1) is referred to as the "elliptic Fourier transformation". Another similarity is that in both cases some nontrivial operators—the derivative and $q$-scaling are converted to the multiplication by a function—the linear and theta functions, respectively.

Let us indicate how the true Fourier transform actually emerges in a particular degeneration limit of (8.1). Take first the limit $p \to 0$ for fixed $q, t$ and $w$. This yields

$$\beta(w, t) = \frac{(q; q)_\infty}{4\pi i} \int_{\mathbb{T}} \frac{(t^2, z^{\pm 2}; q)_\infty}{(tw^{\pm 1}z^{\pm 1}; q)_\infty} \alpha(z, t) \frac{dz}{z}.$$

In the integrand one can write

$$(z^{\pm 2}; q)_\infty = (z^{\pm 1}, -z^{\pm 1}; q)_\infty (q^{1/2}z^{\pm 1}, -q^{1/2}z^{\pm 1}; q)_\infty.$$

We can rewrite the above transform in a renormalized form

$$\tilde{\beta}(w) = \frac{1}{4\pi i} \int_{\mathbb{T}} \frac{(z^{\pm 1}, -z^{\pm 1}; q)_\infty}{(tw^{\pm 1}z^{\pm 1}; q)_\infty} \tilde{\alpha}(z) \frac{dz}{z},$$

where $\tilde{\beta}(w) := \beta(w, t)/(q, t^2; q)_\infty$ and $\tilde{\alpha}(z) := (q^{1/2}z^{\pm 1}, -q^{1/2}z^{\pm 1}; q)_\infty \alpha(z, t)$. Passing to the angular parameter $\theta$, $z = e^{i\theta}$, and introducing a new integration variable $x = \cos\theta$ we obtain $\int_{\mathbb{T}} dz/z = 2i \int_{-1}^{1} dx/\sqrt{1 - x^2}$. Denoting $tw = q^{\alpha + 1/2}$, $tw^{-1} = -q^{\beta + 1/2}$, and using the $q$-binomial limiting relation (4.3) we deduce

$$\lim_{q \to 1^-} \frac{(z^{\pm 1}, -z^{\pm 1}; q)_\infty}{(q^{\alpha + 1/2}z^{\pm 1}, -q^{\beta + 1/2}z^{\pm 1}; q)_\infty} = 2^{\alpha + \beta + 1}(1 - x)^{\alpha + 1/2}(1 + x)^{\beta + 1/2}$$

and come to the integral transform

$$g(\alpha, \beta) = \frac{2^{\alpha + \beta}}{\pi} \int_{-1}^{1} (1 - x)^\alpha (1 + x)^\beta f(x) dx.$$

where we have to assume that $\mathrm{Re}(\alpha)$, $\mathrm{Re}(\beta) > -1$ for convergence of the integral for regular functions $f(x)$. Rescaling in the integral $x \to x/\lambda$ and taking the limit $\lambda \to +\infty$, we obtain asymptotically the transform

$$\frac{2^{\alpha+\beta}}{\pi\lambda} \int_{-\infty}^{\infty} e^{\frac{\beta-\alpha}{\lambda}x - \frac{\alpha+\beta}{2\lambda^2}x^2 + O(\frac{\beta-\alpha}{\lambda^3})} \tilde{f}(x)dx,$$

where $\tilde{f}(x) = \lim_{\lambda\to\infty} f(x/\lambda)$. Demanding that $\alpha + \beta = o(\lambda^2)$ and $\beta - \alpha = iy\lambda$ for some finite variable $y$, we obtain

$$\frac{2^{\alpha+\beta}}{\pi\lambda} \int_{-\infty}^{\infty} e^{iyx} \tilde{f}(x)dx,$$

which is the standard Fourier transformation up to some diverging factor. So, in terms of the original variables, the action of the integral operator (8.1) passes to the ordinary Fourier transformation after setting $p = 0$, proper normalization of the source and image functions, and taking the limit $q \to 1^-$ in the parameterization $w = -iq^{-iy\lambda/2}$, $z + z^{-1} = 2x/\lambda$, $t = iq^c$ with the subsequent limit $\lambda \to +\infty$ and the constraint that $c = (\alpha + \beta + 1)/2$ is an arbitrary parameter which may grow only slower than $\lambda^2$.

The integral Bailey lemma provides an algorithm for constructing infinitely many Bailey pairs from a given one. It is formulated as follows. Let $\alpha(z, t)$ and $\beta(z, t)$ be some functions related by (8.1) for some parameter $t$. Then the functions

$$\alpha'(w, st) = D(s; y, w)\alpha(w, t), \quad D(s; y, w) = \Gamma(\sqrt{pq}s^{-1}y^{\pm 1}w^{\pm 1}; p, q),$$

$$\beta'(w, st) = D(t^{-1}; y, w)M(s)_{wx}D(st; y, x)\beta(x, t), \tag{8.2}$$

where $w \in \mathbb{T}$, $|s|, |t| < 1$, $|\sqrt{pq}y^{\pm 1}| < |st|$, form an integral elliptic Bailey pair with respect to the parameter $st$. Note that the parameters $s$ and $y$ are two new arbitrary variables.

It is necessary to show that $\beta'(w, st) = M(st)_{wz}\alpha'(z, st)$. Substitute in both sides of this equality the definitions (8.2) and use the relation $D(t^{-1}; y, w) = 1/D(t; y, w)$ following from the elliptic gamma function inversion property. This yields the operator identity

$$M(s)_{wx}D(st; y, x)M(t)_{xz} = D(t; y, w)M(st)_{wz}D(s; y, z). \tag{8.3}$$

Substitution of the explicit forms of $M$- and $D$-operators shows that the integral over the variable $x$ on the left-hand side of (8.3) can be computed explicitly using the elliptic beta integral evaluation formula. The resulting expression takes exactly the form given on the right-hand side.

Iterative applications of the maps (8.2) lead to a chain of Bailey pairs satisfying by definition the key relation (8.1). Explicitly this leads to certain nontrivial identities for multiple elliptic hypergeometric integrals. For instance, if the pair

$\alpha$ and $\beta$ is determined from the formula (4.1), then the relation $\beta'(w, st) = M(st)_{wz}\alpha'(z, st)$ yields the key $W(E_7)$-transformation for the $V$-function (5.2).

As shown in [10] the algebraic relations emerging from the described integral Bailey lemma can have the meaning of Coxeter relations for a permutation group. For that interpretation we introduce three operators $S_{1,2,3}(\mathbf{t})$ acting on the functions of two complex variables $f(z_1, z_2)$ as follows

$$[S_1(\mathbf{t})f](z_1, z_2) := M(t_1/t_2)_{z_1z}f(z, z_2),$$

$$[S_2(\mathbf{t})f](z_1, z_2) := D(t_2/t_3; z_1, z_2)f(z_1, z_2),$$

$$[S_3(\mathbf{t})f](z_1, z_2) := M(t_3/t_4)_{z_2z}f(z_1, z),$$

for some complex parameters $\mathbf{t} = (t_1, t_2, t_3, t_4)$. The products of these operators are defined via the cocycle condition

$$S_jS_k := S_j(s_k(\mathbf{t}))S_k(\mathbf{t}),$$

where $s_k$ are elementary transposition operators generating the permutation group $\mathfrak{S}_4$:

$$s_1(\mathbf{t}) = (t_2, t_1, t_3, t_4), \quad s_2(\mathbf{t}) = (t_1, t_3, t_2, t_4), \quad s_3(\mathbf{t}) = (t_1, t_2, t_4, t_3).$$

Now one can check validity of the Coxeter relations

$$S_j^2 = 1, \quad S_iS_j = S_jS_i \text{ for } |i - j| > 1, \quad S_jS_{j+1}S_j = S_{j+1}S_jS_{j+1} \qquad (8.4)$$

as a consequence of properties of the Bailey lemma operator entries. The quadratic relations represent inversion relations for the $M$- and $D$-operators. The cubic relation is equivalent to (8.3) and it is called also the star-triangle relation. A somewhat different application of the operator identity (8.3) is considered in [36]. Extension of the above considerations to the rarefied elliptic beta integral was considered in [59].

Let us replace in (8.3) all variables $z \to e^{iz}$, $x \to e^{ix}$, $y \to e^{iy}$, $w \to e^{iw}$ and denote $s = e^{-\alpha}$, $t = e^{-\beta}$, $pq = e^{-2\eta}$, and pass to the integrations over the line segment $x, z \in [0, 2\pi]$. Applying now this operator identity to the Dirac delta-function $(\delta(z - u) + \delta(z + u))/2$ for some parameter $u$, one comes to formula (4.1) written in the form

$$\int_0^{2\pi} \rho(x)D_{\eta-\alpha}(w, x)D_{\alpha+\beta}(y, x)D_{\eta-\beta}(u, x)dx$$

$$= \chi(\alpha, \beta)D_\beta(y, w)D_{\eta-\alpha-\beta}(w, u)D_\alpha(y, u), \qquad (8.5)$$

where

$$D_\alpha(y, u) = D(e^{-\alpha}; e^{iy}, e^{iu}) = \Gamma(e^{\alpha-\eta\pm iy\pm iu}; p, q) \qquad (8.6)$$

and

$$\rho(u) = \frac{(p; p)_\infty (q; q)_\infty}{2\Gamma(e^{\pm 2iu}; p, q)}, \quad \chi(\alpha, \beta) = \Gamma(e^{-2\alpha}, e^{-2\beta}, e^{2\alpha+2\beta-2\eta}; p, q).$$

In [5] this form of the star-triangle relation was used for building a new two-dimensional integrable lattice model. Namely, one considers a two-dimensional square lattice and ascribes the Boltzmann weight $D_\alpha(x, u)$ to the horizontal edges connecting continuous spins $x$ and $u$ sitting in the neighboring vertices of the lattice. The vertical edges have Boltzmann weights $D_{\eta-\alpha}(x, u)$. Each vertex has the self-interaction energy $\rho(u)$.

Let us substitute in (8.5) $D_\alpha(y, w) = m(\alpha)W_\alpha(y, w)$ and choose the normalization constant $m(\alpha)$ from the condition

$$\frac{m(\alpha)m(\beta)m(\eta - \alpha - \beta)}{m(\eta - \alpha)m(\eta - \beta)m(\alpha + \beta)} \chi(\alpha, \beta) = 1. \tag{8.7}$$

This gives a compact block representation of the elliptic beta integral evaluation

$$\int_0^{2\pi} \rho(x) W_{\eta-\alpha}(w, x) W_{\alpha+\beta}(y, x) W_{\eta-\beta}(u, x) dx = W_\beta(y, w) W_{\eta-\alpha-\beta}(w, u) W_\alpha(y, u).$$

Equality (8.7) holds true, if

$$m(\alpha + \eta) = \Gamma(e^{2\alpha}; p, q)m(-\alpha).$$

In order to compute $m(\alpha)$ it is convenient to consider the function

$$\mu(x; p, q, t) = \frac{\Gamma(xt\sqrt{pqt}; p, q, t^2)}{\Gamma(x^{-1}t\sqrt{pqt}; p, q, t^2)} = \exp\left(\sum_{n\in\mathbb{Z}/\{0\}} \frac{(\sqrt{pqt}x)^n}{n(1 - p^n)(1 - q^n)(1 + t^n)}\right), \tag{8.8}$$

where $\Gamma(z; p, q, t^2)$ is the second order elliptic gamma function with bases $p, q, t^2$. One has the evident reflection equation $\mu(x^{-1}; p, q, t)\mu(x; p, q, t) = 1$. Another easily verifiable functional equation,

$$\mu(x; p, q, t)\mu(t^{-1}x; p, q, t) = \Gamma\left(x\sqrt{\frac{pq}{t}}; p, q\right),$$

becomes equivalent to the equation for $m(\alpha)$ after setting $t = pq$ and denoting $x = e^{2\alpha}$. As a result, we find the normalizing factor of interest

$$m(\alpha) = \frac{\Gamma(e^{2\alpha}(pq)^2; p, q, (pq)^2)}{\Gamma(e^{-2\alpha}(pq)^2; p, q, (pq)^2)}, \quad m(\alpha)m(-\alpha) = 1. \tag{8.9}$$

The partition function of the described lattice model has the form

$$Z = \int \prod_{(ij)} W_\alpha(u_i, u_j) \prod_{(kl)} W_{\eta-\alpha}(u_k, u_l) \prod_m \rho(u_m) du_m,$$

where the product $\prod_{(ij)}$ is taken over the horizontal edges, the product $\prod_{(kl)}$ takes into account vertical edges, and the product in $m$ counts self-energies of all lattice vertices. As argued in [5], for the edge Boltzmann weights $W_\alpha(x, u)$ the free energy per edge vanishes in the thermodynamic limit, i.e. $\lim_{N, M \to \infty} \frac{1}{NM} \log Z = 0$, where $N$ and $M$ are the numbers of edges in the rows and columns of the lattice. As observed in [57], the partition function $Z$ and similar ones describe superconformal indices of four dimensional supersymmetric quiver gauge theories and the integrability conditions represent certain electromagnetic dualities of such theories (see the next section).

The star-triangle relation can be used for constructing $R$-matrices satisfying the Yang–Baxter equation. We skip consideration of this subject, limiting to the statement that the elliptic Fourier transformation operator serves as the intertwining operator of equivalent representations of the Sklyanin algebra [48], emerging from the $RLL$-relation associated with Baxter's 8-vertex model [4]. More precisely, the Sklyanin algebra is generated by four operators $\mathbf{S}^a$ satisfying quadratic relations

$$\mathbf{S}^\alpha \mathbf{S}^\beta - \mathbf{S}^\beta \mathbf{S}^\alpha = i \left( \mathbf{S}^0 \mathbf{S}^\gamma + \mathbf{S}^\gamma \mathbf{S}^0 \right),$$

$$\mathbf{S}^0 \mathbf{S}^\alpha - \mathbf{S}^\alpha \mathbf{S}^0 = i J_{\beta\gamma} \left( \mathbf{S}^\beta \mathbf{S}^\gamma + \mathbf{S}^\gamma \mathbf{S}^\beta \right), \tag{8.10}$$

where the structure constants $J_{\beta\gamma} = (J_\gamma - J_\beta)/J_\alpha$ and $(\alpha, \beta, \gamma)$ is an arbitrary cyclic permutation of $(1, 2, 3)$. An explicit realization of $\mathbf{S}^a(g)$ by finite-difference operators has been found in [48]

$$\mathbf{S}_z^a(g) = e^{\pi i z^2/\eta} \frac{i^{\delta_{a,2}} \theta_{a+1}(\eta|\tau)}{\theta_1(2z|\tau)} \left[ \theta_{a+1} \left( 2z - g + \eta|\tau \right) e^{\eta \partial_z} \right.$$
$$\left. - \theta_{a+1} \left( -2z - g + \eta|\tau \right) e^{-\eta \partial_z} \right] e^{-\pi i z^2/\eta}, \tag{8.11}$$

where $e^{\pm \eta \partial_z}$ denote the shift operators, $e^{\pm \eta \partial_z} f(z) = f(z \pm \eta)$, and the standard theta functions are

$$\theta_2(z|\tau) = \theta_1(z + \tfrac{1}{2}|\tau), \quad \theta_3(z|\tau) = e^{\frac{\pi i \tau}{4} + \pi i z} \theta_2(z + \tfrac{\tau}{2}|\tau), \quad \theta_4(z|\tau) = \theta_3(z + \tfrac{1}{2}|\tau).$$

We added the subindex $z$ to the operators $\mathbf{S}_z^a(g)$ in order to indicate the arguments of the functions which they are acting on. The usual notation for the variable $g$ is $g = \eta(2\ell + 1)$, where $\ell \in \mathbb{C}$ is called the spin. The Casimir operators have the form

$$\mathbf{K}_0 = \sum_{a=0}^{3} \mathbf{S}^a \mathbf{S}^a = 4\theta_1^2(g|\tau), \quad \mathbf{K}_2 = \sum_{\alpha=1}^{3} J_\alpha \mathbf{S}^\alpha \mathbf{S}^\alpha = 4\theta_1(g - \eta|\tau)\theta_1(g + \eta|\tau).$$

They are invariant with respect to the transformation $g \to -g$, i.e. parameters $g$ and $-g$ correspond to equivalent representations of the Sklyanin algebra.

In (8.11) the operators $\mathbf{S}_z^a$ found in [48] are conjugated by exponentials $e^{\pm \pi i z^2/\eta}$, which is done for a special reason. Let us denote $q = e^{4\pi i \eta}$, $p = e^{2\pi i \tau}$, and $t = e^{-2\pi i g}$. Then one has the following intertwining relations [10]:

$$M(t)_{WZ} \mathbf{S}_z^a(g) = \mathbf{S}_w^a(-g) M(t)_{WZ}, \qquad M(t)_{WZ} \tilde{\mathbf{S}}_z^a(g) = \tilde{\mathbf{S}}_w^a(-g) M(t)_{WZ}, \tag{8.12}$$

where $W = e^{2\pi i w}$ and $Z = e^{2\pi i z}$. The operator $M(t)_{WZ}$ is symmetric in $p$ and $q$, and the second relation in (8.12) emerges from the first one after interchanging $p$ and $q$. Operators $\tilde{\mathbf{S}}_z^a(g)$ are thus obtained from (8.11) after permutation of $2\eta$ and $\tau$ and they realize another Sklyanin algebra with different structure constants $\tilde{J}_\alpha$. Jointly these two Sklyanin algebras form the elliptic modular double [56] generalizing Faddeev's modular double for $\mathfrak{sl}_q(2)$ algebra [16]. Intertwining operators of equivalent representations play an important role in the representation theory. In particular, their null spaces are invariant under the action of algebra generators which is helpful for building finite-dimensional irreducible representations.

There are useful recurrence relations for the elliptic Fourier transform operator $M(t)$ [8, 11]:

$$A_k(g) M(t) = M(q^{-1/2} t) \theta_k\left(z | \tfrac{\tau}{2}\right), \qquad B_k(g) M(t) = M\left(p^{-1/2} t\right) \theta_k(z | \eta), \tag{8.13}$$

where $k = 3, 4$ and $A_k(g)$ and $B_k(g)$ are the following difference operators

$$A_k(g) = \frac{e^{\pi i \frac{(z+\eta)^2}{\eta}}}{\theta(e^{4\pi i z}; p)} \left[ \theta_k\left(z + g + \eta | \tfrac{\tau}{2}\right) e^{\eta \partial_z} - \theta_k\left(z - g - \eta | \tfrac{\tau}{2}\right) e^{-\eta \partial_z} \right] e^{-\pi i \frac{z^2}{\eta}},$$

$$B_k(g) = \frac{e^{2\pi i \frac{(z+\tau/2)^2}{\tau}}}{\theta(e^{4\pi i z}; q)} \left[ \theta_k\left(z + g + \tfrac{\tau}{2} | \eta\right) e^{\frac{\tau}{2} \partial_z} - \theta_k\left(z - g - \tfrac{\tau}{2} | \eta\right) e^{-\frac{\tau}{2} \partial_z} \right] e^{-2\pi i \frac{z^2}{\tau}}.$$

In (8.13) we drop coordinate subindices and use the convention that the $z$-coordinate to the right of $M$-operator is the internal integration variable, but to the left—it is a free variable playing the role of $w$ in (8.12).

The initial condition $M(1) = 1$ (the unit operator) is proved by the residue calculus that we used in the proof of the elliptic beta integral (in this case two pairs of poles pinch the integration contour for $t \to 1$). Then for $t = q^{-n/2} p^{-m/2}$, $n, m \in \mathbb{Z}_{\geq 0}$, the recurrence relations can be resolved to yield the finite difference operator

$$M\left(q^{-n/2} p^{-m/2}\right) = A_k(n\eta - \eta + m\tfrac{\tau}{2}) \cdots A_k(\eta + m\tfrac{\tau}{2}) A_k(m\tfrac{\tau}{2})$$

$$\times B_k\left(m\tfrac{\tau}{2} - \tfrac{\tau}{2}\right) \cdots B_k\left(\tfrac{\tau}{2}\right) B_k(0) \theta_k^{-m}(z|\eta) \theta_k^{-n}\left(z|\tfrac{\tau}{2}\right), \tag{8.14}$$

which does not depend on the choice of $k = 3$ or $4$. This is only one of many possible ways to represent $M\left(q^{-n/2}p^{-m/2}\right)$ as a product of $A_k$- and $B_k$-operators.

Finally, we describe the Bailey lemma for $A_n$-root system. Define

$$M(t)_{wz}f(z) := \mu_n \int_{\mathbb{T}^n} \frac{\prod_{j,k=1}^{n+1} \Gamma(tw_j z_k^{-1})f(z)}{\Gamma(t^{n+1}) \prod_{1 \le j < k \le n+1} \Gamma(z_j z_k^{-1}, z_j^{-1} z_k)} \prod_{k=1}^{n} \frac{dz_k}{2\pi i z_k},$$

(8.15)

where $\prod_{k=1}^{n+1} z_k = 1$, $\Gamma(z) := \Gamma(z; p, q)$, and set

$$D(t; u, z) := \prod_{j=1}^{n+1} \Gamma(\sqrt{pq}\, t^{-\frac{n+1}{2}} \frac{u}{z_j}, \sqrt{pq}\, t^{-\frac{n+1}{2}} \frac{z_j}{u}), \quad D(t; u, z)D(t^{-1}; u, z) = 1.$$

(8.16)

For $n = 1$ operator (8.15) coincides with (8.1). For arbitrary $n$ it was defined in [63], where the Fourier type inversion relation $M(t)_{wz}^{-1} = M(t^{-1})_{wz}$ was established for the space of $A_n$-invariant functions under certain constraints on $t$ and $w_j$.

Similar to the univariate case, from a given Bailey pair satisfying $\beta(w, t) = M(t)_{wz}\alpha(z, t)$, the rules

$$\alpha'(w, st) = D(s; t^{-\frac{n-1}{2}}u, w)\alpha(w, t),$$

$$\beta'(w, st) = D(t^{-1}; s^{\frac{n-1}{2}}u, w)M(s)_{wz}D(ts; u, z)\beta(z, t)$$

determine a new Bailey pair with respect to the parameter $st$. From these expressions, the relation $\beta'(w, st) = M(st)_{wz}\alpha'(z, st)$ yields the cubic relation [7]

$$M(s)_{wz}D(st; u, z)M(t)_{zx} = D(t; s^{\frac{n-1}{2}}u, w)M(st)_{wx}D(s; t^{-\frac{n-1}{2}}u, x), \quad (8.17)$$

which holds true due to the elliptic beta integral on the $A_n$ root system (6.4). Although the change of $t \to t^{-1}$ inverts $D$ and $M$ operators, for $n > 1$ it is not possible to give to equality (8.17) a straightforward meaning of the Coxeter relation. A substatially more complicated Bailey lemma based on the multiple $C_n$-elliptic hypergeometric integrals of type II was formulated by Rains in [36].

# 9 Connection with Four Dimensional Superconformal Indices

A completely unexpected development of the theory of elliptic hypergeometric integrals emerged from quantum field theory when Dolan and Osborn [14] have discovered that superconformal indices of four dimensional supersymmetric gauge field theories are expressed in terms of such integrals. This was both the most striking physical application of these integrals and a powerful boost in understanding of their structure. We describe briefly some ingredients of the corresponding construction and refer to surveys [28, 38] for a more detailed account and list of references.

Massless $\mathcal{N} = 1$ supersymmetric field theories on the flat four dimensional space-time have a very large symmetry group $G_{\text{full}} = SU(2, 2|1) \times G \times F$. The superconformal group $SU(2, 2|1)$ contains Lorentz rotations described by $SO(3, 1)$-subgroup which is generated by $J_k, \overline{J}_k, k = 1, 2, 3$. It involves also ordinary translations and their superspace partners generated by $P_\mu, \mu = 0, \ldots, 3$, and $Q_\alpha, \overline{Q}_{\dot\alpha}, \alpha, \dot\alpha = 1, 2$, respectively. Further it includes the special conformal transformations generator $K_\mu$ and its superpartners $S_\alpha, \overline{S}_{\dot\alpha}$. Finally it contains the dilations generated by $H$, and $U(1)_R$-rotations of superpartners generated by the $R$-charge. Other symmetry groups are the local gauge invariance group $G$ and the flavor group $F$ describing global gauge invariance symmetries of matter superfields. Altogether they satisfy a system of supercommutation relations forming a specific super-Lie algebra.

The superconformal index is constructed as a character valued generalization of the Witten index involving generators of a maximal Cartan subalgebra preserving one supersymmetry relation. In particular, for a distinguished pair of supercharges $Q = \overline{Q}_1$ and $Q^\dagger = -\overline{S}_1$, one has the relation

$$QQ^\dagger + Q^\dagger Q = 2\mathcal{H}, \quad Q^2 = (Q^\dagger)^2 = 0, \qquad \mathcal{H} = H - 2\overline{J}_3 - 3R/2. \qquad (9.1)$$

Then, the fermionic generators $Q$ and $Q^\dagger$ commute with the bosonic operators $\mathcal{R} = H - R/2$ and $J_3$ and with the maximal torus generators of the flavor group $F_k$. The latter bosonic operators commute between each other as well. In lagrangian quantum field theory one works with the fields given by irreducible representations of the group $G_{\text{full}}$ which are realized as operators acting in the Hilbert space. All the symmetry generators are then defined as functionals of specific combinations of these fields. In this situation the superconformal index is formally defined as the following trace over the Hilbert space of states [29, 40]

$$I(y; p, q) = \text{Tr}\left((-1)^{\mathcal{F}} p^{\mathcal{R}/2+J_3} q^{\mathcal{R}/2-J_3} \prod_k y_k^{F_k} e^{-\beta\mathcal{H}}\right), \qquad (9.2)$$

where $(-1)^{\mathcal{F}}$ is the $\mathbb{Z}_2$-grading operator for representations of the $SU(2, 2|1)$ supergroup. The variables $p, q, y_k, \beta$ are arbitrary group parameters whose values are restricted by the condition of convergence of (9.2). Presence of the term $(-1)^{\mathcal{F}}$ shows that all eigenstates of $\mathcal{H}$ with non-zero eigenvalues drop out from this trace because of the cancellation of bosonic and fermionic state contributions. It means that the superconformal index is a weighted sum over BPS states which do not form long multiplets, $Q\psi = Q^\dagger\psi = 0$. Because of that there is no $\beta$-dependence in (9.2).

This index was computed heuristically on the basis of physical consideration of theories on curved background $S^3 \times \mathbb{R}$ associated with the radial quantization, or $S^3 \times S^1$ in the Euclidean space. Space-time symmetry group is reduced and conformal invariance is in general absent (it emerges in the infrared fixed points). Still, the meaning of operators entering (9.2) as Cartan generators preserving supersymmetry remains intact.

The field theories of interest may contain the vector superfield which is always in the adjoint representation of the gauge group $G$ with the corresponding character $\chi_{adj}(z)$, and it is invariant with respect to $F$. They involve also a set of chiral superfields transforming as certain irreducible representations of the gauge group with the character $\chi_{R_G,j}(z)$ and of the flavor group $F$ with the characters $\chi_{R_F,j}(y)$ (index $j$ counts such representations). The antichiral fields are described by conjugated representations with the characters $\chi_{\bar{R}_G,j}(z)$ and $\chi_{\bar{R}_F,j}(y)$. The characters depend on the maximal torus variables $z_a$, $a = 1, \ldots, \text{rank } G$, and $y_k$, $k = 1, \ldots, \text{rank } F$.

The final result for the index can be represented in the following explicit form:

$$I(y; p, q) = \int_G d\mu(z) \exp\left(\sum_{n=1}^{\infty} \frac{1}{n} \text{ind}\left(p^n, q^n, z^n, y^n\right)\right), \qquad (9.3)$$

where $d\mu(z)$ is the Haar measure for the gauge group $G$ and

$$\begin{aligned}
\text{ind}(p, q, z, y) &= \frac{2pq - p - q}{(1-p)(1-q)} \chi_{adj_G}(z) \\
&+ \sum_j \frac{(pq)^{R_j/2} \chi_{R_F,j}(y) \chi_{R_G,j}(z) - (pq)^{1-R_j/2} \chi_{\bar{R}_F,j}(y) \chi_{\bar{R}_G,j}(z)}{(1-p)(1-q)}
\end{aligned} \qquad (9.4)$$

with some fractional numbers $R_j$ called $R$-charges. The function (9.4) is called the one-particle states index and the integrand of (9.3) is called the plethystic exponential. Emergence of the integration over $G$ reflects the fact that the trace in (9.2) is taken over the gauge invariant states.

For example, for $G = SU(N)$ one has $z = (z_1, \ldots, z_N)$ with $\prod_{j=1}^{N} z_j = 1$. The gauge group measure for functions depending only on $z_j$ has the form

$$\int_{SU(N)} d\mu(z) = \frac{1}{N!} \int_{\mathbb{T}^{N-1}} \Delta(z)\Delta(z^{-1}) \prod_{a=1}^{N-1} \frac{dz_a}{2\pi i z_a},$$

where $\Delta(z) = \prod_{1 \le a < b \le N}(z_a - z_b)$. The fundamental representation character has the form $\chi_{SU(N),f}(z) = \sum_{k=1}^{N} z_k$, and the adjoint representation character is $\chi_{SU(N),adj}(z) = (\sum_{i=1}^{N} z_i)(\sum_{j=1}^{N} z_j^{-1}) - 1$.

Consider the field theory with $(G = SU(2), F = SU(6))$ containing two representations. The vector superfield transforming as (adj, 1) with the character $\chi_{SU(2),adj}(z) = z^2 + z^{-2} + 1$. The chiral superfield which is described by the fundamental representations of both groups $(f, f)$ with the characters $\chi_{SU(2),f}(z) = z + z^{-1}$ and

$$\chi_{SU(6),f}(y) = \sum_{k=1}^{6} y_k, \quad \chi_{SU(6),\bar{f}}(y) = \sum_{k=1}^{6} y_k^{-1}, \quad \prod_{k=1}^{6} y_k = 1.$$

Let us fix also the chiral field $R$-charge as $R = 1/3$.

*Exercise:* show that after plugging these data into the formula (9.2) and passing from the plethystic exponential to the infinite product form of the integrand, one obtains precisely the left-hand side expression for the elliptic beta integral evaluation formula (4.1) after the identification $t_k = (pq)^{1/6} y_k$.

In this picture the unitarity condition for $SU(6)$ group expressed by the equality $\prod_{k=1}^{6} y_k = 1$ becomes the balancing condition $\prod_{k=1}^{6} t_k = pq$ for the integral which is associated with the hidden ellipticity condition.

Thus, the elliptic beta integral describes the superconformal index $I_E$ of a particular four dimensional gauge field theory. Consider now another field theory without gauge group $G = 1$ and containing only one free chiral superfield transforming as the antisymmetric tensor of the second rank $T_A$ of the same flavor group $F = SU(6)$. The corresponding character is

$$\chi_{SU(6),T_A}(y) = \sum_{1 \leq i < j \leq 6} y_i y_j,$$

and we fix the $R$-charge for this field as $R = 2/3$.

*Exercise:* check that substituting these data to the same formula (9.2) one comes precisely to the right-hand side expression in (4.1).

So, the result of evaluation of the elliptic beta integral yields the superconformal index $I_M$ of a completely differently looking field theory than in the previous case. The two described theories represent the simplest example of the so-called Seiberg duality [47] which states a conjectural equivalence of two models in their infrared fixed points. It is a natural extension of the electromagnetic duality to non-abelian gauge field theories. Therefore the first described model is called the "electric" theory and the second model—the "magnetic" one. The equality of superconformal indices of these two models, $I_E = I_M$, expressed by the evaluation formula (4.1) can be considered as a proof of this duality in the sectors of BPS states which appear to be identical. The physical phenomenon when the theory in the ultraviolet regime with nontrivial gauge interaction becomes in the low energy regime an effective field theory without gauge degrees of freedom is called the confinement. Thus, the process of computation of the elliptic beta integral is equivalent to the transition from high to lower energy physics. From mathematical point of view it describes some group-theoretical duality, when a particular function on characters yields the same result for two different sets of representations of two different groups.

Consider now the full Seiberg electric-magnetic duality [47]. The electric theory has the gauge group $G = SU(N_c)$ and the flavor group $SU(N_f)_l \times SU(N_f)_r \times$

$U(1)_B$ (it enlarges to $SU(2N_f)$ for $N_c = 2$). The representation properties of the fields are described in the table below (where $\tilde{N}_c = N_f - N_c$):

| | $SU(N_c)$ | $SU(N_f)_l$ | $SU(N_f)_r$ | $U(1)_B$ | $U(1)_R$ |
|---|---|---|---|---|---|
| $Q$ | $f$ | $f$ | 1 | 1 | $\tilde{N}_c/N_f$ |
| $\tilde{Q}$ | $\bar{f}$ | 1 | $\bar{f}$ | $-1$ | $\tilde{N}_c/N_f$ |
| $V$ | adj | 1 | 1 | 0 | 1 |

The magnetic theory has different gauge group $G = SU(\tilde{N}_c)$ and the same flavor group. The representation properties of the fields are described in the next table:

| | $SU(\tilde{N}_c)$ | $SU(N_f)_l$ | $SU(N_f)_r$ | $U(1)_B$ | $U(1)_R$ |
|---|---|---|---|---|---|
| $q$ | $f$ | $\bar{f}$ | 1 | $N_c/\tilde{N}_c$ | $N_c/N_f$ |
| $\tilde{q}$ | $\bar{f}$ | 1 | $f$ | $-N_c/\tilde{N}_c$ | $N_c/N_f$ |
| $M$ | 1 | $f$ | $\bar{f}$ | 0 | $2\tilde{N}_c/N_f$ |
| $\tilde{V}$ | adj | 1 | 1 | 0 | 1 |

The first columns of these tables contain usual notation for the fields and last columns contain the abelian group charges—eigenvalues of the generators of $U(1)_B$ and $U(1)_R$ groups. The vector superfields are described in the last rows with all other rows describing some chiral superfields. According to Seiberg's conjecture, these two $\mathcal{N} = 1$ supersymmetric models have identical physical behaviour at their infrared fixed points where superconformal symmetry is fully realized. The suggested consistency checks included the facts that the global anomalies of theories match ('t Hooft anomaly matching conditions) and that the reductions $N_f \to N_f - 1$ match for both theories. Validity of both criteria can be traced from the equality of the electric and magnetic theory indices which we describe now.

Superconformal indices for these general theories were constructed in [14] (see also [60]) and we skip the details of their computation. After passing from maximal torus variables for the flavor group to the canonical elliptic hypergeometric integral parameters, the electric theory index takes the form:

$$I_E = \kappa_{N_c} \int_{\mathbb{T}^{N_c-1}} \frac{\prod_{i=1}^{N_f} \prod_{j=1}^{N_c} \Gamma(s_i z_j, t_i z_j^{-1}; p, q)}{\prod_{1 \leq i < j \leq N_c} \Gamma(z_i z_j^{-1}, z_i^{-1} z_j; p, q)} \prod_{j=1}^{N_c-1} \frac{dz_j}{z_j},$$

where $ST = (pq)^{N_f - N_c}$, $S = \prod_{i=1}^{N_f} s_i$, $T = \prod_{i=1}^{N_f} t_i$, and

$$\prod_{j=1}^{N_c} z_j = 1, \qquad \kappa_{N_c} = \frac{(p; p)_\infty^{N_c-1}(q; q)_\infty^{N_c-1}}{N_c!(2\pi i)^{N_c-1}}.$$

This is a multiple integral for the root system $A_{N_c-1}$, which coincides with (6.4) for $N_f = N_c + 1$ and $n = N_c - 1$.

For the magnetic theory one has:

$$I_M = \kappa_{\tilde{N}_c} \prod_{i,j=1}^{N_f} \Gamma(s_i t_j; p, q) \int_{\mathbb{T}^{\tilde{N}_c-1}} \frac{\prod_{i=1}^{N_f} \prod_{j=1}^{\tilde{N}_c} \Gamma(S^{\frac{1}{\tilde{N}_c}} s_i^{-1} x_j, T^{\frac{1}{\tilde{N}_c}} t_i^{-1} x_j^{-1}; p, q)}{\prod_{1 \le i < j \le \tilde{N}_c} \Gamma(x_i x_j^{-1}, x_i^{-1} x_j; p, q)} \prod_{j=1}^{\tilde{N}_c-1} \frac{dx_j}{x_j},$$

where $\prod_{j=1}^{\tilde{N}_c} x_j = 1$, $\tilde{N}_c = N_f - N_c$.

As observed by Dolan and Osborn [14], the dual indices coincide $I_E = I_M$, since the equality of corresponding elliptic hypergeometric integrals was rigorously established by Rains [32] (for some particular values of the parameters it was proven or conjectured by the author [49, 51]). Evidently, this identity is a multivariable extension of the second $V$-function transformation law (5.3).

In the case when the electric index is explicitly computable, i.e. $N_f = N_c + 1$, one has the confinement of colored particles without chiral symmetry breaking. For $N_f = N_c$ one has the confinement with chiral symmetry breaking which is reflected in the appearance of Dirac delta-functions in the description of indices [62]. In general, equality of dual indices is currently the most rigorous mathematical justification of the Seiberg duality conjecture.

Reduction of the number of chiral fields $N_f \to N_f - 1$ is reached by the restriction of the parameters $s_{N_f} t_{N_f} = pq$. In this case $s_{N_f}$ and $t_{N_f}$ disappear from $I_E$ and the rank of the flavor group of electric theory is reduced by one. In the magnetic theory it is more involved—a number of poles start to pinch the integration contour of $I_M$ and the integral starts to diverge, but the vanishing prefactor $\Gamma(s_{N_f} t_{N_f}; p, q)$ makes the product finite with the effective reduction of ranks of both the magnetic gauge and flavor groups by one, which matches with the physical picture of [47].

As to the 't Hooft anomaly matching conditions, they are described by the modified analogues of the above integrals $I_E$ and $I_M$ [69]. Define for the electric theory

$$I_E^{mod} = \kappa_{N_c}^{mod} \int_{-\omega_3/2}^{\omega_3/2} \frac{\prod_{i=1}^{N_f} \prod_{j=1}^{N_c} G(\alpha_i + u_j, \beta_i - u_j; \omega)}{\prod_{1 \le i < j \le N_c} G(u_i - u_j, -u_i + u_j; \omega)} \prod_{j=1}^{N_c-1} \frac{du_j}{\omega_3}, \qquad (9.5)$$

where $\sum_{j=1}^{N_c} u_j = 0$,

$$\kappa_{N_c}^{mod} = \frac{\kappa(\omega)^{N_c-1}}{N_c!}, \qquad \kappa(\omega) = -\frac{\omega_3}{\omega_2} \frac{(p; p)_\infty (q; q)_\infty (r; r)_\infty}{(\tilde{q}; \tilde{q})_\infty}.$$

and the balancing condition reads

$$\alpha + \beta = (N_f - N_c) \sum_{k=1}^{3} \omega_k, \qquad \alpha = \sum_{i=1}^{N_f} \alpha_i, \qquad \beta = \sum_{i=1}^{N_f} \beta_i.$$

We denoted the products of modified elliptic gamma functions as $G(a, b; \omega) := G(a; \omega)G(b; \omega)$. An analogue of $I_M$ has the form

$$I_M^{mod} = \kappa_{\widetilde{N}_c}^{mod} \prod_{1 \le i,j \le N_f} G(\alpha_i + \beta_j; \omega) \tag{9.6}$$

$$\times \int_{-\omega_3/2}^{\omega_3/2} \frac{\prod_{i=1}^{N_f} \prod_{j=1}^{\widetilde{N}_c} G(\alpha/\widetilde{N}_c - \alpha_i + v_j, \beta/\widetilde{N}_c - \beta_i - v_j; \omega)}{\prod_{1 \le i < j \le \widetilde{N}_c} G(v_i - v_j, -v_i + v_j; \omega)} \prod_{j=1}^{\widetilde{N}_c-1} \frac{dv_j}{\omega_3},$$

where $\widetilde{N}_c = N_f - N_c$ and $\sum_{j=1}^{\widetilde{N}_c} v_j = 0$.

*Exercise:* show that $I_E^{mod} = I_M^{mod}$ under the conditions

$$\text{Im}(\alpha_i/\omega_3), \text{Im}((\alpha/\widetilde{N}_c - \alpha_i)/\omega_3) < 0, \quad \text{Im}(\beta_i/\omega_3), \text{Im}((\beta/\widetilde{N}_c - \beta_i)/\omega_3) < 0,$$

when the integration contour in both integrals can be chosen as the straight line segment connecting $-\omega_3/2$ and $\omega_3/2$. In a sketchy way, this is reached by substitution of the expression (3.13) to (9.5), (9.6) and analysis of the exponential factors $e^{\varphi_E}$ and $e^{\varphi_M}$ containing sums of $B_{3,3}$-Bernoulli polynomials. The phase $\varphi_E$ (or $\varphi_M$) looks like a homogeneous cubic polynomial of the integration variables $u_j$ (or $v_j$) and parameters $\alpha_j, \beta_j, \omega_i$ divided by $\omega_1\omega_2\omega_3$. However, it appears that the integration variables cancel out in both of them. As a result, $I_E^{mod} = e^{\varphi_E}\tilde{I}_E$ and $I_M^{mod} = e^{\varphi_M}\tilde{I}_M$, where the integrals $\tilde{I}_E$ and $\tilde{I}_M$ are obtained from $I_E$ and $I_M$ after the replacements $s_j \to e^{-2\pi i \alpha_j/\omega_3}$, $t_j \to e^{-2\pi i \beta_j/\omega_3}$, $p \to \tilde{p}$, and $q \to \tilde{r}$. Assuming the original parametrization $s_j = e^{2\pi i \alpha_j/\omega_2}$ and $t_j = e^{2\pi i \beta_j/\omega_2}$ this boils down to the modular transformation $(\omega_2, \omega_3) \to (-\omega_3, \omega_2)$ for $I_E$ and $I_M$. Explicit computation shows that $\varphi_E = \varphi_M$ and this proves the required equality.

For dual field theories the coincidence of $\varphi_E$ and $\varphi_M$ describes the 't Hooft anomaly matching. Namely, each coefficient of their numerator cubic polynomials corresponds to a particular triangle Feynman diagram involving fermions and particular gauge or other currents describing global symmetries of the theories. The above consideration shows that the ratio of kernels of particular elliptic hypergeometric integrals corresponding to electric and magnetic superconformal indices has a particular behaviour from the viewpoint of $SL(3, \mathbb{Z})$-group. One can formalize this statement in a general setting by taking the following parametrization for such a ratio

$$\Delta(x_1, \ldots, x_n; p, q) = (p; p)_\infty^{r-}(q; q)_\infty^{r-} \prod_{a=1}^{K} \Gamma\left((pq)^{\frac{R_a}{2}} x_1^{m_1^{(a)}} x_2^{m_2^{(a)}} \ldots x_n^{m_n^{(a)}}; p, q\right)^{\epsilon_a}, \tag{9.7}$$

where $K$ is the total number of independent elliptic gamma functions appearing in this ratio in the integer power $\epsilon_a$ with its own $R$-charge $R_a$ and $m_j^{(a)}$—integer powers of $n$ independent group parameters $x_j$ (playing the role of fugacities $y_j$

in the original definition of the superconformal indices). For the Seiberg duality the integer number $r_-$ is equal to the difference between ranks of the electric and magnetic gauge groups.

Using the parametrization $x_j = e^{2\pi i u_j/\omega_2}$ one can define a modified elliptic gamma function analogue of (9.7)

$$\Delta^{mod}(u_1, \ldots, u_n; \omega) = \kappa(\omega)^{r_-} \prod_{a=1}^{K} G\left(R_a \sum_{k=1}^{3} \frac{\omega_k}{2} + \sum_{j=1}^{n} u_j m_j^{(a)}; \omega\right)^{\epsilon_a}. \tag{9.8}$$

Now one demands validity of an $SL(3, \mathbb{Z})$-modular transformation relation between functions (9.7) and (9.8)

$$\Delta^{mod}(u_1, \ldots, u_n; \omega) = \Delta(e^{-2\pi i u_1/\omega_3}, \ldots, e^{-2\pi i u_n/\omega_3}; \tilde{p}, \tilde{r}). \tag{9.9}$$

There are six independent in form functional combinations of $u_j$ and $\omega_i$ in the sum of $B_{3,3}$-polynomials, appearing after substitution of relation (3.13) in (9.9), and additional terms generated by the Dedekind function modular transformation. The coefficients in front of them should vanish, which yields the following set of equations

$$\sum_{a=1}^{K} \epsilon_a m_i^{(a)} m_j^{(a)} m_k^{(a)} = 0, \tag{9.10}$$

$$\sum_{a=1}^{K} \epsilon_a m_i^{(a)} m_j^{(a)} (R_a - 1) = 0, \tag{9.11}$$

$$\sum_{a=1}^{K} \epsilon_a m_i^{(a)} (R_a - 1)^2 = 0, \tag{9.12}$$

$$\sum_{a=1}^{K} \epsilon_a m_i^{(a)} = 0, \tag{9.13}$$

$$\sum_{a=1}^{K} \epsilon_a (R_a - 1)^3 + r_- = 0, \tag{9.14}$$

$$\sum_{a=1}^{K} \epsilon_a (R_a - 1) + r_- = 0. \tag{9.15}$$

Assuming rationality of $R_a$ we come to a system of Diophantine equations which were not systematically investigated yet from mathematical point of view, although all known physical dualities satisfy them as the 't Hooft anomaly matching conditions. We do not describe the physical meaning of each type of the above equations referring for details to [61]. We only mention that in the context of superconformal indices the combinations of integration variables entering the Bernoulli polynomials must cancel independently for electric and magnetic indices to be able to pull exponentials $e^{\varphi_{E,M}}$ out of the integrals.

*Exercise:* suppose that (9.7) is a kernel of an elliptic hypergeometric integral with $x_1, \ldots, x_r$ being the integration variables, i.e. that it satisfies a set of $r$ $q$-difference equations in these variables with $p$-elliptic function coefficients. Show that this requirement is equivalent to Eqs. (9.10) and (9.11) with $1 \leq i, j \leq r$ together with an extra requirement $\sum_{a=1}^{K} \epsilon_a m_i^{(a)} m_j^{(a)} \in 2\mathbb{Z}$.

In all known dual theories the latter extra evenness condition is automatically satisfied, though it is not clear whether it follows from general Eqs. (9.10)–(9.15). Condition (9.10) for all $1 \leq i, j, k \leq r$ physically corresponds to the demand of absence of the gauge anomalies, which is needed for the consistency of field theories whose indices are described by the corresponding integrals. As we see, it follows from the original definition of the elliptic hypergeometric integrals (2.7) and its multivariable extension, which thus gets an interesting physical interpretation.

As a summary of connections with the superconformal indices, we mention that very many identities for elliptic hypergeometric integrals were found following the physical duality conjectures, and they still require rigorous proofs, see, e.g. [60]. Vice versa, there is a good number of new physical dualities conjectured from proven integral identities. There are also applications of superconformal indices to topological field theories, description of lower and higher dimensional field theories, and some other constructions of mathematical physics [28, 38].

**Acknowledgements** The author is indebted to E. M. Rains and S. O. Warnaar for helpful discussions. This work is supported in part by the Laboratory of Mirror Symmetry NRU HSE, RF government grant, ag. no. 14.641.31.0001.

# References

1. G.E. Andrews, R. Askey, R. Roy, *Special Functions*. Encyclopedia of Mathematics and its Applications, vol. 71 (Cambridge University Press, Cambridge, 1999)
2. N.I. Akhiezer, *Elements of the Theory of Elliptic Functions* (Nauka, Moscow, 1970)
3. E.W. Barnes, On the theory of the multiple gamma function. Trans. Cambridge Phil. Soc. **19**, 374–425 (1904)
4. R.J. Baxter, Partition function of the eight-vertex lattice model. Ann. Phys. **70**, 193–228 (1972)
5. V.V. Bazhanov, S.M. Sergeev, A master solution of the quantum Yang-Baxter equation and classical discrete integrable equations. Adv. Theor. Math. Phys. **16**, 65–95 (2012)
6. S. Bloch, M. Kerr, P. Vanhove, Local mirror symmetry and the sunset Feynman integral. Adv. Theor. Math. Phys. **21**, 1373–1453 (2017)
7. F. Brünner, V.P. Spiridonov, A duality web of linear quivers. Phys. Lett. **B 761**, 261–264 (2016)
8. D. Chicherin, S. Derkachov, D. Karakhanyan, R. Kirschner, Baxter operators with deformed symmetry. Nucl. Phys. B **868**, 652–683 (2013)
9. E. Date, M. Jimbo, A. Kuniba, T. Miwa, M. Okado, Exactly solvable SOS models, II: proof of the star-triangle relation and combinatorial identities. Adv. Stud. Pure Math. **16**, 17–122 (1988)

10. S.E. Derkachov, V.P. Spiridonov, Yang-Baxter equation, parameter permutations, and the elliptic beta integral. Russ. Math. Surv. **68**(6), 1027–1072 (2013)
11. S.E. Derkachov, V.P. Spiridonov, Finite dimensional representations of the elliptic modular double. Theor. Math. Phys. **183**(2), 597–618 (2015)
12. T. Dimofte, Complex Chern-Simons theory at level $k$ via the 3d-3d correspondence. Commun. Math. Phys. **339**, 619–662 (2015)
13. A.L. Dixon, On a generalization of Legendre's formula $KE' - (K - E)K' = \pi/2$. Proc. Lond. Math. Soc. **3**(1), 206–224 (1905)
14. F.A. Dolan, H. Osborn, Applications of the superconformal index for protected operators and $q$-hypergeometric identities to $\mathcal{N} = 1$ dual theories. Nucl. Phys. **B 818**, 137–178 (2009)
15. L.D. Faddeev, Current-like variables in massive and massless integrable models, in *Quantum Groups and Their Applications in Physics, Varenna* (1994), pp. 117–135
16. L.D. Faddeev, Modular double of a quantum group, in *Conference Moshé Flato 1999, vol. I*. Mathematical Physics Studies, vol. 21 (Kluwer, Dordrecht, 2000), pp. 149–156
17. L.D. Faddeev, R.M. Kashaev, A.Y. Volkov, Strongly coupled quantum discrete Liouville theory. 1. Algebraic approach and duality. Commun. Math. Phys. **219**, 199–219 (2001)
18. G. Felder, A. Varchenko, The elliptic gamma function and $SL(3, \mathbb{Z}) \ltimes \mathbb{Z}^3$. Adv. Math. **156**, 44–76 (2000)
19. P.J. Forrester, S.O. Warnaar, The importance of the Selberg integral. Bull. Am. Math. Soc. **45**, 489–534 (2008)
20. I.B. Frenkel, V.G. Turaev, Elliptic solutions of the Yang-Baxter equation and modular hypergeometric functions, in *The Arnold-Gelfand Mathematical Seminars* (Birkhäuser, Boston, 1997), pp. 171–204
21. G. Gasper, M. Rahman, Basic Hypergeometric Series. Encyclopedia of Mathematics and its Applications, vol. 96 (Cambridge University Press, Cambridge, 2004)
22. R.A. Gustafson, Some $q$-beta integrals on $SU(n)$ and $Sp(n)$ that generalize the Askey-Wilson and Nassrallah-Rahman integrals. SIAM J. Math. Anal. **25**, 441–449 (1994)
23. F.H. Jackson, The basic gamma-function and the elliptic functions. Proc. R. Soc. Lond. **A 76**, 127–144 (1905)
24. M. Jimbo, T. Miwa, Quantum KZ equation with $|q| = 1$ and correlation functions of the XXZ model in the gapless regime. J. Phys. A Math. Gen. **29**, 2923–2958 (1996)
25. K. Kajiwara, T. Masuda, M. Noumi, Y. Ohta, Y. Yamada, $_{10}E_9$ solution to the elliptic Painlevé equation. J. Phys. A Math. Gen. **36**, L263–L272 (2003)
26. A.P. Kels, New solutions of the star-triangle relation with discrete and continuous spin variables. J. Phys. A Math. Theor. **48**, 435201 (2015)
27. A.P. Kels, M. Yamazaki, Elliptic hypergeometric sum/integral transformations and supersymmetric lens index. SIGMA **14**, 013 (2018)
28. S. Kim, Superconformal indices and instanton partition functions, in Partition Functions and Automorphic Forms. Lecture Notes of the Dubna Winter School (29.01.2018–02.02.2018, JINR, Dubna, Russia), edited by V.A. Gritsenko, V.P. Spiridonov. Moscow Lectures, vol. 5. Springer, Heidelberg (2020). https://doi.org/10.1007/978-3-030-42400-8_1
29. J. Kinney, J.M. Maldacena, S. Minwalla, S. Raju, An index for 4 dimensional super conformal theories. Commun. Math. Phys. **275**, 209–254 (2007)
30. V. Pasol, W. Zudilin, A study of elliptic gamma function and allies. Res. Math. Sci. **5**, 39 (2018)
31. M. Rahman, An integral representation of a $_{10}\phi_9$ and continuous bi-orthogonal $_{10}\phi_9$ rational functions. Can. J. Math. **38**, 605–618 (1986)
32. E.M. Rains, Transformations of elliptic hypergeometric integrals. Ann. Math. **171**, 169–243 (2010)
33. E.M. Rains, Limits of elliptic hypergeometric integrals. Ramanujan J. **18**(3), 257–306 (2009)
34. E.M. Rains, An isomonodromy interpretation of the hypergeometric solution of the elliptic Painlevé equation (and generalizations). SIGMA **7**, 088 (2011)
35. E.M. Rains, The noncommutative geometry of elliptic difference equations (2016). arXiv:1607.08876

36. E.M. Rains, Multivariate quadratic transformations and the interpolation kernel. SIGMA **14**, 019 (2018)
37. E.M. Rains, V.P. Spiridonov, Determinants of elliptic hypergeometric integrals. Funct. Anal. Appl. **43**(4), 297–311 (2009)
38. L. Rastelli, S.S. Razamat, The supersymmetric index in four dimensions. J. Phys. A Math. Theor. **50**, 443013 (2017)
39. S.S. Razamat, B. Willett, Global properties of supersymmetric theories and the lens space. Commun. Math. Phys. **334**, 661–696 (2015)
40. C. Römelsberger, Counting chiral primaries in $\mathcal{N} = 1$, $d = 4$ superconformal field theories. Nucl. Phys. **B 747**, 329–353 (2006)
41. H. Rosengren, Elliptic hypergeometric functions. Lectures at OPSF-S6, College Park, Maryland, 11–15 July 2016. arXiv:1608.06161
42. H. Rosengren, S.O. Warnaar, Elliptic hypergeometric functions associated with root systems (2017). arXiv:1704.08406
43. S.N.M. Ruijsenaars, First order analytic difference equations and integrable quantum systems. J. Math. Phys. **38**, 1069–1146 (1997)
44. H. Sakai, Rational surfaces associated with affine root systems and geometry of the Painlevé equations. Commun. Math. Phys. **220**, 165–229 (2001)
45. G.A. Sarkissian, V.P. Spiridonov, General modular quantum dilogarithm and beta integrals. Proc. Steklov Institute of Math. **309** (2020), to appear. arXiv:1910.11747 [hep-th]
46. G.A. Sarkissian, V.P. Spiridonov, The endless beta integrals. arXiv:2005.01059 [math-ph]
47. N. Seiberg, Electric–magnetic duality in supersymmetric non-Abelian gauge theories. Nucl. Phys. **B 435**, 129–146 (1995)
48. E.K. Sklyanin, Some algebraic structures connected with the Yang-Baxter equation. Representation of a quantum algebra. Funct. Anal. Appl. **17**(4), 273–284 (1983)
49. V.P. Spiridonov, On the elliptic beta function. Russ. Math. Surveys **56**(1), 185–186 (2001)
50. V.P. Spiridonov, Theta hypergeometric series. *Proceedings of NATO ASI Asymptotic Combinatorics with Applications to Mathematical Physics, St. Petersburg, Russia, July 9–23, 2001* (Kluwer, Dordrecht, 2002), pp. 307–327
51. V.P. Spiridonov, Theta hypergeometric integrals. Algebra i Analiz **15**(6), 161–215 (2003). (St. Petersburg Math. J. **15**(6), 929–967 (2004))
52. V.P. Spiridonov, A Bailey tree for integrals. Theor. Math. Phys. **139**, 536–541 (2004)
53. V.P. Spiridonov, Short proofs of the elliptic beta integrals. Ramanujan J. **13**(1–3), 265–283 (2007)
54. V.P. Spiridonov, Elliptic hypergeometric functions and Calogero-Sutherland type models. Theor. Math. Phys. **150**(2), 266–278 (2007)
55. V.P. Spiridonov, Essays on the theory of elliptic hypergeometric functions. Russ. Math. Surv. **63**(3), 405–472 (2008)
56. V.P. Spiridonov, Continuous biorthogonality of an elliptic hypergeometric function. Algebra i Analiz **20**(5), 155–185 (2008). (St. Petersburg Math. J. **20**(5) 791–812 (2009))
57. V.P. Spiridonov, Elliptic beta integrals and solvable models of statistical mechanics. Contemp. Math. **563**, 181–211 (2012)
58. V.P. Spiridonov, Rarefied elliptic hypergeometric functions. Adv. Math. **331**, 830–873 (2018)
59. V.P. Spiridonov, The rarefied elliptic Bailey lemma and the Yang-Baxter equation. J. Phys. A: Math. and Theor. **52**, 355201 (2019)
60. V.P. Spiridonov, G.S. Vartanov, Elliptic hypergeometry of supersymmetric dualities. Commun. Math. Phys. **304**, 797–874 (2011)
61. V.P. Spiridonov, G.S. Vartanov, Elliptic hypergeometric integrals and 't Hooft anomaly matching conditions. J. High Energy Phys. **06**, 016 (2012)
62. V.P. Spiridonov, G.S. Vartanov, Vanishing superconformal indices and the chiral symmetry breaking. J. High Energy Phys. **06**, 062 (2014)
63. V.P. Spiridonov, S.O. Warnaar, Inversions of integral operators and elliptic beta integrals on root systems. Adv. Math. **207**, 91–132 (2006)

64. V.P. Spiridonov, A.S. Zhedanov, Spectral transformation chains and some new biorthogonal rational functions. Commun. Math. Phys. **210**, 49–83 (2000)
65. F.J. van de Bult, E.M. Rains, J.V. Stokman, Properties of generalized univariate hypergeometric functions. Commun. Math. Phys. **275**, 37–95 (2007)
66. J.F. van Diejen, Integrability of difference Calogero-Moser systems. J. Math. Phys. **35**, 2983–3004 (1994)
67. J.F. van Diejen, V.P. Spiridonov, An elliptic Macdonald-Morris conjecture and multiple modular hypergeometric sums. Math. Res. Letters **7**, 729–746 (2000)
68. J.F. van Diejen, V.P. Spiridonov, Elliptic Selberg integrals. Int. Math. Res. Not. **20**, 1083–1110 (2001)
69. J.F. van Diejen, V.P. Spiridonov, Unit circle elliptic beta integrals. Ramanujan J. **10**(2), 187–204 (2005)
70. S.O. Warnaar, 50 Years of Bailey's lemma, in *Algebraic Combinatorics and Applications* (Springer, Berlin, 2001), pp. 333–347
71. S.O. Warnaar, Summation and transformation formulas for elliptic hypergeometric series. Constr. Approx. **18**, 479–502 (2002)

# Feynman Integrals and Mirror Symmetry

## Pierre Vanhove

## Contents

1   Introduction.................................................................................... 320
2   Feynman Integrals........................................................................... 321
   2.1   The Parametric Representation..................................................... 322
   2.2   Maximal Cut ......................................................................... 324
   2.3   The Differential Equations ......................................................... 325
3   Toric Geometry and Feynman Graphs ..................................................... 330
   3.1   Toric Polynomials and Feynman Graphs ......................................... 331
   3.2   The GKZ Approach : A Review.................................................... 333
   3.3   Hypergeometric Functions and GKZ System..................................... 336
   3.4   The Massive One-Loop Graph ..................................................... 337
   3.5   The Two-Loop Sunset .............................................................. 342
   3.6   The Generic Case ................................................................... 347
   3.7   Toward a Systematic Approach to Differential Equations ....................... 352
4   Analytic Evaluations for Sunset Integral.................................................. 353
   4.1   The Sunset Integral as an Elliptic Dilogarithm .................................. 353
   4.2   The Sunset Integral as a Trilogarithm............................................. 356
   4.3   Mirror Symmetry and Sunset Integral ............................................ 359
5   Conclusion..................................................................................... 362
References ........................................................................................ 363

**Abstract**  In this text we describe various approaches to the computation of Feynman integrals. One approach uses toric geometry to derive differential equations

IPHT-t18/095

P. Vanhove (✉)
CEA, DSM, Institut de Physique Théorique, IPhT, CNRS, MPPU, URA2306, Gif-sur-Yvette, France

National Research University, Higher School of Economics, Moscow, Russia
e-mail: pierre.vanhove@ipht.fr

© Springer Nature Switzerland AG 2020
V. A. Gritsenko, V. P. Spiridonov (eds.), *Partition Functions and Automorphic Forms*, Moscow Lectures 5,
https://doi.org/10.1007/978-3-030-42400-8_7

satisfied by the imaginary part of the Feynman integrals. We then discuss how this can be used to obtain the full differential equation acting on the integral. In a second part of this text we explain how Calabi–Yau geometry is naturally associated to Feynman integrals and that mirror symmetry plays some role in evaluating some particular Feynman integrals.

# 1   Introduction

Scattering amplitudes are fundamental quantities used to understand fundamental interactions and the elementary constituents in Nature. It is well known that scattering amplitudes are used in particle physics to compare the theoretical predictions to experimental measurements in particle colliders (see [1] for instance). More recently the use of modern developments in scattering amplitudes has been extended to gravitational physics like unitarity methods to gravitational wave physics [2–6].

The $l$-loop scattering amplitude $A_{n,l}^D(\underline{s}, \underline{m}^2)$ between $n$ fields in $D$ dimensions is a function of the kinematics invariants $\underline{s} = \{s_{ij} = (p_i + p_j)^2, 1 \leq i, j \leq n\}$ where $p_i$ are the incoming momenta of the external particle, and the internal masses $\underline{m}^2 = (m_1, \ldots, m_r)$.

We focus on the questions : what kind of function is a Feynman integral? What is the best analytic representation?

The answer to these questions depends very strongly on which properties one wants to display. An analytic representation suitable for an high precision numerical evaluation may not be the one that displays the mathematical nature of the integral.

For instance the two-loop sunset integral has received many different, but equivalent, analytical expressions: hypergeometric and Lauricella functions [7, 8], Bessel integral representation [9–11], Elliptic integrals [12, 13], Elliptic polylogarithms [14–20] and trilogarithms [20].

The approach that we will follow here will be guided by the geometry of the graph polynomial using the parametric representation of the Feynman integral. In Sect. 2 we review the description of the Feynman integral $I_\Gamma$ for a graph $\Gamma$ in parametric space. We focus on the properties of the second Symanzik polynomial as a preparation for the toric approach used in Sect. 3. In Sect. 2.2 we show that the maximal cut $\pi_\Gamma$ of a Feynman integral has a parametric representation similar to the one of the Feynman integral $I_\Gamma$ where the only difference is the cycle of integration. The toric geometry approach is described in Sect. 3. In Sect. 3.2 we explain that the maximal cut integral is an hypergeometric series from the Gel'fand-Kapranov-Zelevinski (GKZ) construction. In Sect. 3.4.2, we show on examples how to derive the minimal differential operator annihilating the maximal cut integral. In Sect. 4 we review the evaluation of the two-loop sunset integral in two space-time dimensions.

In Sect. 4.1 we give its expression as an elliptic dilogarithm

$$I_{\ominus}(p^2, \xi_1^2, \xi_2^2, \xi_3^2) \propto \varpi \sum_{i=1}^{6} c_i \sum_{n \geq 1} (\text{Li}_2(q^n z_i) - (\text{Li}_2(-q^n z_i))) \tag{1}$$

where $\varpi$ is a period of the elliptic curve defined by the graph polynomial, $q$ the nome function of the external momentum $p^2$ and internal masses $\xi_i^2$ for $i = 1, 2, 3$. In Sect. 4.2 we show that the sunset integral evaluates to

$$I_{\ominus}(p^2, \xi_1^2, \xi_2^2, \xi_3^2) \propto \varpi \left( \log Q \left( \frac{d}{d \log Q} \right)^2 - \frac{d}{d \log Q} \right) \mathscr{F}(p^2, \xi_1^2, \xi_2^2, \xi_3^2).$$

$$\tag{2}$$

where $\mathscr{F}(p^2, \xi_1^2, \xi_2^2, \xi_3^2)$ is a sum of trilogarithm functions in (167)

$$\mathscr{F}(p^2, \xi_1^2, \xi_2^2, \xi_3^2) = -(\log Q)^3 + \sum_{(k_1, k_2, k_3) \geq 0} n_{k_1, k_2, k_3} \text{Li}_3 \left( \prod_{i=1}^{3} \xi_i^{2n_i} Q^{n_i} \right). \tag{3}$$

In Sect. 4.3 we show that the equivalence between these two expressions is the result of a local mirror map, $q \leftrightarrow Q$ in (169), for the non-compact Calabi–Yau threefold obtained as the anti-canonical bundle over the del Pezzo 6 surface defined by the sunset graph polynomial. The free energy $\mathscr{F}(p^2, \xi_1^2, \xi_2^2, \xi_3^2)$ has been computed in [21, §6.6] as an application of the computation of local refined BPS numbers for toric del Pezzo surfaces. It is remarkable that the sunset Feynman integral is expressed in terms of the genus zero local Gromov-Witten numbers [20] and $\mathscr{F}(p^2, \xi_1^2, \xi_2^2, \xi_3^2)$ is the local genus 0 Gromov–Witten prepotential described in Sect. 4.3. This result generalises to the whole family of multi-loop sunset integrals as [22, 23]. Therefore this provides a natural application for Batyrev's mirror symmetry techniques [24]. One remarkable fact is that the computation can be done using the existing technology of mirror symmetry developed in other physical [21, 25, 26] or mathematical [27] contexts.

## 2 Feynman Integrals

A connected Feynman graph $\Gamma$ is determined by the number $n$ of propagators (internal edges), the number $l$ of loops, and the number $v$ of vertices. The Euler characteristic of the graph relates these three numbers as $v = n - l + 1$, therefore only the number of loops $l$ and the number $n$ of propagators are needed.

In a momentum representation an $l$-loop with $n$ propagators Feynman graph reads

$$I_\Gamma(\underline{s}, \underline{\xi}^2, \underline{v}, D) := \frac{(\mu^2)^\omega}{\pi^{\frac{lD}{2}}} \frac{\prod_{i=1}^n \Gamma(v_i)}{\Gamma(\omega)} \int_{(\mathbb{R}^{1,D-1})^l} \frac{\prod_{i=1}^l d^D \ell_i}{\prod_{i=1}^n (q_i^2 - m_i^2 + i\varepsilon)^{v_i}}, \quad (4)$$

where $D$ is the space-time dimension, and we set $\omega := \sum_{i=1}^n v_i - lD/2$ and $q_i$ is the momentum flowing in between the vertices $i$ and $i+1$. With $\mu^2$ a scale of dimension mass squared. From now on we set $m_i^2 = \xi_i^2 \mu^2$ and $p_i \to p_i \mu$, with these new variables the $\mu^2$ dependence disappear. The internal masses are positive $\xi_i^2 \geq 0$ with $1 \leq i \leq n$. Finally $+i\varepsilon$ with $\varepsilon > 0$ is the Feynman prescription for the propagators for a space-time metric of signature $(+ - \cdots -)$. The arguments of the Feynman integral are $\underline{\xi}^2 := \{\xi_1^2, \ldots, \xi_n^2\}$ and $\underline{v} := \{v_1, \ldots, v_n\}$ and $\underline{s} :=$ $\{s_{ij} = (p_i + p_j)^2\}$ with $p_i$, $i = 1, \ldots, v_e$ and $0 \leq v_e \leq v$ the external momenta subject to the momentum conservation condition $p_1 + \cdots + p_{v_e} = 0$. There are $n$ internal masses $\xi_i^2$ with $1 \leq i \leq n$, $v_e$ is the number of external momenta, we have $v_e$ external masses $p_i^2$ with $1 \leq i \leq v_e$ (some of the masses could vanish but we do a generic counting here), and $\frac{v_e(v_e-3)}{2}$ independent kinematics invariants $s_{ij} = (p_i + p_j)^2$. The total number of kinematic parameters is

$$N_\Gamma(n, l) = n + \frac{v_e(v_e - 1)}{2} \leq N_\Gamma(n, l)^{max} = n + \frac{(n - l + 1)(n - l)}{2}. \quad (5)$$

We set

$$I_\Gamma(\underline{s}, \underline{m}, D) := I_\Gamma(\underline{s}, \underline{m}, 1, \ldots, 1, D), \quad (6)$$

and for $v_i$ positive integers we have

$$I_\Gamma(\underline{s}, \underline{m}, \underline{v}, D) = \prod_{i=1}^n \left(\frac{\partial}{\partial(\xi_i^2)}\right)^{v_i} I_\Gamma(\underline{s}, \underline{m}, D). \quad (7)$$

## 2.1  The Parametric Representation

Introducing the variables $x_i$ with $1 \leq i \leq n$ such that

$$\sum_{i=1}^n x_i(q_i^2 - \xi_i^2) = (\ell_1^\mu, \ldots, \ell_l^\mu) \cdot \Omega \cdot (\ell_1^\mu, \ldots, \ell_l^\mu)^T + (\ell_1^\mu, \ldots, \ell_l^\mu) \cdot (Q_1^\mu, \ldots, Q_l^\mu) - J,$$

$$(8)$$

and performing standard Gaussian integrals on the $x_i$ (see [28] for instance) one obtains the equivalent parametric representation that we will use in these notes

$$I_\Gamma(\underline{s}, \underline{\xi}, \underline{v}, D) = \int_{\Delta_n} \Omega_\Gamma, \tag{9}$$

the integrand is the $n - 1$-form

$$\Omega_\Gamma = \prod_{i=1}^{n} x_i^{v_i - 1} \frac{\mathcal{U}^{\omega - \frac{D}{2}}}{\mathcal{F}^\omega} \Omega_0, \tag{10}$$

where $\Omega_0$ is the differential $n - 1$-form on the real projective space $\mathbb{P}^{n-1}$

$$\Omega_0 := \sum_{j=1}^{n} (-1)^{j-1} x_j \, dx_1 \wedge \cdots \wedge \widehat{dx_j} \wedge \cdots \wedge dx_n, \tag{11}$$

where $\widehat{dx_j}$ means that $dx_j$ is omitted in this sum. The domain of integration $\Delta_n$ is defined as

$$\Delta_n := \{[x_1, \cdots, x_n] \in \mathbb{P}^{n-1} | x_i \in \mathbb{R}, x_i \geq 0\}. \tag{12}$$

The second Symanzik polynomial $\mathcal{F} = \mathcal{U}\left((Q_1^\mu, \ldots, Q_l^\mu) \cdot \Omega^{-1} \cdot (Q_1^\mu, \ldots, Q_l^\mu)^T - J\right)$, takes the form

$$\mathcal{F}(\underline{s}, \underline{\xi}^2, x_1, \ldots, x_n) = \mathcal{U}(x_1, \ldots, x_n) \left(\sum_{i=1}^{n} \xi_i^2 x_i\right) - \sum_{1 \leq i \leq j \leq n} s_{ij} \mathcal{G}_{ij}(x_1, \ldots, x_n) \tag{13}$$

where the first Symanzik polynomial $\mathcal{U}(x_1, \ldots, x_n) = \det \Omega$ and $\mathcal{G}_{ij}(x_1, \ldots, x_n)$ are polynomial in the $x_i$ variables only.

- The first Symanzik polynomial $\mathcal{U}(x_1, \ldots, x_n)$ is an homogeneous polynomial of degree $l$ in the Feynman parameters $x_i$ and it is at most linear in each of the $x_i$ variables. It does not depend on the physical parameters. This polynomial is also known as the Kirchhoff polynomial of graph $\Gamma$. Which is as well the determinant of the Laplacian of the graph, see [29, eq (35)] for a definition.
- The polynomial $\mathcal{U}(x_1, \ldots, x_n)$ can be seen as the determinant of the period matrix $\Omega$ of the punctured Feynman graph [28], i.e. the graph with amputated external legs. Or equivalently it can be obtained by considering the degeneration limit of a genus $l$ Riemann surfaces with $n$ punctures. This connection plays an important role in understanding the quantum field theory Feynman integrals as the $\alpha' \to 0$ limit of the corresponding string theory integrals [30, 31].

- The graph polynomial $\mathscr{F}$ is homogeneous of degree $l+1$ in the variables $(x_1, \ldots, x_n)$. This polynomial depends on the internal masses $\xi_i^2$ and the kinematic invariants $s_{ij} = (p_i \cdot p_j)/\mu^2$. The polynomials $\mathscr{G}_{ij}$ are at most linear in all the variables $x_i$ since this is given by the spanning 2-trees [29]. Therefore if all internal masses are vanishing then $\mathscr{F}$ is linear in the Feynman parameters $x_i$.
- The $\mathscr{U}$ and $\mathscr{F}$ are independent of the dimension of space-time. The space-time dimension enters only in the powers of $\mathscr{U}$ and $\mathscr{F}$ in the parametric representation for the Feynman graphs. Therefore one can see the Feynman integral as a meromorphic function of $(\underline{v}, D)$ in $\mathbb{C}^{1+n}$ as discussed in [32].
- All the physical parameters, the internal masses $\xi_i^2$ and the kinematic variables $s_{ij} = (p_i \cdot p_j)/\mu^2$ (that includes the external masses) enter linearly. This will be important for the toric approach described in Sect. 3.

## 2.2 Maximal Cut

We show that the maximal cut of a Feynman graph has a nice parametric representation. Let us consider the maximal cut

$$\pi_\Gamma(\underline{s}, \underline{\xi}^2, D) := \frac{1}{\Gamma(\omega)(2i\pi)^n \pi^{\frac{lD}{2}}} \int_{(\mathbb{R}^{1,D-1})^L} \prod_{i=1}^{l} d^D \ell_i \prod_{i=1}^{n} \delta(q_i^2 - m_i^2 + i\varepsilon), \tag{14}$$

of the Feynman integral $I_\Gamma(\underline{s}, \underline{\xi}^2, D)$ which is obtained from the Feynman integral in (4) by replacing all propagators by a delta-function

$$\frac{1}{d^2} \rightarrow \frac{1}{2i\pi}\delta(d^2). \tag{15}$$

Using the representation of the $\delta$-function

$$\delta(x) = \int_{-\infty}^{+\infty} dw e^{iwx}, \tag{16}$$

we obtain that the integral is

$$\pi_\Gamma(\underline{s}, \underline{m}, D) := \frac{1}{\Gamma(\omega)(2i\pi)^n \pi^{\frac{lD}{2}}} \int_{\mathbb{R}^{(1,D-1)L}} e^{-i \sum_{i=1}^{n} x_i(\ell_i^2 + m_i^2 - i\varepsilon)} \prod_{i=1}^{l} d^D \ell_i \prod_{i=1}^{n} dx_i. \tag{17}$$

At this stage the integral is similar to the one leading to the parametric representation with the replacement $x_r \to i x_r$ with $x_r \in \mathbb{R}$. Setting $\tilde{x}_r = i x_r$ and performing the Gaussian integrals over the loop momenta, we get

$$\pi_n(\underline{s}, \xi^2, D) := \frac{1}{(2i\pi)^n} \int_{i\mathbb{R}^n} \frac{\tilde{\mathscr{U}}^{\omega - \frac{D}{2}}}{\tilde{\mathscr{F}}^\omega} \prod_{i=1}^{n} \delta\left(1 - \sum_{i=1}^{n} \tilde{x}_i\right) d\tilde{x}_i . \tag{18}$$

Using the projective nature of the integrand we have $\frac{\tilde{\mathscr{U}}^{\omega - D/2}}{\tilde{\mathscr{F}}^\omega} = i^{-n} \frac{\mathscr{U}^{\omega - D/2}}{\mathscr{F}^\omega}$ and the integral can be rewritten as the torus integral

$$\pi_\Gamma(\underline{s}, \xi^2, D) := \frac{1}{(2i\pi)^n} \int_{|x_1| = \cdots = |x_{n-1}| = 1} \frac{\mathscr{U}^{\omega - D/2}}{\mathscr{F}^\omega} \prod_{i=1}^{n-1} dx_i . \tag{19}$$

This integral shares the same integrand with the Feynman integral $I_\Gamma$ in (9) but the cycle of integration differs since we are integrating over a $n$-torus. We show in Sect. 3.2 that this maximal cut arises naturally from the toric formalism.

## 2.3 The Differential Equations

In general a Feynman integral $I_\Gamma(\underline{s}, \xi^2, \underline{\nu}, D)$ satisfies an inhomogeneous system of differential equations

$$\mathscr{L}_\Gamma I_\Gamma = \mathscr{S}_\Gamma, \tag{20}$$

where the inhomogeneous term $\mathscr{S}_\Gamma$ essentially arises from boundary terms corresponding to reduced graph topologies where internal edges have been contracted. Knowing the maximal cut integral allows one to determine the differential operators $\mathscr{L}_\Gamma$

$$\mathscr{L}_\Gamma \pi_\Gamma(\underline{s}, \xi^2, D) = 0. \tag{21}$$

This fact has been exploited in [33–36] to obtain the minimal order differential operator. The important remark in this construction is to use that the only difference between the Feynman integral $I_\Gamma$ and the maximal cut $\pi_\Gamma$ is the choice of the cycle of integration. Since the Picard–Fuchs operator $\mathscr{L}_\Gamma$ acts as

$$\mathscr{L}_\Gamma \pi_\Gamma(\underline{s}, \xi^2, D) = \int_{\gamma_n} \mathscr{L}_\Gamma \Omega_F = \int_{\gamma_n} d(\beta_\Gamma) = 0, \tag{22}$$

this integral vanishes because the cycle $\gamma_n = \{|x_1| = \cdots = |x_n| = 1\}$ has no boundaries $\partial \gamma_n = \emptyset$. In the case of the Feynman integral $I_\Gamma$ this is not longer true as

$$\mathscr{L}_\Gamma I_\Gamma (\underline{s}, \xi^2, D) = \int_{\Delta_n} d(\beta_\Gamma) = \int_{\partial \Delta_n} \beta_\Gamma = \mathscr{S}_\Gamma \neq 0. \tag{23}$$

The boundary contributions arises from the configuration with some of the Schwinger coordinate $x_i = 0$ vanishing which corresponds to the so-called reduced topologies that are known to arise when applying the integration-by-part algorithm (see [37–39] for instance).

We illustrate this logic on some elementary examples of differential equations for multi-valued integrals relevant for the one- and two-loop massive sunset integrals discussed in this text.

### 2.3.1 The Logarithmic Integral

We consider the integral

$$I_1(t) = \int_a^b \frac{dx}{x(x-t)}, \tag{24}$$

and its cut integral

$$\pi(t) = \int_\gamma \frac{dx}{x(x-t)}, \tag{25}$$

where $\gamma$ is a cycle around the point $x = t$. Clearly we have

$$\frac{d}{dx}\left(\frac{1}{t-x}\right) = \frac{1}{x(x-t)} + t\frac{d}{dt}\left(\frac{1}{x(x-t)}\right), \tag{26}$$

therefore the integral $\pi(t)$ satisfies the differential equation

$$t\frac{d}{dt}\pi(t) + \pi(t) = \int_\gamma \frac{d}{dx}\left(\frac{1}{t-x}\right) = 0, \tag{27}$$

and the integral $I_1(t)$ satisfies

$$t\frac{d}{dt}I_1(t) + I_1(t) = \int_a^b \frac{d}{dx}\left(\frac{1}{t-x}\right) = \frac{1}{b(b-t)} - \frac{1}{a(a-t)}. \tag{28}$$

Changing variables from $t$ to $p^2$ or an internal mass will give the familiar differential equation for the one-loop bubble that will be commented further in Sect. 3.4.

## 2.3.2 Elliptic Curve

The second example is the differential equation for the period of an elliptic curve $\mathscr{E} : y^2 z = x(x - z)(x - tz)$ which is the geometry of the two-loop sunset integral. Consider the differential of the first kind on the elliptic curve

$$\omega = \frac{dx}{\sqrt{x(x-1)(x-t)}}, \tag{29}$$

this form can be seen as a residue evaluated on the elliptic curve $\omega = \mathrm{Res}_{\mathscr{E}} \Omega$ of the form on the projective space $\mathbb{P}^2$

$$\Omega = \frac{\Omega_0}{y^2 z - x(x-z)(x-tz)}, \tag{30}$$

where $\Omega_0 = z dx \wedge dy + y dz \wedge dx + x dy \wedge dz$ is the natural top form on the projective space $[x : y : z]$. A systematic way of deriving Picard–Fuchs operators for elliptic curve is given by Griffith's algorithm [40]. Consider the second derivative with respect to the parameter $t$

$$\frac{d^2}{dt^2} \Omega = 2 \frac{x^2 (x-z)^2 z^2}{(y^2 z - x(x-z)(x-tz))^2} \Omega_0. \tag{31}$$

The numerator belongs to the Jacobian ideal[1] of the polynomial $p(x, y, z) := y^2 z - x(x - z)(x - tz)$, $J_1 = \langle \partial_x p(x, y, z) = -3x^2 + 2(t+1)xz - tz^2 2, \partial_y p(x, y, z) = 2yz, \partial_z p(x, y, z) = (t+1)x^2 + y^2 - 2txz \rangle$, since

$$x^2 (x-z)^2 z^2 = m_x^1 \partial_x p(x, y, z) + m_y^1 \partial_y p(x, y, z) + m_z^1 \partial_z p(x, y, z). \tag{32}$$

---

[1]An ideal $I$ of a ring $R$, is the subset $I \subset R$, such that 1) $0 \in I$, 2) for all $a, b \in I$ then $a + b \in I$, 3) for $a \in I$ and $b \in R$, $a \cdot b \in R$. For $P(x_1, \ldots, x_n)$ an homogeneous polynomial in $R = \mathbb{C}[x_1, \ldots, x_n]$ the Jacobian ideal of $P$ is the ideal generated by the first partial derivative $\{\partial_{x_i} P(x_1, \ldots, x_n)\}$ [41]. Given a multivariate polynomial $P(x_1, \ldots, x_n)$ its Jacobian ideal is easily evaluated using $\mathtt{Singular}$ command $\mathtt{jacob(P)}$. The hypersurface $P(x_1, \cdots, x_n) = 0$ for an homogeneous polynomial, like the Symanzik polynomials, is of codimension 1 in the projective space $\mathbb{P}^{n-1}$. The singularities of the hypersurface are determined by the irreducible factors of the polynomial. This determines the cohomology of the complement of the graph hypersurface and the number of independent master integrals as shown in [42].

This implies that

$$
\frac{d^2}{dt^2}\Omega = \frac{\partial_x m_x^1 + \partial_y m_y^1 + \partial_z m_z^1}{(y^2 z - x(x-z)(x-tz))^2}\Omega_0
$$
$$
+ d\left(\frac{(ym_z^1 - zm_y^1)dx + (zm_x^1 - xm_z^1)dy + (xm_y^1 - ym_x^1)dz}{(y^2 z - x(x-z)(x-tz))^2}\right). \tag{33}
$$

Therefore

$$
\frac{d^2}{dt^2}\Omega + p_1(t)\frac{d}{dt}\Omega = \frac{-p_1(t)x(x-z)z + \partial_x m_x^1 + \partial_y m_y^1 + \partial_z m_z^1}{(y^2 z - x(x-z)(x-tz))^2}\Omega_0
$$
$$
+ d\left(\frac{(ym_z^1 - zm_y^1)dx + (zm_x^1 - xm_z^1)dy + (xm_y^1 - ym_x^1)dz}{(y^2 z - x(x-z)(x-tz))^2}\right). \tag{34}
$$

One easily derives that $\partial_x m_x^1 + \partial_y m_y^1 + \partial_z m_z^1$ is in the Jacobian ideal generated by $J_1$ and $x(x-z)z$ with the result that

$$
\partial_x m_x^1 + \partial_y m_y^1 + \partial_z m_z^1 = m_x^2 \partial_x p(x, y, z) + m_y^2 \partial_y p(x, y, z) + m_z^2 \partial_z p(x, y, z)
$$
$$
+ \frac{2t-1}{t(t-1)}x(x-z)z, \tag{35}
$$

therefore $p_1(t) = \frac{2t-1}{t(t-1)}$ and the Picard–Fuchs operator reads

$$
\frac{d^2}{dt^2}\Omega + \frac{2t-1}{t(t-1)}\frac{d}{dt}\Omega - \frac{\partial_x m_x^2 + \partial_y m_y^2 + \partial_z m_z^2}{(y^2 z - x(x-z)(x-tz))^2}\Omega_0 =
$$
$$
d\left(\frac{(ym_z^1 - zm_y^1)dx + (zm_x^1 - xm_z^1)dy + (xm_y^1 - ym_x^1)dz}{(y^2 z - x(x-z)(x-tz))^2}\right)
$$
$$
+ d\left(\frac{(ym_z^2 - zm_y^2)dx + (zm_x^2 - xm_z^2)dy + (xm_y^2 - ym_x^2)dz}{y^2 z - x(x-z)(x-tz)}\right). \tag{36}
$$

Since $\partial_x m_x^2 + \partial_y m_y^2 + \partial_z m_z^2 = -\frac{1}{4t(t-1)}$ we have that

$$
\left(4t(t-1)\frac{d^2}{dt^2} - 4(2t-1)\frac{d}{dt} + 1\right)\omega = -2\partial_x\left(\frac{y}{(x-t)^2}\right). \tag{37}
$$

For $\alpha$ and $\beta$ a (sympletic) basis of $H_1(\mathscr{E}, \mathbb{Z})$ the period integrals $\varpi_1(t) := \int_\alpha \omega$ and $\varpi_2(t) := \int_\beta \omega$ both satisfy the differential equation

$$\left( 4t(t-1)\frac{d^2}{dt^2} - 4(1-2t)\frac{d}{dt} + 1 \right) \varpi_i(t) = 0 \,. \tag{38}$$

Again this differential operator acting on an integral with a different domain of integration can lead to an homogeneous terms as this is the case for the two-loop sunset Feynman integral.

The whole procedure is easily implemented in Singular [43] with the following set of commands

```
In [1]: // Griffith-Dwork method for
deriving the Picard--Fuchs operator for the elliptic curve
 y^2z=x(x-z)(x-tz)

In [2]: ring A=(0,t),(x,y,z),dp;

In [3]: poly f=y^2*z-x*(x-z)*(x-t*z);

In [4]: ideal I1=jacob(f); I1

Out[4]: I1[1]=-3*x2+(2t+2)*xz+(-t)*z2
        I1[2]=2*yz
        I1[3]=(t+1)*x2+y2+(-2t)*xz

In [5]: matrix M1=lift(I1,x^2*(x-z)^2*z^2); M1

Out[5]: M1[1,1]=2/(3t+3)*xz3
        M1[2,1]=-1/(2t+2)*x2yz+1/(6t+6)*yz3
        M1[3,1]=1/(t+1)*x2z2-1/(3t+3)*z4

In [6]: // checking the decomposition
x^2*(x-z)^2*z^2-M1[1,1]*I[1]-M1[2,1]*I[2]-M1[3,1]*I[3]

Out[6]: 0

In [7]:  poly dC1=diff(M[1,1],x)+diff(M[2,1],y)
+diff(M[3,1],z);
dC1

Out[7]:  dC1=3/(t+1)*x2z-1/(t+1)*z3

In [8]:  ideal I2=jacob(f),x*(x-z)*z;

In [9]:  matrix M2=lift(I2,dC1); M2

Out[9]: M2[1,1]=1/(2t2+2t)*z
        M2[2,1]=1/(4t2-4t)*y
        M2[3,1]=-1/(2t2-2t)*z
        M2[4,1]=(2t-1)/(t2-t)
```

```
In [10] : // checking the decomposition
dC1-M2 [1,1] *I [1] -M2 [2,1] *I [2] -M2 [3,1] *I [3]
-M2 [4,1] *x* (x-z) *z
```

```
Out [10] : 0
```

```
In [11] :   poly dC2=diff (M2 [1,1] ,x) +diff (M2 [2,1] ,y)
+diff (M2 [3,1] ,z) ;
dC2
```

```
Out [11] : -1/ (4t2-4t)
```

## 3   Toric Geometry and Feynman Graphs

We will show how the toric approach provides a nice way to obtain this maximal cut integral. The maximal cut integral $\pi_\Gamma(\underline{s}, \xi^2, D)$ is the particular case of generalised Euler integrals

$$\int_\sigma \prod_{i=1}^r P_i(x_1, \ldots, x_n)^{\alpha_i} \prod_{i=1}^n x_i^{\beta_i} dx_i \tag{39}$$

studied by Gel'fand, Kapranov and Zelevinski (GKZ) in [44, 45]. There $P_i(x_1, \ldots, x_n)$ are Laurent polynomials, $\alpha_i$ and $\beta_i$ are complex numbers and $\sigma$ is a cycle. The cycle entering the maximal cut integral in (19) is the product of circles $\sigma = \{|x_1| = |x_2| = \cdots = |x_n| = 1\}$. But other cycles arise when considering different cuts of Feynman graphs. The GKZ approach provides a totally combinatorial approach to differential equation satisfied by these integrals.

As well in the case when $P(\underline{x}, \underline{z}) = \sum_i z_{i_1, \ldots, i_r} \prod_{i=1}^n x_i^{\alpha_i}$ is the Laurent polynomial defining a Calabi–Yau hypersurface $\{P(\underline{x}, \underline{z}) = 0\}$, Batyrev showed that there is one canonical period integral [46, 47]

$$\Pi(\underline{z}) := \frac{1}{(2i\pi)^n} \int_{|x_1|=\cdots=|x_n|=1} \frac{1}{P(\underline{x}, \underline{z})} \prod_{i=1}^n \frac{dx_i}{x_i}. \tag{40}$$

This corresponds to the maximal cut integral (19) in the case where $\omega = D/2 = 1$ which is satisfied by the $(n-1)$-loop sunset integral in $D = 2$ dimensions. The graph hypersurface of the $(n-1)$-loop sunset (see (48)) is always a Calabi–Yau $(n-1)$-fold. See for more comments about this at the end of Sect. 4.3.1. We refer to the reviews [48, 49] for some introduction to toric geometry for physicists.

## 3.1 Toric Polynomials and Feynman Graphs

The second Symanzik polynomial $\mathscr{F}(\underline{s}, \xi^2, x_1, \ldots, x_n)$ defined in (13) is a specialisation of the homogeneous (toric) polynomial[2] of degree $l + 1$ at most quadratic in each variable of in the projective variables $(x_1, \ldots, x_n) \in \mathbb{P}^{n-1}$

$$\mathscr{F}_l^{toric}(\underline{z}, x_1, \ldots, x_n) = \mathscr{U}_l^{tor}(x_1, \ldots, x_n) \left( \sum_{i=1}^{n} \xi_i^2 x_i \right) - \mathscr{V}_l^{tor}(x_1, \ldots, x_n),$$

(41)

where for $l \leq n$

$$\mathscr{U}_l^{tor}(x_1, \ldots, x_n) := \sum_{\substack{0 \leq r_i \leq 1 \\ r_1 + \cdots + r_n = l}} u_{i_1, \ldots, i_n} \prod_{i=1}^{n} x_i^{r_i},$$

(42)

where the coefficients $u_{i_1, \ldots, i_n} \in \{0, 1\}$. The expression in (41) is the most generic form compatible with the properties of the Symanzik polynomials listed in Sect. 2.1.

There are $\frac{n!}{(n-l)! l!}$ independent coefficient in the polynomial $\mathscr{U}_l^{tor}(x_1, \ldots, x_n)$. Of course this is a huge over counting, as this does not take into account the symmetries of the graphs and the constraints on the non-vanishing of some coefficients. This will be enough for the toric description we are using here. In order to keep most of the combinatorial power of the toric approach we will only do the specialisation of the toric coefficients with the physical slice corresponding to the Feynman graph polynomial at the end on solutions. This will avoid having to think on the constrained system of differential equations which is a difficult problem discussed recently in [42].

---

[2]Consider an homogeneous polynomial of degree $d$

$$P(\underline{z}, \underline{x}) = \sum_{\substack{0 \leq r_i \leq n \\ r_1 + \cdots + r_n = d}} z_{i_1, \ldots, i_n} \prod_{i=1}^{n} x_i^{r_i}$$

this is called a *toric polynomial* if it is invariant under the following actions

$$z_i \to \prod_{j=1}^{n} t_i^{\alpha_{ij}} z_i; \qquad x_i \to \prod_{j=1}^{n} t_i^{\beta_{ij}} x_i$$

for $(t_1, \ldots, t_n) \in \mathbb{C}^n$ and $\alpha_{ij}$ and $\beta_{ij}$ integers. The second Symanzik polynomial has a natural torus action acting on the mass parameters and the kinematic variables as we will see on some examples below. We refer to the book [41] for more details.

The kinematics part has the toric polynomial

$$\mathcal{V}_l^{tor}(x_1, \ldots, x_n) := \sum_{\substack{0 \le r_i \le 1 \\ r_1 + \cdots + r_n = l+1}} z_{i_1, \ldots, i_n} \prod_{i=1}^{n} x_i^{r_i}, \tag{43}$$

where the coefficients $z_{i_1, \cdots, i_n} \in \mathbb{C}$. The number of independent toric variables $z_i$ in $\mathcal{V}^{tor}(x_1, \ldots, x_n)$ is $\frac{n!}{(n-l-1)!(l+1)!}$.

### 3.1.1 Some Important Special Cases

There are a few important special cases.

- The one-loop order $l = 1$ and the number of independent toric variables in $\mathcal{V}^{tor}(x_1, \ldots, x_n)$ is exactly the number of independent kinematics for an $n$-gon one-loop amplitude

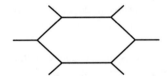

In this case the most general toric one-loop polynomial is

$$\mathscr{F}_1^{tor}(x_1, \cdots, x_n) = \left(\sum_{i=1}^{n} x_i\right)\left(\sum_{i=1}^{n} \xi_i^2 x_i\right) - \mathcal{V}_1^{tor}(x_1, \cdots, x_n). \tag{44}$$

- For $l = n$ there is only one vertex the graph is $n$-bouquet which is a product of $n$ one-loop graphs. These graphs contribute to the reduced topologies entering the determination of the inhomogeneous term $\mathscr{S}_\Gamma$ of the Picard–Fuchs equation (20). They don't contribute to the maximal cut $\pi_\Gamma$ for $l > 1$.

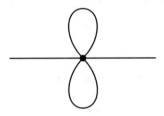

- The case $l = n - 1$ corresponds to the $(n - 1)$-loop two-point sunset graphs

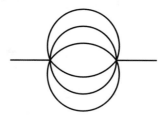

In that case the kinematic polynomial is just

$$\mathcal{V}_{n-1}^{tor}(x_1, \ldots, x_n) = z_{1,\ldots,1} x_1 \cdots x_n, \tag{45}$$

and the toric polynomial

$$\mathcal{F}_{n-1}^{tor}(x_1, \ldots, x_n) = x_1 \cdots x_n \left( \sum_{i=1}^{n} \frac{u_{1,\ldots,0,\ldots,1}}{x_i} \right) \left( \sum_{i=1}^{n} \xi_i^2 x_i \right) - z_{1,\ldots,1} x_1 \cdots x_n, \tag{46}$$

where the index 0 in $u_{1,\ldots,0,\ldots,1}$ is at the position $i$. Actually by redefining the parameter $z_{1,\ldots,1}$ the generic toric polynomial associated to the sunset graph becomes

$$\mathcal{F}_{\ominus}^{tor}(x_1, \ldots, x_n) = x_1 \cdots x_n \left( \sum_{\substack{1 \le i, j \le n \\ i \neq j}} z_{ij} \frac{x_i}{x_j} - z_0 \right), \tag{47}$$

where $z_{ij} \in \mathbb{C}$ and $z_0 \in \mathbb{C}\backslash\{0\}$. This polynomial has $1 - n + n^2$ parameters whereas the sunset graph has $n + 1$ physical parameters given by $n$ masses and one kinematics invariant

$$\mathcal{F}_{\ominus}^{l}(p^2, \underline{\xi}^2, \underline{x}) = x_1 \cdots x_{l+1} \left( \sum_{i=1}^{l+1} \frac{1}{x_i} \right) \left( \sum_{i=1}^{l+1} \xi_i^2 x_i \right) - p^2 x_1 \cdots x_{l+1}. \tag{48}$$

So there are too many parameters from $n \ge 3$ but this generalisation will be useful for the GKZ description used in the next sections.

## 3.2   The GKZ Approach : A Review

In this section we briefly review the GKZ construction based on [44, 45], see as well [50]. We consider the Laurent polynomial of $n - 1$ variables $P(z_1, \ldots, z_r) =$

$\mathscr{F}^{tor}\underline{z}, x_1, \ldots, x_n/(x_1 \cdots x_n)$ from the toric polynomial of Sect. 3.1. The coefficients of monomials are $z_i$ (by homogeneity we set $x_n = 1$)

$$P(z_1, \ldots, z_r) = \sum_{\mathbf{a}=(a_1, \ldots, a_{n-1}) \in \mathbf{A}} z_{\mathbf{a}} \prod_{i=1}^{n-1} x_i^{a_i}, \tag{49}$$

with $\mathbf{a} = (a_1, \ldots, a_{n-1})$ is an element of $\mathbf{A} = (\mathbf{a}_1, \ldots, \mathbf{a}_r)$ a finite subset of $\mathbb{Z}^{n-1}$. The number of elements in $A$ is $r$ the number of monomials in $P(z_1, \ldots, z_r)$.

We consider the natural fundamental period integral [51]

$$\Pi(\underline{z}) := \frac{1}{(2i\pi)^{n-1}} \int_{|x_1|=\cdots=|x_{n-1}|=1} P(z_1, \ldots, z_r)^m \prod_{i=1}^{n-1} \frac{dx_i}{x_i}, \tag{50}$$

which is the same as maximal cut $\pi_\Gamma$ in (19) for $D = 2\omega = -m$. The derivative with respect to $z_{\mathbf{a}}$ reads

$$\frac{\partial}{\partial z_{\mathbf{a}}} \Pi(\underline{z}) = \frac{1}{(2i\pi)^{n-1}} \int_{|x_1|=\cdots=|x_{n-1}|=1} m P(z_1, \ldots, z_r)^{m-1} \prod_{i=1}^{n-1} x_i^{a_i} \frac{dx_i}{x_i}, \tag{51}$$

therefore for every vector $\boldsymbol{\ell} = (\ell_1, \ldots, \ell_r) \in \mathbb{Z}^{n-1}$ such that

$$\ell_1 + \cdots + \ell_r = 0, \qquad \ell_1 \mathbf{a}_1 + \cdots + \ell_r \mathbf{a}_r = \boldsymbol{\ell} \cdot \mathbf{A} = 0, \tag{52}$$

there holds the differential equation

$$\left( \prod_{l_i > 0} \partial_{z_i}^{l_i} - \prod_{l_i < 0} \partial_{z_i}^{-l_i} \right) \Pi(\underline{z}) = 0. \tag{53}$$

Introduce the so-called $\mathscr{A}$-hypergeometric functions[3] $\Phi_{\mathbb{L}, \boldsymbol{\gamma}}(z_1, \ldots, z_r)$ of $r$ complex variables $(z_1, \ldots, z_r) \in \mathbb{C}^r$

$$\Phi_{\mathbb{L}, \boldsymbol{\gamma}}(z_1, \ldots, z_r) = \sum_{(\ell_1, \ldots, \ell_r) \in \mathbb{L}} \prod_{j=1}^{r} \frac{z_j^{\gamma_j + \ell_j}}{\Gamma(\gamma_j + \ell_j + 1)}, \tag{54}$$

---

[3]The convergence of these series is discussed in [52, §3-2] and [50, §5.2].

depending on the complex parameters $\boldsymbol{\gamma} := (\gamma_1, \ldots, \gamma_r) \in \mathbb{C}^r$ and the lattice

$$\mathbb{L} := \{(\ell_1, \ldots, \ell_r) \in \mathbb{Z} | \sum_{i=1}^{r} \ell_i \mathbf{a}_i = 0, \ell_1 + \cdots + \ell_r = 0\}, \tag{55}$$

with $r$ elements $\{\mathbf{a}_1, \ldots, \mathbf{a}_r\} \in \mathbb{Z}^n$. These functions are solutions of the so-called $\mathscr{A}$-hypergeometric system of differential equations given by a vector $\mathbf{c} \in \mathbb{C}^n$ and :

- For every $\boldsymbol{\ell} = (\ell_1, \ldots, \ell_r) \in \mathbb{L}$ there are one differential operator

$$\Box_{\boldsymbol{\ell}} := \prod_{\ell_i > 0} \partial_{z_i}^{\ell_i} - \prod_{\ell_i < 0} \partial_{z_i}^{-\ell_i}, \tag{56}$$

such that $\Box_{\boldsymbol{\ell}} \Phi_{\mathbb{L},\gamma}(z_1, \ldots, z_r) = 0$,
- and $n$ differential operators $\mathbf{E} := (E_1, \ldots, E_{n-1})$

$$\mathbf{E} := \mathbf{a}_1 z_1 \frac{\partial}{\partial z_1} + \cdots + \mathbf{a}_r z_r \frac{\partial}{\partial z_r}, \tag{57}$$

such that for $\mathbf{c} = (c_1, \ldots, c_{n-1})$ we have

$$(\mathbf{E} - \mathbf{c}) \Phi_{\mathbb{L},\gamma}(z_1, \ldots, z_r) = 0. \tag{58}$$

Notice that $E_1 = \sum_{i=1}^{n} z_i \frac{\partial}{\partial z_i}$ is the Euler operator and $c_1$ is the degree of homogeneity of the hypergeometric function.

These operators satisfy the commutation relations

$$\mathbf{z}^{\mathbf{u}} \mathbf{E} - \mathbf{E} \mathbf{z}^{\mathbf{u}} = -(\mathbf{A} \cdot \mathbf{u}) \mathbf{z}^{\mathbf{u}},$$
$$\partial_z^{\mathbf{u}} \mathbf{E} - \mathbf{E} \partial_z^{\mathbf{u}} = (\mathbf{A} \cdot \mathbf{u}) \partial_z^{\mathbf{u}}, \tag{59}$$

with $\mathbf{z}^{\mathbf{u}} := \prod_{i=1}^{r} z_r^{u_r}$ and $\partial_z^{\mathbf{u}} := \prod_{i=1}^{r} \partial_{z_r}^{u_r}$.

Using the GKZ construction one can easily derive a system of differential operator annihilating the maximal cut of any Feynman integral after identification of the toric variables with the physical parameters. The system of differential operators obtained from the GKZ system can be massaged into a set of Picard–Fuchs differential operators in a spirit similar to the one used in mirror symmetry [25, 41, 53].

Since it is rather complicated to restrict differential operators but it is easier to restrict functions, it is therefore preferable to determine the $\mathscr{A}$-hypergeometric representation of the maximal cut integral and derive the minimal differential operator annihilating this integral. For well chosen vector $\boldsymbol{\ell} \in \mathbb{L}$ the differential operator factorises with a factor being given by the minimal (Picard–Fuchs) differential operator acting on the Feynman integral.

An important remark is that the maximal cut integral

$$\pi_\Gamma = \int_{|x_1|=\cdots=|x_{n-1}|=1} \frac{1}{\mathscr{F}_\Gamma} \prod_{i=1}^{n-1} dx_i, \tag{60}$$

is a particular case of the fundamental period $\Pi(\underline{z})$ in (50) with $m = -1$ and therefore is given by an $\mathscr{A}$-hypergeometric function once we have identified the toric variables $z_i$ with the physical parameters.

In the next section we illustrate this approach on some simple but fundamental examples.

## 3.3  Hypergeometric Functions and GKZ System

The relation between hypergeometric functions and the GKZ differential system can be simply understood as follows (see [50, 54, 55]).

### 3.3.1  The Gauß Hypergeometric Series

Consider the case of $\mathbb{L} = (1, 1, -1, -1)\mathbb{Z} \subset \mathbb{Z}^4$ and the vector $\gamma = (0, c - 1, -a, -b) \in \mathbb{C}^4$ and $c$ a positive integer. The GKZ hypergeometric function is

$$\Phi_{\mathbb{L},\gamma}(u_1, \ldots, u_4) = \sum_{n\in\mathbb{Z}} \frac{u_1^n u_2^{1-c+n} u_3^{-a-n} u_4^{-b-n}}{\Gamma(1+n)\Gamma(c+n)\Gamma(1-n-a)\Gamma(1-n-b)}, \tag{61}$$

which can be rewritten as

$$\Phi_{\mathbb{L},\gamma}(u_1, \ldots, u_4) = \frac{u_2^{c-1} u_3^{-a} u_4^{-b}}{\Gamma(c)\Gamma(1-a)\Gamma(1-b)}\, {}_2F_1\left(\begin{matrix} a, & b \\ & c \end{matrix}\,\bigg|\, \frac{u_1 u_2}{u_3 u_4}\right). \tag{62}$$

The GKZ system is

$$\left(\frac{\partial^2}{\partial u_1 \partial u_2} - \frac{\partial^2}{\partial u_3 \partial u_4}\right) \Phi_{\mathbb{L},\gamma}(u_1, \ldots, u_4) = 0,$$

$$\left(u_1 \frac{\partial}{\partial u_1} - u_2 \frac{\partial}{\partial u_2} + 1 - c\right) \Phi_{\mathbb{L},\gamma}(u_1, \ldots, u_4) = 0,$$

$$\left(u_1 \frac{\partial}{\partial u_1} + u_3 \frac{\partial}{\partial u_3} + a\right) \Phi_{\mathbb{L},\gamma}(u_1, \ldots, u_4) = 0,$$

$$\left(u_1 \frac{\partial}{\partial u_1} + u_4 \frac{\partial}{\partial u_4} + b\right) \Phi_{\mathbb{L},\gamma}(u_1, \ldots, u_4) = 0. \tag{63}$$

By differentiating we find

$$\left( u_2 \frac{\partial^2}{\partial u_1 \partial u_2} - u_1 \frac{\partial^2}{\partial u_1^2} + c \frac{\partial}{\partial u_1} \right) \Phi_{\mathbb{L},\gamma}(u_1, \ldots, u_4) = 0,$$

$$\left( u_3 u_4 \frac{\partial^2}{\partial u_3 \partial u_4} - \left( u_1 \frac{\partial}{\partial u_1} + a \right) \left( u_1 \frac{\partial}{\partial u_1} + b \right) \right) \Phi_{\mathbb{L},\gamma}(u_1, \ldots, u_4) = 0. \qquad (64)$$

combining these equations one finds

$$\left( u_1^2 \frac{\partial}{\partial u_1} + (1 + a + b) u_1 \frac{\partial}{\partial u_1} + ab \right) \Phi_{\mathbb{L},\gamma}(u_1, \ldots, u_4)$$

$$= \frac{u_3 u_4}{u_2} \left( u_1 \frac{\partial^2}{\partial u_1^2} + c \frac{\partial}{\partial u_1} \right) \Phi_{\mathbb{L},\gamma}(u_1, \ldots, u_4). \qquad (65)$$

Setting $F(z) = \Gamma(c)\Gamma(1-a)\Gamma(1-b)\Phi_{\mathbb{L},\gamma}(z, 1, 1, 1)$ gives that $F(z) = {}_2F_1\left({}^{a\ b}_{\ c} | z\right)$ satisfies the Gauß hypergeometric differential equation

$$z(z-1) \frac{d^2 F(z)}{dz^2} + ((a + b + 1)z - c) \frac{dF(z)}{dz} + abF(z) = 0. \qquad (66)$$

## 3.4   The Massive One-Loop Graph

In this section we show how to apply the GKZ formalism on the one-loop bubble integral

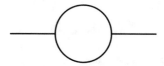

### 3.4.1   Maximal Cut

The one-loop sunset (or bubble) graph has the graph polynomial

$$\mathscr{F}_o(x_1, x_2, t, \xi_1^2, \xi_2^2) = p^2 x_1 x_2 - (\xi_1^2 x_1 + \xi_2^2 x_2)(x_1 + x_2). \qquad (67)$$

The most general toric degree two polynomial in $\mathbb{P}^2$ with at most degree two monomial is given by

$$\mathscr{F}_\circ^{tor}(x_1, x_2, z_1, z_2, z_3) = z_1 x_1^2 + z_2 x_2^2 + z_3 x_1 x_2 \,. \qquad (68)$$

This toric polynomial has three parameters which is exactly the number of independent physical parameters. The identification of the variables is given by

$$z_1 = -\xi_1^2, \qquad z_2 = -\xi_2^2, \qquad z_3 = p^2 - (\xi_1^2 + \xi_2^2) \,. \qquad (69)$$

We consider the equivalent toric Laurent polynomial

$$P(x_1, x_2) = \frac{\mathscr{F}_\circ^{tor}}{x_1 x_2} = \sum_{i=1}^{3} z_i x_1^{a_i^1} x_2^{a_i^2} \,, \qquad (70)$$

so that $p^2$ in (67) corresponds to the constant term (or the origin the Newton polytope) and setting $\mathbf{a}_i = (1, a_i^1, a_i^2)$ we have

$$\mathbf{A}_\circ = \begin{pmatrix} \mathbf{a}_1 \\ \mathbf{a}_2 \\ \mathbf{a}_3 \end{pmatrix} = \begin{pmatrix} 1 & -1 & 1 \\ 1 & 1 & -1 \\ 1 & 0 & 0 \end{pmatrix} \,. \qquad (71)$$

The lattice is defined by

$$\mathbb{L}_\circ := \{\boldsymbol{\ell} := (\ell_1, \ell_2, \ell_3) \in \mathbb{Z}^3 | \ell_1 \mathbf{a}_1 + \ell_2 \mathbf{a}_2 + \ell_3 \mathbf{a}_3 = \boldsymbol{\ell} \cdot \mathbf{A}_\circ = 0\} \,. \qquad (72)$$

This means that the elements of $\mathbb{L}_\circ$ are in the kernel of $\mathbf{A}_\circ$. This lattice in $\mathbb{Z}^3$ has rank one

$$\mathbb{L}_\circ = (1, 1, -2)\mathbb{Z} \,. \qquad (73)$$

Notice that all the elements automatically satisfy the condition $\ell_1 + \ell_2 + \ell_3 = 0$.

Because the rank is one the GKZ system of differential equations is given by

$$e_1 := \frac{\partial^2}{\partial z_1 \partial z_2} - \frac{\partial^2}{(\partial z_3)^2},$$

$$d_1 := \sum_{r=1}^{3} z_r \frac{\partial}{\partial z_r},$$

$$d_2 := z_1 \frac{\partial}{\partial z_1} - z_2 \frac{\partial}{\partial z_2}, \qquad (74)$$

By construction for $\alpha \in \mathbb{C}$

$$e_1(\mathscr{F}_\circ^{tor})^\alpha = 0,$$
$$d_1(\mathscr{F}_\circ^{tor})^\alpha = \alpha \, (\mathscr{F}_\circ^{tor})^\alpha, \tag{75}$$

and

$$d_2(\mathscr{F}_\circ^{tor})^\alpha = \frac{1}{2} \left( \partial_{x_1}(x_1(\mathscr{F}_\circ^{tor})^\alpha) - \partial_{x_2}(x_2(\mathscr{F}_\circ^{tor})^\alpha) \right), \tag{76}$$

therefore the action of the derivative $d_2$ vanishes on the integral but not the integrand

$$d_2 \int_\gamma (\mathscr{F}_\circ^{tor})^\alpha = 0 \qquad \text{for} \qquad \partial \gamma = \emptyset. \tag{77}$$

The GKZ hypergeometric series is defined as for $\gamma_i \notin \mathbb{Z}$

$$\Phi^\circ_{\mathbb{L},\gamma} = \sum_{\ell \in \mathbb{L}_\circ} \prod_{i=1}^3 \frac{z_i^{l_i + \gamma_i}}{\Gamma(l_i + \gamma_i + 1)}, \tag{78}$$

in this sum we have $\ell = n(1, 1, -2)$ with $n \in \mathbb{Z}$, and the condition $\sum_{i=1}^3 \gamma_i \mathbf{a}_i = (0, 0, -1)$ which can be solved using $\boldsymbol{\gamma} = (\gamma_1, \gamma_2, \gamma_3) = \gamma(1, 1, -2) + (0, 0, -1)$, leading to

$$\Phi^\circ_{\mathbb{L},\gamma} = \frac{1}{z_3} \sum_{n \in \mathbb{Z}} \frac{u_1^n}{\Gamma(n + \gamma + 1)^2 \Gamma(-2n + \gamma)}, \tag{79}$$

where we have introduced the new toric coordinate

$$u_1 := \frac{z_1 z_2}{z_3^2} = \frac{\xi_1^2 \xi_2^2}{\left(p^2 - (\xi_1^2 + \xi_2^2)\right)^2}. \tag{80}$$

This is the natural coordinate dictated by the invariance of the period integral under the transformation $(x_1, x_2) \to (\lambda x_1, \lambda x_2)$ and $(z_1, z_2, z_3) \to (z_1/\lambda, z_2/\lambda, z_3/\lambda)$.

This GKZ hypergeometric function is a combination of $_3F_2$ hypergeometric functions

$$\Phi^\circ_{\mathbb{L},\gamma} = \frac{1}{z_3^{1-2\gamma}} \left( \frac{u_1^{\gamma-1}}{\Gamma(\gamma)\Gamma(\gamma+2)} \, _3F_2 \left( \begin{matrix} 1, 1-\gamma, 1-\gamma \\ 1+\frac{\gamma}{2}, \frac{3}{2}+\frac{\gamma}{2} \end{matrix} \middle| \frac{1}{4u_1} \right) \right.$$

$$\left. + \frac{u_1^\gamma}{\Gamma(\gamma+1)^2} \, _3F_2 \left( \begin{matrix} 1, \frac{1}{2}-\frac{\gamma}{2}, \frac{1}{2}-\frac{\gamma}{2} \\ 1+\gamma, 1+\gamma \end{matrix} \middle| 4u_1 \right) \right). \tag{81}$$

For $\gamma = 0$ the series is trivially zero as the system is resonant and needs to be regularised [52, 56] . The regularisation is to use the functional equation for the $\Gamma$-function $\Gamma(z)\Gamma(1 - z) = \pi / \sin(\pi z)$ to replace the pole term by

$$\lim_{\epsilon \to 0} \frac{\Gamma(\epsilon)}{\Gamma(-2n + \epsilon)} = \Gamma(1 + 2n), \qquad n \in \mathbb{Z}\backslash\{0\}, \tag{82}$$

and write the associated regulated period as

$$\pi_\circ = \lim_{\epsilon \to 0} \frac{1}{z_3} \sum_{n \in \mathbb{N}} \frac{u_1^n \Gamma(\epsilon)}{\Gamma(n + 1)^2 \Gamma(-2n + \epsilon)}, \tag{83}$$

which is easily shown to be

$$\pi_\circ(z_1, z_2, z_3) = \frac{1}{z_3} {}_2F_1 \left( \begin{matrix} \frac{1}{2} & 1 \\ & 1 \end{matrix} \middle| 4u_1 \right) = \frac{1}{\sqrt{z_3^2 - 4z_1 z_2}}$$

$$= \frac{1}{\sqrt{(p^2 - (\xi_1 + \xi_2)^2)(p^2 - (\xi_1 - \xi_2)^2)}}. \tag{84}$$

This expression of course matches the expression for the maximal cut (19) integral $\pi_\circ(p^2, \xi_1^2, \xi_2^2, 2)$ in two dimensions

$$\pi_\circ(p^2, \xi_1^2, \xi_2^2, 2) = \frac{1}{(2i\pi)^2} \int_{|x_1|=|x_2|=1} \frac{dx_1 dx_2}{\mathscr{F}_\circ(x_1, x_2)}. \tag{85}$$

### 3.4.2 The Differential Operator

From the expression of the maximal cut $\pi_\circ$ in (84) as an hypergeometric series, which satisfies a second order differential equation (66), we can extract a differential operator with respect to $p^2$ or the masses $\xi_i^2$ annihilating the maximal cut. This differential equation is not the minimal one as it can be factorised leaving minimal order differential operators annihilating the maximal cut such that $L_{PF,(1)}^\circ \pi_\circ(p^2, \xi_1^2, \xi_2^2) = 0$ and $L_{PF,(2)}^\circ \pi_\circ(p^2, \xi_1^2, \xi_2^2) = 0$ with

$$L_{PF,(1)}^\circ = p^2 \frac{d}{dp^2} + \frac{p^2(p^2 - \xi_1^2 - \xi_2^2)}{(p^2 - (\xi_1 + \xi_2)^2)(p^2 - (\xi_1 - \xi_2)^2)}, \tag{86}$$

and

$$L_{PF,(2)}^\circ = \xi_1^2 \frac{d}{d\xi_1^2} - \frac{\xi_1^2(p^2 - \xi_1^2 + \xi_2^2)}{(p^2 - (\xi_1 + \xi_2)^2)(p^2 - (\xi_1 - \xi_2)^2)}, \tag{87}$$

with of course a similar operator with the exchange of $\xi_1$ and $\xi_2$. These operators do not annihilate the integrand but lead to total derivatives

$$L^\circ_{PF,(1)} \frac{1}{\mathscr{F}_\circ(\underline{x}, p^2, \underline{\xi}^2)} = \partial_{x_1} \left( \frac{p^2(2\xi_2^2 - (p^2 - (\xi_1^2 + \xi_2^2))x_1)}{(p^2 - (\xi_1 + \xi_2)^2)(p^2 - (\xi_1 - \xi_2)^2)\mathscr{F}_2(x_1, 1, p^2, \underline{\xi}^2)} \right),$$

(88)

and

$$L^\circ_{PF,(2)} \frac{1}{\mathscr{F}_\circ(\underline{x}, p^2, \underline{\xi}^2)} = \partial_{x_1} \left( \frac{((p^2 - \xi_2^2)^2 - \xi_1^2(p^2 + \xi_2^2))x_1 - \xi_2^2(p^2 + \xi_1^2 - \xi_2^2)}{(p^2 - (\xi_1 + \xi_2)^2)(p^2 - (\xi_1 - \xi_2)^2)\mathscr{F}_2(x_1, 1, p^2, \underline{\xi}^2)} \right).$$

(89)

These operators can be obtained from the operator $td/dt + 1$ derived in Sect. 2.3.1 and the change of variables $t = \frac{\sqrt{(p^2 - \xi_1^2 - \xi_2^2)^2 - 4\xi_1^2 \xi_2^2}}{\xi_1^2}$. For the boundary term one needs to pay attention that the shift induces a dependence on the physical parameters in the domain of integration.

### 3.4.3 The Massive One-Loop Sunset Feynman Integral

Having determined the differential operators acting on the maximal cut it is now easy to obtain the action of these operators on the one-loop integral. The action of the Picard–Fuchs operators on the Feynman integral $I_\circ(p^2, \xi_1^2, \xi_2^2, 2)$ are given by

$$L^\circ_{PF,(1)} I_\circ(p^2, \xi_1^2, \xi_2^2, 2) = -\frac{2}{(p^2 - (\xi_1 + \xi_2)^2)(p^2 - (\xi_1 - \xi_2)^2)},$$

(90)

and

$$L^\circ_{PF,(2)} I_\circ(p^2, \xi_1^2, \xi_2^2, 2) = \frac{\xi_1^2 - \xi_2^2 - p^2}{(p^2 - (\xi_1 + \xi_2)^2)(p^2 - (\xi_1 - \xi_2)^2)}.$$

(91)

It is then easy to obtain that in $D = 2$ dimensions the one-loop massive bubble evaluates to

$$I_\circ(p^2, \xi_1^2, \xi_2^2) = \frac{1}{\sqrt{(p^2 - (\xi_1 + \xi_2)^2)(p^2 - (\xi_1 - \xi_2)^2)}}$$

$$\times \log \left( \frac{p^2 - (\xi_1^2 + \xi_2^2) - \sqrt{(p^2 - (\xi_1 + \xi_2)^2)(p^2 - (\xi_1 - \xi_2)^2)}}{p^2 - (\xi_1^2 + \xi_2^2) + \sqrt{(p^2 - (\xi_1 + \xi_2)^2)(p^2 - (\xi_1 - \xi_2)^2)}} \right).$$

(92)

## 3.5   The Two-Loop Sunset

The sunset graph polynomial is the most general cubic in $\mathbb{P}^2$ with maximal order two degree for each variable

$$\mathscr{F}_{\ominus}(x_1, x_2, x_3, t, \underline{\xi}^2) = x_1 x_2 x_3 \left( p^2 - (\xi_1^2 x_1 + \xi_2^2 x_2 + \xi_3^2 x_3) \left( \frac{1}{x_1} + \frac{1}{x_2} + \frac{1}{x_3} \right) \right), \tag{93}$$

which corresponds to the toric polynomial

$$\mathscr{F}_{\ominus}^{tor} = x_1 x_2 x_3 \left( \frac{x_3 z_1}{x_1} + \frac{x_2 z_2}{x_1} + \frac{x_3 z_3}{x_2} + \frac{x_1 z_4}{x_3} + \frac{x_2 z_5}{x_3} + \frac{x_1 z_6}{x_2} + z_7 \right). \tag{94}$$

Contrary to the one-loop case there are more toric parameters $z_i$ than physical variables. The identification of the physical variables is

$$-\xi_1^2 = z_4 = z_6, \quad -\xi_2^2 = z_2 = z_5, \quad -\xi_3^2 = z_1 = z_3, \quad p^2 - (\xi_1^2 + \xi_2^2 + \xi_3^2) = z_7. \tag{95}$$

As before writing the toric polynomial as

$$P_{\ominus} = \sum_{i=1}^{7} z_i x_1^{a_i^1} x_2^{a_i^2} x_3^{a_i^3}, \tag{96}$$

and setting $\mathbf{a}_i = (1, a_i^1, a_i^2, a_i^3)$ we have

$$\mathbf{A}_{\ominus} = \begin{pmatrix} \mathbf{a}_1 \\ \vdots \\ \mathbf{a}_7 \end{pmatrix} = \begin{pmatrix} 1 & -1 & 0 & 1 \\ 1 & -1 & 1 & 0 \\ 1 & 0 & -1 & 1 \\ 1 & 1 & 0 & -1 \\ 1 & 0 & 1 & -1 \\ 1 & 1 & -1 & 0 \\ 1 & 0 & 0 & 0 \end{pmatrix}, \tag{97}$$

The lattice is now defined by

$$\mathbb{L}_{\ominus} := \{ \boldsymbol{\ell} := (\ell_1, \ldots, \ell_7) \in \mathbb{Z}^7 | \ell_1 \mathbf{a}_1 + \cdots + \ell_7 \mathbf{a}_7 = \boldsymbol{\ell} \cdot \mathbf{A}_{\ominus} = 0 \}. \tag{98}$$

This lattice in $\mathbb{Z}^7$ has rank four $\mathbb{L}_{\ominus} = \oplus_{i=1}^4 L_i \mathbb{Z}$ with the basis

$$\begin{pmatrix} L_1 \\ \vdots \\ L_4 \end{pmatrix} = \begin{pmatrix} 1 & 0 & 0 & 0 & 1 & 1 & -3 \\ 0 & 1 & 0 & 0 & 0 & 1 & -2 \\ 0 & 0 & 1 & 0 & 1 & 0 & -2 \\ 0 & 0 & 0 & 1 & -1 & -1 & 1 \end{pmatrix}, \tag{99}$$

From this we derive the sunset GKZ system

$$
e_1 := \frac{\partial^3}{\partial z_1 \partial z_5 \partial z_6} - \frac{\partial^3}{(\partial z_7)^3},
$$

$$
e_2 := \frac{\partial^2}{\partial z_2 \partial z_6} - \frac{\partial^2}{(\partial z_7)^2},
$$

$$
e_3 := \frac{\partial^2}{\partial z_3 \partial z_5} - \frac{\partial^2}{(\partial z_7)^2},
$$

$$
e_4 := \frac{\partial^2}{\partial z_4 \partial z_7} - \frac{\partial^2}{\partial z_5 \partial z_6} \tag{100}
$$

by construction $e_i (\mathscr{F}_\ominus^{tor})^\alpha = 0$ with $\alpha \in \mathbb{C}$ for $1 \leq i \leq 4$. We have as well this second set of operators from the operators

$$
d_1 := \sum_{r=1}^{7} z_r \frac{\partial}{\partial z_r},
$$

$$
d_2 := z_1 \frac{\partial}{\partial z_1} + z_2 \frac{\partial}{\partial z_2} - z_4 \frac{\partial}{\partial z_4} - z_6 \frac{\partial}{\partial z_6},
$$

$$
d_3 := z_2 \frac{\partial}{\partial z_2} - z_3 \frac{\partial}{\partial z_3} + z_5 \frac{\partial}{\partial z_5} - z_6 \frac{\partial}{\partial z_6},
$$

$$
d_4 := z_1 \frac{\partial}{\partial z_1} + z_3 \frac{\partial}{\partial z_3} - z_4 \frac{\partial}{\partial z_4} - z_5 \frac{\partial}{\partial z_5} \tag{101}
$$

The interpretation of these operators is the following

- The Euler operator $d_1 \mathscr{F}_{tor}^\alpha = \alpha \, \mathscr{F}_{tor}^\alpha$ for $\alpha \in \mathbb{C}$.
- To derive the action of these operators on the maximal cut period integral

$$
\pi_\ominus^{tor}(z_1, \ldots, z_7) = \frac{1}{(2i\pi)^3} \int_\gamma \frac{1}{\mathscr{F}_\ominus^{tor}} \prod_{i=1}^{3} dx_i , \tag{102}
$$

we remark that if $\mathscr{F}_\ominus^{tor} = x_1 x_2 x_3 P_\ominus$ we have

$$
d \left( \frac{1}{P_\ominus} \frac{dx_1}{x_1} \right) = \frac{-z_1 x_1/x_2 + z_3 x_2 + z_4 x_2/x_1 - z_6/x_2}{P_\ominus^2} \frac{dx_1}{x_1} \wedge \frac{dx_2}{x_2},
$$

$$
d \left( \frac{1}{P_\ominus} \frac{dx_1}{x_1} \right) = -\frac{z_1 x_1/x_2 + z_2 x_1 - z_4 x_2/x_1 - z_5/x_1}{P_\ominus^2} \frac{dx_1}{x_1} \wedge \frac{dx_2}{x_2}, \tag{103}
$$

therefore since the cycle $\gamma$ has no boundary

$$d_2\pi_\ominus^{tor} = \int_\gamma d\left(\frac{1}{P_\ominus}\frac{dx_1}{x_1}\right) = 0,$$

$$d_3\pi_\ominus^{tor} = -\int_\gamma d\left(\frac{1}{P_\ominus}\frac{dx_2}{x_2}\right) = 0,$$

$$d_4\pi_\ominus^{tor} = \int_\gamma d\left(\frac{1}{P_\ominus}\left(\frac{dx_1}{x_1} + \frac{dx_2}{x_2}\right)\right) = 0. \tag{104}$$

- The natural toric coordinates are

$$u_1 := \frac{z_1 z_5 z_6}{z_7^3}, \qquad u_2 := \frac{z_2 z_6}{z_7^2}, \qquad u_3 := \frac{z_3 z_5}{z_7^2}, \qquad u_4 := \frac{z_4 z_7}{z_5 z_6}, \tag{105}$$

which reads in terms of the physical parameters

$$u_2 = \frac{\xi_1^2 \xi_2^2}{\left(p^2 - (\xi_1^2 + \xi_2^2 + \xi_3^2)\right)^2}, \qquad u_3 = \frac{\xi_2^2 \xi_3^2}{\left(p^2 - (\xi_1^2 + \xi_2^2 + \xi_3^2)\right)^2},$$

$$u_4 = \frac{p^2 - (\xi_1^2 + \xi_2^2 + \xi_3^2)}{\xi_2^2}, \qquad u_1 = u_2 u_3 u_4. \tag{106}$$

They are the natural variables associated with the toric symmetries of the period integral

$$(x_1, x_2) \to (\lambda x_1, x_2), \qquad (z_1, z_2, z_3, z_4, z_5, z_6, z_7) \to (z_1/\lambda, z_2/\lambda, z_3, z_4\lambda, z_5\lambda, z_6, z_7),$$

$$(x_1, x_2) \to (x_1, \lambda x_2), \qquad (z_1, z_2, z_3, z_4, z_5, z_6, z_7) \to (z_1\lambda, z_2, z_3/\lambda, z_4/\lambda, z_5, z_6\lambda, z_7),$$

$$(x_1, x_2) \to (\lambda x_1, \lambda x_2), \qquad (z_1, z_2, z_3, z_4, z_5, z_6, z_7) \to (z_1, z_2/\lambda, z_3/\lambda, z_4, z_5\lambda, z_6\lambda, z_7). \tag{107}$$

The sunset GKZ hypergeometric series is defined as for $\gamma_i \notin \mathbb{Z}$ with $1 \leq i \leq 7$

$$\Phi_{\mathbb{L},\gamma}^\ominus(z_1, \ldots, z_7) = \sum_{\ell \in \mathbb{L}} \prod_{i=1}^7 \frac{z_i^{l_i + \gamma_i}}{\Gamma(l_i + \gamma_i + 1)}, \tag{108}$$

in this sum we have $\ell = \sum_{i=1}^4 n_i L_i$ with $n_i \in \mathbb{Z}$, and the condition $\sum_{i=1}^7 \gamma_i \mathbf{a}_i = (-1, 0, 0, 0)$ which can be solved using $\gamma = (\gamma_1, \ldots, \gamma_7) = \sum_{i=1}^4 \gamma_i \mathscr{L}_i + (0, \ldots, 0, -1)$. Using toric variables the solution reads

$$\Phi_{\mathbb{L},\gamma}^\ominus(z_1, \ldots, z_7) = \frac{1}{z_7} \sum_{(n_1,\ldots,n_4)\in\mathbb{Z}} \frac{u_1^{n_1+\gamma_1} u_2^{n_2+\gamma_2} u_3^{n_3+\gamma_3} u_4^{n_4+\gamma_4}}{\prod_{i=1}^4 \Gamma(n_i + \gamma_i + 1)} \times$$

$$\times \; \frac{1}{\Gamma(n_1 + n_2 - n_4 + \gamma_1 + \gamma_2 - \gamma_4 + 1)\Gamma(n_1 + n_3 - n_4 + \gamma_1 + \gamma_3 - \gamma_4 + 1)}$$

$$\times \; \frac{1}{\Gamma(-3n_1 - 2n_2 - 2n_3 + n_4 - 3\gamma_1 - 2\gamma_2 - 2\gamma_3 + \gamma_4)} \,. \tag{109}$$

With $\gamma = (0,0,0,0,0,0,0)$ the series is trivially zero as being resonant. The resolution is to regularise the term with a zero by using for $\ell_7 < 0$

$$\lim_{\epsilon \to 0} \frac{\Gamma(\epsilon)}{\Gamma(\ell_7 + \epsilon)} = (-1)^{\ell_7}\Gamma(1 - \ell_7)\,, \tag{110}$$

and write the associated regulated period as

$$\pi_{\ominus}^{(2)}(p^2, \underline{\xi}^2) = \lim_{\epsilon \to 0} \sum_{(n_1, n_2, n_3, n_4) \in \mathbb{N}} \frac{(\xi_1^2)^{n_1 + n_2}(\xi_2^2)^{n_1 + n_2 + n_3 - n_4}(\xi_3^2)^{n_1 + n_3}}{\prod_{i=1}^{4} \Gamma(1 + n_i)}$$

$$\times \; \frac{(p^2 - (\xi_1^2 + \xi_2^2 + \xi_3^2))^{-3n_1 - 2n_2 - 2n_3 + n_4 - 1}(-1)^{-3n_1 - 2n_2 - 2n_3 + n_4}\Gamma(\epsilon)}{\Gamma(1 + n_1 + n_2 - n_4)\Gamma(1 + n_1 + n_3 - n_4)\Gamma(-3n_1 - 2n_2 - 2n_3 + n_4 + \epsilon)} \,. \tag{111}$$

One can expand this expression as a series near $t = \infty$ to get that

$$\pi_{\ominus}^{(2)}(p^2, \xi_1^2, \xi_2^2, \xi_3^2) = \sum_{n \geq 0}(p^2)^{-n-1} \sum_{n_1 + n_2 + n_3 = n} \left(\frac{n!}{n_1! n_2! n_3!}\right)^2 \xi_1^{2n_1} \xi_2^{2n_2} \xi_3^{2n_3}\,, \tag{112}$$

which is the series expansion of the maximal cut integral

$$\pi_{\ominus}^{(2)}(p^2, \underline{\xi}^2) = \frac{1}{(2i\pi)^3} \int_{\gamma} \frac{1}{\mathscr{F}_{\ominus}} \prod_{i=1}^{3} dx_i\,, \tag{113}$$

where $\gamma = \{|x_1| = |x_2| = |x_3| = 1\}$. The construction generalises easily to the case of multi-loop sunset integrals in an easy way [22].

### 3.5.1 The Differential Operators

Now that we have the expression for the maximal cut it is easy to derive the minimal order differential operator annihilating this period. There are various methods to derive the Picard–Fuchs operator from the maximal cut. One method is to use the series expansion of the period around $s = 1/t = 0$. Another method is to reduce the GKZ system of differential operators in similar fashion as shown for the hypergeometric function in Sect. 3.3.1. This method leads to a fourth order differential operator which factorises to a minimal second order operator. We notice that this approach is similar to the integration-by-part based approach

The minimal order differential operator is of second order

$$
\mathscr{L}_{PF}^{\ominus} = \left(p^2 \frac{d}{dp^2}\right)^2 + q_1(p^2, \underline{\xi}^2)\left(p^2 \frac{d}{dp^2}\right) + q_0(p^2, \underline{\xi}^2),
\tag{114}
$$

with the coefficients given in [20, 57]. The action of this differential operator on the maximal cut is given by

$$
\mathscr{L}_{PF}^{\ominus} \pi_{\ominus}^{(2)} = \frac{1}{(2i\pi)^3} \int_{\gamma} \mathscr{L}_{PF}^{\ominus} \frac{1}{\mathscr{F}_{\ominus}} \prod_{i=1}^{3} dx_i = \frac{1}{(2i\pi)^3} \int_{\gamma} \left(\sum_{i=1}^{3} \partial_i \beta_i\right) \prod_{i=1}^{3} dx_i = 0.
\tag{115}
$$

As to the action of this operator on the Feynman integral we find that full differential operator action on the two-loop sunset integral is given by

$$
\mathscr{L}_{PF}^{\ominus} I_{\ominus}(p^2, \underline{\xi}^2) = \int_{\substack{x_1 \geq 0 \\ x_2 \geq 0}} \left(\sum_{i=1}^{3} \partial_i \beta_i\right) \delta(x_3 = 1) \prod_{i=1}^{3} dx_i = \mathscr{S}_{\ominus},
\tag{116}
$$

where the inhomogeneous term reads

$$
\mathscr{S}_{\ominus} = \mathscr{Y}_{\ominus}(p^2, \underline{\xi}^2) + c_1(p^2, \underline{\xi}^2) \log\left(\frac{m_1^2}{m_3^2}\right) + c_2(p^2, \underline{\xi}^2) \log\left(\frac{m_2^2}{m_3^2}\right),
\tag{117}
$$

with the Yukawa coupling[4]

$$
\mathscr{Y}_{\ominus}(p^2, \underline{\xi}^2) = \frac{6(p^2)^2 - 4p^2(\xi_1^2 + \xi_2^2 + \xi_3^2) - 2\prod_{i=1}^{4} \mu_i}{(p^2)^2 \prod_{i=1}^{4}(p^2 - \mu_i^2)},
\tag{118}
$$

where $(\mu_1, \ldots, \mu_4) = ((-\xi_1 + \xi_2 + \xi_3)^2, (\xi_1 - \xi_2 + \xi_3)^2, (\xi_1 + \xi_2 - \xi_3)^2, (\xi_1 + \xi_2 + \xi_3)^2)$. A geometric interpretation has the integral [20]

$$
\mathscr{Y}_{\ominus}(p^2, \underline{\xi}^2) = \int_{\mathscr{E}_{\ominus}} \Omega_{\ominus} \wedge p^2 \frac{d}{p^2} \Omega_{\ominus},
\tag{119}
$$

---

[4]This quantity is the usual Yukawa coupling of particle physics and string theory compactification. The Yukawa coupling is determined geometrically by the integral of the wedge product of differential forms over particular cycles [58]. The Yukawa couplings which depend non-trivially on the internal geometry appear naturally in the differential equations satisfied by the periods of the underlying geometry as explained for instance in these reviews [48, 59].

where $\Omega_\ominus$ is the sunset residue differential form

$$\Omega_\ominus = \operatorname{Res}_{\mathscr{E}_\ominus=0} \frac{x_1 dx_2 \wedge dx_3 + x_3 dx_1 \wedge dx_2 + x_2 dx_3 \wedge dx_1}{\mathscr{F}_\ominus}, \tag{120}$$

on the sunset elliptic curve

$$\mathscr{E}_\ominus := \{p^2 x_1 x_2 x_3 - (\xi_1^2 x_1 + \xi_2^2 x_2 + \xi_3^2 x_3)(x_1 x_2 + x_1 x_3 + x_2 x_3)|(x_1, x_2, x_3) \in \mathbb{P}^2\}. \tag{121}$$

The Yukawa coupling satisfies the differential equation

$$p^2 \frac{d}{p^2} \mathscr{Y}_\ominus(t) = (2 - q_1(p^2, \underline{\xi}^2)) \mathscr{Y}_\ominus(p^2, \underline{\xi}^2). \tag{122}$$

The coefficients $c_1$ and $c_2$ in (117) are the integrals of the residue one form between the marked points on $Q_1 = [0, -\xi_3^2, \xi_2^2]$, $Q_2 = [-\xi_3^2, 0, \xi_1^2]$ and $Q_3 = [-\xi_2^2, \xi_1^2, 0]$ on the elliptic curve [20]

$$c_1(p^2, \underline{\xi}^2) := p^2 \frac{d}{p^2} \int_{Q_1}^{Q_3} \Omega_\ominus, \qquad c_2(p^2, \underline{\xi}^2) := p^2 \frac{d}{p^2} \int_{Q_2}^{Q_3} \Omega_\ominus. \tag{123}$$

## 3.6 The Generic Case

In this section we show how to determine the differential equation for the $l$-loop sunset integral from the knowledge of the maximal cut. The maximal cut of the $l$-loop sunset integral is given by

$$\pi_\ominus^{(l)}(p^2, \underline{\xi}^2) = \sum_{n \geq 0} t^{-n-1} A_\ominus(l, n, \xi_1^2, \dots, \xi_{l+1}^2), \tag{124}$$

with

$$A_\ominus(l, n, \xi_1^2, \cdots, \xi_{l+1}^2) := \sum_{r_1 + \cdots + r_{l+1} = n} \left(\frac{n!}{r_1! \cdots r_{l+1}!}\right)^2 \prod_{i=1}^{l+1} \xi_i^{2r_i}. \tag{125}$$

### 3.6.1 The All Equal Mass Case

For the all equal mass case one can easily determine the differential equation to all orders [28] using the Bessel integral representation of [9]. We present here a different derivation.

For all equal masses the coefficient of the maximal cut satisfies a nice recursion [60]

$$\sum_{k\geq 0}\left( n^{l+2} \sum_{\substack{1\leq i\leq k \\ a_i+b_i=l+2 \\ 1<a_{i+1}+1<a_i\leq l+1}} \prod_{i=1}^{k}(-a_ib_i)\left(\frac{n-i}{n-i+1}\right)^{a_i-1}\right) A_{\ominus}(l,n-k,\underline{1})=0,$$

(126)

where $a_i \in \mathbb{N}$. Standard method gives that the associated differential operator acting on $t\pi_{\ominus}^{l}(t,1,\ldots,1)=\sum_{n\geq 0}(p^2)^{-n}A(l+1,n,1,\ldots,1)$ reads

$$\mathscr{L}_{PF,\ominus}^{(l),1mass}=\sum_{k\geq 0}(p^2)^k \sum_{\substack{1\leq i\leq k \ a_i+b_i=l+2,a_{k+1}=0 \\ 1<a_{i+1}+1<a_i\leq l+1}} \left(k-p^2\frac{d}{p^2}\right)^{l+2-a_1}$$

$$\times \prod_{i=1}^{k}(-a_ib_i)\left(k-i-p^2\frac{d}{dp^2}\right)^{a_i-a_{i+1}}.$$

(127)

This operator has been derived in [28, §9] using different method.

They are differential operators of order $l$, the loop order, in $d/dp^2$ and the coefficients are polynomials of degree $l+1$

$$\mathscr{L}_{PF}^{(l),1mass}=(-p^2)^{\lceil l/2\rceil-1}\prod_{i=1}^{\lfloor l/2\rfloor+1}(p^2-\mu_i^2)\left(\frac{d}{dp^2}\right)^l+\cdots$$

(128)

where $\mu_i^2:=(\pm 1\pm 1\cdots\pm 1)^2$ is the set of the different thresholds. The operator $\mathscr{L}_{PF}^{(2),1mass}$ is the Picard–Fuchs operator of the family of elliptic curves for $\Gamma_1(6)$ for the all equal mass sunset [61], the operator $\mathscr{L}_{PF}^{(3),1mass}$ of the family of $K3$ surfaces [62]. Having determined the Picard–Fuchs operator it is not difficult to derive its action on the Feynman integral with the result that [28]

$$\mathscr{L}_{PF}^{(l),1mass}(I_{\ominus}(p^2,1,\ldots,1))=-(l+1)!.$$

(129)

### 3.6.2 The General Mass Case

For unequal masses the recursion relation does not close only on the coefficients (125) and no simple closed formula is known for the differential operator on the maximal cut. The minimal differential operator annihilating the $\pi_{\ominus}^{(l)}(t,\xi^2)$ can be obtained using the GKZ hypergeometric function discussed in the previous section.

For the $l$-loop sunset integral the GKZ lattice has rank $l^2$, $\mathbb{L} = \sum_{i=1}^{l^2} n_i L_i$. For instance for the three-loop sunset the regulated hypergeometric series representation of the maximal cut reads

$$\pi_{\ominus}^{(3)}(p^2, \underline{\xi}^2) = -\lim_{\epsilon \to 0} \sum_{(n_1,\ldots,n_9) \in \mathbb{N}^9} \frac{(\xi_1^2)^{n_1+n_2+n_3} (\xi_2^2)^{n_1+n_3+n_4+n_6-n_7-n_8+n_9}}{\prod_{i=1}^{9} \Gamma(1+n_i)}$$

$$\times \frac{(\xi_3^2)^{n_2+n_5+n_8} (\xi_4^2)^{n_1+n_4+n_6}}{\Gamma(n_1 + n_4 + n_6 - n_7 - n_8 + 1)\Gamma(n_2 + n_5 - n_6 + n_8 - n_9 + 1)}$$

$$\times \frac{1}{\Gamma(n_1 + n_3 - n_5 + n_6 - n_7 - n_8 + n_9 + 1)}$$

$$\times \frac{(-p^2 + \xi_1^2 + \xi_2^2 + \xi_3^2 + \xi_4^2)^{-3n_1-2n_2-2n_3-2n_4-n_5-2n_6+n_7-n_9-1}\Gamma(\epsilon)}{\Gamma(-3n_1 - 2n_2 - 2n_3 - 2n_4 - n_5 - 2n_6 + n_7 - n_9 + \epsilon)}. \quad (130)$$

The minimal order differential operator annihilating the maximal cut $p^2 \pi_{\ominus}^{(3)}(p^2, \underline{\xi}^2)$ with generic mass configurations, $\xi_1 \neq \xi_2 \neq \xi_3 \neq \xi_4$ and all the masses non vanishing, is an operator of order 6, with polynomial coefficients $c_k(t)$ of degree up to 29

$$L_{PF,\ominus}^3 = \sum_{k=0}^{6} c_k(t) \left(t\frac{d}{t}\right)^k. \quad (131)$$

For instance the differential operator for the mass configuration $\xi_i = i$ with $1 \leq i \leq 4$ is given by

$$c_6 = (t - 100)(t - 36)(t - 64)(t - 4)^2(t - 16)^2$$

$$\times \left(345t^{12} - 10275t^{11} + 243243t^{10} + 700860t^9 - 289019444t^8 + 9517886160t^7\right.$$

$$- 169244843904t^6 + 2163112875520t^5 - 24375264125952t^4$$

$$+ 198627459010560t^3 - 8965173122170088t^2$$

$$+ 1570362910310400t - 1192050032640000\big), \quad (132)$$

and

$$c_5 = (t - 4)(t - 16)\big(7245t^{17} - 1461150t^{16} + 108842709t^{15} - 4073021820t^{14}$$

$$+ 79037467036t^{13} + 706049613520t^{12} - 1229771149488000t^{11}$$

$$+ 4897976525794560t^{10} - 118057966435402752t^9$$

$$+ 2042520337021317120t^8 - 28129034886941589504t^7$$

$$+ 321784682881513881600t^6 - 28775225280576592281160t^5$$

$$+ 17978948962533528043520t^4 - 69950845277551433089024t^3$$
$$+ 15117855778012806512640 0t^2 - 1822506963713182924800 00t$$
$$+ 966762112871301120 00000),$$

(133)

and

$$c_4 = 2\left(23460t^{19} - 4086975t^{18} + 273974766t^{17} - 9833465295t^{16}\right.$$
$$+ 173874227860t^{15} + 3780156754180t^{14}$$
$$- 419091386081744t^{13} + 16647873781420800t^{12}$$
$$- 425729411677916160t^{11} + 8098824799795968000t^{10}$$
$$- 125136842089603031040t^9 + 1631034274362173030400t^8$$
$$+ 17364390414642101354496t^7 + 140612615518097533829120t^6$$
$$- 80786806001514379214848 0t^5 + 3100095209313936311582720t^4$$
$$- 7563751451192001262780416t^3 + 11448586013594218187980800t^2$$
$$- 9812428506034109153280000t + 33748786485689057280 00000),$$   (134)

and

$$c_3 = 12\left(8970t^{19} - 1147050t^{18} + 56442264t^{17} - 1477273050t^{16} - 447578647t^{15}\right.$$
$$+ 2416587481200t^{14} - 130189239609348t^{13} + 4001396495500560t^{12}$$
$$- 86975712270293184t^{11} + 1511724058206439680t^{10}$$
$$- 22690173944998831104t^9 + 289974679497600921600t^8$$
$$- 2900762618196498137088t^7 + 20882244400635484241920t^6$$
$$- 101090327023260610854912t^5 + 308760428925736546467840t^4$$
$$- 559057237244267332632576t^3 + 533177283118109609164800t^2$$
$$- 133034777312420167680000t - 140619943690371072000000),$$

(135)

and

$$c_2 = 24\left(3105t^{19} - 260100t^{18} + 8740695t^{17} - 121279200t^{16} - 8982728081t^{15}\right.$$
$$+ 771645247175t^{14} - 29786960482306t^{13} + 741851366254700t^{12}$$

$$- 14140682364004072t^{11} + 237224880534337760t^{10}$$

$$- 3605462277123620992t^9 + 44725169880349560320t^8$$

$$- 405767142088142927872t^7 + 2549108215435181793280t^6$$

$$- 11307241496864563101696t^5 + 40972781273200446013440t^4$$

$$- 141797614014479525216256t^3 + 363118631232748702924800t^2$$

$$- 41518049060871733248000t + 21092991553555660800000),$$

(136)

and

$$c_1 = 24 \left(345t^{19} - 15000t^{18} + 345675t^{17} + 7323600t^{16} - 3165461083t^{15}\right.$$

$$+ 184943420750t^{14} - 5084383561348t^{13} + 91042473303800t^{12}$$

$$- 1344824163401536t^{11} + 17444484465759680t^{10}$$

$$- 146155444722244096t^9 - 426434786380119040t^8$$

$$+ 3179868308848698 9824t^7 - 4884830766562838 93760t^6$$

$$+ 5136134162164414021632t^5 - 40834519838668015534080t^4$$

$$+ 222597043391679285952512t^3 - 685074395310881085849600t^2$$

$$+ 830360981217434664960000t - 421859831071113216000000),$$

(137)

and

$$c_0 = 1728 \left(21908444t^{15} - 1482071825t^{14} + 40507170144t^{13} - 668436089250t^{12}\right.$$

$$+ 8209054542408t^{11} - 65000176183240t^{10} - 503218239747392t^9$$

$$+ 31962708303867520t^8 - 619576476284137472t^7 + 7554395788685281280t^6$$

$$- 73455221906789646336t^5 + 571135922816871792640t^4$$

$$- 3095113137012548304896t^3 + 9514922157095570636800t^2$$

$$- 11532791405797703680000t + 585916432043212800 0000\right).$$

(138)

A systematic study of the differential operators for the multiloop sunset integral will appear in [22].

## 3.7 Toward a Systematic Approach to Differential Equations

In the previous section we have presented a way of deriving a differential equation for the maximal cut of a Feynman integral based on the knowledge of its representation as an hypergeometric function. One can as well use the series expansion for determining the differential operator acting on this series. The Griffith method presented in Sect. 2.3.2 using the reduction of polynomial in the cohomology unfortunately does not work when there are non-isolated singularities. This is unfortunately a rather generic case for Feynman integrals beyond the special case of the two-loop sunset. In principle one can resolve the singularity by deforming the graph polynomial by introduction a finite number of deformation parameters $\underline{\lambda} = \{\lambda_i\}$ until one has only isolated singularities. The Picard–Fuchs operator $L^{\underline{\lambda}}$. Taking the limit $\underline{\lambda} \to \underline{0}$ would give a differential operator $L^{\underline{0}}$ that would factorise on the minimal order Picard–Fuchs operator $L$ acting on the Feynman integral. Unfortunately this approach is not very useful in practice as the deformation increases a lot the complexity of the computation.

At more systematic approach is the use of the telescoping method introduced by Zeilberger [63] and developed by Chyzak [64, 65] and Koutschan [66]. One wants to derive an annihilating operator $L$ for the maximal cut integral $\pi_\Gamma$ in (14).

Let first illustrate the idea of the method on a multi-parameter one dimensional integral

$$I(\underline{\xi}) = \int_a^b f(x, \underline{\xi}) . \tag{139}$$

We want to construct an operator $P$ of the form

$$P := T(\underline{\xi}, \partial_{\underline{\xi}}) + \frac{d}{dx} C(x, \underline{\xi}, \partial_{\underline{\xi}}) \tag{140}$$

This clearly implies that

$$T(\underline{\xi}, \partial_{\underline{\xi}}) I(\underline{\xi}) = - \int_a^b \frac{d}{dx} C(x, \underline{\xi}, \partial_{\underline{\xi}}) f(x, \underline{\xi}) = -C(x, \underline{\xi}, \partial_{\underline{\xi}}) f(x, \underline{\xi}) \Big|_{x=a}^{x=b} . \tag{141}$$

In the case that the Griffith method works this gives the same answer but the advantage of the method is that the algorithm works as well when there are non-isolated singularities. For the case of multidimensional integrals one just applies the same method iteratively.

Because we know that Feynman integrals are annihilated by a finite order differential operator because the dimension of master integrals (or the system of Gauss–Manin connection) is finite [42, 67] the creative telescoping algorithm in [63, 64, 64, 66] will finish in a finite time although this may take a long time for complicated Feynman graphs.

This provides a direct approach for finding the differential equation for Feynman integrals.

# 4 Analytic Evaluations for Sunset Integral

In this section we give different analytic expressions for the two-loop sunset integral. In one form the two-loop sunset integral is given by an elliptic dilogarithm as review in Sect. 4.1 or as a ordinary trilogarithm as reviewed in Sect. 4.1. In Sect. 4.3 we explain that the equivalence between the two expressions is a manifestation of the mirror symmetry proven in [20].

## 4.1 The Sunset Integral as an Elliptic Dilogarithm

The geometry of the graph hypersurface is a family of elliptic curves

$$\mathscr{E}_\ominus := \{p^2 x_1 x_2 x_3 - (\xi_1^2 x_1 + \xi_2^2 x_2 + \xi_3^2 x_3)(x_1 x_2 + x_1 x_3 + x_2 x_3) | (x_1, x_2, x_3) \in \mathbb{P}^2\}.$$

$$(142)$$

One can use the information from the geometry of the graph polynomial and use a parameterisation of the physical variables making the geometry of the elliptic curve explicit.

The elliptic curve $\mathscr{E}_\ominus$ can be represented as $\mathbb{C}^\times/q^{\mathbb{Z}}$ where $q = \exp(2i\pi\tau)$ and $\tau$ is the period ratio of the elliptic curve. There are six special points on the elliptic curve $\mathscr{E}_\ominus$ the three points that intersect the domain of integration

$$P_1 := [1, 0, 0], \qquad P_2 := [0, 1, 0], \qquad P_3 := [0, 0, 1], \tag{143}$$

and three other points outside the domain of integration

$$Q_1 := [0, -\xi_3^2, \xi_2^2], \qquad Q_2 := [-\xi_3^2, 0, \xi_1^2], \qquad Q_3 := [-\xi_1^2, \xi_2^2, 0]. \tag{144}$$

If one denotes by $x(P_i)$ the image of the point $P_i$ in $\mathbb{C}^\times/q^{\mathbb{Z}}$ and $x(Q_i)$ the image of the point $Q_i$ we have $x(P_i) = -x(Q_i)$ with $i = 1, 2, 3$

$$\left(\frac{\theta_1(x(P_i)/x(P_j))}{\theta_c(x(P_i)/x(P_j))}\right)^2 = \frac{\xi_k}{\sqrt{t\xi_i\xi_j}}, \tag{145}$$

with $(i, j, k)$ a permutation of $(1, 2, 3)$ and $c$ a permutation of $(2, 3, 4)$.[5] It was shown in [20] that the sunset Feynman integral is given by

$$I_\ominus(p^2, \underline{\xi}^2) \equiv \frac{i\varpi_r}{\pi} \left( \hat{E}_2 \left( \frac{x(P_1)}{x(P_2)} \right) + \hat{E}_2 \left( \frac{x(P_2)}{x(P_3)} \right) + \hat{E}_2 \left( \frac{x(P_3)}{x(P_1)} \right) \right) \quad \text{mod periods},$$
(146)

where $\hat{E}_2(x)$ is the elliptic dilogarithm

$$\hat{E}_2(x) = \sum_{n \geq 0} \left( \mathrm{Li}_2 \left( q^n x \right) - \mathrm{Li}_2 \left( -q^n x \right) \right) - \sum_{n \geq 1} \left( \mathrm{Li}_2 \left( q^n / x \right) - \mathrm{Li}_2 \left( -q^n / x \right) \right).$$
(147)

The $J$-invariant of the sunset elliptic curve is

$$J_\ominus = 256 \frac{(3 - u_\ominus^2)^3}{4 - u_\ominus^2},$$
(148)

where the Hauptmodul is

$$u_\ominus = \frac{(p^2 - \xi_1^2 - \xi_2^2 - \xi_3^2)^2 - 4(\xi_1^2 \xi_2^2 + \xi_1^2 \xi_3^2 + \xi_2^2 \xi_3^2)}{\sqrt{16 t \xi_1^2 \xi_2^2 \xi_3^2}},$$
(149)

given in term of Jacobi theta functions

$$u_\ominus^{3,4} = \frac{\theta_3^4 + \theta_4^4}{\theta_3^2 \theta_4^2}, \quad u_\ominus^{2,3} = -\frac{\theta_3^4 + \theta_2^4}{\theta_3^2 \theta_2^2}, \quad u_\ominus^{2,4} = i \frac{\theta_2^4 - \theta_4^4}{\theta_2^2 \theta_4^2},$$
(150)

and the period is given for each pair $(a, b) = (3, 4), (2, 3), (2, 4)$ by

$$\varpi_r = \frac{t^{\frac{1}{4}} \pi \theta_a \theta_b}{(\xi_1^2 \xi_2^2 \xi_3^2)^{\frac{1}{4}}},$$
(151)

is the elliptic curve period which is real on the line $t < (\xi_1 + \xi_2 + \xi_3)^2$.

By using the dilogarithm functional equations one can bring the expression (146) in a form similar to the one used in [68]

$$\sum_{i=1}^{3} \sum_{n \in \mathbb{Z}} \mathrm{Li}_2(q^n x_i).$$
(152)

---

[5]The Jacobi theta functions are defined by $\theta_2(q) := 2q^{\frac{1}{8}} \prod_{n \geq 1} (1 - q^n)(1 + q^n)^2$, $\theta_3(q) := \prod_{n \geq 1} (1 - q^n)(1 + q^{n-\frac{1}{2}})^2$ and $\theta_4(q) := \prod_{n \geq 1} (1 - q^n)(1 - q^{n-\frac{1}{2}})^2$.

This representation needs to be properly regularised as discussed in [68] whereas the representation in (147) is a converging sum. An equivalent representation used multiple elliptic polylogarithms [14–19] this representation has the advantage of generalising to other graphs [69–74].

For the all equal masses case, $1 = \xi_1 = \xi_2 = \xi_3$, the family of elliptic curves

$$\mathscr{E}_{\ominus} := \{p^2 x_1 x_2 x_3 - (x_1 + x_2 + x_3)(x_1 x_2 + x_1 x_3 + x_2 x_3) = 0 | (x_1, x_2, x_3) \in \mathbb{P}^2\},$$
(153)

defines a pencil of elliptic curves in $\mathbb{P}^2$ corresponding to a modular family of elliptic curves $f : \mathscr{E}_{\ominus} \to X_1(6) = \{\tau \in \mathbb{C} | \Im m(\tau) > 0\}/\Gamma_1(6)$ (see [61]). When all the masses are equal the map is easier since the elliptic curve is a modular curve for $\Gamma_1(6)$ and the coordinates of the points are mapped to sixth root of unity $x(P_r) = e^{\frac{2i\pi r}{6}}$ and $x(Q_r) = -e^{\frac{2i\pi r}{6}}$ with $r = 1, 2, 3$.

The integral is expressed as the following combination of elliptic dilogarithms

$$I_{\ominus}(p^2, 1, 1, 1) = \varpi_r(t)(i\pi - \log q) - 6\frac{\varpi_r(p^2)}{\pi} E_{\ominus}(q),$$
(154)

where the Hauptmodul

$$p^2 = 9 + 72\frac{\eta(q^2)}{\eta(q^3)}\left(\frac{\eta(q^6)}{\eta(q)}\right)^5,$$
(155)

and the real period for $p^2 < \xi_1^2 + \xi_2^2 + \xi_3^2$

$$\varpi_r(p^2) = \frac{\pi}{\sqrt{3}}\frac{\eta(q)^6\eta(q^6)}{\eta(q^2)^3\eta(q^3)^2}.$$
(156)

In this case the elliptic dilogarithm is given by

$$E_{\ominus}(q) = -\frac{1}{2i}\sum_{n\geq 0}\left(\text{Li}_2\left(q^n\zeta_6^5\right) + \text{Li}_2\left(q^n\zeta_6^4\right) - \text{Li}_2\left(q^n\zeta_6^2\right) - \text{Li}_2\left(q^n\zeta_6\right)\right)$$

$$+ \frac{1}{4i}\left(\text{Li}_2\left(\zeta_6^5\right) + \text{Li}_2\left(\zeta_6^4\right) - \text{Li}_2\left(\zeta_6^2\right) - \text{Li}_2\left(\zeta_6\right)\right).$$
(157)

which we can write as a $q$-expansion

$$E_{\ominus}(q) = \frac{1}{2}\sum_{k\in\mathbb{Z}\backslash\{0\}}\frac{(-1)^{k-1}}{k^2}\frac{\sin(\frac{n\pi}{3}) + \sin(\frac{2n\pi}{3})}{1 - q^k}.$$
(158)

## 4.2   The Sunset Integral as a Trilogarithm

In this section we evaluate the sunset two-loop integral in a different way, leading to an expression in terms of trilogarithms. We leave the interpretation of the equivalence of two with the previous evaluation to Sect. 4.3 where we explain that these results are a manifestation of local mirror symmetry.

We introduce the quantity — the logarithmic Mahler measure $R_0(p^2, \underline{\xi}^2)$

$$R_0(p^2, \underline{xi}^2) = -i\pi + \int_{|x|=|y|=1} \log(p^2 - (\xi_1^2 x + \xi_2^2 y + \xi_3^2)(x^{-1} + y^{-1} + 1)) \frac{d\log x\, d\log y}{(2\pi i)^2},$$
(159)

which evaluates to

$$R_0 = \log(-p^2) - \sum_{n \geq 1} \frac{(p^2)^{-n}}{n} A_{\ominus}(2, n, \xi_1^2, \xi_2^2, \xi_3^2),$$
(160)

where $A_{\ominus}(2, n, \xi_1^2, \xi_2^2, \xi_3^2)$ is defined in (125). Differentiating with respect to $p^2$ leads to maximal cut

$$\frac{d}{dp^2} R_0(p^2, \xi_1^2, \xi_2^2, \xi_3^2) = \pi_{\ominus}^{(2)}(p^2, \xi_1^2, \xi_2^2, \xi_3^2),$$
(161)

where $\pi_{\ominus}^{(2)}(p^2, \xi_1^2, \xi_2^2, \xi_3^2)$ is defined in (124). It was shown in [20] that the sunset integral has the expansion

$$I_{\ominus}(p^2, \underline{\xi}^2) = -2i\pi\, \pi_{\ominus}^{(2)}(t, \underline{\xi}^2) \left( 3R_0^3 + \sum_{\substack{\ell_1 + \ell_2 + \ell_3 = \ell > 0 \\ (\ell_1, \ell_2, \ell_3) \in \mathbb{N}^3 \setminus (0,0,0)}} \ell(1 - \ell R_0) N_{\ell_1, \ell_2, \ell_3} \prod_{i=1}^{3} \xi_i^{2\ell_i} e^{\ell_i R_0} \right),$$
(162)

where the invariant numbers $N_{\ell_1, \ell_2, \ell_3}$ can be computed from the Yukawa coupling (119) using [20, proposition 7.6]

$$6 - \sum_{\substack{\ell_1 + \ell_2 + \ell_3 = \ell > 0 \\ (\ell_1, \ell_2, \ell_3) \in \mathbb{N}^3 \setminus (0,0,0)}} \ell^3 N_{\ell_1, \ell_2, \ell_3} R_0^\ell \prod_{i=1}^{3} \xi_i^{2\ell_i} = \frac{(6(p^2)^2 - 4p^2(\xi_1^2 + \xi_2^2 + \xi_3^2) + 2\mu_1 \cdots \mu_4)}{p^2 \prod_{i=1}^{4}(p^2 - \mu_i^2)\, (\pi_{\ominus}^{(2)}(p^2, \underline{\xi}^2))^3}.$$
(163)

These quantities can be expressed in terms of the virtual integer numbers of rational curves of degree $\ell = \ell_1 + \ell_2 + \ell_3$ by the covering formula

$$N_{\ell_1, \ell_2, \ell_3} = \sum_{d | \ell_1, \ell_2, \ell_3} \frac{1}{d^3} n_{\frac{\ell_1}{d}, \frac{\ell_2}{d}, \frac{\ell_3}{d}}.$$
(164)

A first few Gromov–Witten numbers are given by (these invariants are symmetric in their indices so we list only one representative)

| $(\ell_1, \ell_2, \ell_3)$ | (100) | $k>0$ (k00) | (110) | (210) | (111) | (310) | (220) | (211) | (221) |
|---|---|---|---|---|---|---|---|---|---|
| $N_{\ell_1,\ell_2,\ell_3}$ | 2 | $2/k^3$ | $-2$ | 0 | 6 | 0 | $-1/4$ | $-4$ | 10 |
| $n_{\ell_1,\ell_2,\ell_3}$ | 2 | 0 | $-2$ | 0 | 6 | 0 | 0 | $-4$ | 10 |

(165)

| $(\ell_1, \ell_2, \ell_3)$ | (410) | (320) | (311) | (510) | (420) | (411) | (330) | (321) | (222) |
|---|---|---|---|---|---|---|---|---|---|
| $N_{\ell_1,\ell_2,\ell_3}$ | 0 | 0 | 0 | 0 | 0 | 0 | $-2/27$ | $-1$ | $-189/4$ |
| $n_{\ell_1,\ell_2,\ell_3}$ | 0 | 0 | 0 | 0 | 0 | 0 | 0 | $-1$ | $-48$ |

(166)

Introducing the variables $Q_i = \xi_i^2 e^{R_0}$ we can rewrite the sunset integral as

$$\frac{I_\ominus(p^2, \underline{\xi}^2)}{2i\pi\pi_\ominus^{(2)}(p^2, \underline{\xi}^2)} = \left( \log Q \left( \frac{d}{d\log Q} \right)^2 - \frac{d}{d\log Q} \right) \mathscr{F}(p^2, \xi_1^2, \xi_2^2, \xi_3^2),$$

(167)

where $\mathscr{F}$ is the local Gromov–Witten prepotential for the Calabi–Yau threefold with given as the total space of the anti-canonical line bundle of the $dP_6$ del Pezzo surface computed in [21]

$$\mathscr{F}(p^2, \xi_1^2, \xi_2^2, \xi_3^2) = -(\log Q)^3 + \sum_{(k_1,k_2,k_3)\geq 0} n_{k_1,k_2,k_3} \mathrm{Li}_3 \left( \prod_{i=1}^{3} \xi_i^{2n_i} Q^{n_i} \right).$$

(168)

and $\mathrm{Li}_3 = \sum_{n\geq 1} x^n/n^3$ is the trilogarithm. How the $dP_6$ del Pezzo surface arises from the sunset graph is discussed in Sect. 4.3.1. In Sect. 4.3 we will explain that these numbers are local Gromov–Witten numbers $N_{\ell_1,\ell_2,\ell_3}$ and the sunset Feynman integral is the Legendre transformation of the local prepotential as shown [20].

Using the relation between the complex structure of $2i\pi\tau = \log q$ of the elliptic curve and $R_0$ (see [20, proposition 7.6] and Sect. 4.3)

$$\log q = 2 \sum_{i=1}^{3} \log(Q_i^2) - \sum_{\substack{\ell_1+\ell_2+\ell_3=\ell>0 \\ (\ell_1,\ell_2,\ell_3)\in\mathbb{N}^3\setminus(0,0,0)}} \ell^2 N_{\ell_1,\ell_2,\ell_3} \prod_{i=1}^{3} Q_i^{\ell_i},$$

(169)

one can check the equivalence between the expressions (146) and (162).

### 4.2.1 The All Equal Masses Case

In this section we compute the local invariants for the all equal masses case $\xi_1 = \xi_2 = \xi_3 = 1$ the sunset integral reads

$$I_\ominus(p^2, 1, 1, 1) = \pi_\ominus^{(2)}(p^2, 1, 1, 1) \left( 3R_0^3 + \sum_{\substack{\ell_1+\ell_2+\ell_3=\ell>0 \\ (\ell_1,\ell_2,\ell_3)\in\mathbb{N}^3\setminus(0,0,0)}} \ell(1 - \ell \log Q) N_{\ell_1,\ell_2,\ell_3} \, Q_0^\ell \right). \tag{170}$$

with $Q_0 = \exp(R_0)$ where

$$R_0 = -\log(-p^2) + \sum_{\ell>0} \frac{(p^2)^{-\ell}}{\ell} \sum_{p_1+p_2+p_3=\ell} \left( \frac{\ell!}{p_1!p_2!p_3!} \right)^2, \tag{171}$$

and using the expression for $p^2$ in (155) we have that

$$R_0(q) = i\pi + \log q - \sum_{n\geq 1} (-1)^{n-1} \left( \frac{-3}{n} \right) n \operatorname{Li}_1 (q^n), \tag{172}$$

where $\left( \frac{-3}{n} \right) = 0, 1, -1$ for $n \equiv 0, 1, 2 \mod 3$. The maximal cut in (112) reads

$$p^2 \pi_\ominus^{(2)}(p^2, 1, 1, 1) = \frac{\eta(q^2)^6 \eta(q^3)}{\eta(q)^3 \eta(q^6)^2}. \tag{173}$$

We recall the $p^2$ is the hauptmodul in (155). The Gromov–Witten invariant $N_\ell$ can be computed using [20, proposition 7.6]

$$6 - \sum_{\ell\geq 1} \ell^3 N_\ell Q^\ell = \frac{6}{p^2(p^2-1)(p^2-9)\,(\pi_\ominus^{(2)}(q))^3}. \tag{174}$$

Introducing the virtual numbers $n_\ell$ of degree $\ell$

$$N_\ell = \sum_{d|\ell} \frac{1}{d^3} n_{\frac{\ell}{d}}, \tag{175}$$

we have

$$n_k/6 = 1, -1, 1, -2, 5, -14, 42, -136, 465, -1655, 6083, -22988, \tag{176}$$
$$88907, -350637, 1406365, -5724384, 23603157, -98440995,$$
$$414771045, -1763651230, 7561361577, -32661478080,$$

$$142046490441, -621629198960, 2736004885450,$$

$$- 12105740577346, 53824690388016, \ldots$$

The relation between $Q$ and $q$

$$Q = -q \prod_{n \geq 1} (1 - q^n)^{n\delta(n)}; \qquad \delta(n) := (-1)^{n-1} \left( \frac{-3}{n} \right), \qquad (177)$$

which we will interpret as a mirror map in Sect. 4.3, in the expansion in (170) gives the dilogarithm expression in (154).

## 4.3 Mirror Symmetry and Sunset Integral

In this section we review the result of [20] where it was shown that the sunset two-loop integral is the Legendre transform of the local Gromov–Witten prepotential and that the equivalence between the elliptic dilogarithm expression and the trilogarithm expansion of the previous section is a manifestation of local mirror symmetry. The techniques used in this section are standard in the study of mirror symmetry in string theory. We refer to the physicists oriented reviews [48, 49] for some presentation of the mathematical notions used in this section.

### 4.3.1 The Sunset Graph Polynomial and del Pezzo Surface

To the sunset Laurent polynomial

$$\phi_\ominus(p^2, \underline{\xi}^2, x_1, x_2, x_3) = p^2 - (x_1\xi_1^2 + x_2\xi_2^2 + x_3\xi_3^2) \left( \frac{1}{x_1} + \frac{1}{x_2} + \frac{1}{x_3} \right), \quad (178)$$

we associate the Newton polyhedron in Fig. 1. The vertices of the polyhedron are the powers of the monomial in $x_1$ and $x_2$ with $x_3 = 1$.

This corresponds to a maximal toric blow-up of three points in $\mathbb{P}^2$ leading to a del Pezzo surface of degree 6 $\mathscr{B}_3$.[6] The hexagon in Fig. 1 resulted from the blow-up (in red on the figure) of a triangle at the points $P_1 = [1 : 0 : 0]$, $P_2 = [0 : 1 : 0]$ and

---

[6] A del Pezzo surface is a two-dimensional Fano variety. A Fano variety is a complete variety whose anti-canonical bundle is ample. The anti-canonical bundle of a non-singular algebraic variety of dimension $n$ is the line bundle defined as the $n$th exterior power of the inverse of the cotangent bundle. An ample line bundle is a bundle with enough global sections to set up an embedding of its base variety or manifold into projective space.

**Fig. 1** The Newton polyhedron associated with the sunset second Symanzik polynomial. The coordinates $(a, b)$ of the vertices give the powers of $x^a y^b$ and we give the value of the coefficient in $\phi_\ominus(p^2, \underline{\xi}^2, x, y, 1)$

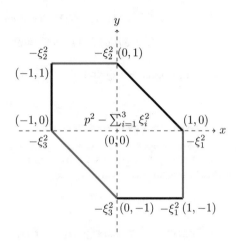

$P_3 = [0 : 0 : 1]$ by the mass parameters see [61, §6] and [20, §4]. The del Pezzo 6 surfaces are rigid.[7]

Notice that the external momentum $p^2$ appears only in the centre of the Newton polytope making this variable special.

One can construct a non-compact Calabi–Yau threefold $\mathcal{M}_\ominus$ defined as the anti-canonical hypersurface over the del Pezzo surface $\mathcal{B}_3$. This non-compact threefold is obtained as follows [20, §5]. Consider the Laurent polynomial

$$F_\ominus = a + bu^2 v^{-1} + cu^{-1}v + u^{-1}v^{-1}\phi_\ominus(p^2, \underline{\xi}^2, x_1, x_2, x_3), \qquad (179)$$

with $a, b, c \in \mathbb{C}^*$. Its Newton polytope $\Delta$ is the convex hull of $\{(0, 0, 2, -1), (0, 0, -1, 1), \Delta_\ominus \times (-1, -1)\}$ where $\Delta_\ominus$ is the Newton polytope given by the hexagon in Fig. 1. The newton polytope $\Delta$ is reflexive because its polar polytope $\Delta^\circ := \{y \in \mathbb{R}^4 | \langle y, x \rangle \geq -1, \forall x \in \Delta\} = $ convex hull$\{(0, 0, 1, 0), (0, 0, 0, 1), 6\Delta_\ominus^\circ \times (-2, -3)\}$ is integral. Notice that for the sunset polytope is self-dual $\Delta_\ominus = \Delta_\ominus^\circ$. A triangulation of $\Delta$ gives a complete toric fan[8] on $\Delta^\circ$, which then provides Fano variety $\mathbb{P}_\Delta$ of dimension four [76]. For general $a, b, c$ and the generic physical parameters $p^2, \xi_1^2, \xi_2^2, \xi_3^2$ in the sunset graph polynomial, the singular compactification $\mathcal{M}_\ominus := \overline{\{F = 0\}}$ is a smooth Calabi–Yau threefold. This non-compact Calabi–Yau threefold can be seen as a limit of compact Calabi–Yau threefold following the approach of [26] to local mirror symmetry. One can consider a semi-stably degenerating a family of elliptically-fibered Calabi–Yau threefolds $\mathcal{M}_z$ to a singular compactification $\mathcal{M}_\ominus$ for $z = 0$ and to compare the asymptotic

---

[7] The graph polynomial (48) for higher loop sunset graphs defines Fano variety, which is as well a Calabi–Yau manifold.

[8] The fan of a toric variety is defined in the standard reference [75] and the review oriented to a physicists audience in [49].

Hodge theory[9] of this B-model to that of the mirror (elliptically fibered) A-model Calabi–Yau $\mathcal{M}_{\ominus}^{\circ}$. Both $\mathcal{M}_{\ominus}$ and $\mathcal{M}_{\ominus}^{\circ}$ are elliptically fibered over the del Pezzo of degree 6 $\mathcal{B}_3$. Under the mirror map we have the isomorphism of A- and B-model $\mathbb{Z}$-variation of Hodge structure [20]

$$H^3(\mathcal{M}_{z0}) \cong H^{even}(\mathcal{M}_{q0}^{\circ}). \tag{180}$$

This situation is not unique to the two-loop sunset. The sunset graph have a reflexive polytopes containing the origin. The origin of the polytope is associated with the coefficient $p^2 - \sum_{i=1}^n \xi_i^2$, and plays a very special role. The ambient space of the sunset polytope defines a Calabi–Yau hypersurfaces (the anti-canonical divisor defines a Gorenstein toric Fano variety). Therefore they are a natural home for Batyrev's mirror symmetry techniques [24].

### 4.3.2 Local Mirror Symmetry

Putting this into practise means recasting the computation in Sect. 4.2 and the mirror symmetry description in [20, §7] in the language of [21], matching the computation of the Gromov–Witten prepotential in [21, §6.6].

The first step is to remark that the holomorphic $(3, 0)$ period of Calabi–Yau threefold $\mathcal{M}_{\ominus}$ reduces to the third period $R_0$ once integrated on a vanishing cycle [78, Appendix A], [79, §4] and [20, §5.7]

$$\int_{\text{vanishing cycle}} \mathrm{Res}_{F_{\ominus}=0}\left(\frac{1}{F_{\ominus}} \frac{du \wedge dv \wedge \wedge dx_1 \wedge dx_2}{uvx_1 x_2}\right) \propto R_0(p^2, \underline{\xi}^2), \tag{181}$$

where $F_{\ominus}$ is given in (179) and $R_0(p^2, \underline{\xi}^2)$ is given in (159). This second period is related to the analytic period near $p^2 = \infty$ by $\pi_{\ominus}^{(2)}(p^2, \underline{\xi}^2) = \frac{d}{dp^2} R_0(p^2, \underline{\xi}^2)$.[10]

The Gromov–Witten invariant evaluated in (165) Sect. 4.2 are actually the BPS numbers for the del Pezzo 6 case evaluated in [21, §6.6] since

$$\sum_{\substack{\ell_1+\ell_2+\ell_3=\ell>0 \\ (\ell_1,\ell_2,\ell_3)\in\mathbb{N}^3\setminus(0,0,0)}} N_{\ell_1,\ell_2,\ell_3} R_0^{\ell} \prod_{i=1}^{3} \xi_i^{2\ell_i} = \sum_{(\tilde{\ell}_1,\tilde{\ell}_2,\tilde{\ell}_3)\in\mathbb{N}^3\setminus(0,0,0)} n_{\tilde{\ell}_1,\tilde{\ell}_2,\tilde{\ell}_3} \mathrm{Li}_3(\prod_{i=1}^{3} \xi_i^{2\tilde{\ell}_i} e^{\tilde{\ell}_i R_0}),$$

$$\tag{182}$$

---

[9]Feynman integrals are period integrals of mixed Hodge structures [28, 77]. At a singular point some cycles of integration vanish, the so-called vanishing cycles, and the limiting behaviour of the period integral is captured by the asymptotic behaviour of the cohomological Hodge theory. The asymptotic Hodge theory inherits some filtration and weight structure of the original Hodge theory.

[10]It has been already noticed in [80] the special role played by the Mahler measure and mirror symmetry.

where we used the covering relation (164). With the following identifications[11] $Q_1 = 1$, $Q_2 = \xi_1^2 e^{R_0}$, $Q_3 = \xi_2^2 e^{R_0}$ and $Q_4 = \xi_3^2 e^{R_0}$, the expression in (182) reproduces the local genus 0 prepotential $F_0 = F_0^{\text{class}} + \sum_{\beta \in H^2(\mathcal{M}, \mathbb{Z})} n_0^{\boldsymbol{\beta}} \text{Li}_3(\prod_{r=1}^4 Q_r^{\beta_r})$ computed in [21, eq.(6.51)] with $F_0^{\text{class}} = \prod_{i=1}^3 (R_0 + \log(\xi_i^2))$ in our case. A proof that the all loop order sunset Feynman integrals compute the genus 0 relative Gromov–Witten invariants will be given in [22].

From the complex structure of the elliptic curve we define the dual period $\pi_1(p^2, \underline{\xi}^2) = 2i\pi \tau \pi_\ominus^{(2)}(p^2, \underline{\xi}^2)$ one the other homology cycle. Which gives the dual third period $R_1$, such that $\pi_1^{(2)}(p^2, \underline{\xi}^2) = \frac{d}{dp^2} R_1(p^2, \underline{\xi}^2)$. This dual period $R_1$ is therefore identified with the derivative of local prepotential $F_0$

$$2i\pi R_1 = \frac{\partial}{\partial R_0} F_0 \tag{183}$$

$$= \sum_{1 \le i < j \le 3} (R_0 + \log(\xi_i^2))(R_0 + \log(\xi_j^2)) - \sum_{\substack{\ell_1 + \ell_2 + \ell_3 = \ell > 0 \\ (\ell_1, \ell_2, \ell_3) \in \mathbb{N}^3 \setminus \{0,0,0\}}} \ell N_{\ell_1, \ell_2, \ell_3} \prod_{i=1}^3 \xi_i^{2\ell_i} e^{\ell_i R_0},$$

as shown in [20, theorem 6.1] and [20, Corollary 6.3]. With these identifications it is not difficult to see that the sunset Feynman integral is actually given by the Legendre transform of $R_1$

$$I_\ominus(p^2, \underline{\xi}^2) = -2i\pi \pi_\ominus^{(2)}(p^2, \underline{\xi}^2) \left( \frac{\partial R_1}{\partial R_0} R_0 - R_1 \right). \tag{184}$$

This shows the relation between the sunset Feynman integral and the local Gromov–Witten prepotential. The local mirror symmetry map $Q \leftrightarrow q$ given in the relations (169) and (177) maps the B-model expression, where the sunset Feynman integral is an elliptic dilogarithm function of the complex structure $\log(q)/(2i\pi)$ of the elliptic curve and the A-model expansion in terms of the Kähler moduli $Q_i$.

# 5 Conclusion

We have reviewed the toric approach to the evaluation of the maximal cut of Feynman integrals and the derivation of the minimal order differential operator acting on the Feynman integral. We have explained that recent algorithms based on Creative Telescoping provide a direct way of deriving differential equations for Feynman integrals. On the particular example of the sunset integral we have

---

[11]We would like to thank Albrecht Klemm for discussions and communication that helped clarifying the link between the work in [20] and the analysis in [21].

shown that the Feynman integral can take two different but equivalent forms. One form is an elliptic polylogarithm but it can be as well expressed as the standard trilogarithm. We have explained that mirror symmetry can be used to evaluate around the point where $p^2 = \infty$. The expressions there make explicit all the mass parameters. One remarkable fact is that the computation can be done using the existing technology of mirror symmetry developed in other physical [21, 25, 26] or mathematical [27] contexts. This analysis extends naturally to the higher loop sunset integrals [22, 23]. The elliptic polylogarithm representation generalises to other two-loop integrals like the kite integral [81–83] or the all equal masses three-loop sunset [62]. This representation leads to fast numerical evaluation [82]. But it has the disadvantage of hiding all the physical parameters in the geometry of the elliptic curve. The expression using the trilogarithm has the advantage of making all the mass parameters explicit and generalising to all loop orders since the expansion of the higher-loop sunset graphs around $p^2 = \infty$ makes explicit the Kähler parameters which are proportional to the mass parameters [22].

It is natural to ask how generic is the appearance of Calabi–Yau geometry and the role of mirror symmetry in the evaluation of Feynman integrals? It is still too early to give a precise answer but various recent works have pointed out that Calabi–Yau manifold arises more often than expected from graph polynomials [84, 85] following an earlier argument by Francis Brown in [86, §12.7]. As we explained in Sect. 3.1 the graph polynomial is an homogeneous polynomial of degree $L + 1$ in $n$ variables. The sunset family where $n = L + 1$ seems special as this is naturally associated with a Calabi–Yau geometry as explained in Sect. 4.3.1. It was actually argued in [84, 85] that for Feynman graphs with massless internal propagators, one can perform at most $n - 1 - (L + 1)D/2$ integrations and bring the denominator of the integrand to a form of a product of powers of homogeneous polynomials $P_i(x)$ with a degree matching the number of variables, therefore leading to the Calabi–Yau condition of [86, §12.7]. We leave the investigation of this intriguing property of Feynman integrals to a further study.

**Acknowledgements** It is a pleasure to thank Charles Doran and Albrecht Klemm for discussions. The research of P. Vanhove has received funding the ANR grant "Amplitudes" ANR-17- CE31-0001-01, and is partially supported by Laboratory of Mirror Symmetry NRU HSE, RF Government grant, ag. N° 14.641.31.0001.

# References

1. J.R. Andersen et al., Les Houches 2017: physics at TeV colliders standard model working group report (2018). arXiv:1803.07977 [hep-ph]
2. D. Neill, I.Z. Rothstein, Classical space-times from the S matrix. Nucl. Phys. B **877**, 177 (2013). https://doi.org/10.1016/j.nuclphysb.2013.09.007 [arXiv:1304.7263 [hep-th]]
3. N.E.J. Bjerrum-Bohr, J.F. Donoghue, P. Vanhove, On-shell techniques and universal results in quantum gravity. J. High Energy Phys. **1402**, 111 (2014). https://doi.org/10.1007/JHEP02(2014)111 [arXiv:1309.0804 [hep-th]]

4. F. Cachazo, A. Guevara, Leading singularities and classical gravitational scattering (2017). arXiv:1705.10262 [hep-th]
5. A. Guevara, Holomorphic classical limit for spin effects in gravitational and electromagnetic scattering (2017). arXiv:1706.02314 [hep-th]
6. N.E.J. Bjerrum-Bohr, P.H. Damgaard, G. Festuccia, L. Planté, P. Vanhove, General relativity from scattering amplitudes (2018). arXiv:1806.04920 [hep-th]
7. O.V. Tarasov, Hypergeometric representation of the two-loop equal mass sunrise diagram. Phys. Lett. B **638**, 195 (2006). https://doi.org/10.1016/j.physletb.2006.05.033 [hep-ph/0603227]
8. S. Bauberger, F.A. Berends, M. Bohm, M. Buza, Analytical and numerical methods for massive two loop selfenergy diagrams. Nucl. Phys. B **434**, 383 (1995). https://doi.org/10.1016/0550-3213(94)00475-T [hep-ph/9409388]
9. D.H. Bailey, J.M. Borwein, D. Broadhurst, M.L. Glasser, Elliptic integral evaluations of bessel moments. J. Phys. A **41**, 205203 (2008). https://doi.org/10.1088/1751-8113/41/20/205203 [arXiv:0801.0891 [hep-th]]
10. D. Broadhurst, Elliptic integral evaluation of a bessel moment by contour integration of a lattice green function (2008). arXiv:0801.4813 [hep-th]
11. D. Broadhurst, Feynman integrals, L-series and kloosterman moments. Commun. Num. Theor. Phys. **10**, 527 (2016). https://doi.org/10.4310/CNTP.2016.v10.n3.a3 [arXiv:1604.03057 [physics.gen-ph]]
12. M. Caffo, H. Czyz, E. Remiddi, The pseudothreshold expansion of the two loop sunrise selfmass master amplitudes. Nucl. Phys. B **581**, 274 (2000). https://doi.org/10.1016/S0550-3213(00)00274-1 [hep-ph/9912501]
13. S. Laporta, E. Remiddi, Analytic treatment of the two loop equal mass sunrise graph. Nucl. Phys. B **704**, 349 (2005). [hep-ph/0406160]
14. L. Adams, C. Bogner, S. Weinzierl, The two-loop sunrise graph with arbitrary masses in terms of elliptic dilogarithms (2014). arXiv:1405.5640 [hep-ph]
15. L. Adams, C. Bogner, S. Weinzierl, The two-loop sunrise integral around four space-time dimensions and generalisations of the Clausen and Glaisher functions towards the elliptic case. J. Math. Phys. **56**(7), 072303 (2015). https://doi.org/10.1063/1.4926985 [arXiv:1504.03255 [hep-ph]].
16. L. Adams, C. Bogner, S. Weinzierl, The iterated structure of the all-order result for the two-loop sunrise integral. J. Math. Phys. **57**(3), 032304 (2016). https://doi.org/10.1063/1.4944722 [arXiv:1512.05630 [hep-ph]]
17. L. Adams, C. Bogner, S. Weinzierl, A walk on sunset boulevard. PoS RADCOR **2015**, 096 (2016). https://doi.org/10.22323/1.235.0096 [arXiv:1601.03646 [hep-ph]]
18. L. Adams, S. Weinzierl, On a class of feynman integrals evaluating to iterated integrals of modular forms (2018). arXiv:1807.01007 [hep-ph]
19. L. Adams, E. Chaubey, S. Weinzierl, From elliptic curves to Feynman integrals. arXiv:1807.03599 [hep-ph]
20. S. Bloch, M. Kerr, P. Vanhove, Local mirror symmetry and the sunset Feynman integral. Adv. Theor. Math. Phys. **21**, 1373 (2017). https://doi.org/10.4310/ATMP.2017.v21.n6.a1 [arXiv:1601.08181 [hep-th]]
21. M.X. Huang, A. Klemm, M. Poretschkin, Refined stable pair invariants for E-, M- and [$p$, $Q$]-strings. J. High Energy Phys. **1311**, 112 (2013). https://doi.org/10.1007/JHEP11(2013)112 [arXiv:1308.0619 [hep-th]]
22. C. Doran, A. Novoseltsev, P. Vanhove, Mirroring towers: the Calabi-Yau geometry of the multiloop sunset Feynman integrals (to appear)
23. P. Vanhove, Mirroring towers of Feynman integrals: fibration and degeneration in Feynman integral Calabi-Yau geometries. (String Math 2019). https://www.stringmath2019.se/wp-content/uploads/sites/39/2019/07/Vanhove_StringMath2019.pdf
24. V.V. Batyrev, Dual polyhedra and mirror symmetry for CalabiYau hypersurfaces in toric varieties. J. Algebr. Geom. **3**, 493–535 (1994)

25. S. Hosono, A. Klemm, S. Theisen, S.T. Yau, Mirror symmetry, mirror map and applications to Calabi-Yau hypersurfaces. Commun. Math. Phys. **167**, 301 (1995). https://doi.org/10.1007/BF02100589 [hep-th/9308122]

26. T.-M. Chiang, A. Klemm, S.-T. Yau, E. Zaslow, Local mirror symmetry: calculations and interpretations. Adv. Theor. Math. Phys. **3**, 495 (1999). https://doi.org/10.4310/ATMP.1999.v3.n3.a3 [hep-th/9903053]

27. C.F. Doran, M. Kerr, Algebraic K-theory of toric hypersurfaces. Commun. Number Theory Phys. **5**(2), 397–600 (2011)

28. P. Vanhove, The physics and the mixed hodge structure of Feynman integrals. Proc. Symp. Pure Math. **88**, 161 (2014). https://doi.org/10.1090/pspum/088/01455 [arXiv:1401.6438 [hep-th]]

29. C. Bogner, S. Weinzierl, Feynman graph polynomials. Int. J. Mod. Phys. A **25**, 2585 (2010). [arXiv:1002.3458 [hep-ph]]

30. P. Tourkine, Tropical amplitudes (2013). arXiv:1309.3551 [hep-th]

31. O. Amini, S. Bloch, J.I.B. Gil, J. Fresan, Feynman amplitudes and limits of heights. Izv. Math. **80**, 813 (2016). https://doi.org/10.1070/IM8492 [arXiv:1512.04862 [math.AG]]

32. E.R. Speer, *Generalized Feynman Amplitudes*. Annals of Mathematics Studies, vol. 62 (Princeton University Press, New Jersey, 1969)

33. A. Primo, L. Tancredi, On the maximal cut of Feynman integrals and the solution of their differential equations. Nucl. Phys. B **916**, 94 (2017). https://doi.org/10.1016/j.nuclphysb.2016.12.021 [arXiv:1610.08397 [hep-ph]]

34. A. Primo, L. Tancredi, Maximal cuts and differential equations for Feynman integrals. An application to the three-loop massive banana graph. Nucl. Phys. B **921**, 316 (2017). https://doi.org/10.1016/j.nuclphysb.2017.05.018 [arXiv:1704.05465 [hep-ph]]

35. J. Bosma, M. Sogaard, Y. Zhang, Maximal cuts in arbitrary dimension. J. High Energy Phys. **1708**, 051 (2017). https://doi.org/10.1007/JHEP08(2017)051 [arXiv:1704.04255 [hep-th]]

36. H. Frellesvig, C.G. Papadopoulos, Cuts of Feynman integrals in Baikov representation. J. High Energy Phys. **1704**, 083 (2017). https://doi.org/10.1007/JHEP04(2017)083 [arXiv:1701.07356 [hep-ph]]

37. K.G. Chetyrkin, F.V. Tkachov, Integration by parts: the algorithm to calculate beta functions in 4 loops. Nucl. Phys. B **192**, 159 (1981). https://doi.org/10.1016/0550-3213(81)90199-1

38. O.V. Tarasov, Generalized recurrence relations for two loop propagator integrals with arbitrary masses. Nucl. Phys. B **502**, 455 (1997). https://doi.org/10.1016/S0550-3213(97)00376-3 [hep-ph/9703319]

39. O.V. Tarasov, Methods for deriving functional equations for Feynman integrals. J. Phys. Conf. Ser. **920**(1), 012004 (2017). https://doi.org/10.1088/1742-6596/920/1/012004 [arXiv:1709.07058 [hep-ph]]

40. P. Griffiths, On the periods of certain rational integrals: I. Ann. Math. **90**, 460 (1969)

41. D. Cox, S. Katz, Mirror symmetry and algebraic geometry. Mathematical Surveys and Monographs, vol. 68. (American Mathematical Society, Providence, 1999). https://doi.org/10.1090/surv/068

42. T. Bitoun, C. Bogner, R.P. Klausen, E. Panzer, Feynman integral relations from parametric annihilators (2017). arXiv:1712.09215 [hep-th]

43. W. Decker, G.-M. Greuel, G. Pfister, H. Schönemann, SINGULAR 4-1-1 — a computer algebra system for polynomial computations (2018). http://www.singular.uni-kl.de

44. I.M. Gelfand, M.M. Kapranov, A.V. Zelevinsky, Generalized Euler integrals and a-hypergeometric functions. Adv. Math. **84**, 255–271 (1990)

45. I.M. Gelfand, M.M. Kapranov, A.V. Zelevinsky, *Discriminants, Resultants and Multidimensional Determinants* (Birkhäuser, Boston, 1994)

46. V.V. Batyrev, Variations of the mixed hodge structure of affine hypersurfaces in algebraic tori. Duke Math. J. **69**(2), 349–409 (1993)

47. V.V. Batyrev, D.A. Cox, On the hodge structure of projective hypersurfaces in toric varieties. Duke Math. J. **75**(2), 293–338 (1994)

48. S. Hosono, A. Klemm, S. Theisen, Lectures on mirror symmetry. Lect. Notes Phys. **436**, 235 (1994). https://doi.org/10.1007/3-540-58453-6_13 [hep-th/9403096]

49. C. Closset, Toric geometry and local Calabi-Yau varieties: an introduction to toric geometry (for physicists) (2009). arXiv:0901.3695 [hep-th]

50. J.S. Jan, GKZ hypergeometric structures (2005). arXiv:math/0511351

51. V.V. Batyrev, D. van Straten, Generalized hypergeometric functions and rational curves on Calabi-Yau complete intersections in toric varieties. Commun. Math. Phys. **168**, 493 (1995). https://doi.org/10.1007/BF02101841 [alg-geom/9307010]

52. S. Hosono, *GKZ Systems, Gröbner Fans, and Moduli Spaces of Calabi-Yau Hypersurfaces* (Birkhäuser, Boston, 1998)

53. S. Hosono, A. Klemm, S. Theisen, S.T. Yau, Mirror symmetry, mirror map and applications to complete intersection Calabi-Yau spaces. Nucl. Phys. B **433**, 501 (1995). [AMS/IP Stud. Adv. Math. **1** (1996) 545] https://doi.org/10.1016/0550-3213(94)00440-P [hep-th/9406055]

54. E. Cattani, Three lectures on hypergeometric functions (2006). https://people.math.umass.edu/~cattani/hypergeom_lectures.pdf

55. F. Beukers, Monodromy of A-hypergeometric functions. J. Reine Angew. Math. **718**, 183–206 (2016)

56. J. Stienstra, Resonant hypergeometric systems and mirror symmetry, in *Proceedings of the Taniguchi Symposium 1997 "Integrable Systems and Algebraic Geometry"* (1998) [alg-geom/9711002]

57. L. Adams, C. Bogner, S. Weinzierl, The two-loop sunrise graph with arbitrary masses (2013). arXiv:1302.7004 [hep-ph]

58. P. Candelas, X.C. de la Ossa, P.S. Green, L. Parkes, A pair of Calabi-Yau manifolds as an exactly soluble superconformal theory. Nucl. Phys. B **359**, 21 (1991). [AMS/IP Stud. Adv. Math. **9** (1998) 31]. https://doi.org/10.1016/0550-3213(91)90292-6

59. D.R. Morrison, Picard-Fuchs equations and mirror maps for hypersurfaces. AMS/IP Stud. Adv. Math. **9**, 185 (1998). [hep-th/9111025]

60. H.A. Verrill, Sums of squares of binomial coefficients, with applications to Picard-Fuchs equations (2004). arXiv:math/0407327

61. S. Bloch, P. Vanhove, The elliptic dilogarithm for the sunset graph. J. Number Theory **148**, 328 (2015). https://doi.org/10.1016/j.jnt.2014.09.032 [arXiv:1309.5865 [hep-th]]

62. S. Bloch, M. Kerr, P. Vanhove, A Feynman integral via higher normal functions. Compos. Math. **151**(12), 2329 (2015). https://doi.org/10.1112/S0010437X15007472 [arXiv:1406.2664 [hep-th]]

63. D. Zeilberger, The method of creative telescoping. J. Symb. Comput. **11**(3), 195–204 (1991)

64. F. Chyzak, An extension of Zeilberger's fast algorithm to general holonomic functions. Discret. Math. **217**(1–3), 115–134 (2000)

65. F. Chyzak, The ABC of creative telescoping — algorithms, bounds, complexity. Symbolic Computation [cs.SC]. Ecole Polytechnique X (2014)

66. C. Koutschan, Advanced applications of the holonomic systems approach. ACM Commun. Comput. Algebra **43**, 119 (2010)

67. A.V. Smirnov, A.V. Petukhov, The number of master integrals is finite. Lett. Math. Phys. **97**, 37 (2011). https://doi.org/10.1007/s11005-010-0450-0 [arXiv:1004.4199 [hep-th]].

68. F.C.S. Brown, A. Levin, Multiple elliptic polylogarithms (2011). arXiv:1110.6917

69. J. Broedel, C. Duhr, F. Dulat, L. Tancredi, Elliptic polylogarithms and iterated integrals on elliptic curves. Part I: general formalism. J. High Energy Phys. **1805**, 093 (2018). https://doi.org/10.1007/JHEP05(2018)093 [arXiv:1712.07089 [hep-th]]

70. J. Broedel, C. Duhr, F. Dulat, L. Tancredi, Elliptic polylogarithms and iterated integrals on elliptic curves II: an application to the sunrise integral. Phys. Rev. D **97**(11), 116009 (2018). https://doi.org/10.1103/PhysRevD.97.116009 [arXiv:1712.07095 [hep-ph]]

71. J. Broedel, C. Duhr, F. Dulat, B. Penante, L. Tancredi, Elliptic symbol calculus: from elliptic polylogarithms to iterated integrals of Eisenstein series (2018). arXiv:1803.10256 [hep-th]

72. J. Broedel, C. Duhr, F. Dulat, B. Penante, L. Tancredi, From modular forms to differential equations for Feynman integrals (2018). arXiv:1807.00842 [hep-th]

73. J. Broedel, C. Duhr, F. Dulat, B. Penante, L. Tancredi, Elliptic polylogarithms and two-loop Feynman integrals (2018). arXiv:1807.06238 [hep-ph]

74. E. Remiddi, L. Tancredi, An elliptic generalization of multiple polylogarithms. Nucl. Phys. B **925**, 212 (2017). https://doi.org/10.1016/j.nuclphysb.2017.10.007 [arXiv:1709.03622 [hep-ph]]

75. W. Fulton, *Introduction to Toric Varieties*. Annals of Mathematics Studies (Princeton University Press, Princeton, 1993)

76. D.A. Cox, J.B. Little, H.K. Schenck, *Toric Varieties*. Graduate Studies in Mathematics (Book 124) (American Mathematical Society, Providence, 2011)

77. S. Bloch, H. Esnault, D. Kreimer, On motives associated to graph polynomials. Commun. Math. Phys. **267**, 181 (2006). https://doi.org/10.1007/s00220-006-0040-2 [math/0510011 [math-ag]]

78. S. Hosono, Central charges, symplectic forms, and hypergeometric series in local mirror symmetry, in *Mirror Symmetry V*, ed. by N. Yui, S.-T. Yau, J. Lewis (American Mathematical Society, Providence, 2006), pp. 405–439

79. S.H. Katz, A. Klemm, C. Vafa, Geometric engineering of quantum field theories. Nucl. Phys. B **497**, 173 (1997). https://doi.org/10.1016/S0550-3213(97)00282-4 [hep-th/9609239]

80. J. Stienstra, Mahler measure variations, Eisenstein series and instanton expansions, in *Mirror Symmetry V*, ed. by N. Yui, S.-T. Yau, J.D. Lewis. AMS/IP Studies in Advanced Mathematics, vol. 38 (International Press & American Mathematical Society, Providence, 2006), pp. 139–150. [arXiv:math/0502193]

81. L. Adams, C. Bogner, A. Schweitzer, S. Weinzierl, The kite integral to all orders in terms of elliptic polylogarithms. J. Math. Phys. **57**(12), 122302 (2016). https://doi.org/10.1063/1.4969060 [arXiv:1607.01571 [hep-ph]]

82. C. Bogner, A. Schweitzer, S. Weinzierl, Analytic continuation and numerical evaluation of the kite integral and the equal mass sunrise integral. Nucl. Phys. B **922**, 528 (2017). https://doi.org/10.1016/j.nuclphysb.2017.07.008 [arXiv:1705.08952 [hep-ph]]

83. C. Bogner, A. Schweitzer, S. Weinzierl, Analytic continuation of the kite family (2018). arXiv:1807.02542 [hep-th]

84. J.L. Bourjaily, Y.H. He, A.J. Mcleod, M. Von Hippel, M. Wilhelm, Traintracks through Calabi-Yau manifolds: scattering amplitudes beyond elliptic polylogarithms. Phys. Rev. Lett. **121**(7), 071603 (2018). https://doi.org/10.1103/PhysRevLett.121.071603 [arXiv:1805.09326 [hep-th]]

85. J.L. Bourjaily, A.J. McLeod, M. von Hippel, M. Wilhelm, A (Bounded) bestiary of Feynman integral Calabi-Yau geometries (2018). arXiv:1810.07689 [hep-th]

86. F.C.S. Brown, On the periods of some Feynman integrals (2009). arXiv:0910.0114 [math.AG]

# Theory and Applications of the Elliptic Painlevé Equation

**Yasuhiko Yamada**

## Contents

1  Introduction.................................................................... 370
2  "Integrability" of Painlevé Equations ......................................... 370
   2.1   The Differential Painlevé Equations ..................................... 370
   2.2   Painlevé Test ......................................................... 372
   2.3   Isomonodromic Deformation.............................................. 373
   2.4   Space of Initial Conditions ........................................... 374
   2.5   QRT System............................................................. 375
   2.6   "Integrable" Deautonomization ......................................... 378
   2.7   Sakai Scheme .......................................................... 379
   2.8   Short Summary on Cubic Curve and Elliptic Function .................... 380
3  Affine Weyl Groups ............................................................ 383
   3.1   Affine Weyl Group and Its Translations ................................ 383
   3.2   Birational Representations............................................. 385
   3.3   Weyl Group Actions on Borel Subgroups ................................. 387
   3.4   Weyl Group Actions on Point Configurations ............................ 389
   3.5   The Elliptic Painlevé Equation in $E_{10}$ Form........................ 391
   3.6   Picard Lattice......................................................... 393
   3.7   $q$-Painlevé Equation of Type $E_8^{(1)}$ ............................. 395
   3.8   Toric-Like Form........................................................ 398
4  Results on the Elliptic Painlevé Equation ..................................... 401
   4.1   $\tau$ Functions....................................................... 401
   4.2   Lax Formalism ......................................................... 404
   4.3   Hypergeometric Solution ............................................... 408
References ...................................................................... 412

**Abstract** This note is intended to provide an introduction to the theory of discrete Painlevé equations focusing mainly on the elliptic difference case. The elliptic Painlevé equation is the master case of the continuous/discrete Painlevé equations in two variables and has the largest affine Weyl group symmetry $W(E_8^{(1)})$. Various

Y. Yamada (✉)
Department of Mathematics, Kobe Univ. Rokko, Kobe, Japan
e-mail: yamaday@math.kobe-u.ac.jp

© Springer Nature Switzerland AG 2020
V. A. Gritsenko, V. P. Spiridonov (eds.), *Partition Functions and Automorphic Forms*, Moscow Lectures 5,
https://doi.org/10.1007/978-3-030-42400-8_8

"integrable" nature such as the symmetry of equation, bilinear form, Lax pair and special solutions are explained from a geometric point of view.

# 1 Introduction

The Painlevé equations now arise in many areas in mathematics/physics, and we can access them through various approaches: isomonodromic deformations, soliton equations, matrix models, affine Weyl groups, geometry of algebraic curves and surfaces, etc. In this note, I will take a geometric approach, and explain various aspects from this view point.

Contents of the paper is as follows. In Sect. 2, starting with differential cases, we explain "integrability" of Painlevé equations and how it can be extended to discrete cases. In Sect. 3, we discuss the affine Weyl groups and their birational representations as the main technical tool to construct discrete Painlevé equations. In Sect. 4, we present fundamental results on the elliptic Painlevé equations: bilinear form, Lax pair and hypergeometric solutions.

There are many important materials which are not discussed in this note. We only give a few comments here. (1) *Multivariate generalizations.* In this paper, we will consider the discrete isomonodromic systems in two variables. There are various studies on multivariate generalizations, see [37, 52, 57, 60] for example in $q$-difference case. Some generalizations in elliptic case are also known [45, 66]. (2) *Kiev formula.* Using the connection to the conformal field theory, an explicit formula of the $\tau$-function for generic solutions is established [8]. The result is generalized in various way, e.g. to $q$-difference case [3, 16], to higher rank [9, 10], etc. Their application to Painlevé/gauge correspondence [5, 6, 35] is also very important.

# 2 "Integrability" of Painlevé Equations

## 2.1 The Differential Painlevé Equations

Let us start with differential case. Classically there are six (or eight[1]) differential Painlevé equations $P_J$:

$$P_{VI} \to P_V \to P_{III_1} \to (P_{III_2}) \to (P_{III_3})$$
$$\searrow \qquad \searrow \qquad \searrow$$
$$P_{IV} \to P_{II} \to P_I \qquad\qquad (1)$$

---

[1]From a geometric point of view [51], the third Painlevé equation $P_{III}$ is further classified into $P_{III_1} = P_{III}^{D_6^{(1)}}$, $P_{III_2} = P_{III}^{D_7^{(1)}}$ and $P_{III_3} = P_{III}^{D_8^{(1)}}$.

Each equation $P_J$ can be written as a *non-autonomous* Hamiltonian system

$$\frac{dq}{dt} = \frac{\partial H_J}{\partial p}, \quad \frac{dp}{dt} = -\frac{\partial H_J}{\partial q}. \tag{2}$$

The Hamiltonians are given as

$$H_{VI} = \frac{q(q-1)(q-t)}{t(t-1)}\left\{p^2 - \left(\frac{\alpha_0 - 1}{q-t} + \frac{\alpha_3}{q-1} + \frac{\alpha_4}{q}\right)p\right\} + \frac{(q-t)\alpha_2(\alpha_1 + \alpha_2)}{t(t-1)},$$

$$\alpha_0 + \alpha_1 + 2\alpha_2 + \alpha_3 + \alpha_4 = 1,$$

$$H_V = \frac{1}{t}\left\{q(q-1)p(p+t) - (\alpha_1 + \alpha_3)qp + \alpha_1 p + \alpha_2 tq\right\},$$

$$H_{III_1} = \frac{1}{t}\left\{p(p-1)q^2 + (\alpha_1 + \alpha_2)qp + tp - \alpha_2 q\right\},$$

$$H_{III_2} = \frac{1}{t}(p^2q^2 + q + pt + \alpha_1 pq),$$

$$H_{III_3} = \frac{1}{t}(p^2q^2 + pq + q + \frac{t}{q}),$$

$$H_{IV} = qp(p - q - t) - \alpha_1 p - \alpha_2 q,$$

$$H_{II} = \frac{p^2}{2} - (q^2 + \frac{t}{2})p - aq,$$

$$H_I = \frac{p^2}{2} - 2q^3 - tq.$$

$$\tag{3}$$

Let us consider the autonomous limit.

*Example 2.1* $P_{IV}$ case. By a rescaling $(p, q, t, \alpha_i) \to (\delta^{-\frac{1}{2}}p, \delta^{-\frac{1}{2}}q, \delta^{-\frac{1}{2}}t, \delta^{-1}\alpha_i)$ and putting $s = \delta^{-1}t$, the $P_{IV}$ equation can be written as

$$q' = q(2p - q - t) - \alpha_1, \quad p' = p(2q + t - p) + \alpha_2, \quad t' = \delta, \quad \alpha_i' = 0. \quad ('= \frac{d}{ds}) \tag{4}$$

Then a limit $\delta \to 0$ gives the *autonomous* $P_{IV}$ where $t$ is a constant and $H_{IV}$ is conserved. The conservation low $H_{IV}(q, p) = C$ define a elliptic curve and the system can be solved in terms of elliptic functions.

A relation between the curves $H_J = C$ and the Seiberg-Witten curves was observed in [22]. For non-autonomous case, the Hamiltonians $H_J$ are no longer conserved. Nevertheless, we will see that the Painlevé equations are still "integrable" in certain sense.

## 2.2  Painlevé Test

Solutions of a nonlinear differential equation may have singularities whose locations depend on the initial values. Such singularity is called *movable*. For instance, the solution $y = \sqrt{t - c}$ of the equation $2y\frac{dy}{dt} = 1$ has a movable branch cut. The requirement that all the movable singularities to be poles is called the *Painlevé property*. The differential equations satisfied by elliptic functions give the typical examples of such equations. For instance the equation

$$\frac{d^2 y}{dt^2} = 6y^2 - \frac{g_2}{2},\tag{5}$$

has the general solution in Laurent series in $z = t - t_0$

$$y = z^{-2} + \frac{g_2}{20}z^2 + \frac{g_3}{28}z^4 + \frac{g_2^2}{1200}z^6 + \frac{3g_2 g_3}{6160}z^8 + \cdots,\tag{6}$$

admitting two free parameters $g_3, t_0$. This solution is the Weierstrass $\wp$-function $y = \wp(z; g_2, g_3)$ whose all singularities are poles of order two.

For a given equation, searching for a Laurent series solution admitting sufficient number of free parameters is a very efficient way to find "integrable" cases. This method (Painlevé test) was first successfully used by S. Kovalevskaya in her famous work on the rotation of a rigid body around a fixed point.

*Example 2.2 (Kovalevskaya's Top)*  The equation of motion is

$$\frac{d}{dt}\mathbf{L} + \boldsymbol{\omega} \times \mathbf{L} = \mathbf{a} \times \boldsymbol{\gamma}, \quad \frac{d}{dt}\boldsymbol{\gamma} = \boldsymbol{\gamma} \times \boldsymbol{\omega},\tag{7}$$

where $\boldsymbol{\gamma}, \boldsymbol{\omega}$ are the dynamical variables, $\mathbf{L} = \mathrm{diag}(I_1, I_2, I_3)\boldsymbol{\omega}$, and $\mathbf{a}, I_1, I_2, I_3$ are constants. In general, this system has three integrals: $E = \frac{1}{2}\boldsymbol{\omega} \cdot \mathbf{L} - \mathbf{a} \cdot \boldsymbol{\gamma}, \ell = \boldsymbol{\gamma} \cdot \mathbf{L}$ and $k = \boldsymbol{\gamma}^2$. Before Kovalevskaya, two special cases (Euler, Lagrange) with one additional integral were known. In both cases the solutions are given by elliptic functions and hence have no singular points other than pole. She asked herself, "Is this property preserved in general case?". By looking for Laurent series solutions, she found a new integrable case: the Kovalevskaya's top[2]

|  | Conditions | Additional integral |
|---|---|---|
| Euler | $\mathbf{a} = 0$ | $\mathbf{L}^2$ |
| Lagrange | $I_1 = I_2, a_1 = a_2 = 0$ | $L_3$ |
| Kovalevskaya | $I_1 = I_2 = 2I_3, a_2 = a_3 = 0$ | $(\omega_1^2 - \omega_2^2 + \frac{a_1}{I_3}\gamma_1)^2 + (2\omega_1\omega_2 + \frac{a_1}{I_3}\gamma_2)^2$ |

This method is now widely used as a very efficient way to find integrable cases.

---

[2] Solving the Kovalevskaya's top is much more difficult than the Euler/Lagrange cases because one need hyperelliptic functions.

In case of the first Painlevé equation $P_I$, the equation

$$\frac{d^2q}{dt^2} = 6q^2 + t, \tag{8}$$

admits a series solution similar to (6)

$$q = z^{-2} - \frac{t_0}{10}z^2 - \frac{1}{6}z^3 + cz^4 + \frac{t_0^2}{300}z^6 + \frac{t_0}{150}z^7 + \frac{(5 - 36ct_0)}{1320}z^8 + \cdots, \tag{9}$$

where $z = t - t_0$ and $t_0$, $c_6$ are free parameters. Hence, the $P_I$ equation passes the Painlevé test. This indicates that $P_I$ is a good deautonomization of the Eq. (5).

## 2.3 Isomonodromic Deformation

Consider a linear differential equation on $\mathbb{P}^1$

$$L: \quad Y''(x) + a(x)Y'(x) + b(x)Y(x) = 0. \tag{10}$$

We allow the coefficient $a(x), b(x)$ to depend on some deformation parameter $t$, hence the solution and its monodromy also may depend on $t$. If the monodromy is independent of $t$, such deformation is called the *isomonodromic deformation*. This condition is equivalent to the compatibility of $L$ with a deformation equation

$$B: \quad \frac{\partial}{\partial t}Y(x) = r(x)Y'(x) + s(x)Y(x), \tag{11}$$

where $r(x), s(x)$ are rational functions in $x$ depending also on $t$ in nontrivial way. Borrowing the term in integrable systems, the equations $L, B$ is called the *Lax pair*. The Lax formulation is a cornerstone for "integrability" from which various results can be obtained.

*Example 2.3 (Lax Pair for $P_{VI}$)* The $P_{VI}$ equation is the isomonodromic deformation of the 2nd order equation $L$ (10) on $\mathbb{P}^1$ with following properties:

- It is a Fuchsian equation with the following local exponents:

| $x$ | 0 | 1 | $t$ | $\infty$ | $q$ |
|------|---------|---------|---------|--------------------|---|
| exp. | 0 | 0 | 0 | $\alpha_2$ | 0 |
| | $\alpha_4$ | $\alpha_3$ | $\alpha_0$ | $\alpha_1 + \alpha_2$ | 2 |

$$\tag{12}$$

- $x = q$ is an apparent singularity, namely the solutions are regular there.

These conditions determine the coefficients $a(x), b(x)$ in (10) as

$$a(x) = \frac{1-\alpha_4}{x} + \frac{1-\alpha_3}{x-1} + \frac{1-\alpha_0}{x-t} + \frac{-1}{x-q},$$

$$b(x) = \frac{1}{x(x-1)}\left\{\frac{q(q-1)p}{x-q} - \frac{t(t-1)H_{VI}(q,p)}{x-t} + \alpha_2(\alpha_1+\alpha_2)\right\}. \tag{13}$$

The isomonodromic deformation of $L$ is given by

$$B: \quad \frac{t(t-1)}{q-t}\frac{\partial}{\partial t}Y(x) + \frac{x(x-1)}{q-x}Y'(x) + \frac{pq(q-1)}{x-q}Y(x) = 0. \tag{14}$$

The compatibility condition of $L, B$ is equivalent to $P_{VI}$.

## 2.4   Space of Initial Conditions

In [43] Okamoto defined a surface $X_J$ which parametrize all the solutions for each equation $P_J$. Then in [33, 53], some surfaces $X_J$ are constructed as a union of several charts patched together by birational symplectic transformations.

*Example 2.4* For $P_{IV}$, the surface $X_{IV}$ can be defined as a union of four charts $\mathbb{C}^2$

$$X_{IV} = \{(q,p)\} \cup \{(q_1,p_1)\} \cup \{(q_2,p_2)\} \cup \{(q_2,p_2), \tag{15}$$

where

$$\begin{aligned}(q,p) &= (\alpha_1 p_1 + q_1 p_1^2, \frac{1}{p_1}) \\ &= (\frac{1}{q_2}, -\alpha_2 q_2 + q_2^2 p_2) \\ &= (\frac{1}{q_3}, \frac{1}{q_3} + t - \alpha_0 q_3 - q_3^2 p_3) \quad (\alpha_0+\alpha_1+\alpha_2 = 1).\end{aligned} \tag{16}$$

The construction goes as follows. Since the $P_{IV}$ equation

$$q' = q(2p-t-q) - \alpha_1, \quad p' = p(2q+t-p) + \alpha_2. \tag{17}$$

is a polynomial system, one naively expect that the solutions can be parametrized by its initial values $(q, p) \in \mathbb{C}^2$ at $t = t_0$. However this is not enough since there may be a solution such as $q \to \infty$ and/or $p \to \infty$ for $t \to t_0$. To include the solution with $p \to \infty$, we put $(q, p) = (x, \frac{1}{y})$ and get

$$x' = \frac{2x}{y} - \alpha_1 - x(t+x), \quad y' = 1 - (1+2x)y - \alpha_2 y^2. \tag{18}$$

We see $x = 0$ for $y = 0$, so we put $(x, y) = (uy, y)$ (a blow-up) and get

$$u' = \frac{u - \alpha_1}{y} + (\alpha_2 + u)uy, \quad y' = 1 - ty - (\alpha_2 + 2u)y. \tag{19}$$

Again we see $u = \alpha_1$ for $y = 0$, hence we further put $(u, y) = (\alpha_1 + vy, y)$ and get

$$\begin{aligned} v' &= \alpha_1(\alpha_1 + \alpha_2) + (t + 2(2\alpha_1 + \alpha_2)y)v + 3v^2 y^2, \\ y' &= 1 - ty - (2\alpha_1 + \alpha_2)y^2 - 2vy^3. \end{aligned} \tag{20}$$

Thus, the singular solutions (with $p \to \infty$) can be parametrized by the initial values of $(v, y)$ where $q = (\alpha_1 + vy)y$, $p = \frac{1}{y}$. This gives the transformation to $(q_1, p_1)$ chart. In a similar way, one can construct a surface $X_{\mathrm{IV}}$ (15) which parametrize all the solutions including singular ones at $t \to t_0$.

Since the patching transformations are rational symplectic, the equation extended to each chart is also a Hamiltonian system whose transformed Hamiltonians may have poles. It is remarkable that the Hamiltonians are still polynomial, and moreover, the equation with such holomorphy is unique [33, 53]. This result shows that the geometry knows Painlevé equations! Sakai extended this idea to discrete Painlevé equations. In the following, we will consider some examples of discrete Painlevé equations. First, we look at the autonomous case.

## 2.5 QRT System

The QRT system [47, 48] is an important class of birational mapping on $\mathbb{C}^2$ having algebraic integral. Starting with suitable integral $h$, the system can be easily constructed as follows [59, 61]. Let $h(x, y) = \frac{f(x,y)}{g(x,y)}$ be a rational function of bidegree $(2, 2)$, namely $f(x, y)$, $g(x, y)$ are polynomials of degree 2 in each variable $x$, $y$. The equation for $x' : h(x', y) = h(x, y)$ has two solutions. One is trivial $x' = x$. Using the other solution $x'(x, y)$ we define a rational transformation $\iota_1(x) = x'(x, y)$. Similarly we define a transformation $\iota_2(y) = y'(x, y)$ from $h(x, y') = h(x, y)$. By construction, the transformations $\iota_1, \iota_2$ are involutions and birational. The iteration of the composition $T = \iota_1 \circ \iota_2$ (or $T^{-1} = \iota_2 \circ \iota_1$) gives the discrete dynamical system for which $h(x, y)$ gives the integral. Here and in the followings, the transformations are considered as automorphisms on a field of rational functions, and trivial actions such as $\iota_1(y) = y$ will be omitted.

For a "fixed" curve $C$ of bidegree $(2, 2)$, the construction above has been classically known. The important thing here is that we have a mapping not only on the fixed curve $C$ but also for the whole plane $\mathbb{C}^2$ (or its compactification $\mathbb{P}^1 \times \mathbb{P}^1$). This is achieved by considering one parameter family of curves $f(x, y) - hg(x, y) = 0$. Where the parameter $h$ is determined by the initial data $(x_0, y_0)$ as $h = h(x_0, y_0)$.

By the Bézout's theorem, a "pencil" (one parameter family) of bidegree $(2, 2)$ curve $f(x, y) - hg(x, y) = 0$ has 8 intersection points $\{f(x, y) = 0\} \cap \{g(x, y) = 0\}$. These 8 points are in a special position since any 7 of them determine the rest one. A projection from $\mathbb{P}^1 \times \mathbb{P}^1$ to $\mathbb{P}^1$ given by $(x, y) \mapsto h = \frac{f(x,y)}{g(x,y)}$ define a elliptic fiblation $X$ over $\mathbb{P}^1$. Where $X$ is a blowing up of $\mathbb{P}^1 \times \mathbb{P}^1$ at the 8 points. Note that these blowing ups are needed to make the projection well defined.

*Example 2.5 Autonomos $q$-$P(D_5^{(1)})$.* This case arises from the following 8 points:

$$(x, y) = (a_1, 0), (a_2, 0), (a_3, \infty), (a_4, \infty), (0, a_5), (0, a_6), (\infty, a_7), (\infty, a_8),$$

sitting on the four lines. These points admit a pencil of curve $f(x, y) - hg(x, y) = 0$ where

$$\begin{aligned} f(x, y) &= (x - a_3)(x - a_4)y^2 - ((a_7 + a_8)x^2 + a_3 a_4(a_5 + a_6))y \\ &\quad + a_7 a_8(x - a_1)(x - a_2), \\ g(x, y) &= xy, \end{aligned} \tag{21}$$

iff a constraint $a_1 a_2 a_7 a_8 = a_3 a_4 a_5 a_6$ is satisfied.

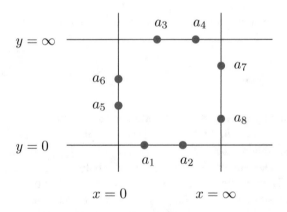

The corresponding QRT system is defined by $\iota_1(x) = x'$, $\iota_2(y) = y'$, where

$$yy' = a_7 a_8 \frac{(x - a_1)(x - a_2)}{(x - a_3)(x - a_4)}, \qquad xx' = a_3 a_4 \frac{(y - a_5)(y - a_6)}{(y - a_7)(y - a_8)}. \tag{22}$$

*Example 2.6 Autonomos $q$-$P(E_8^{(1)})$.* Consider the 8 points

$$(x, y) = \left(\psi_{\kappa_1}(v_i), \psi_{\kappa_2}(v_i)\right)_{i=1}^{8}, \qquad \psi_\kappa(u) = u + \frac{\kappa}{u}. \tag{23}$$

This configuration of points admits a pencil of bidegree $(2, 2)$ curve iff we have a constraint $\prod_{i=1}^8 v_i = \kappa_1^2 \kappa_2^2$. The pencil is given by $f(x, y) - h g(x, y) = 0$ where

$$
\begin{aligned}
f(x, y) = m_1 & \frac{xy(\kappa_1 y - \kappa_2 x) + \kappa_2^2 x - \kappa_1^2 y}{\kappa_1 - \kappa_2} + m_2 \frac{(\kappa_1 y - \kappa_2 x)^2}{(\kappa_1 - \kappa_2)^2} + m_3 \frac{\kappa_1 y - \kappa_2 x}{\kappa_1 - \kappa_2} + m_4 \\
+ m_5 & \frac{x - y}{\kappa_1 - \kappa_2} + m_6 \frac{(x - y)^2}{(\kappa_1 - \kappa_2)^2} + m_7 \frac{xy(x - y) - \kappa_1 x + \kappa_2 y}{\kappa_1 \kappa_2 (\kappa_1 - \kappa_2)} \\
+ & (x^2 - 2\kappa_1)(y^2 - 2\kappa_2) - \kappa_1^2 - \kappa_2^2,
\end{aligned}
$$

$$
g(x, y) = (x - y)(\kappa_2 x - \kappa_1 y) + (\kappa_1 - \kappa_2)^2.
$$

$$(24)$$

In fact, we have $f(\psi_{\kappa_1}(u), \psi_{\kappa_2}(u)) = U(u)$ and $g(\psi_{\kappa_1}(u), \psi_{\kappa_2}(u)) = 0$ where $U(u) = u^4 + \sum_{i=1}^7 m_i u^{4-i} + \frac{\kappa_1^2 \kappa_2^2}{u^4} = u^{-4} \prod_{i=1}^8 (u - v_i)$. Corresponding QRT map is given by involutions $\iota_1(x) = x'$ and $\iota_2(y) = y'$, where the rational functions $x' = x'(x, y)$, $y' = y'(x, y)$ are determined by

$$
\begin{aligned}
\frac{x' - \psi_{\kappa_1}(\frac{\kappa_2}{u})}{x' - \psi_{\kappa_1}(u)} \frac{x - \psi_{\kappa_1}(\frac{\kappa_2}{u})}{x - \psi_{\kappa_1}(u)} &= \frac{U(\frac{\kappa_2}{u})}{U(u)} \quad \text{for} \quad y = \psi_{\kappa_2}(u), \\
\frac{y' - \psi_{\kappa_2}(\frac{\kappa_1}{u})}{y' - \psi_{\kappa_2}(u)} \frac{y - \psi_{\kappa_2}(\frac{\kappa_1}{u})}{y - \psi_{\kappa_2}(u)} &= \frac{U(\frac{\kappa_1}{u})}{U(u)} \quad \text{for} \quad x = \psi_{\kappa_1}(u).
\end{aligned}
$$

$$(25)$$

This kind of simple expressions using auxiliary variable $u$ were first obtained in non-autonomous situation [41].

*Example 2.7 Autonomos $q$-$P(E_1^{(1)})$.* As a very degenerate case, Consider a function

$$
h = x + \frac{1}{y} + \frac{1}{xy} + ay.
$$

$$(26)$$

The corresponding involutions and the composition $T = \iota_1 \circ \iota_2$ are given by

$$\iota_1(x) = \frac{1}{xy}, \quad \iota_2(y) = \frac{1+x}{axy}, \tag{27}$$

$$T(x) = \frac{1}{xy}, \quad T(y) = \frac{1+xy}{ay}. \tag{28}$$

The following figure (Left) is a plot of the $10^4$ points in $(\log|x|, \log|y|)$ coordinates with initial condition $(x, y) = (3 + i, 4)$ and $a = 2$.

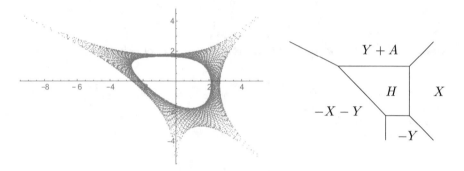

In the *ultra discrete limit* [30, 31, 58] $x = e^{\frac{X}{\epsilon}}$, $y = e^{\frac{Y}{\epsilon}}$, $a = e^{\frac{A}{\epsilon}}$, $h = e^{\frac{H}{\epsilon}}$ and $\epsilon \to 0$, we have a piecewise linear system $T(x) = -X - Y$, $T(Y) = \max(0, X + Y) - A - Y$, whose integral $H = \max(X, -Y, -X - Y, A + Y)$ gives a tropical curve (Right) [46].

## 2.6 "Integrable" Deautonomization

Many discrete Painlevé equations has been discovered by the *singularity confinement*, which is the discrete analog of the Painlevé test (see the review [11, 12, 50] and references therein). The singularity confinement demands that the singularity must disappear after some iteration steps.

*Example 2.8* Consider an deautonomization of the mapping in (28):

$$x_{n+1} = \frac{1}{x_n y_n}, \quad y_{n+1} = \frac{1 + x_n y_n}{a_n y_{n-1}}. \tag{29}$$

For initial value $(x_0, y_0)$ with $\epsilon = x_0 y_0 + 1 \to 0$, we observe successive singularities for $x_2, y_2, y_3, x_4, y_5, x_7, y_8, \cdots$. Since such infinite sequence of poles is a discrete analog of movable branch cut, it is natural to demand that the singularities will terminate after certain finite steps. This is the singularity confinement condition. In our case, we find

$$y_5 = -\frac{a_1(a_3^2 - a_2 a_4)}{x^2 a_2 a_3 a_5 \epsilon} + O(\epsilon^0), \tag{30}$$

and the pole in $y_5$ (and successive poles in $x_7, y_8, \cdots$) is cancelled iff $a_n = q^n a$. Thus we obtain deautonomized system

$$T(x) = \frac{1}{xy}, \quad T(y) = \frac{1+xy}{ay}, \quad T(a) = qa. \tag{31}$$

Though the system (31) does not admit an algebraic integral, it still has certain "integrability". For instance, the degrees of the rational functions $T^n(x)$ and $T^n(y)$ increase quadratic in $n$, similarly to the addition formula of elliptic functions [32]. We note that the system (31) is also given by the composition $T = \iota_1 \circ \iota_2$ of two involutions: $\iota_1 : x \mapsto \frac{1}{xy}, a \mapsto \frac{a}{q}, q \mapsto \frac{1}{q}$ and $\iota_2 : y \mapsto \frac{q(1+x)}{axy}, a \mapsto \frac{a}{q^2}, q \mapsto \frac{1}{q}$. Thus, the singularity confinement (or degree growth condition) provides a practical method to find "integrable" case and it was successfully applied to construct various discrete Painlevé equations. Similarly, "integrable" deautonomization of (22) gives the $q$-$P_{VI}$:

$$yT(y) = a_7 a_8 \frac{(x - a_1)(x - a_2)}{(x - a_3)(x - a_4)}, \quad xT^{-1}(x) = a_3 a_4 \frac{(y - a_5)(y - a_6)}{(y - a_7)(y - a_8)}, \tag{32}$$

where $T(a_1, \ldots, a_8) = (qa_1, qa_2, a_3, a_4, qa_5, qa_6, a_7, a_8)$ and $qa_1 a_2 a_7 a_8 = a_3 a_4 a_5 a_6$. This system was first obtained as the compatibility of linear $q$-difference equations [15].

## 2.7  Sakai Scheme

The Sakai's list [51] gives a classification of the surfaces associated with 2nd order (discrete) Painlevé equations where each surface is labeled by a root system representing the intersection of the components of the curve (or divisors) characterizing the configuration.

elliptic    $A_0^{(1)}$

$$A_7^{(1)'}$$

multiplicative $A_0^{(1)} \to A_1^{(1)} \to A_2^{(1)} \to A_3^{(1)} \to A_4^{(1)} \to A_5^{(1)} \to A_6^{(1)} \to A_7^{(1)} \to A_8^{(1)}$

additive   $A_0^{(1)} \to A_1^{(1)} \to A_2^{(1)} \qquad \to \qquad D_4^{(1)} \to D_5^{(1)} \to D_6^{(1)} \to D_7^{(1)} \to D_8^{(1)}$

$$E_6^{(1)} \to E_7^{(1)} \to E_8^{(1)}$$

Sakai's list (labeled by surface type)

For each surface, there exist an action of affine Weyl group given as follows:

elliptic    $E_8^{(1)}$

$$E_1^{(1)'}$$

multiplicative $E_8^{(1)} \to E_7^{(1)} \to E_6^{(1)} \to E_5^{(1)} \to E_4^{(1)} \to E_3^{(1)} \to E_2^{(1)} \to E_1^{(1)} \to E_0^{(1)}$

additive   $E_8^{(1)} \to E_7^{(1)} \to E_6^{(1)} \qquad \to \qquad D_4^{(1)} \to D_3^{(1)} \to D_2^{(1)} \to D_1^{(1)} \to D_0^{(1)}$

$$A_2^{(1)} \to A_1^{(1)} \to A_0^{(1)}$$

where $E_5, \ldots, E_0$ means $D_5, A_4, A_2+A_1, A_1+A_1, A_1, A_0$, and $D_3, \ldots, D_0$ means $A_3, A_1+A_1, A_1, A_0$.

Sakai's list (labeled by symmetry type)

The Eqs. (22), (25) and (28) are of multiplicative type corresponding to the surface type $A_3^{(1)}, A_0^{(1)}, A_7^{(1)}$ and the symmetry type $D_5^{(1)}, E_8^{(1)}, E_1^{(1)}$ respectively.

## 2.8 Short Summary on Cubic Curve and Elliptic Function

Geometrically, the three classes elliptic/multiplicative/additive correspond to the group structure of the associated cubic curve. We will give a quick summary on this. Consider a plane cubic curve

$$C : F(x, y) = 0, \tag{33}$$

where $F(x, y)$ is a polynomial in $x, y$ of degree 3; a linear combination of ten monomials $x^i y^j$ ($i + j \leq 3$). The most important property of the cubic is its Abelian group structure defined as follows. Chose a point, say $O \in C$, which will play the role of the unit element of the group. For given points $P, Q \in C$, consider a line $\ell_{PQ}$ passing through them (if $P = Q$, the line $\ell_{PQ}$ means the tangent of $C$ at $P$).

Since $C$ is a cubic, it intersect with the line $\ell_{PQ}$ at three points: $P$, $Q$ and the third one denoted by $P * Q$. Then the sum $P + Q$ and the inverse element $-P$ is defined by $P + Q = (P * Q) * O$ and $-P = P * (O * O)$. It is easy to see that $P + O = P$, $P + (-P) = O$, $P + Q = Q + P$. Moreover

**Proposition 2.9** *The associativity* $(Q + P) + R = Q + (P + R)$ *holds.*

*Proof* It is enough to show that $(P + Q) * R = (P + R) * Q$, namely, the two lines $f_3 = 0$, $g_1 = 0$ and the cubic $C : F(x, y) = 0$ meat at a point $X$. Here $f_i = 0$ and $g_i = 0$ are equations of lines defined as in the figure.

Consider eight points $O$, $P$, $Q$, $R$, $P * Q$, $P * R$, $P + Q$, $P + R$ on $C$. In generic case, a cubic polynomial vanishes at the eight points has $10 - 8 = 2$ free parameters. For such polynomials, we have two distinguished basis $f_1 f_2 f_3$ and $g_1 g_2 g_3$. Hence $F(x, y) = \lambda f_1 f_2 f_3 + \mu g_1 g_2 g_3$. ($\lambda, \mu \in \mathbb{C}$) and we have $F(X) = 0$ for $X = (f_3 = 0) \cap (g_1 = 0)$ as desired.                    □

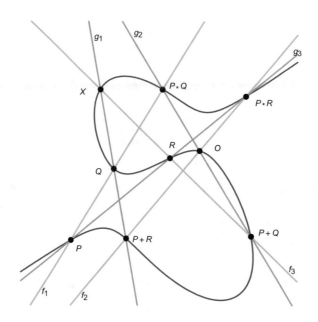

Though the proof is well known but we presented it here since the argument is useful later. This proof shows that any cubic passing through generic eight points also passes additional ninth point. This *ninth point* plays important role in the elliptic Painlevé equation.

The special case where the cubic degenerates to a conic and a line is known as Pascal's *mystic hexagon theorem*: " For any six points $p_1, \ldots, p_6$ on a conic, three intersections $P = \ell_{p_1 p_2} \cap \ell_{p_4 p_5}$, $Q = \ell_{p_2 p_3} \cap \ell_{p_5 p_6}$ and $R = \ell_{p_3 p_4} \cap \ell_{p_6 p_1}$ are co-linear". Further special case where the conic degenerates to two lines is due to Pappus.

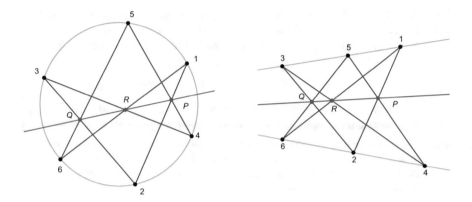

Introduce a parameterization $x = x(t)$, $y = y(t)$ of the cubic curve $C$ : $F(x, y) = 0$ by the solution of the differential equation:

$$\frac{dx}{dt} = \frac{\partial F}{\partial y}, \quad \frac{dy}{dt} = -\frac{\partial F}{\partial x}. \tag{34}$$

Then the group structure on $C$ can be interpreted as the standard addition on parameter $t \in \mathbb{C}$. Namely

**Proposition 2.10** *If three points $p_i = (x(t_i), y(t_i))$ $(i = 1, 2, 3)$ are on a line $\ell$, then the sum $t_1 + t_2 + t_3$ is a constant independent of $\ell$.*

The following geometric proof is due to H. Abel [62, p442].

***Proof*** Let $\ell : y = ax + b$ be the line passing through $p_1, p_2, p_3$. We will show that the sum $t_1 + t_2 + t_3$ is invariant under the deformation of the coefficients $a, b$. Since $x_1, x_2, x_3$ are the roots of $F(x, ax + b)$, we have an identity $F(x, ax + b) = k(x - x_1)(x - x_2)(x - x_3)$. By a deformation $\delta a$, $\delta b$

$$\frac{\partial F}{\partial y}(x, ax+b)(x\delta a+\delta b) = \delta k \prod_{i=1}^{3}(x-x_i) - k\{\delta x_1(x-x_2)(x-x_3)+\text{cyclic}\}, \tag{35}$$

and putting $x = x_1$ we have $\dfrac{\partial F}{\partial y}(p_1)(x_1\delta a + \delta b) = -k\delta x_1(x_1 - x_2)(x_1 - x_3)$. Together with the similar relations for $x = x_2, x_3$, we obtain the desired result

$$-k\sum_{i=1}^{3}\delta t_i = -k\sum_{i=1}^{3}\frac{\delta x_i}{\dfrac{\partial F}{\partial y}(p_i)} = \frac{x_1\delta a + \delta b}{(x_1 - x_2)(x_1 - x_3)} + \text{cyclic} = 0. \tag{36}$$

The last equality follows from the Lagrange's interpolation identity.                    □

In generic case, the cubic $C$ is a smooth elliptic curve and the parametrization $x(t)$, $y(t)$ are given by elliptic functions. For instance, when $F(x, y) = y^2 - (4x^3 - g_2 x - g_3)$, one can parametrize it as $x(t) = \wp(t)$, $y(t) = \wp'(t)$ where $\wp(t) = \wp(t; g_2, g_3)$ is the Weierstrass $\wp$ function. By the construction, we have the addition formulae

$$\wp(t_1 + t_2) = -\wp(t_1) - \wp(t_2) + \frac{1}{4} \left\{ \frac{\wp'(t_1) - \wp'(t_2)}{\wp(t_1) - \wp(t_2)} \right\}^2,$$

$$\wp'(t_1 + t_2) = -\frac{\wp'(t_1) - \wp'(t_2)}{\wp(t_1) - \wp(t_2)} \{\wp(t_1 + t_2) - \wp(t_1)\} - \wp'(t_1). \tag{37}$$

When the cubic $C$ become singular (nodal: $g_2^3 - 27g_3^2 = 0$ or cuspidal: $g_2 = g_3 = 0$), the parametrization $x(t)$, $y(t)$ degenerate to trigonometric or rational functions.

To give the explicit expression of the elliptic functions, it is convenient to introduce holomorphic quasi-periodic functions. The simplest one is

$$\vartheta(z) = (z)_\infty (q/z)_\infty, \quad (a)_\infty = \prod_{i=0}^{\infty} (1 - aq^i), \tag{38}$$

which satisfies $\vartheta(qz) = \vartheta(z^{-1}) = -z^{-1}\vartheta(z)$. Any elliptic function $f(x)$ of degree $d$ with periods 1 and $\tau$ can be written as $f(x) = c \prod_{i=1}^{d} \frac{\vartheta(a_i z)}{\vartheta(b_i z)}$, where $\prod_{i=1}^{d} \frac{a_i}{b_i} = 1$, $z = e^{2\pi i x}$, $q = e^{2\pi i \tau}$. An important identity for $\vartheta$-function is the Riemann relation:

$$a^{-1}\vartheta(ab)\vartheta(a/b)\vartheta(cz)\vartheta(z/c) + (abc \text{ cyclic}) = 0, \tag{39}$$

which can be proved by using the fact that $\frac{\text{(LHS)}}{\vartheta(cz)\vartheta(z/c)}$ is a holomorphic elliptic function in $x$ hence a constant. The extra factor such as $a^{-1}$ can be absorbed if we use $[z] = z^{-\frac{1}{2}}\vartheta(z)$ which is odd in $x$ (multiplicatively odd in $z$: $[\frac{1}{z}] = -[z]$).

## 3 Affine Weyl Groups

### 3.1 Affine Weyl Group and Its Translations

Let $A = (a_{ij})_{i,j \in I}$ be a Cartan matrix, namely

$$a_{ii} = 2, \quad a_{ij} \in \mathbb{Z}_{\leq 0}, \quad a_{ij} = 0 \Leftrightarrow a_{ji} = 0 \quad (i \neq j). \tag{40}$$

Define a linear actions $s_i$ $(i \in I)$ on parameters $\{\alpha_i (i \in I)\}$ as

$$s_i : \alpha_j \to \alpha_j - \alpha_i a_{ij}. \tag{41}$$

These are involutions $s_i^2 = 1$ and satisfy the relations $(s_i s_j)^{m_{ij}} = 1$ where $m_{ij} = 2, 3, 4, 6$ for $a_{ij} a_{ji} = 0, 1, 2, 3$ respectively. This means that the actions $s_i$ give a representation of Weyl group $W(A)$ associated to the Cartan matrix $A$ (or associated root system).

Let $V$ be a vector space with an inner product $\langle *, * \rangle$ and $\alpha_i (i \in I)$ are basis of $V$ such that $\frac{2\langle \alpha_i, \alpha_j \rangle}{\langle \alpha_i, \alpha_i \rangle} = a_{ij}$. Then the reflections $r_{\alpha_i}(x) = x - \frac{2\langle \alpha_i, x \rangle}{\langle \alpha_i, \alpha_i \rangle}\alpha_i$ w.r.t. the hyperplane orthogonal to $\alpha_i$ give a linear representation of the Weyl group $W(A)$. The Eq. (41) arises in this way, by viewing $\alpha_i$ as variables.

For the 2nd order (discrete) Painlevé equations, the relevant Weyl groups are of affine ADE type. They are given by the following Cartan matrix $(a_{ij})$ with index set $I = \{0, 1, \ldots, n\}$:

$$
A_{n(\geq 2)}^{(1)} : \begin{bmatrix} 2 & -1 & & & -1 \\ -1 & 2 & -1 & & \\ & \ddots & \ddots & \ddots & \\ & & -1 & 2 & -1 \\ -1 & & & -1 & 2 \end{bmatrix}, \qquad A_1^{(1)} : \begin{bmatrix} 2 & -2 \\ -2 & 2 \end{bmatrix},
$$

$$
D_{n(\geq 4)}^{(1)} : \begin{bmatrix} 2 & & -1 & & & & \\ & 2 & -1 & & & & \\ -1 & -1 & 2 & -1 & & & \\ & & \ddots & \ddots & \ddots & & \\ & & & -1 & 2 & -1 & -1 \\ & & & & -1 & 2 & \\ & & & & -1 & & 2 \end{bmatrix}, \qquad E_{n(=6,7,8)}^{(1)} : \begin{bmatrix} 2 & & & -1 & & & & \\ & 2 & -1 & & & & & \\ & -1 & 2 & -1 & & & & \\ -1 & & -1 & 2 & -1 & & & \\ & & & -1 & 2 & -1 & & \\ & & & & \ddots & \ddots & \ddots & \\ & & & & & -1 & 2 & -1 \\ & & & & & & -1 & 2 \end{bmatrix}.
$$

$$(42)$$

An important fact for us is that the affine Weyl group contains a translation subgroup.

*Example 3.1* The case $W(A_2^{(1)})$ corresponds to the following Coxeter-Dynkin diagram:

The action of $s_i$ is given by $s_i(\alpha_{i \pm 1}) = \alpha_{i \pm 1} + \alpha_i$, $s_i(\alpha_i) = -2\alpha_i$, where indices $\{0, 1, 2\}$ are considered as in $\mathbb{Z}/(3\mathbb{Z})$. $W(A_2^{(1)})$ can be extended to $\tilde{W}(A_2^{(1)})$ by including the diagram automorphisms $\pi(\alpha_0, \alpha_1, \alpha_2) = (\alpha_1, \alpha_2, \alpha_0)$. The translations $T_1 = s_2 s_1 \pi$, $T_2 = s_1 \pi s_2$ and $T_0 = \pi s_2 s_1 \in \tilde{W}(A_2^{(1)})$ act as

$$T_1(\alpha_0, \alpha_1, \alpha_2) = (\alpha_0, \alpha_1, \alpha_2) + (-1, 0, 1)\delta,$$

$$T_2(\alpha_0, \alpha_1, \alpha_2) = (\alpha_0, \alpha_1, \alpha_2) + (0, 1, -1)\delta, \qquad (43)$$

$$T_0(\alpha_0, \alpha_1, \alpha_2) = (\alpha_0, \alpha_1, \alpha_2) + (1, -1, 0)\delta,$$

where $\delta = \alpha_0 + \alpha_1 + \alpha_2$. The translations $T_1, T_2, T_0 = (T_1 T_2)^{-1}$ generate the weight lattice, and the translations in $W(A_2^{(1)})$ generated without using $\pi$ (such as $T_1 T_2^{-1} = s_2 s_1 s_0 s_1$) generate the root lattice.

*Example 3.2* The case $W(E_8^{(1)})$ corresponding to the following Coxeter-Dynkin diagram:

$$\qquad (44)$$

We have no diagram automorphism and the translations generate the root lattice. There are 120 shortest directions (and their inverses), all of them can be given as product of 58 simple reflections such as

$$T_{78} = s_7 s_6 s_5 s_4 s_3 s_0 s_2 s_1 s_3 s_2 s_4 s_3 s_0 s_5 s_4 s_3 s_2 s_1 s_6 s_5 s_4 s_3 s_2 s_8 s_7 s_6 s_5 s_4 s_3$$
$$\times s_0 s_3 s_2 s_1 s_4 s_3 s_2 s_5 s_4 s_3 s_0 s_6 s_5 s_4 s_3 s_2 s_7 s_6 s_5 s_4 s_3 s_0 s_1 s_2 s_3 s_4 s_5 s_6 s_8. \qquad (45)$$

The action of $T_{78}$ on variables $\alpha_i$ is

$$T_{78}(\alpha_0, \alpha_1, \ldots, \alpha_8) = (\alpha_0, \alpha_1, \ldots, \alpha_8) + (0, \ldots, 0, -1, 2, -1)\delta. \qquad (46)$$

where $\delta = 3\alpha_0 + 2\alpha_1 + 4\alpha_2 + 6\alpha_3 + 5\alpha_4 + 4\alpha_5 + 3\alpha_6 + 2\alpha_7 + \alpha_8$.

## 3.2 Birational Representations

Many discrete Painlevé equations are obtained as the translation part of affine Weyl groups. Nontrivial equations arise from birational representations. Typical examples of such representations are given as the symmetry (Bäcklund transformation) of differential Painlevé equations. The equations in the diagram (1) have the following symmetry:

$$\tilde{W}(D_4^{(1)}) \rightarrow \tilde{W}(A_3^{(1)}) \rightarrow \tilde{W}((A_1 + A_1)^{(1)}) \rightarrow \tilde{W}(A_1^{(1)}) \rightarrow \mathfrak{S}_2$$
$$\searrow \qquad\qquad \searrow \qquad\qquad \searrow \qquad \qquad . \qquad (47)$$
$$\tilde{W}(A_2^{(1)}) \qquad\qquad \rightarrow \tilde{W}(A_1^{(1)}) \rightarrow \{1\}$$

*Example 3.3 ($\tilde{W}(A_2^{(1)})$ Symmetry of $P_{IV}$)* Consider the equation in the form (4)

$$\delta q' = q(2p - q - t) - \alpha_1, \quad \delta p' = p(2q + t - p) + \alpha_2, \quad t' = \delta, \quad \alpha_i' = 0. \quad (48)$$

On the field of rational functions $K = \mathbb{C}(q, p, t, \alpha_0, \alpha_1, \alpha_2)$, the Eq. (48) define a derivation (a linear map satisfying the Leibniz rule). In this setting, a Bäcklund transformation can be simply described as an algebra homomorphism $w$ on the field $K$ preserving the derivation: $\{w(f)\}' = w(f')$. Such transformations are given as follows

$$s_1(p) = p - \frac{\alpha_1}{q}, \quad s_1(q) = q, \quad s_1(\alpha_1) = -\alpha_1, \quad s_1(\alpha_2) = \alpha_1 + \alpha_2,$$
$$\pi(p) = -p + q + t, \quad \pi(q) = -p, \quad \pi(\alpha_1) = \alpha_2, \quad \pi(\alpha_2) = \delta - \alpha_1 - \alpha_2.$$
$$(49)$$

Putting $s_2 = \pi s_1 \pi^{-1}$, $s_0 = \pi s_2 \pi^{-1}$, we have

$$s_i^2 = \pi^3 = 1, \quad s_i s_{i+1} s_i = s_{i+1} s_i s_{i+1}, \quad \pi s_i = s_{i+1} \pi. \quad (i \in \mathbb{Z}/(3\mathbb{Z})) \quad (50)$$

This means that the actions (49) give a birational representation of the affine Weyl group $W(A_2^{(1)})$ (or extended affine Weyl group $\tilde{W}(A_2^{(1)})$) generated by $s_1, s_2$ (and $\pi$).

For the representation $\tilde{W}(A_2^{(1)})$ in Eq. (49), the translation $T_2 = s_1 \pi s_2$ gives the following birational map (a Schlesinger transformation of $P_{IV}$):

$$T_2(\alpha_1) = \alpha_1 + \delta, \quad T_2(\alpha_2) = \alpha_2 - \delta,$$
$$T_2(p) = -p + q + t + \frac{\alpha_1}{q}, \quad T_2^{-1}(q) = p - q - t - \frac{\alpha_2}{p}. \quad (51)$$

As a discrete mapping, this system (51) is known as one of the $d$-$P_{II}$ equations. Its autonomization ($\delta \rightarrow 0$) is an integrable mapping where the Hamiltonian $H_{IV}$ is the integral. In fact, the Eq. (51) (for $\delta = 0$) can be obtained as the QRT system associated to the curve $H_{IV} = C$.

There are at least two general methods[3] to construct birational representations of the Weyl groups, namely the actions (1) on Borel subgroups or (2) on point configurations. We will explain these in the following subsections.

## 3.3  Weyl Group Actions on Borel Subgroups

Let $M = (m_{i,j})$ be an $n \times n$ upper triangular matrix: i.e. $m_{i,j} = 0$, $(i > j)$. We put $G_i(u) = 1 + uE_{i+1,i}$ where $E_{i,j}$ is the matrix unit with $(k, l)$-element $\delta_{ik}\delta_{jl}$. The matrix

$$\tilde{M} = G_i(u)MG_i(u)^{-1}, \tag{52}$$

is upper triangular if $u = 0$ or $u = \frac{m_{i,i} - m_{i+1,i+1}}{m_{i,i+1}}$. In the latter case, the transformation $M \mapsto \tilde{M}$ defines a birational automorphism $s_i$ on the field $K = \mathbb{C}(\{m_{i,j}\}_{1 \le i < j \le n})$. We can check that $s_1, \cdots, s_{n-1}$ give a birational representation of the Weyl group $W(A_{n-1}) = \mathfrak{S}_n$. This construction can be generalized to affine Weyl group $\tilde{W}(A_{n-1}^{(1)})$ as follows.

Consider a $n \times n$ matrix $M(z) = (m_{i,j}(z))$ of functions in $z$. We call $M(z)$ affine upper triangular if $M_{i,j}(z)$ is regular at $z = 0$ and $m_{i,j}(0) = 0$ for $i > j$, namely the expansion of $M(z)$ at $z = 0$ takes the form

$$M(z) = M_0 + M_1 z + M_2 z^2 + \cdots \tag{53}$$

where $M_0$ is upper triangular. The actions $s_1, s_2, \ldots, s_n$ are defined in the same way as above. To define $s_0$, we put $G_0(u) = 1 + \frac{u}{z}E_{1,n}$. Then $\tilde{M}(z) = G_i(u)M(z)G_i(u)^{-1}$ is affine upper triangular if $u = 0$ or $u = \frac{(M_0)_{n,n} - (M_0)_{1,1}}{(M_1)_{n,1}}$, and for the latter case the transformation $M(z) \to \tilde{M}(z)$ gives the action of $s_0$. Similar construction for general Kac-Moody algebras are given in [40].

For $x = (x_1, x_2, \cdots, x_n) \in \mathbb{C}^n$, consider $n \times n$ matrix of the form

$$X(x) = \begin{bmatrix} x_1 & 1 & & & \\ & x_2 & 1 & & \\ & & \ddots & \ddots & \\ & & & x_{n-1} & 1 \\ z & & & & x_n \end{bmatrix}. \tag{54}$$

---

[3]Recently a method using the cluster algebra is also developed. See [4, 44] for example.

The matrix has the exchange relation [63]

$$X(x)X(y) = X(x')X(y'),$$  (55)

where

$$x_i' = y_i \frac{Q_{i-1}}{Q_i}, \quad y_i' = x_i \frac{Q_i}{Q_{i-1}}, \quad Q_i = \sum_{k=1}^{n}\Big(\prod_{a=1}^{k-1} x_{i+a} \prod_{a=i+1}^{n} y_{i+a}\Big),$$  (56)

$x_{i+n} = x_i, y_{i+n} = y_i$. Note that this transformation is subtraction-free and hence it has combinatorial analog through ultra-discretization "$+ \to \max, \times \to +$".[30, 31, 58]

On the product

$$M(z) = X_1 X_2 \cdots X_m, \quad X_i = X(x_{i1}, x_{i2}, \ldots, x_{in}),$$  (57)

we have two affine Weyl group actions: $\tilde{W}(A_{m-1}^{(1)})$ and $\tilde{W}(A_{m-1}^{(1)})$. The first one comes from the exchange of the factors $X_i$, and the second one arises from $M \to G_j M G_j^{-1}$ actions explained above. These actions have the following interesting properties.

- The actions of $\tilde{W}(A_{m-1}^{(1)})$ and $\tilde{W}(A_{m-1}^{(1)})$ are commutative.
- The spectral curve $f(\lambda, z) = |\lambda - M(z)| = 0$ is invariant under the actions of $\tilde{W}(A_{m-1}^{(1)})$ and $\tilde{W}(A_{m-1}^{(1)})$.
- Under the transformation $x_{ij} \leftrightarrow x_{ji}$, the actions of $\tilde{W}(A_{m-1}^{(1)})$ and $\tilde{W}(A_{m-1}^{(1)})$ are transformed with each other.
- The combinatorial analog [38] has an interpretation in $B_{\ell_1} \otimes \cdots \otimes B_{\ell_m}$ where $B_\ell$ is the $A_{n-1}^{(1)}$-crystal with single low.[4]

In [17, 18], the corresponding discrete dynamical systems and their deautonomization was studied. The case $(m, n) = (2, 3)$ or $(3, 2)$ gives the $q$-$P_{IV}$ and others are considered as certain generalization of it. Recently a slightly modified version of $(m, n) = (2N + 2, 2)$ cases are studied in relation with the $q$-Garnier system of $2N$ variables.[37]

*Example 3.4 The Case $N = 1$, $(m, n) = (4, 2)$.* Consider the following $2 \times 2$ Lax pair

$$\Psi(qz) = A(z)\Psi(z), \quad T(\Psi(z)) = B(z)\Psi(z),$$  (58)

---

[4]For the application to integrable systems, see the review [14] and references therein.

where

$$A(z) = \frac{1}{(z - a_1)(z - a_3)} \begin{bmatrix} d_1 & 0 \\ 0 & d_2 q \end{bmatrix} \begin{bmatrix} x_1 & 1 \\ z & \frac{a_1}{x_1} \end{bmatrix} \begin{bmatrix} x_2 & 1 \\ z & \frac{a_1}{x_2} \end{bmatrix} \begin{bmatrix} x_3 & 1 \\ z & \frac{a_1}{x_3} \end{bmatrix} \begin{bmatrix} x_4 & 1 \\ z & \frac{a_1}{x_4} \end{bmatrix},$$

$$B(z) = \frac{1}{(z - a_3)} \begin{bmatrix} x_3 & 1 \\ z & \frac{a_1}{x_3} \end{bmatrix} \begin{bmatrix} x_4 & 1 \\ z & \frac{a_1}{x_4} \end{bmatrix}.$$

$$(59)$$

The compatibility $T(A(z))B(z) = B(qz)A(z)$ gives the $q$-$P_{VI}$ equation

$$T : (a_1, a_2, a_3, a_4, d_1, d_2, d_3, d_4) \mapsto (a_1, a_2, a_3/q, a_4/q, d_1, d_2, qd_3, qd_4),$$
$$T^{-1}(g)g = d_1 d_2 \frac{(f - a_1)(f - a_2)}{(f - a_3)(f - a_4)}, \quad T(f)f = \frac{a_3 a_4}{q} \frac{(g - d_3)(g - d_4)}{(g - d_1)(g - d_2)}.$$

$$(60)$$

Where variables $x_1, \ldots, x_4$ and $d_3, d_4, f, g, w$ ($\frac{a_1 a_2 d_1 d_2}{a_3 a_4 d_3 d_4} = q$) are related by

$$\lim_{z \to 0} A(z) = \frac{a_4}{a_1} \begin{bmatrix} d_3 & * \\ 0 & d_4 \end{bmatrix}, \quad A(z)_{12} = \frac{w d_1 (z - f)}{(z - a_1)(z - a_3)}, \quad A(f)_{11} = g \frac{f - a_4}{f - a_1}.$$

$$(61)$$

## 3.4 Weyl Group Actions on Point Configurations

Consider configurations of $n$ points $p_1, \ldots, p_n$ on projective space $\mathbb{P}^{m-1}$. A naive moduli space of the configuration is given by the coset

$$\mathcal{M}_{m,n} = \mathrm{GL}(m) \setminus \left\{ \begin{bmatrix} x_{11} & x_{12} & \cdots & x_{1n} \\ x_{21} & x_{22} & \cdots & x_{2n} \\ \vdots & \vdots & \cdots & \vdots \\ x_{m1} & x_{m2} & \cdots & x_{mn} \end{bmatrix} \right\} / (\mathbb{C}^\times)^n, \qquad (62)$$

where the multiplications of $\mathrm{GL}(m)$ and $(\mathbb{C}^\times)^n$ correspond the projective transformation and the scaling of the homogeneous coordinates $p_j = (x_{1j}, x_{2j}, \ldots, x_{mj})^T$. A generic configuration $[p_1, \cdots, p_n] \in \mathcal{M}_{m,n}$ can be represented by the following canonical form:

$$[p_1, \cdots, p_n] \simeq \begin{bmatrix} 1 & \cdots & 0 & 0 & 1 & u_{1,m+2} & \cdots & u_{1,n} \\ \vdots & \ddots & \vdots & \vdots & \vdots & \vdots & \cdots & \vdots \\ 0 & \cdots & 1 & 0 & 1 & u_{m-1,m+2} & \cdots & u_{m-1,n} \\ 0 & \cdots & 0 & 1 & 1 & 1 & \cdots & 1 \end{bmatrix}, \qquad (63)$$

where the first $m$ by $m$ block is the identity matrix and

$$u_{i.j} = \frac{d_{i,j}d_{m,m+1}}{d_{m,j}d_{i,m+1}}, \quad d_{i,j} = \det[p_1, \cdots, \widehat{p_i}, \cdots, , p_m, p_j]. \tag{64}$$

The formula for $u_{i,j}$ follows from its invariance under the $GL(m) \times (\mathbb{C}^\times)^n$-action.

Let $W_{m,n}$ be the Weyl group generated by involutions $s_0, s_1, \ldots, s_n$ with fundamental relations given by the following Coxeter-Dynkin diagram

$$\tag{65}$$

$$
\begin{array}{c}
s_0 \\
\bullet \\
| \\
\bullet - \bullet - \cdots - \bullet - \bullet - \cdots - \bullet \\
s_1 \quad s_2 \qquad s_m \quad s_{m+1} \qquad s_{n-1}
\end{array}
$$

**Theorem 3.5 ([7])** *There is a natural birational action of $W_{m,n}$ on $\mathcal{M}_{m,n}$.*

**Proof** For $i = 1, \ldots, n-1$, the action $s_i$ is given by the permutation of two points $p_i$, $p_{i+1}$. The action $s_0$ is given by the standard Cremona transformation centered at $p_1, \ldots, p_m$. Their explicit actions in coordinates $u_{i,j}$ are given as follows.

$$s_0 : u_{i,j} \mapsto \frac{1}{u_{i,j}},$$

$$s_i : u_{i,j} \leftrightarrow u_{i+1,j} \quad (1 \le i \le m-2),$$

$$s_{m-1} : u_{m-1,j} \mapsto \frac{1}{u_{m-1,j}}, \quad u_{i,j} \mapsto \frac{u_{i,j}}{u_{m-1,j}} \quad (1 \le i \le m-2), \tag{66}$$

$$s_m : u_{i,j} \mapsto 1 - u_{i,j},$$

$$s_{m+1} : u_{i,m+2} \mapsto \frac{1}{u_{i,m+2}}, \quad u_{i,j} \mapsto \frac{u_{i,j}}{u_{i,m+2}} \quad (m+3 \le j \le n),$$

$$s_j : u_{i,j} \leftrightarrow u_{i,j+1} \quad (m+2 \le j \le n-1),$$

where $1 \le i \le m-1, m+2 \le j \le n$ unless otherwise stated. The Coxeter relations can be checked directly.                                                                                        □

The birational action (66) can be linearized under certain specialization. Define linear actions $s_i \in W_{m,n}$ on a set of parameters $\epsilon_0, \ldots, \epsilon_n$ as

$$s_0 : \epsilon_i \mapsto \epsilon_i + \alpha_0 \quad (0 \le i \le m), \tag{67}$$

$$s_i : \epsilon_i \leftrightarrow \epsilon_{i+1} \quad (1 \le i \le n-1),$$

where $\alpha_0 = (m-2)\epsilon_0 - \sum_{i=1}^{m} \epsilon_i$. Then we have

**Proposition 3.6 ([23])** *The actions* (66), (67) *are compatible with the following parametrization*

$$u_{i,j} = \frac{[\alpha_0 + \epsilon_{m,m+1}][\epsilon_{i,m+1}][\alpha_0 + \epsilon_{i,j}][\epsilon_{m,j}]}{[\alpha_0 + \epsilon_{i,m+1}][\epsilon_{m,m+1}][\alpha_0 + \epsilon_{m,j}][\epsilon_{i,j}]}, \quad (1 \le i \le m-1, m+2 \le j \le n),$$

(68)

*where $\epsilon_{i,j} = \epsilon_i - \epsilon_j$ and $[x]$ is the odd theta function.*

**Proof** The compatibility for $s_i$ ($i \neq m$) can be checked only by the oddness $[-x] = -[x]$. The compatibility for $s_m$ is satisfied thanks to the Riemann relation:

$$\begin{aligned}
&[x_1 - x_2][x_3 - x_4][a - x_1 - x_2][a - x_3 - x_4] \\
&-[x_1 - x_3][x_2 - x_4][a - x_1 - x_3][a - x_2 - x_4] \\
&+[x_1 - x_4][x_2 - x_3][a - x_1 - x_4][a - x_2 - x_3] = 0.
\end{aligned}$$

(69)

where $x_1 = \epsilon_i, x_2 = \epsilon_m, x_3 = \epsilon_{m+1}, x_4 = \epsilon_j, a = \alpha + \epsilon_i + \epsilon_m$. $\qquad \square$

## 3.5 The Elliptic Painlevé Equation in $E_{10}$ Form

Consider the case of $(m, n) = (3, 10)$. We take variables $f_i, g_i$ ($i = 5, \ldots, 10$) as

$$[p_1, \cdots, p_{10}] \simeq \begin{bmatrix} 1 & 0 & 0 & 1 & f_5 & \cdots & f_{10} \\ 0 & 1 & 0 & 1 & g_5 & \cdots & g_{10} \\ 0 & 0 & 1 & 1 & 1 & \cdots & 1 \end{bmatrix}.$$

(70)

Then the following actions $s_0, \ldots, s_9$ on $\mathbb{C}(f_5, g_5, \ldots, f_{10}, g_{10})$

$$\begin{aligned}
s_0 &: \quad f_i \to \frac{1}{f_i}, \quad g_i \to \frac{1}{g_i}, \\
s_1 &: \quad f_i \to g_i, \quad g_i \to f_i, \\
s_2 &: \quad f_i \to \frac{f_i}{g_i}, \quad g_i \to \frac{1}{g_i}, \\
s_3 &: \quad f_i \to 1 - f_i, \quad g_i \to 1 - g_i, \\
s_4 &: \quad f_5 \to \frac{1}{f_5}, \quad g_5 \to \frac{1}{g_5}, \quad f_i \to \frac{f_i}{f_5}, \quad g_i \to \frac{g_i}{g_5} \quad (i \ge 6), \\
s_j &: \quad f_{j+1} \leftrightarrow f_j, \quad g_{j+1} \leftrightarrow g_j \qquad (j = 5, \ldots, 9).
\end{aligned}$$

(71)

give a birational representation of the Coxeter-Weyl group $W_{3,10} = W(E_{10})$:

$$s_0$$
$$\bullet$$
$$\vert$$
$$\tag{72}$$
$$\bullet - \bullet - \bullet - \bullet - \cdots - \bullet$$
$$s_1 \quad s_2 \quad s_3 \quad s_4 \qquad\quad s_9$$

In the context of the discrete Painlevé equation [35], the points $p_1, \ldots, p_9 \in \mathbb{P}^2$ play the role of parameters, while the point $p_{10}$ corresponds to the unknown variables. In view of this, we will consider the affine Weyl group $W(E_8^{(1)}) \subset W(E_{10})$ generated by $s_0, \ldots, s_8$ omitting $s_9$.

*Example 3.7* Consider the Translation $T_{78} \in W(E_8^{(1)})$ in (45). Since some of the expressions of $w(f_i), w(g_i) \in \mathbb{C}(f_5, g_5, \ldots, f_{10}, g_{10})$ are too big, we give their numerical values. For instance, specializing to

$$\begin{bmatrix} f_5 & \cdots & f_{10} \\ g_5 & \cdots & g_{10} \end{bmatrix} = \begin{bmatrix} 7 & 9 & 5 & 6 & 2 & 3 \\ 2 & 3 & 4 & 9 & 11 & 8 \end{bmatrix}, \tag{73}$$

we have

$$\begin{bmatrix} w(f_5) & \cdots & w(f_{10}) \\ w(g_5) & \cdots & w(g_{10}) \end{bmatrix} = \begin{bmatrix} 7 & 9 & \frac{25726500}{34417211} & \frac{15539377}{667873009} & 2 & \frac{219878505}{1788517354} \\ 2 & 3 & \frac{66717390}{7202521} & \frac{38499785890}{4540144829} & 11 & \frac{364794359225}{40085905392} \end{bmatrix}. \tag{74}$$

From (68), the parametrization of the variables $f_i, g_i$ $(i > 4)$ is given by

$$f_i = \frac{[\epsilon_{23i}][\epsilon_{14}][\epsilon_{124}][\epsilon_{3i}]}{[\epsilon_{234}][\epsilon_{1i}][\epsilon_{12i}][\epsilon_{34}]}, \quad g_i = \frac{[\epsilon_{13i}][\epsilon_{24}][\epsilon_{124}][\epsilon_{3i}]}{[\epsilon_{134}][\epsilon_{2i}][\epsilon_{12i}][\epsilon_{34}]}, \tag{75}$$

where $\epsilon_{ijl} = \epsilon_0 - \epsilon_i - \epsilon_j - \epsilon_k$ and $\epsilon_{ij} = \epsilon_i - \epsilon_j$. This is equivalent to

$$p_i = \left( \frac{[\epsilon_{23i}]}{[\epsilon_{1i}]}, \frac{[\epsilon_{13i}]}{[\epsilon_{2i}]}, \frac{[\epsilon_{12i}]}{[\epsilon_{3i}]} \right). \tag{76}$$

This parametrization can be used for $p_1, \ldots, p_9$ without loss of generality, since there exist unique cubic curve $C_0$ passing through generic 9 points. However, the parametrization for $p_{10}$ is available only for the special case such as $p_{10} \in C_0$.

Note that $T_{78}$ in (45) can be written as

$$T_{78} = s_{78}s_{127}s_{347}s_{569}s_{347}s_{127},$$
$$s_{ij} = \{\epsilon_i \leftrightarrow \epsilon_j\}, \quad s_{ijk} = \{\epsilon_l \to \epsilon_l + (\epsilon_0 - \epsilon_i - \epsilon_j - \epsilon_k) \quad (l = 0, i, j, k)\}. \tag{77}$$

Then we have

$$
\begin{aligned}
T_{78}(\epsilon_0) &= \epsilon_0 + 3(\delta + \epsilon_7 - \epsilon_8), \quad T_{78}(\epsilon_i) = \epsilon_i + (\delta + \epsilon_7 - \epsilon_8) \quad (i \neq 0, 7, 8), \\
T_{78}(\epsilon_7) &= \epsilon_7 + \delta + (\delta + \epsilon_7 - \epsilon_8), \quad T_{78}(\epsilon_8) = \epsilon_8 - \delta + (\delta + \epsilon_7 - \epsilon_8),
\end{aligned}
\tag{78}
$$

hence, for $p_i$ parametrized as (76), we obtain

$$
\begin{aligned}
T_{78}(p_i) &= p_i, \quad (i \neq 7, 8, 10) \quad T_{78}(p_7) = p_7|_{\epsilon_7 \to \epsilon_7 + \delta}, \\
T_{78}(p_8) &= p_8|_{\epsilon_8 \to \epsilon_8 - \delta}, \quad T_{78}(p_{10}) = p_{10}|_{\epsilon_{10} \to \epsilon_{10} - \epsilon_7 + \epsilon_8 - \delta}.
\end{aligned}
\tag{79}
$$

**Theorem 3.8 ([19])** *For the translation* $T_{ij} = s_{ij} s_{ia_1 a_2} s_{ia_3 a_4} s_{a_5 a_6 a_7} s_{ia_3 a_4} s_{ia_1 a_2}$, *($\{i, j\} \cup \{a_1, \ldots, a_7\} = \{1, 2, \ldots, 9\}$), the discrete evolution $\overline{x} = T_{ij}(x)$ has the following geometric description.*

$$
\begin{aligned}
\overline{p_k} &= p_k, \quad (k \neq 7, 8, 10), \\
\overline{p_i} &= C(\widehat{p_{10}}) \cap C(\widehat{p_i}) \setminus \{p_1, \ldots, p_9\}, \\
\overline{p_i} + \overline{p_j} &= p_i + p_j, \quad \text{on} \quad C(\widehat{p_{10}}) \\
\overline{p_i} + \overline{p_{10}} &= p_j + p_{10}, \quad \text{on} \quad C(\widehat{p_i})
\end{aligned}
\tag{80}
$$

*where $C(\widehat{p_i})$ is the cubic curve passing through the nine points $p_1, p_2, \ldots, \widehat{p_i}, \ldots, p_{10}$.*

**Proof** If $p_{10} \in C_0$, the relations (80) are obvious from Eq. (79). The case $p_{10} \notin C_0$ can be reduced to the case $p_{10} \in C_0$, using the fact that $\overline{p_{10}}$ are independent of $p_7$. $\qquad \square$

One can also prove this more geometric way by using the result of [32].

## 3.6 Picard Lattice

The linear representation (67) has a natural geometric meaning. We will explain in case of $W_{3,n}$. Consider a lattice

$$
\Lambda = \mathbb{Z}\mathcal{E}_0 \oplus \mathbb{Z}\mathcal{E}_1 \oplus \cdots \oplus \mathbb{Z}\mathcal{E}_n,
\tag{81}
$$

with inner product

$$
\langle \mathcal{E}_0, \mathcal{E}_0 \rangle = -1, \quad \langle \mathcal{E}_i, \mathcal{E}_j \rangle = \delta_{i,j} \ (i > 0).
\tag{82}
$$

The reflections for simple roots

$$
\alpha_0 = \mathcal{E}_0 - \mathcal{E}_1 - \mathcal{E}_2 - \mathcal{E}_3, \quad \alpha_i = \mathcal{E}_i - \mathcal{E}_{i+1} \quad (1 \leq i \leq n - 1).
\tag{83}
$$

give the representation (67) by the identification $\mathcal{E}_i \leftrightarrow \epsilon_i$.

Geometrically, the lattice $\Lambda$ arises as the Picard lattice $\mathrm{Pic}(X)$ of the surface $X = \mathrm{Bl}_n(\mathbb{P}^2)$ obtained as $n$ points blowing up of $\mathbb{P}^2$. $\mathcal{E}_0$ denotes the class of the total transform of the line in $\mathbb{P}^2$ and $\mathcal{E}_i$ $(i > 0)$ are exceptional divisors. $x \cdot y = -\langle x, y \rangle$ is the intersection pairing.

Similarly, for a surface $Y = \mathrm{Bl}_{n-1}(\mathbb{P}^1 \times \mathbb{P}^1)$ obtained as $n - 1$ times blowing up of $\mathbb{P}^1 \times \mathbb{P}^1$, we have

$$\mathrm{Pic}(Y) = \mathbb{Z}H_1 \oplus \mathbb{Z}H_2 \oplus \mathbb{Z}E_1 \oplus \cdots \oplus \mathbb{Z}E_{n-1}, \tag{84}$$

where $H_1$ (resp. $H_2$) denotes the class associated with the line of degree $(1, 0)$ (resp. $(0, 1)$), hence the intersections are

$$H_1 \cdot H_2 = H_2 \cdot H_1 = 1, \quad E_i \cdot E_j = -\delta_{i,j}. \tag{85}$$

For $n \geq 2$, the surfaces $X$ and $Y$ are birationally equivalent. An isomorphism of their Picard lattices $\mathrm{Pic}(X) \simeq \mathrm{Pic}(Y)$ is given by

$$\mathcal{E}_0 = H_1 + H_2 - E_1, \quad \mathcal{E}_1 = H_1 - E_1, \quad \mathcal{E}_2 = H_2 - E_1, \quad \mathcal{E}_{i+1} = E_i, (i \geq 2) \tag{86}$$

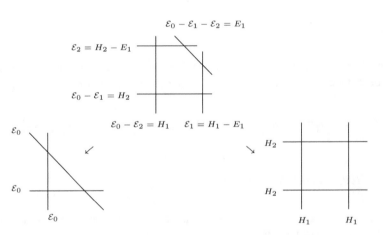

In the $\mathbb{P}^1 \times \mathbb{P}^1$ picture, the root system (83) is written as

$$\alpha_0 = E_1 - E_2, \quad \alpha_1 = H_1 - H_2,$$
$$\alpha_2 = H_2 - E_1 - E_2, \quad \alpha_i = E_i - E_{i+1}, \quad (i \geq 3) \tag{87}$$

The actions of corresponding simple reflections are explicitly given as

$$s_0 : \quad E_1 \leftrightarrow E_2,$$

$$s_1 : \quad H_1 \leftrightarrow H_2,$$

$$s_2 : \quad E_1 \rightarrow H_2 - E_2, \quad E_2 \rightarrow H_2 - E_1, \quad H_1 \rightarrow H_1 + H_2 - E_1 - E_2,$$

$$s_j : \quad E_{j-1} \leftrightarrow E_j \quad (j \geq 3).$$

$$(88)$$

Also, we introduce the set of parameters $h_1, h_2, e_1, \ldots, e_8$ and define the actions on them by identifying $H_i$ with $h_i$ and $E_i$ with $e_i$. The parameters $\epsilon_i$ for $\mathbb{P}^2$ and $h_i, e_i$ for $\mathbb{P}^1 \times \mathbb{P}^1$ are also related by the same way as (86). In both cases, the null root parameter

$$\delta = 3\epsilon_0 - \sum_{i=1}^{9} \epsilon_i = 2h_1 + 2h_2 - \sum_{i=1}^{8} e_i, \tag{89}$$

is $W(E_8^{(1)})$-invariant and will play the role of the interval for additive difference.

A canonical form of the $n - 1$ points configuration of $\mathbb{P}^1 \times \mathbb{P}^1$ can be taken as

$$[p_1, \ldots, p_{n-1}] \simeq \begin{bmatrix} \infty & 0 & 1 & x_4 & \cdots & x_{n-1} \\ \infty & 0 & 1 & y_4 & \cdots & y_{n-1} \end{bmatrix}. \tag{90}$$

Then the birational representation of $W_{3,n}$ is the same as (71) by the identification:

$$f_i = x_{i+1}, \quad g_i = y_{i+1} \quad (i \geq 4). \tag{91}$$

Corresponding parametrization of 8 points are given by

$$(x_i, y_i) = ( \frac{[e_{2i}][e_{1|2i}][e_{13}][e_{1|13}]}{[e_{1i}][e_{1|1i}][e_{23}][e_{1|23}]}, \frac{[e_{2i}][e_{2|2i}][e_{13}][e_{2|13}]}{[e_{1i}][e_{2|1i}][e_{23}][e_{2|23}]} ), \tag{92}$$

where $e_{i|jk} = h_i - e_j - e_k, e_{ij} = e_i - e_j$.


## 3.7 q-Painlevé Equation of Type $E_8^{(1)}$

We will derive a $q$-Painlevé equation from a representation of $W(E_8^{(1)})$ and give its relation to Eq. (25). In this section, the parameters $h_1, h_2, e_1, \ldots, e_8$ are multiplicative ones, and we use the notation

$$f_i = \psi_{h_1}(e_i), \quad g_i = \psi_{h_2}(e_i), \quad \psi_h(u) = u + \frac{h}{u}. \tag{93}$$

On the field of rational functions $K = \mathbb{C}(h_1, h_2, e_1, \ldots, e_8, f, g)$, we define the automorphisms $s_{ij}, s_{k|ij}, c$ as follows

$$s_{ij} = \{e_i \leftrightarrow e_j\}, \quad c = \{h_1 \leftrightarrow h_2, f \leftrightarrow g\},$$
$$s_{2|ij} = \{e_1 \mapsto \frac{h_2}{e_2}, e_2 \mapsto \frac{h_2}{e_1}, h_1 \mapsto \frac{h_1 h_2}{e_1 e_2}, f \mapsto s_{2|ij}(f)\}, \quad s_{1|ij} = c s_{2|ij} c, \tag{94}$$

where $s_{2|ij}(f)$ is determined by the following relation [36]

$$s_{2|ij}\left(\frac{f - f_i}{f - f_j}\right) = \frac{f - f_i}{f - f_j} \frac{g - g_j}{g - g_i}. \tag{95}$$

We note that

$$s_{2|ij}\left(\frac{f - \psi_{h_1}(t)}{f - \psi_{h_1}(\frac{h_2}{t})}\right) = \frac{f - \psi_{h_1}(t)}{f - \psi_{h_1}(\frac{h_2}{t})} \frac{t^2}{h_2} \frac{e_i - \frac{h_2}{t}}{e_i - t} \frac{e_j - \frac{h_2}{t}}{e_j - t} \quad \text{for} \quad g = \psi_{h_2}(t), \tag{96}$$

where $t \in \mathbb{C}$ is a constant.

The actions $s_0 = s_{12}, s_1 = c, s_2 = s_{2|12}, s_3 = s_{23}, \cdots, s_8 = s_{78}$, give a representation of affine Weyl group $W(E_8^{(1)})$. Since $q = \frac{h_1^2 h_2^2}{e_1 \cdots e_8}$ is a central element, it will be considered as a constant $q \in \mathbb{C}$. Our parametrization of the eight points $p_i = (f_i, g_i)$ is Weyl group covariant in the following sense. For $w \in W(E_8^{(1)})$, if $w(h_1) = h_1^{d_1} h_2^{d_2} \prod_{i=1}^{8} e_i^{-m_i}$, then $w(f)$ is a ratio of polynomials in $(f, g)$ of bidegree $(d_1, d_2)$ vanishing at $p_i$ with multiplicity $m_i$ ($i = 1, \ldots, 8$). Using this vanishing conditions, the rational expression of $w(f) = F(f, g)$ can be determined by the condition $F(\psi_{h_1}(t), \psi_{h_2}(t)) = w(\psi_{h_1}(t))$. Similar relation also holds for $w(h_2)$ and $w(g)$.

Consider the involutions $w_1 = s_{12} s_{1|12} s_{34} s_{1|34} s_{56} s_{1|56} s_{78} s_{1|78}$ and $w_2 = c w_1 c$, acting as

$$w_1 : (h_1, h_2, e_i, e, f, g) \mapsto (h_1, \frac{h_1^4 h_2}{e_1 \cdots e_8}, \frac{h_1}{e_i}, \frac{h_1}{e}, f, w_1(g)),$$
$$w_2 : (h_1, h_2, e_i, e, f, g) \mapsto (\frac{h_1 h_2^4}{e_1 \cdots e_8}, h_2, \frac{h_2}{e_i}, \frac{h_2}{e}, w_2(f), g). \tag{97}$$

Here, we included an extra parameter $e$ which transforms similarly to $e_i$. Then the composition $T = w_2 w_1$ gives a translation acting on parameters as

$$T : (h_1, h_2, e_i, e) \mapsto (\frac{h_1}{q}\xi^2, qh_2\xi^2, e_i\xi, e\xi), \quad \xi = \frac{qh_2}{h_1}. \tag{98}$$

For a constant $t \in \mathbb{C}$ such that $g = \psi_{h_2}(t)$, we have from Eq. (96) that

$$w_2\left(\frac{f - \psi_{h_1}(t)}{f - \psi_{h_1}(\frac{h_2}{t})}\right)\frac{f - \psi_{h_1}(\frac{h_2}{t})}{f - \psi_{h_1}(t)} = \frac{t^8}{h_2^4}\prod_{i=1}^{8}\frac{e_i - \frac{h_2}{t}}{e_i - t}. \tag{99}$$

Noting that $T(f) = w_2(f)$, $T(h_1) = w_2(h_1)$ and $Tw_1(h_2) = w_2(h_2)$, this means

$$T\left(\frac{f - \psi_{h_1}(t)}{f - \psi_{h_1}(\frac{w_1(h_2)}{t})}\right)\frac{f - \psi_{h_1}(\frac{h_2}{t})}{f - \psi_{h_1}(t)} = \frac{t^8}{h_2^4}\prod_{i=1}^{8}\frac{e_i - \frac{h_2}{t}}{e_i - t}, \quad \text{for} \quad g = \psi_{h_2}(t), \tag{100}$$

and similarly we have

$$T^{-1}\left(\frac{g - \psi_{h_2}(t)}{g - \psi_{h_2}(\frac{w_2(h_1)}{t})}\right)\frac{g - \psi_{h_2}(\frac{h_1}{t})}{g - \psi_{h_2}(t)} = \frac{t^8}{h_1^4}\prod_{i=1}^{8}\frac{e_i - \frac{h_1}{t}}{e_i - t}, \quad \text{for} \quad f = \psi_{h_1}(t). \tag{101}$$

The Eqs. (100) and (101) look similar but different from the QRT system (25). They can be almost related through a rescaling of variables

$$(h_i\lambda^2, e_i\lambda, e\lambda, f\lambda, g\lambda) = (\kappa_i, \upsilon_i, \upsilon, x, y), \tag{102}$$

where the factor $\lambda$ transforms as $w_i(\lambda) = \frac{h_i}{\lambda}$, $T(\lambda) = \xi\lambda = \frac{qh_2}{h_1}\lambda$. Then the parameters transform in a suitable way

$$T : (\kappa_1, \kappa_2, \upsilon_i, \upsilon) \mapsto (\frac{\kappa_1}{q}, q\kappa_2, \upsilon_i, \upsilon). \tag{103}$$

To make a connection to (25), we replace the *constant* $t$ by the *parameter* $e$. This change is needed since the factor $\lambda$ is $T$-dependent. Using the relations

$$\left\{w_2\left(\frac{f - \psi_{h_1}(t)}{f - \psi_{h_1}(\frac{h_2}{t})}\right)\right\}_{t \to e} = \left\{\frac{w_2(f) - \psi_{w_2(h_1)}(t)}{w_2(f) - \psi_{w_2(h_1)}(\frac{h_2}{t})}\right\}_{t \to e} = w_2\left(\frac{f - \psi_{h_1}(\frac{h_2}{e})}{f - \psi_{h_1}(e)}\right),$$

$$T\left(\frac{f - \psi_{h_1}(\frac{h_2}{qe})}{f - \psi_{h_1}(e)}\right) = w_2 w_1\left(\frac{f - \psi_{h_1}(\frac{h_2}{qe})}{f - \psi_{h_1}(e)}\right) = w_2\left(\frac{f - \psi_{h_1}(\frac{h_1}{h_2 e})}{f - \psi_{h_1}(\frac{h_1}{e})}\right) = w_2\left(\frac{f - \psi_{h_1}(\frac{h_2}{e})}{f - \psi_{h_1}(e)}\right), \tag{104}$$

the equation $\{\text{Eq.}\,(99)\}_{t \to e}$ can be written as

$$T\left(\frac{f - \psi_{h_1}(\frac{h_2}{qe})}{f - \psi_{h_1}(e)}\right)\frac{f - \psi_{h_1}(\frac{h_2}{e})}{f - \psi_{h_1}(e)} = \frac{e^8}{h_2^4}\prod_{i=1}^{8}\frac{e_i - \frac{h_2}{e}}{e_i - e}, \quad \text{for} \quad g = \psi_{h_2}(e). \quad (105)$$

Hence, by the rescaling (102), we have

$$T\left(\frac{x - \psi_{\kappa_1}(\frac{\kappa_2}{qv})}{x - \psi_{\kappa_1}(v)}\right)\frac{x - \psi_{\kappa_1}(\frac{\kappa_2}{v})}{x - \psi_{\kappa_1}(v)} = \frac{v^8}{\kappa_2^4}\prod_{i=1}^{8}\frac{v_i - \frac{\kappa_2}{v}}{v_i - v}, \quad \text{for} \quad y = \psi_{\kappa_2}(v). \quad (106)$$

Similarly we have

$$T^{-1}\left(\frac{y - \psi_{\kappa_2}(\frac{\kappa_1}{qv})}{y - \psi_{\kappa_2}(v)}\right)\frac{y - \psi_{\kappa_2}(\frac{\kappa_1}{v})}{y - \psi_{\kappa_2}(v)} = \frac{v^8}{\kappa_1^4}\prod_{i=1}^{8}\frac{v_i - \frac{\kappa_1}{v}}{v_i - v}, \quad \text{for} \quad x = \psi_{\kappa_1}(v).$$
$$(107)$$

We see that Eqs. (106) and (107) are de-autonomization of Eq. (25).

## 3.8   Toric-Like Form

There is a correspondence between gauge theories and Painlevé to equations [35], where additive/multiplicative/elliptic Painlevé equations corresponds to 4d/5d/6d gauge/string theories respectively. In recent studies, more precise connection has been studied (see [5, 6, 13] for example and references therein). For instance, in [6], a remarkable connection between the $\tau$-functions of $q$-Painlevé equations and the grand canonical topological string partition functions is established. In this correspondence, the conserved curve[5] of the autonomous limit of discrete Painlevé equations are identified with the mirror curve of the corresponding geometry. It seems to be difficult to generalize these results to $E_7^{(1)}$ and $E_8^{(1)}$ cases since the corresponding geometry is not realized as toric geometry. Here, we will give a construction of the toric like form as the integral curve for autonomous $q$-$P(E_8^{(1)})$ system.

We take the following 9 points on a nodal cubic in $\mathbb{P}^2$

$$p_i = (u_i : \frac{u_0 - u_i^3}{u_i} : 1), \quad (i = 1, \ldots, 9) \quad (108)$$

---

[5]The conserved curve and the spectral curve can be identified for genus 1 case.

with multiplicative parameters $u_i = e^{\epsilon_i}$. The $W(E_8^{(1)})$ actions on $p_1, \ldots, p_9$ and $p_{10} = (x, y, 1)$ are computed as follows:

$$s_i(u_i) = u_{i+1}, \quad s_i(u_{i+1}) = u_i, \quad (i = 1, \ldots, 8)$$

$$s_0(u_0) = \frac{u_0}{u_1 u_2 u_3}, \quad s_0(u_1) = \frac{u_0}{u_2 u_3}, \quad s_0(u_2) = \frac{u_0}{u_1 u_3}, \quad s_0(u_3) = \frac{u_0}{u_1 u_2},$$

$$s_0(x) = \frac{x^2 u_0 k_1 - u_0(u_0 - k_3) + xy u_0 - x u_0 k_2}{y k_3 + x u_0 k_1 - u_0 k_2 - x^2(u_0 - k_3)},$$

$$s_0(y) = \frac{y^2 u_0 k_3 - y u_0^2 k_2 + u_0^2 k_1(u_0 - k_3) - x^2 u_0 k_2(u_0 - k_3) + xy u_0 k_1 k_3 + x u_0(u_0 - k_3)^2}{k_3(y k_3 + x u_0 k_1 - u_0 k_2 - x^2(u_0 - k_3))},$$

$$\tag{109}$$

$k_1 = u_1 + u_2 + u_3$, $k_2 = u_1 u_2 + u_1 u_3 + u_2 u_3$, $k_3 = u_1 u_2 u_3$. These actions can be transformed into "tropical" form (subtraction free and hence admitting the combinatorial analog [46])

$$s_i(b_j) = b_j b_i^{-a_{ij}}, \quad s_0(f) = f b_0^{-1}, \quad s_2(g) = b_2 g,$$

$$s_3(f) = f \frac{b_3 + b_3 f + g}{1 + f + g}, \quad s_3(g) = g \frac{1 + b_3 f + g}{b_3(1 + f + g)}. \tag{110}$$

The relations of the variables $(u_i, x, y)$ and $(b_i, f, g)$ are given by

$$u_3(1 + f) + xg = 0,$$

$$(x^3 + xy - u_0)u_1 u_2 u_3 f + (x - u_1)(x - u_2)(x - u_3)u_0 = 0 \tag{111}$$

$$b_0 = \frac{u_0}{u_1 u_2 u_3}, \quad b_i = \frac{u_i}{u_{i+1}}, \quad (i > 0).$$

We note that

$$\frac{df \wedge dg}{fg} = \frac{dx \wedge dy}{u_0^3 - xy - x^3}. \tag{112}$$

When $u_0^3 = u_1 u_2 \cdots u_9$ (autonomous case), we have a pencil of cubic passing through the nine points, which is given by

$$m_8(x^2 y + y^2 + xu_0)u_0^{-2} - m_7(x^2 + y)u_0^{-1} + m_6 - m_5 x + m_4 x^2 - m_3 x^3$$
$$-m_2 x(xy - u_0) - m_1(xy^2 + x^2 u_0 - yu_0) - m_9(y^3 - 3x^3 u_0 + 3u_0^2)u_0^{-3}$$
$$+u_0 H(x^3 + xy - u_0) = 0,$$

$$\tag{113}$$

where $\prod_{i=1}^{9}(1 + u_i z) = \sum_{i=0}^{9} m_i z^i$.

By the variable change (111), we obtain the integral curve for $b_0 b_1^2 b_2^4 b_3^6 b_4^5 b_5^4 b_6^3$ $b_7^2 b_8 = 1$

$$H = \frac{1}{f^2 g^3}\left(\sum_{i=0}^{6}\sum_{j=0}^{6-i} c_{ij} f^i g^j\right). \tag{114}$$

Where

$$c_{0,0} = \frac{k}{l_3^2}, \quad c_{0,1} = \frac{2kl_1}{l_3^2}, \quad c_{0,2} = \frac{k(l_1^2 + 2l_2)}{l_3^2}, \quad c_{0,3} = \frac{2k(l_1 l_2 + l_3)}{l_3^2},$$

$$c_{0,4} = \frac{k(l_2^2 + 2l_1 l_3)}{l_3^2}, \quad c_{0,5} = \frac{2kl_2}{l_3}, \quad c_{0,6} = k, \quad c_{1,0} = \frac{3(k + l_3)}{l_3^2},$$

$$c_{1,1} = \frac{4kl_1 + 4l_3 l_1 + m_1 l_3}{l_3^2},$$

$$c_{1,2} = \frac{l_1^2 k^3 + 3l_2 k^3 + l_1^2 l_3 k^2 + m_1 l_1 l_3 k^2 + 3l_2 l_3 k^2 + m_5 l_3^2}{k^2 l_3^2},$$

$$c_{1,3} = \frac{l_1 l_2 k^3 + 3l_3 k^3 + 3l_3^2 k^2 + m_1 l_2 l_3 k^2 + l_1 l_2 l_3 k^2 + m_5 l_1 l_3^2}{k^2 l_3^2},$$

$$c_{1,4} = \frac{l_1 k^3 + m_1 l_3 k^2 + l_1 l_3 k^2 + m_5 l_2 l_3}{k^2 l_3}, \quad c_{1,5} = \frac{m_5 l_3}{k^2},$$

$$c_{2,0} = \frac{3(k^2 + 3l_3 k + l_3^2)}{kl_3^2}, \quad c_{2,1} = \frac{2(l_1 k^2 + m_1 l_3 k + 4l_1 l_3 k + m_1 l_3^2 + l_1 l_3^2)}{kl_3^2},$$

$$c_{2,2} = \frac{l_2 k^4 + l_1^2 l_3 k^3 + m_1 l_1 l_3 k^3 + 4l_2 l_3 k^3 + m_2 l_3^2 k^2 + m_1 l_1 l_3^2 k^2 + l_2 l_3^2 k^2 + m_5 l_3^2 k + m_5 l_3^3}{k^3 l_3^2},$$

$$c_{2,3} = 6, \quad c_{2,4} = \frac{m_4 l_3}{k^2}, \quad c_{3,0} = \frac{(k + l_3)(k^2 + 8l_3 k + l_3^2)}{k^2 l_3^2},$$

$$c_{3,1} = \frac{m_1 k^2 + 4l_1 k^2 + 4m_1 l_3 k + 4l_1 l_3 k + m_1 l_3^2}{k^2 l_3},$$

$$c_{3,2} = \frac{l_2 k^3 + m_2 l_3 k^2 + m_1 l_1 l_3 k^2 + l_2 l_3 k^2 + m_2 l_3^2 k + m_5 l_3^2}{k^3 l_3},$$

$$c_{3,3} = \frac{m_3 l_3}{k^2}, \quad c_{4,0} = \frac{3(k^2 + 3l_3 k + l_3^2)}{k^2 l_3}, \quad c_{4,1} = \frac{2(km_1 + l_3 m_1 + kl_1)}{k^2},$$

$$c_{4,2} = \frac{m_2 l_3}{k^2},$$

$$c_{5,0} = \frac{3(k + l_3)}{k^2}, \quad c_{5,1} = \frac{m_1 l_3}{k^2}, \quad c_{6,0} = \frac{l_3}{k^2},$$

$$\tag{115}$$

and

$$\begin{aligned}
&k = b_0 b_1 b_2^2, \\
&(1 + z)(1 + b_2 z)(1 + b_1 b_2 z) = 1 + l_1 z + l_2 z^2 + l_3 z^3, \\
&\left(1 + \frac{z}{b_3}\right)\left(1 + \frac{z}{b_3 b_4}\right)\cdots\left(1 + \frac{z}{b_3 b_4 \cdots b_8}\right) = 1 + m_1 z + \cdots + m_6 z^6.
\end{aligned} \tag{116}$$

Some comments are in order for the curve (114).

- The curve corresponds to the "Tuned $T_6$ curve" in [28] (see also [2, 29]).
- The curve is totally positive. Its quantization will admit good spectral property.
- The curve is rather complicated. Its characterization by the invariance under the simple actions (110) will be important.

# 4 Results on the Elliptic Painlevé Equation

## 4.1 $\tau$ Functions

For the elliptic Painlevé equation, the bilinear equations in terms of the $\tau$-functions on $E_8$ lattice is known [42]. We will explain it from a geometric point of view.

For each element $\lambda$ of the Picard lattice $\Lambda = \mathrm{Pic}(\mathrm{Bl}_8(\mathbb{P}^1 \times \mathbb{P}^1))$ of the form

$$\lambda = d_1 H_1 + d_2 H_2 - m_1 E_1 - \cdots - m_8 E_8 \in \Lambda, \quad d_i, \, m_i \in \mathbb{Z}, \tag{117}$$

one can associate a class of curves $|\lambda|$ on $\mathbb{P}^1 \times \mathbb{P}^1$ of bidegree $(d_1, d_2)$ passing through the blowing-up points $p_i$ with multiplicity $\geq m_i$ ($i = 1, \ldots, 8$). For example, the null root $\delta$ given by

$$\delta = 2H_1 + 2H_2 - \sum_{i=1}^{8} E_i, \tag{118}$$

corresponds to the class of curves of bidegree $(2, 2)$ passing through the eight points. For generic configurations of eight points, it is classically known that the *dimension* $d(\lambda)$ of the family $|\lambda|$ and the *genus* $g(\lambda)$ of generic member of the family are given by

$$d(\lambda) = (d_1 + 1)(d_2 + 1) - \sum_{i=1}^{8} \frac{m_i(m_i + 1)}{2} - 1,$$

$$g(\lambda) = (d_1 - 1)(d_2 - 1) - \sum_{i=1}^{8} \frac{m_i(m_i - 1)}{2}, \tag{119}$$

and hence it follows that

$$2d(\lambda) = \lambda \cdot (\lambda + \delta), \quad 2g(\lambda) - 2 = \lambda \cdot (\lambda - \delta), \tag{120}$$

respectively. Note that the space $L(\lambda)$ of defining equation of the curves in $|\lambda|$ is a vector space of dimension $\dim L(\lambda) = \dim|\lambda| + 1$.

Suggested by the relations $E_i \cdot E_i = -1$ and $E_i \cdot \delta = 1$, we define the subset $M \subset \Lambda$ by

$$M = \{\lambda \in \Lambda \mid \lambda \cdot \lambda = -1, \ \lambda \cdot \delta = 1\}. \tag{121}$$

Then $d(\lambda) = 0$ and $g(\lambda) = 0$ for $\lambda \in M$. The elements of $M$ are called the *exceptional classes*. Typical elements of $M$ are as follows:

| $\lambda \in M$ | Geometric meaning |
| --- | --- |
| $E_i$ | Exceptional curve |
| $H_i - E_j$ | Line passing through $p_j$ |
| $H_1 + H_2 - E_i - E_j - E_k$ | Curve of bidegree $(1,1)$ passing through $p_i, p_j, p_k$ |

For each $\lambda = d_1 H_1 + d_2 H_2 - \sum_{i=1}^{8} m_i E_i \in M$, the function $\tau(\lambda) = \tau(\lambda; x, y) \in L(\lambda)$ is unique up to normalization. We fix the normalization as follows. Let $p(u)$ be the parametrization of the bidegree $(2, 2)$ curve $C_0$ passing through $p_1, \ldots, p_8$, such that

$$p_i = p(e_i), \quad p(u) = \left( \frac{[b - u][h_1 - b - u]}{[a - u][h_1 - a - u]}, \ \frac{[b - u][h_2 - b - u]}{[a - u][h_2 - a - u]} \right), \tag{122}$$

where $a, b$ are some constants (or parameters). Then we put the normalization condition as

$$\tau(\lambda) \Big|_u = \left[ d_1 h_1 + d_2 h_2 - \sum_{i=1}^{8} m_i e_i - u \right], \tag{123}$$

where, $*\big|_u$ means a specialization of $(x, y)$ to the point $p(u)$.

Then we can derive various bilinear relations as follows. For instance, the polynomials of the form $\tau(E_i)\tau(H_1 - E_i)$ $(i = 1, \ldots, 8)$ belong to the vector space $L(H_1)$ of dimension 2. As a result, there exists a linear relation among any three of such polynomials,

$$c_{1|23}\tau(E_1)\tau(H_1 - E_1) + c_{1|31}\tau(E_2)\tau(H_1 - E_2) + c_{1|12}\tau(E_3)\tau(H_1 - E_3) = 0,$$

$$c_{k|ij} = [e_i - e_j][h_k - e_i - e_j], \tag{124}$$

where the coefficients are fixed by the normalization condition (123).

There are infinitely many bilinear relations. They are consistent system though extremely overdetermined. The consistency is due to the following symmetry under the Weyl group $W(E_8^{(1)})$. Namely we have

**Theorem 4.1 ([24])** *The following actions $s_i$ on the parameters $\{h_1, h_2, e_1, \ldots, e_8\}$ and $\tau$-variables $\{\tau_1, \ldots, \tau_8, \tau_{1,1}, \tau_{1,2}, \tau_{2,1}, \tau_{2,2}\}$ gives a birational representation*

*of* $W(E_8^{(1)})$.

$$s_0: \quad e_i \leftrightarrow e_2, \quad \tau_1 \leftrightarrow \tau_2, \quad \tau_{k,1} \leftrightarrow \tau_{k,2} \quad (k=1,2),$$

$$s_1: \quad h_1 \leftrightarrow h_2, \quad \tau_{1,i} \leftrightarrow \tau_{2,i},$$

$$s_2: \quad e_1 \mapsto h_2 - e_2, \quad e_2 \mapsto h_2 - e_1, \quad h_1 \mapsto h_1 + h_2 - e_1 - e_2,$$

$$\qquad \tau_1 \leftrightarrow \tau_{2,2}, \quad \tau_2 \leftrightarrow \tau_{2,1},$$

$$s_3: \quad e_2 \leftrightarrow e_3, \quad \tau_2 \leftrightarrow \tau_3, \quad \tau_{k,2} \to \frac{c_{k|32}\tau_1\tau_{k,1} + c_{k|13}\tau_2\tau_{k,2}}{c_{k|12}\tau_3} \quad (k=1,2),$$

$$s_j: \quad e_{j-1} \leftrightarrow e_j, \quad \tau_{j-1} \leftrightarrow \tau_j \quad (j=4,\ldots,8), \tag{125}$$

*where* $c_{k|ij} = [e_i - e_j][h_k - e_i - e_j]$.

**Proof** Almost all the relations follows by the oddness $[-x] = -[x]$. Only nontrivial relation is $s_3s_4s_3(\tau_{k,2}) = s_4s_3s_4(\tau_{k,2})$ $(k = 1,2)$ which is equivalent to

$$c_{k|12}c_{k|34} - c_{k|13}c_{k|24} + c_{k|14}c_{k|23} = 0. \tag{126}$$

This reduces to the Riemann relation for $[x]$.                                         $\square$

From the $\tau$ variables, the variables $f_i, g_i$ in (70) are recovered by

$$f_9 = \frac{c_{1|13}\tau(E_2)\tau(H_1 - E_2)}{c_{1|23}\tau(E_1)\tau(H_1 - E_1)}, \quad g_9 = \frac{c_{2|13}\tau(E_2)\tau(H_2 - E_2)}{c_{2|23}\tau(E_1)\tau(H_2 - E_1)}, \tag{127}$$

and $(f_i, g_i) = (f_9, g_9)\big|_{u=e_i}$ $(i = 1, \ldots, 8)$.

Thanks to the bilinear relations, one can define variables $(x, y)$ consistently as

$$\frac{[a - e_i][h_1 - a - e_i]}{[a - e_j][h_1 - a - e_j]} \frac{x - x_i}{x - x_j} = \frac{\tau(E_i)\tau(H_1 - E_i)}{\tau(E_j)\tau(H_1 - E_j)},$$

$$\frac{[a - e_i][h_2 - a - e_i]}{[a - e_j][h_2 - a - e_j]} \frac{y - y_i}{y - y_j} = \frac{\tau(E_i)\tau(H_2 - E_i)}{\tau(E_j)\tau(H_2 - E_j)}, \tag{128}$$

where $(x_i, y_i) = p(e_i)$ in Eq. (122). Since the 8 points are treated in symmetric manner in this coordinate,[6] many of the $s_i$ act on $(x, y)$ trivially. The non-trivial

---

[6]The representation (70) in variables $(f_i, g_i)$ is algebraic but not $\mathfrak{S}_8$-symmetric, while the representation (129) in variables $(e_i, h_i, x, y)$ is $\mathfrak{S}_8$-symmetric but not algebraic. There exist $\mathfrak{S}_8$-symmetric and algebraic representations [1] which are, interestingly, related to the quadrirational Yang-Baxter maps.

transformations are $s_1(x)$, $s_1(y)$, $s_2(x)$ only, and given by

$$s_1(x) = y, \quad s_1(y) = x, \quad s_2\left(\frac{x - x_1}{x - x_2}\right) = \frac{x - x_1}{x - x_2}\frac{y - y_2}{y - y_1}. \tag{129}$$

As a similar way to the $q$-case Sect. 3.7, we can derive a simple explicit form of the elliptic Painlevé equation along the direction $\alpha_1 = H_1 - H_2$. Using the decomposition $T_{\alpha_1} = w_2 w_1$,

$$\begin{aligned}
&w_1 = r_{78}r_{1|78}r_{56}r_{1|56}r_{34}r_{1|34}r_{12}r_{1|12}, \\
&r_{ij} = r_{E_i - E_j}, \quad r_{i|jk} = r_{H_i - E_j - E_k}, \quad w_2 = w_1|_{H_1 \leftrightarrow H_2},
\end{aligned} \tag{130}$$

we have

**Theorem 4.2 ([24, 41])** *The evolutions $T_{\alpha_1}(x) = \overline{x}$ and $T_{\alpha_1}^{-1}(y) = \underline{y}$ are given by*

$$\frac{[\overline{h}_1 - a - t]\,\overline{x} - \overline{\varphi}(t)}{[\overline{h}_1 - a - s]\,\overline{x} - \overline{\varphi}(s)} = \frac{[h_1 - a - t]\,x - \varphi(t)}{[h_1 - a - s]\,x - \varphi(s)} \prod_{i=1}^{8}\frac{[e_i - s]}{[e_i - t]}, \quad \text{for} \quad y = \psi(s) = \psi(t), \tag{131}$$

$$\frac{[\underline{h}_2 - a - t]\,\underline{y} - \psi(t)}{[\underline{h}_2 - a - s]\,\underline{y} - \psi(s)} = \frac{[h_2 - a - t]\,y - \psi(t)}{[h_2 - a - s]\,y - \psi(s)} \prod_{i=1}^{8}\frac{[e_i - s]}{[e_i - t]}, \quad \text{for} \quad x = \varphi(s) = \varphi(t), \tag{132}$$

*where $(\varphi(u), \psi(u)) = p(u)$ is the parametrization of the curve $C_0$ in Eq. (122).*

**Proof** The equation for $\overline{x}$ (131) follows from

$$r_{H_2 - E_i - E_j}\left(\frac{[h_1 - a - t]\,x - \varphi(t)}{[h_1 - a - s]\,x - \varphi(s)}\right) = \frac{[h_1 - a - t]\,x - \varphi(t)\,[e_i - s]\,[e_j - s]}{[h_1 - a - s]\,x - \varphi(s)\,[e_i - t]\,[e_j - t]}, \tag{133}$$

where $y = \psi(s) = \psi(t)$. Equation for $\underline{y}$ (132) is similar.                                    □

### 4.2   Lax Formalism

Throughout this section, we use the multiplicative parameters, $\kappa_1, \kappa_2, v_1, \ldots, v_8$, with $\kappa_1^2\kappa_2^2 = q\prod_{i=1}^{8}v_i$. The direction of time evolution $T$ is chosen as $T : \kappa_1 \mapsto \frac{\kappa_1}{q}, \kappa_2 \mapsto q\kappa_2$ and we use the notations $\overline{X} = T(X)$, $\underline{X} = T^{-1}(X)$.

Since the elliptic Painlevé equation was formulated by geometric method, its isomonodromic origin has been a fundamental problem. It was solved in [49, 64]. The construction in [64] is based on the geometric characterization of the Lax equation. Later the result is simplified using contiguity type Lax pair [41]. To see the ideas of these constructions, we recall the scalar Lax form of $q$-$P_{VI}$.

*Example 4.3 (Scalar Lax Form for $q$-$P_{VI}$)* Consider the $q$-$P_{VI}$ equation of the form

$$\overline{f} f = v_3 v_4 \frac{(g - \frac{v_5}{\kappa_2})(g - \frac{v_6}{\kappa_2})}{(g - \frac{1}{v_1})(g - \frac{1}{v_2})}, \quad \underline{g} g = \frac{1}{v_1 v_2} \frac{(f - \frac{\kappa_1}{v_7})(f - \frac{\kappa_1}{v_8})}{(f - v_3)(f - v_4)}, \tag{134}$$

where $\overline{\kappa_1} = \kappa_1/q, \overline{\kappa_2} = q\kappa_2$. This can be derived as the compatibility of Lax pair:

$$L_1 : \left\{ \frac{z}{qg} \prod_{i=1}^{2} (gv_i - 1) - \frac{1}{fg} \prod_{i=1}^{4} v_i \prod_{i=5}^{6} \left( g - \frac{v_i}{\kappa_2} \right) \right\} Y(z)$$

$$+ \frac{v_1 v_2 \prod_{i=3}^{4} \left( \frac{z}{q} - v_i \right)}{f - \frac{z}{q}} \left\{ gY(z) - Y(\frac{z}{q}) \right\} + \frac{\prod_{i=7}^{8} \left( \frac{\kappa_1}{v_i} - z \right)}{q(f - z)} \left\{ Y(qz) - \frac{1}{g} Y(z) \right\} = 0,$$

$$L_2 : \left( 1 - \frac{f}{z} \right) \overline{Y}(z) + Y(qz) - \frac{1}{g} Y(z) = 0. \tag{135}$$

From this, we observe the following interesting facts:

- As an algebraic curve in $(f, g)$ the equation $L_1$ is uniquely characterized by the conditions: (1) it is of bidegree $(3, 2)$, and (2) passing through the following 12 points:

$$\left( \infty, \frac{1}{v_i} \right)_{i=1}^{2}, \quad \left( v_i, \infty \right)_{i=3}^{4}, \quad \left( 0, \frac{v_i}{\kappa_2} \right)_{i=5}^{6}, \quad \left( \frac{\kappa_1}{v_i}, 0 \right)_{i=7}^{8},$$

$$(z, \infty), \quad \left( \frac{z}{q}, 0 \right), \quad \left( z, \frac{Y(z)}{Y(qz)} \right), \quad \left( \frac{z}{q}, \frac{Y(z/q)}{Y(z)} \right). \tag{136}$$

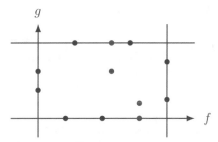

- Compatibility of $L_1, L_2$ can be equivalently written as the compatibility of simpler equations of contiguity type: namely $L_2$ and

$$L_3 : \quad qz(\overline{f} - \frac{z}{q}) \prod_{i=1}^{2} (v_i - \frac{1}{g}) y(z) + \frac{1}{g} \prod_{i=7}^{8} (z - \frac{\kappa_1}{v_i}) \overline{y}(z) - v_1 v_2 \prod_{i=3}^{4} (qv_i - z) \overline{y}(\frac{z}{q}) = 0. \tag{137}$$

Using the above observations as a hint, we can construct the Lax pair for the elliptic Painlevé equation as follows. The most fundamental object in the Lax formulation is the linear difference equation $L_1$ among $y(zq)$, $y(z)$ and $y(z/q)$. One can determine the equation $L_1$ by the geometric condition [64]. The explicit form of the equation $L_1$ is, however, rather complicated. So, we start with the following contiguity type equations:

$$L_2 : G\left(g, \frac{\kappa_1}{z}\right) y(qz) - G(g, z) y(z) - \left[\frac{\kappa_1}{z^2}\right] F(f, z) \overline{y}(z) = 0, \qquad (138)$$

$$L_3 : G\left(g, \frac{\kappa_1}{qz}\right) U(z) \overline{y}(z) - G(g, z) U\left(\frac{\kappa_1}{qz}\right) \overline{y}(qz) - \left[\frac{\kappa_1}{qz^2}\right] w\overline{F}(\overline{f}, z) y(qz) = 0, \qquad (139)$$

where $f$, $g$, $\overline{f}$ and $w$ are variables independent of $z$, and $[z]$ is a multiplicative odd theta function satisfying $[z^{-1}] = -[z]$, $[pz] = -z^{-1}p^{-1/2}[z]$. Other notations are given by

$$F(f, z) = \psi_{\kappa_1, a}(z) f - \psi_{\kappa_1, b}(z), \qquad G(g, z) = \psi_{\kappa_2, a}(z) g - \psi_{\kappa_2, b}(z),$$

$$U(z) = \prod_{i=1}^{8}\left[\frac{v_i}{z}\right], \qquad \psi_{\kappa, a}(z) = \left[\frac{a}{z}\right]\left[\frac{\kappa}{az}\right]. \qquad (140)$$

Parameters $a$, $b$ are introduced to specify the coordinates $(f, g) \in \mathbb{P}^1 \times \mathbb{P}^1$.

As the necessary condition for the compatibility of the Eqs. (138) and (139), one can easily derive elliptic Painlevé equation as follows. When $G(g, z) = 0$, the equations $L_2$ and $L_3$ give

$$G\left(g, \frac{\kappa_1}{z}\right) y(qz) = \left[\frac{\kappa_1}{z^2}\right] F(f, z) \overline{y}(z),$$

$$G\left(g, \frac{\kappa_1}{qz}\right) U(z) \overline{y}(z) = \left[\frac{\kappa_1}{qz^2}\right] w\overline{F}(\overline{f}, z) y(qz), \qquad (141)$$

from which we obtain

$$\frac{w F(f, z)\overline{F}(\overline{f}, z)}{U(z)} = \frac{G\left(g, \frac{\kappa_1}{z}\right) G\left(g, \frac{\kappa_1}{qz}\right)}{\left[\frac{\kappa_1}{z^2}\right]\left[\frac{\kappa_1}{qz^2}\right]} = \left[\frac{\kappa_1}{\kappa_2}\right]\left[\frac{\kappa_1}{q\kappa_2}\right]\left(\frac{g_a(b)}{g_a(z)}\right)^2, \qquad (142)$$

for $G(g, z) = 0$. Since $G(g, z) = G(g, \frac{\kappa_2}{z})$, another relation obtained from (142) by replacing $z$ with $\frac{\kappa_2}{z}$ also holds. Then by taking the ratio of these two expressions, we have

$$\frac{F(f, \frac{\kappa_2}{z})\overline{F}(\overline{f}, \frac{\kappa_2}{z})U(z)}{F(f, z)\overline{F}(\overline{f}, z)U(\frac{\kappa_2}{z})} = 1, \quad \text{for } G(g, z) = 0. \tag{143}$$

On the other hand, putting $F(f, z) = 0$ in $L_2$ and $\underline{L_3}$, we have

$$\frac{G(g, \frac{\kappa_1}{z})\underline{G}(\underline{g}, \frac{\kappa_1}{z})U(z)}{G(g, z)\underline{G}(\underline{g}, z)U(\frac{\kappa_1}{z})} = 1, \quad \text{for } F(f, z) = 0. \tag{144}$$

Equations (143) and (144) are the elliptic Painlevé equation.[7] Eliminating the auxiliary variable $z$, one can show.

**Proposition 4.4 ([41])** *Variables $\overline{f}$ and $\underline{g}$ determined by the Eqs. (143) and (144) have the following characteristic properties: (i) $\overline{f}$ and $\underline{g}$ are rational functions of $(f, g)$ of bidegree $(1, 4)$ and $(4, 1)$ with the points of indeterminacy at $(f(v_i), g(v_i))$ $(i = 1, \ldots, 8)$. (ii) If $(f, g) = (f(z), g(z))$ then $\overline{f} = \overline{f}(\frac{\kappa_2}{z})$ and $\underline{g} = \underline{g}(\frac{\kappa_1}{z})$.*

Here $f(z)$ and $g(z)$ is defined by $F(f(z), z) = 0$ and $G(g(z), z) = 0$. For the proof, see [24, section7] for example. Furthermore we have

**Theorem 4.5 ([41])** *The relations (143), (144) and (142) are sufficient for the compatibility of Eqs. (138) and (139).*

The idea of the proof is as follows. We construct the $L_1$ equation among $y(qz), y(z), y(z/q)$ by eliminating $\overline{y}(z)$ and $\overline{y}(z/q)$ from $L_2, L_2|_{z \to z/q}$ and $L_3|_{z \to z/q}$.

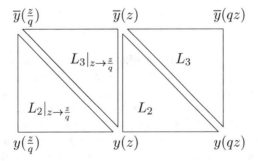

Similarly, by eliminating $y(z)$ and $y(qz)$ from $L_3, L_3|_{z \to z/q}$ and $L_2$, we obtain $L_4$ equation among $\overline{y}(qz), \overline{y}(z), \overline{y}(z/q)$.

---

[7]The relation to the Eqs. (131) and (132) is given by similar way as the $q$-case Sect. 3.7.

The compatibility means $L_1 \propto T^{-1}(L_4)$. This is proved by showing the following geometric properties [64] for the equations $L_1$ and $L_4$. The equation $L_1$ is the equation of the bidegree $(3, 2)$ curve in $(f, g)$ passing through the following 12 points:

$$\left(f(v), g(v)\right)_{v=v_1,\ldots,v_8,z,\frac{q\kappa_1}{z}}, \quad \left(f(x), \gamma_x\right)_{x=z,\frac{z}{q}}, \quad \text{where} \quad \frac{G(\gamma_x, \frac{\kappa_1}{x})}{G(\gamma_x, x)} = \frac{y(x)}{y(qx)}. \tag{145}$$

Similarly $L_4$ is the equation for the curve of bidegree $(3, 2)$ in $(\overline{f}, g)$ passing through

$$\left(\overline{f}(v), g(v)\right)_{v=v_1,\ldots,v_8,\frac{z}{q},\frac{\kappa_1}{qz}}, \quad \left(\overline{f}(x), \gamma'_x\right)_{x=z,\frac{z}{q}}, \quad \text{where} \quad \frac{G(\gamma'_x, x)}{G(\gamma'_x, \frac{\kappa_1}{qx})} \frac{U(\frac{\kappa_1}{qx})}{U(x)} = \frac{\overline{y}(x)}{\overline{y}(qx)}. \tag{146}$$

See [24, section 7] for more details.

## 4.3 Hypergeometric Solution

For some special parameters and initial values, the Painlevé equations sometimes can be reduced to Riccati equations and solved in terms of hypergeometric functions.

*Example 4.6* For the equation in Eq. (1) the hypergeometric solutions are

$$\begin{array}{ccccccc} \text{Gauss} & \to & \text{Kummer} & \to & \text{Bessel} & \to & \text{(none)} & \to & \text{(none)} \\ & \searrow & & \searrow & & \searrow & & & \\ & & \text{Hermite} & \to & \text{Airy} & \to & \text{(none)} & & \end{array} \tag{147}$$

The discrete analog of Riccati equation is a nonautonomous fractional linear transformation $\overline{y} = \frac{ay+b}{cy+d}$ on $\mathbb{P}^1$ which can be linearized easily. In general discrete settings, the existence of the special solutions reducible to Riccati equation is explained geometrically as follows [19]. Consider a root $\beta \in \Lambda, \delta \cdot \beta = 0, \beta^2 = -2$. Since the genus $g$ and dimension $d$ of the linear system $|\beta|$ is $g = 0$ and $d = -1$, the corresponding rational curve $C_\beta = \mathbb{P}^1$ exists only for a special parameter of co-dimension 1. On the Picard lattice $\Lambda$, the action of translation $T_\alpha$ along the direction $\alpha \in \Lambda$ is given by Kac formula

$$T_\alpha(\lambda) = \lambda - (\delta \cdot \lambda)\alpha + \left\{(\alpha \cdot \lambda) - \frac{\alpha^2}{2}(\delta \cdot \lambda)\right\}\delta. \tag{148}$$

If $\alpha \cdot \beta = 0$, we have

$$T_\alpha(\beta) = \beta, \tag{149}$$

and the curve $C_\beta$ is called invariant divisor. In this situation, the Painlevé equation $T_\alpha$ admits a restriction on $C_\beta$ and can be reduced to a Riccati equation.

As an example, one has the following equation in $\mathbb{P}^1 \times \mathbb{P}^1$ formulation.

**Theorem 4.7 ([21, 36])** *Assume that four points $p_1, p_2, p_3, p_4$ are on a $(1, 1)$ curve $C_\beta$: $d_{1234} = 0$, and the ninth point $p_9$ is also on $C_\beta$: $d_{1239} = 0$. Then the time evolution $T_{H_1 - H_2}(f_9) = \overline{f_9}$ is determined by $d_{1239} = d'_{6789} = 0$. Moreover, in variable $y = f_{19}/f_{89}$, the equation is the following Riccati equation explicitly given as*

$$\overline{y} = \frac{ay + b}{cy + d},$$

$$b = f_{12}f_{13}g_{23}d'_{1678}, \quad c = \overline{f_{78}}\,\overline{f_{68}}g_{67}d_{8123}, \quad d = -f_{12}f_{13}\overline{f_{68}}\,\overline{f_{78}}g_{23}g_{18}g_{67},$$

$$\Delta = ad - bc = f_{12}f_{13}f_{23}f_{18}\overline{f_{67}}\,\overline{f_{68}}\,\overline{f_{78}}f_{18}g_{12}g_{13}g_{23}g_{67}g_{68}g_{78},$$

$$f_{ij} = f_i - f_j, \quad g_{ij} - g_i - g_j,$$

$$d_{ijkl} = \det\left[(1, f_s, g_s, f_s g_s)_{s=i,j,k,l}\right], \quad d'_{ijkl} = \det\left[(1, \overline{f_s}, g_s, \overline{f_s}g_s)_{s=i,j,k,l}\right].$$

$$\tag{150}$$

*Proof* From $d_{1234} = 0$ we have $\kappa_1\kappa_2 = v_1v_2v_3v_4$, and hence $\kappa_1\kappa_2 = qv_5v_6v_7v_8$. By the assumption that $(f, g) = (f_9, g_9) \in C_\beta$, we have

$$\frac{F(f, \frac{\kappa_2}{t})}{F(f, t)} = \prod_{j=1}^{4} \frac{\left[\frac{v_j t}{\kappa_2}\right]}{\left[\frac{v_j}{t}\right]} \quad \text{for } g = g(t), \qquad \frac{G(g, \frac{\kappa_1}{s})}{G(g, s)} = \prod_{j=1}^{4} \frac{\left[\frac{v_j s}{\kappa_1}\right]}{\left[\frac{v_j}{s}\right]} \quad \text{for } f = f(s). \tag{151}$$

Then, from the evolution equations (143) and (144), we have

$$\frac{\overline{F}(\overline{f}, \frac{\kappa_2}{t})}{\overline{F}(\overline{f}, t)} = \prod_{j=5}^{8} \frac{\left[\frac{v_j t}{\kappa_2}\right]}{\left[\frac{v_j}{t}\right]} \quad \text{for } g = g(t), \qquad \frac{G(g, \frac{\kappa_1}{s})}{G(g, s)} = \prod_{j=5}^{8} \frac{\left[\frac{v_j s}{\kappa_1}\right]}{\left[\frac{v_j}{s}\right]} \quad \text{for } f = f(s). \tag{152}$$

The first equation shows $d'_{6789} = 0$. Eliminating $g_9$ from $d_{1239} = d'_{6789} = 0$, we obtain the Riccati equation (150). $\qquad \square$

In general, the discrete Riccati equation of the form $\overline{y} = \frac{ay+b}{cy+d}$ can be linearized by putting $y = \frac{d}{c}\frac{\overline{\Phi}-\Phi}{\Phi}$ as

$$\underline{c}\underline{d}d(\overline{\Phi}-\Phi) - \underline{b}\underline{c}c\Phi + \underline{\Delta}c(\underline{\Phi}-\Phi) = 0. \tag{153}$$

In case of (150) we have

$$\frac{f_{12}f_{13}f_{68}f_{78}g_{23}g_{18}g_{67}g_{18}}{d_{1238}}(\overline{\Phi}-\Phi) + d'_{1678}\Phi + \frac{f_{67}f_{18}f_{23}f_{18}g_{12}g_{13}g_{68}g_{78}}{d_{1238}}(\underline{\Phi}-\Phi) = 0. \tag{154}$$

Using the parametrization

$$f_i = \frac{[a/v_i][\kappa_1/(av_i)]}{[b/v_i][\kappa_1/(bv_i)]}, \quad g_i = \frac{[a/v_i][\kappa_1/(av_i)]}{[b/v_i][\kappa_2/(bv_i)]}, \tag{155}$$

and change the parameters as

$$a_0 = \frac{v_1 v_8}{v_2 v_3}, \ a_1 = \frac{q\kappa_1}{v_2 v_3}, \ a_2 = \frac{\kappa_2}{v_2 v_3}, \ a_3 = \frac{v_1}{v_5}, \ a_4 = \frac{\cdot v_1}{v_6}, \ a_5 = \frac{v_1}{v_7}, \ a_6 = \frac{v_8}{v_2}, \ a_7 = \frac{v_8}{v_3}, \tag{156}$$

$(q^2 a_0^3 = a_1 a_2 \cdots a_7)$, the Eq. (154) is written as the linear 3-term equation

$$A(\overline{\Phi}-\Phi) + \Phi + A'(\underline{\Phi}-\Phi) = 0, \tag{157}$$

where $\overline{\Phi} = \Phi_{a_1^- a_2^+} = \Phi|_{a_1 \to a_1/q, a_2 \to q a_2}$, $\underline{\Phi} = \Phi|_{a_1^+ a_2^-}$ and

$$A = \frac{[a_2][a_0/a_2][qa_0/a_2]}{[a_2/a_1][qa_2/a_1][qa_0/a_1 a_2]} \prod_{j=3}^{7} \frac{[qa_0/a_1 a_j]}{[a_j]}, \quad A' = A\Big|_{a_1 \leftrightarrow a_2}. \tag{158}$$

If one of the parameters $a_1, \ldots, a_6$ is $q^{-N}$ ($N \in \mathbb{Z}_{\geq 0}$), we have a solution $\Phi$ expressed by terminating very-well poised elliptic hypergeometric series [54, 55]:

$$\Phi = {}_{12}V_{11}(a_0; a_1, \ldots, a_7) = \sum_{k=0}^{N} \frac{[q^{2k}a_0][a_0]_k}{[a_0][q]_k} \prod_{i=1}^{7} \frac{[a_i]_k}{[qa_0/a_i]_k}, \tag{159}$$

where $[a]_k = [a][qa] \cdots [q^{k-1}a]$. Together with the three term equation (157) this function $\Phi = {}_{12}V_{11}$ satisfy the contiguity relation

$$
\overline{\Phi} - \Phi = -\frac{[qa_0][q^2a_0][qa_2/a_1][qa_0/a_1a_2]}{[qa_0/a_1][q^2a_0/a_1][a_0/a_2][qa_0/a_2]} \prod_{i=3}^{7} \frac{[a_i]}{[qa_0/a_i]}
$$

$$
\times \, \Phi(q^2a_0; a_1, qa_2, qa_3, \ldots, qa_7). \tag{160}
$$

For non-terminating case, the solution can be expressed in terms of the elliptic hypergeometric integral [54, 55]:

$$
V(t_1, \ldots, t_8|p, q) = \frac{(p; p)_\infty (q; q)_\infty}{4\pi\sqrt{-1}} \int_C \frac{\prod_{j=1}^{8} \Gamma(t_j z^{\pm 1}; p, q)}{\Gamma(z^{\pm 2}; p, q)} \frac{dz}{z}, \tag{161}
$$

where $t_1 \cdots t_8 = p^2q^2$, $\Gamma(az^{\pm 1}; p, q) = \Gamma(az; p, q)\Gamma(a/z; p, q)$,

$$
\Gamma(z; p, q) = \frac{(pq/z; p, q)_\infty}{(z; p, q)_\infty}, \quad (z; p, q)_\infty = \prod_{i,j=0}^{\infty} (1 - p^i q^j z), \tag{162}
$$

and the integration contour $C$ can be chosen as $|z| = 1$ when $|t_k| < 1$. The integral $V$ satisfy the following three term equation

$$
\mathcal{A}(U_{t_6^+ t_7^-} - U) + U + \mathcal{A}'(U_{t_6^- t_7^+} - U) = 0, \tag{163}
$$

where

$$
\mathcal{A} = \frac{\theta(t_6/qt_8, t_8t_6, t_8/t_6; p)}{\theta(t_6/t_7, t_7/qt_6, t_6t_7/q; p)} \prod_{k=1}^{5} \frac{\theta(t_7t_k/q; p)}{\theta(t_8t_k; p)},
$$

$$
\mathcal{A}' = \mathcal{A}|_{t_6 \leftrightarrow t_7}, \quad U = \frac{V}{\prod_{k=6}^{7} \Gamma(t_k t_8^{\pm 1}; p, q)}. \tag{164}
$$

Similarly, in $\mathbb{P}^2$ formulation, the hypergeometric solution can be obtained by a geometric method as follows. As an example, we consider the case $\alpha = \mathcal{E}_8 - \mathcal{E}_9$ and $\beta = \mathcal{E}_0 - \mathcal{E}_5 - \mathcal{E}_6 - \mathcal{E}_7$ on $\mathrm{Bl}_9(\mathbb{P}^2)$. The invariant divisor $C_\beta$ is a line on $\mathbb{P}^2$ passing through the three points $p_5, p_6, p_7$ which exist iff $\epsilon_0 - \epsilon_5 - \epsilon_6 - \epsilon_7 = 0$.

**Theorem 4.8 ([19, 20])** *If $p_{10} \in C_\beta$ then $T^n(p_{10}) \in C_\beta$ and for any permutation $(i, j, k, l)$ of $(1, 2, 3, 4)$ we have*

$$
\frac{d_{5,6,8}d_{k,i,8}d_{j,k,9}d_{i,j,\bar{9}}}{d_{i,8,\bar{9}}}\left\{\frac{d_{i,k,\bar{9}}}{d_{i,k,8}}\lambda_l d_{j,k,\overline{10}} - d_{j,k,10}\right\}
$$
$$
+\frac{d_{5,6,9}d_{k,i,9}d_{j,k,8}d_{i,j,\underline{8}}}{d_{i,\underline{8},9}}\left\{\frac{d_{i,k,\underline{8}}}{d_{i,k,9}}\mu_l d_{j,k,\underline{10}} - d_{j,k,10}\right\} = d_{5,6,j}d_{k,8,9}d_{i,j,k}d_{j,k,10}.
$$

$$(165)$$

*Where $d_{i,j,l} = \det(p_i, p_j, p_k)$, $d_{i,j,\bar{k}} = \det(p_i, p_j, \overline{p}_k)$ etc. The factors $\lambda_l, \mu_l$ depend on the gauge and can be chosen as $\lambda_4 = 1$, then $\lambda_3 = \frac{c_{1,4}}{c_{1,3}}$, $\lambda_2 = \frac{c_{1,4}}{c_{1,2}}$, $\lambda_1 = \frac{c_{2,4}}{c_{2,1}}$, $c_{i,j} = \frac{d_{i,j,8}}{d_{i,j,\bar{9}}}$ and $\mu_l = \lambda_l\big|_{8\leftrightarrow9,\bar{9}\leftrightarrow\underline{8}}$.*

For the proof and identification with the terminating hypergeometric series, see [19, 20].

The solution obtained above is sometimes called a "seed" solution. One can obtain more solutions by applying Bäcklund transformations to "seed" solution. The solutions obtained in this way are usually given by the $\tau$ functions in determinant form (see [39] and references therein). "Padé method" provides a convenient way to obtain such solutions [37, 41, 65], which gives the equation, Lax pair and special solutions simultaneously from suitable Padé approximation/interpolation problems. It is interesting to compare these solutions with the determinant representation of the Nekrasov functions given in [34].

Recently, an interesting "lens generalization" of the bilinear elliptic Painlevé equation is proposed [27], which has special solutions in terms of the elliptic hypergeometric sum/integral [26] (see also [25, 56, 67]). Its geometric interpretation along the lines of Sakai's theory [51] and relations to gauge theories and integrable lattice models will be interesting problems.

**Acknowledgements** The author would like to thank his coworkers on various collaborations reviewed in this paper. He is also grateful to M. Bershtein, A. Grassi, A. Kels, V.P. Spiridonov, T. Suzuki, M. Yamazaki for interesting discussions. This work was supported by JSPS Kakenhi Grant (B) 26287018.

# References

1. J. Atkinson, Y. Yamada, Quadrirational Yang-Baxter maps and the elliptic Cremona system (2018). arXiv:1804.01794 [nlin.SI]
2. F. Benini, S. Benvenuti, Y. Tachikawa, Webs of five-branes and $N = 2$ superconformal field theories. J. High Energ. Phys. **2009**(09), 052 (2009)
3. M.A. Bershtein, A.I. Shchechkin, $q$-deformed Painlevé tau function and $q$-deformed conformal blocks. J. Phys. A: Math. Theor. **50**, 085202 (2017)
4. M. Bershtein, P. Gavrylenko, A. Marshakov, Cluster integrable systems, $q$-Painlevé equations and their quantization. J. High Energ. Phys. **2018**, 77 (2018)

5. G. Bonelli, O. Lisovyy, K. Maruyoshi, A. Sciarappa, A. Tanzini, On Painlevé/gauge theory correspondence. Lett. Math. Phys. **107**, 2359 (2017)

6. G. Bonelli, A. Grassi, A. Tanzini, Quantum curves and $q$-deformed Painlevé equations. Lett. Math. Phys. **109**, 1961–2001 (2019). arXiv: 1710.11603 [hep-th]

7. A.B. Coble, Points sets and allied Cremona groups (part I), Trans. Amer. Math. Soc. **16**, 155–198 (1915); – (part II). Ibid. **17** 345–385 (1916).

8. O. Gamayun, N. Iorgov, O. Lisovyy, Conformal field theory of Painlevé VI. J. High Energ. Phys. **10**, 038 (2012)

9. P. Gavrylenko, N. Iorgov, O. Lisovyy, Higher rank isomonodromic deformations and $W$-algebras. Lett. Math. Phys. **110**, 327–364 (2019). arXiv:1801.09608 [hep-th]

10. P. Gavrylenko, N. Iorgov, O. Lisovyy, On solutions of the Fuji-Suzuki-Tsuda system. Symmetry, Integr. Geom. Methods Appl. **14**, 123 (2018). arXiv:1806.08650 [hep-th].

11. B. Grammaticos, F. Nijhoff, A. Ramani, Discrete Painlevé equations, in *The Painlevé Property*. CRM Series in Mathematical Physics (Springer, New York, 1999), pp. 413–516

12. B. Grammaticos, A. Ramani, R. Willox, J. Satsuma, Multiplicative equations related to the affine Weyl group $E_8$. J. Math. Phys. **58**, 083502 (9pp) (2017)

13. A. Grassi, Y. Hatsuda, M. Marino, Topological strings from quantum mechanics. Ann. Henri Poincaré **17**, 3177 (2016)

14. R. Inoue, A. Kuniba, T. Takagi, Integrable structure of box-ball systems: crystal, Bethe ansatz, ultradiscretization and tropical geometry. J. Phys. A Math. Theor. **45**, 073001 (64pp) (2012)

15. M. Jimbo, H. Sakai, A $q$-analog of the sixth Painlevé equation. Lett. Math. Phys. **38**, 145–154 (1996)

16. M. Jimbo, H. Nagoya, H. Sakai, CFT approach to the $q$-Painlevé VI equation. J. Integr. Syst. **2**(1) (2017). arXiv:1706.01940 [hep-th]

17. K. Kajiwara, M. Noumi, Y. Yamada, Discrete dynamical systems with $W(A_{m-1}^{(1)} \times A_{n-1}^{(1)})$ symmetry. Lett. Math. Phys. **60**, 211–219 (2002)

18. K. Kajiwara, M. Noumi, Y. Yamada, $q$-Painlevé systems arising from $q$-KP hierarchy. Lett. Math. Phys. **62**, 259–268 (2002)

19. K. Kajiwara, T. Masuda, M. Noumi, Y. Ohta, Y. Yamada, $_{10}E_9$ solution to the elliptic Painlevé equation. J. Phys. A Math. Gen. **36**, L263–L272 (2003)

20. K. Kajiwara, T. Masuda, M. Noumi, Y. Ohta, Y. Yamada, Hypergeometric solutions to the $q$-Painlevé equations. Int. Math. Res. Not. **2004**, 2497–2521 (2004)

21. K. Kajiwara, T. Masuda, M. Noumi, Y. Ohta, Y. Yamada, Construction of hypergeometric solutions to the $q$-Painlevé equations. Int. Math. Res. Not. **2004**, 1439–1453 (2005)

22. K. Kajiwara, T. Masuda, M. Noumi, Y. Ohta, Y. Yamada, Cubic pencils and Painlevé Hamiltonians. Funkcial. Ekvac. **48**, 147–160 (2005)

23. K. Kajiwara, T. Masuda, M. Noumi, Y. Ohta, Y. Yamada, Point configurations, Cremona transformations and the elliptic difference Painlevé equation. Sémin. Congr. **14**, 169–198 (2006)

24. K. Kajiwara, M. Noumi, Y. Yamada, Geometric aspects of Painlevé equations. J. Phys. A: Math. Theor. **50**, 073001 (2017)

25. A.P. Kels, New solutions of the star-triangle relation with discrete and continuous spin variables. J. Phys. A Math. Theor. **48**, 435201 (2015)

26. A.P. Kels, M. Yamazaki, Elliptic hypergeometric sum/integral transformations and supersymmetric lens index. Symmetry, Integr. Geomet. Methods Appl. **14**, 013 (2018)

27. A.P. Kels, M. Yamazaki, Lens generalisation of $\tau$-functions for elliptic discrete Painlevé equation (2019). arXiv:1810.12103 [nlin.SI]

28. S.-S. Kim, F. Yagi, $5d$ $E_n$ Seiberg-Witten curve via toric-like diagram. J. High Energ. Phys. **06**, 082 (2015)

29. S.-S. Kim, M. Taki, F. Yagi, Tao probing the end of the world. Prog. Theor. Exper. Phys. **8**, 1 (2015)

30. G. Lusztig, *Introduction to Quantum Groups*. Progress in Mathematics, vol. 110 (Birkhäuser, Basel, 1993)

31. G. Lusztig, Total positivity in reductive groups, in *Lie Theory and Geometry*. Progress in Mathematics, vol. 123 (Birkhäuser, Basel, 1994), pp. 531–568
32. J.I. Manin, The Tate height of points on an abelian variety. Its variants and applications. Izv. Akad. Nauk SSSR Ser. Mat. **28**, 1363–1390 (1964); AMS Transl. **59**(2), 82–110 (1966)
33. T. Matano, A. Matumiya, K. Takano, On some Hamiltonian structures of Painlevé systems. II. J. Math. Soc. Japan **51**, 843–866 (1999)
34. A. Mironov, A. Morozov, On determinant representation and integrability of Nekrasov functions. Phys. Lett. **B773**, 34–46 (2017)
35. S. Mizoguchi, Y. Yamada, $W(E_{10})$ symmetry, $M$ theory and Painlevé equations. Phys. Lett. **B537**, 130–140 (2002)
36. M. Murata, H. Sakai, J. Yoneda, Riccati solutions of discrete Painlevé equations with Weyl group symmetry of type $E_8^{(1)}$. J. Math. Phys. **44**, 1396–1414 (2003)
37. H. Nagao, Y. Yamada, Study of $q$-Garnier system by Padé method. Funkcial. Ekvac. **61**, 109–133 (2018)
38. A. Nakayashiki, Y. Yamada, Kostka polynomials and energy functions in solvable lattice models. Sel. math. New Ser. **3**, 547–599 (1997)
39. M. Noumi, Remarks on $\tau$-functions for the difference Painlevé equations of type $E_8$, in *Representation Theory, Special Functions and Painlevé Equations—RIMS 2015* (Mathematical Society of Japan, Tokyo, 2018), pp. 1–65. arXiv:1604.04686
40. M. Noumi, Y. Yamada, Birational Weyl group action arising from a nilpotent Poisson algebra, in *Physics and Combinatorics 1999 (Nagoya)*, ed. by A.N. Kirillov, A. Tsuchiya, H. Umemura (World Scientific, Singapore, 2001)
41. M. Noumi, S. Tsujimoto, Y. Yamada, Padé interpolation problem for elliptic Painlevé equation, in *Symmetries, Integrable Systems and Representations*, ed. by K. Iohara, et al. Springer Proceedings in Mathematics & Statistics, vol. 40 (Springer, Berlin, 2013), pp. 463–482
42. Y. Ohta, A. Ramani, B. Grammaticos, An affine Weyl group approach to the eight parameter discrete Painlevé equation. J. Phys. A Math. Gen. **34**, 10523 (2001)
43. K. Okamoto, Sur les feuilletages associés aux équations du second ordre à points critiques fixés de P. Painlevé. Jpn. J. Math. **5**, 1–79 (1979)
44. N. Okubo, T. Suzuki, Generalized $q$-Painlevé VI system of type $(A_{2n+1} + A_1 + A_1)^{(1)}$ arising from cluster algebra (2018). arXiv:1810.03252 [math-ph]
45. C.M. Ormerod, E.M. Rains, An elliptic Garnier system. Commun. Math. Phys. **355**(2), 741–766 (2017). arXiv:1607.07831 [math-ph]
46. C.M. Ormerod, Y. Yamada, From polygons to ultradiscrete painlevé equations. Symmetry, Integr. Geomet. Methods Appl. **11**, 056, 36 (2015)
47. G.R.W. Quispel, J.A.G. Roberts, C.J. Thompson, Integrable mappings and soliton equations. Phys. Lett. **A126**, 419–421 (1988)
48. G.R.W. Quispel, J.A.G. Roberts, C.J. Thompson, Integrable mappings and soliton equations II. Physica **D34**, 183–192 (1989)
49. E.M. Rains, An isomonodromy interpretation of the hypergeometric solution of the elliptic Painlevé equation (and generalizations). Symmetry, Integr. Geomet. Methods Appl. **7**, 088 (2011)
50. A. Ramani, B. Grammaticos, Singularity analysis for difference Painlevé equations associated with affine Weyl group $E_8$. J. Phys. **A 50**, 055204 (18pp) (2017)
51. H. Sakai, Rational surfaces associated with affine root systems and geometry of the Painlevé equations. Comm. Math. Phys. **220**, 165–229 (2001)
52. H. Sakai, A $q$-analog of the Garnier system. Funkcial. Ekvac. **48**, 273–297 (2005)
53. T. Shioda, K. Takano, On some Hamiltonian structures of Painlevé systems. I. Funkcial. Ekvac. **40**, 271–291 (1997)
54. V.P. Spiridonov, Classical elliptic hypergeometric functions and their applications, in *Elliptic Integrable Systems*, ed. by M. Noumi, K. Takasaki. Rokko Lectures in Mathematics, vol. 18 (2005), pp. 253–287

55. V.P. Spiridonov, Essays on the theory of elliptic hypergeometric functions. Russ. Math. Surv. **63**(3), 405–472 (2008)
56. V.P. Spiridonov, Rarefied elliptic hypergeometric functions. Adv. Math. **331**, 830–873 (2018)
57. T. Suzuki, A reformulation of generalized $q$-Painlevé VI system with $W(A_{2n+1}^{(1)})$ symmetry. J. Integrable Syst. **2**, xyw017 (2017)
58. T. Tokihiro, D. Takahashi, J. Matsukidaira, J. Satsuma, From soliton equations to integrable cellular automata through a limiting procedure. Phys. Rev. Lett. **76**, 3247–3250 (1996)
59. T. Tsuda, Integrable mappings via rational elliptic surfaces. J. Phys. A Math. Gen. **37**, 2721–2730 (2004)
60. T. Tsuda, On an integrable system of $q$-difference equations satisfied by the universal characters: its Lax formalism and an application to $q$-Painlevé equations. Comm. Math. Phys. **293**, 347–359 (2010)
61. A.P. Veselov, Integrable maps. Russ. Math. Surv. **46**(5), 1–51 (1991)
62. E.T. Whittaker, G.N. Watoson, *A Course of Modern Analysis* (Cambridge University Press, Cambridge, 1927)
63. Y. Yamada, A birational representation of Weyl group, combinatorial R matrix and discrete Toda equation, in *Physics and Combinatorics 2000 (Nagoya)*, ed. by A.N. Kirillov, N. Liskova (World Scientific, Singapore, 2001), pp. 305–319
64. Y. Yamada, A Lax formalism for the elliptic difference Painlevé equation. Symmetry, Integr. Geomet. Methods Appl. **5**, 042 (2009)
65. Y. Yamada, Padé method to Painlevé equations. Funkcial. Ekvac. **52**, 83–92 (2009)
66. Y. Yamada, An elliptic Garnier system from interpolation, Symmetry, Integr. Geomet. Methods Appl. **13**, 069 (2017)
67. M. Yamazaki, Integrability as duality: the Gauge/YBE correspondence (2018). arXiv:1808.04374 [hep-th]

Printed in the United States
by Baker & Taylor Publisher Services